Journal
of the American Ceramic Society

T0087595

Discover this journal online at
WILEY ONLINE LIBRARY
...ry.com/journal/jace

Number S1
June 2011

JOURNAL OF THE AMERICAN CERAMIC SOCIETY (ISSN: 0002-7820 [print]; ISSN: 1551-2916 [online]), is published monthly on behalf of the American Ceramic Society by Wiley Subscription Services, Inc., a Wiley Company, 111 River St., Hoboken, NJ 07030-5774.

Periodical postage paid at Hoboken NJ and additional offices.

Postmaster: Send all address changes to JOURNAL OF THE AMERICAN CERAMIC SOCIETY, Journal Customer Services, John Wiley & Sons Inc., 350 Main St., Malden, MA 02148-5020.

Information for Subscribers: Journal of the American Ceramic Society is published in 12 issues per year.
Members of the American Ceramic Society (ACerS) should contact ACerS Customer Services (customerservice@ceramics.org; 866 721 3322; or 240 646 7054) to subscribe to the print edition of the Journal.

(Combined subscription with International Journal of Applied Ceramic Technology and International Journal of Applied Glass Science)

Institutional subscription prices for 2011 are:
Print & Online: US$2,269 (US), US$2,479 (Rest of World), €1,608 (Europe), £1,265 (UK). Prices are exclusive of tax. Asia-Pacific GST, Canadian GST and European VAT will be applied at the appropriate rates. For more information on current tax rates, please go to www.wileyonlinelibrary.com/tax-vat. The price includes online access to the current and all online back files to January 1st 2007, where available. For other pricing options, including access information and terms and conditions, please visit www.wileyonlinelibrary.com/access.

Delivery Terms and Legal Title Where the subscription price includes print issues and delivery is to the recipient's address, delivery terms are Delivered Duty Unpaid (DDU); the recipient is responsible for paying any import duty or taxes. Title to all issues transfers FOB our shipping point, freight prepaid. We will endeavour to fulfil claims for missing or damaged copies within six months of publication, within our reasonable discretion and subject to availability.

Publisher: Journal of the American Ceramic Society is published by Wiley Periodicals Inc., 350 Main St., Malden MA 02148-5020.

Journal Customer Services: For ordering information, claims and any enquiry concerning your journal subscription please go to www.wileycustomerhelp.com/ask or contact your nearest office.

Americas: Email: cs-journals@wiley.com; Tel: +1 781 388 8598 or +1 800 835 6770 (toll free in the USA & Canada).

Europe, Middle East and Africa: Email: cs-journals@wiley.com; Tel: +44 (0) 1865 778315.

Asia Pacific: Email: cs-journals@wiley.com; Tel: +65 6511 8000.

Japan: For Japanese speaking support, Email: cs-japan@wiley.com; Tel: +65 6511 8010 or Tel (toll-free): 005 316 50 480.

Visit our Online Customer Get-Help available in 6 languages at www.wileycustomerhelp.com.

Production Editor: Josh Martino (Email: JACE@wiley.com)

Advertising: Karl Franz (email: kfranz@wiley.com)

Commercial Reprints: Lydia Supple-Pollard (email: lsupple@wiley.com)

Back Issues: Single issues from current and recent volumes are available at the current single issue price from cs-journals@wiley.com. Earlier issues may be obtained from Periodicals Service Company, 11 Main Street, Germantown, NY 12526, USA. Tel: +1 518 537 4700, Fax: +1 518 537 5899, Email: psc@periodicals.com

Introduction

Feature

Review

Original Articles

Contents

J. Am. Ceram. Soc., **94** [S1] S1–S2 (2011)
DOI: 10.1111/j.1551-2916.2011.04592.x
© 2011 The American Ceramic Society

journal

A Tribute to Anthony G. Evans:
Materials Scientist and Engineer December; 4, 1942–September 9, 2009

John W. Hutchinson
January 4, 2011

Tʜɪs special issue of the Journal of the American Ceramics Society brings together papers written by some of the many colleagues who have worked with Tony Evans over his career. The issue itself reflects the extraordinary breadth of Evans' scientific interests. This brief Tribute is an attempt to capture in a few words Tony Evans' remarkable influence and contribution to materials science and more broadly to engineering science. The task would be more daunting were it not for the testimony to Tony Evans in celebration of his 65th birthday by A. Heuer (Int. J. Mat. Res. **98** (2007) 1168–1169) and an obituary by N. A. Fleck, which appeared in the November 29, 2009 issue of the Guardian newspaper.

Anthony G. Evans was one of the most influential materials scientists and materials engineers of his generation. He had no rival when it came to the grasp of the underlying fundamentals of material behavior coupled with an extraordinary ability to focus his attention and to inspire and lead collaborative efforts. Evans was born and raised in Porthcawl, Wales to William Glyn and Annie May Evans. He met his wife, Trisha, in their hometown and they were married in 1964. Trisha and their daughters and grandchildren survive Tony. After obtaining BSc (1964) and PhD (1967) degrees in metallurgy at Imperial College, London, Evans began work in 1967 as a ceramist at the Atomic Energy Research

Establishment, Harwell. Following a sabbatical period at UCLA, Evans worked at the National Bureau of Standards from 1971 to 1974 and then served as a group leader at the Rockwell International Science Center from 1974 to 1978. In 1978 he joined the faculty of the Department of Materials Science and Mineral Engineering, University of California, Berkeley, where he remained until 1985. During these years the emphasis of much of Evans' research was on ceramics and he began his long association with the American Ceramics Society. In 1985 he moved to the Santa Barbara campus of the University of California as the Alcoa Professor. Evans was the founding chair of the Materials Department at UCSB (1985–1991) which would become one of the leading materials departments in the world. Those of us who conducted research with Evans during this period were largely unaware of his efforts as department chair—he later attributed this to the fact that he allocated the period from 7:30 to 9 in the morning each day to his departmental duties, finishing before his colleagues had an opportunity to perturb the process. This was also the period that Evans established himself as a research leader par excellence heading major projects on ceramic matrix composites, toughening of ceramics, and thin films and multilayers. Evans made two more academic moves before completing a circle back to UCSB in 2002. From 1994 until 1998 Evans was the Gordon McKay Professor of Materials Engineering in the Division of Engineering and Applied Sciences at Harvard University and from 1998 to 2002 he served as the Gordon Wu Professor in the Department of Mechanical and Aerospace Engineering at Princeton University. The pull of Santa Barbara and UCSB remained strong, however, and he returned in 2002 to a joint appointment in the Departments of Materials and Mechanical Engineering where he focused primarily on teaching and research.

Any tribute to Tony Evans must begin with the impact of his work. Evans is the most highly cited materials scientist with almost 35 000 citations to over 650 published journal papers. His h-index will soon pass 100. A short list of subjects to which Evans has made major contributions includes micro-cracking and transformation toughening of ceramics, ceramic matrix composites and metal matrix composites, thin film mechanics, interface mechanics, thermal barrier coatings, metallic foams, morphing structures, aerospace materials with special thermomechanical properties, lightweight lattice materials, and superior blast and ballistic resistant structural materials. On each of these problem areas, Evans brought to bear a fundamental understanding of material behavior at all scales together with innovative experiments in the laboratory. The experimental work he and his collaborators performed more often than not focused on observation of micromechanical behavior and new phenomena rather than on refined measurement of material properties.

The synergy between Evans' grasp of theory and his insightful exploitation of experiment, combined with his love of subject and legendary ability to focus, would have been enough, by themselves, to establish his primacy. However, there is more. Evans' skills at assembling, inspiring and leading interdisciplinary teams of engineers and scientists to tackle challenging technological problems is the additional component of his approach which truly set him apart. Those who have had the experience of working with Tony Evans on one or more of the teams he put together will be aware that he had no match as a technology leader in the arena of structural materials. These include his former students, post-doctoral fellows, and a large cadre of colleagues which he brought together from many academic, government and industrial institutions, here and abroad. Particularly notable for each major project that Evans led were the workshops where ongoing work was reviewed and new work was planned with criticism and input from experts from industry and government labs. These workshops, which were always enlivened by Evans' active participation in every detail of the research, were exceptional in identifying the challenges and moving the research forward. Skepticism about the effectiveness of shifting research funding from smaller projects to relatively fewer large projects, a trend that has taken place over the past several decades in the US and is now spreading around the world, would be far less warranted if more large projects were led by individuals with the abilities of an Evans.

Tony Evans provided leadership in the materials community throughout his career in other ways as well. As already noted, he provided critical leadership in founding the Materials Department at UCSB for nearly six years and then later for the Princeton Materials Institute for four years, in each case without appearing to break step in his own research activities. Starting his service in 1974, Evans became the longest serving member of the Defense Sciences Research Council (formerly the Materials Research Council) of DARPA which played a major role in setting the agenda for research in advanced materials in the US. Simultaneous with all his other activities, during his entire career, Evans was a highly engaged consultant to many companies, not only in the materials industry but also in aerospace and electronics. To those of us who worked closely with Evans,

it was clear that much of his research emphasis was motivated by problems that surfaced through his consulting activities. There is little wonder that his work has had, and will continue to have, such major impact.

In his testimony in celebration of Tony Evans' 65th birthday, Arthur Heuer noted that Evans' exceptional generosity to students and colleagues with his ideas and time was one of the keys to his success. Quoting Heuer directly: "It is his incredible ability to focus, his "nose" for important problems to work on, and his generosity in collaborative research, that have led to his revered status in our field." Throughout his career, Evans enjoyed working closely with his many students and post-docs who are now spread far and wide in academia, industry and government. These former junior colleagues, together with all his other collaborators, will keep Evans' legacy alive for years to come.

Tony Evans' contributions have been recognized in many ways—a few of them are listed below. He was a member of both the National Academy of Engineering and the National Academy of Sciences. He was also a Fellow of both the Royal Society (FRS) and the Royal Academy of Engineering in the UK. He was a Distinguished Life Member of the American Ceramics Society and received essentially every major award of this Society. He won the Henry Marion Howe Medal of ASM International, the Turnbull Award of the Materials Research Society, the Griffith Medal and Prize and Mellor Memorial Lectureship of the Institute of Materials, UK, and the Nadai Medal of ASME. He was an Alexander von Humboldt Senior Scientist. In 2008, Evans was especially pleased to receive the highest award granted by his alma mater, Fellow of Imperial College, London.   □

J. Am. Ceram. Soc., **94** [S1] S3–S14 (2011)
DOI: 10.1111/j.1551-2916.2011.04559.x
© 2011 The American Ceramic Society

journal

Hybrid Materials to Expand the Boundaries of Material-Property Space

Mike Ashby[†]

Engineering Department, Cambridge University, CB2 1PZ Cambridge, U.K.

The materials we use today for mechanical design are the outcome of at least 3000 years of development, much of it empirical but much the outcome of systematic science. Both approaches have been motivated by the desire for stiffer, stronger, more durable, and lighter structures, progressively populating material property "space". We first examine the extent to which this space is now filled and estimate the ultimate constraints on this filling. Strategies for expanding the filled regions further include *hybrid material design*—making materials by combining two or more monolithic materials in chosen configurations and connectivity. Here we explore two classes of hybrid—lattice materials and sandwich structures—developing ways of comparing their properties with those of conventional monolithic materials. The comparison reveals the potential of hybrids. This paper builds on ideas that have grown from a long and rewarding collaboration with Tony Evans with whom it has been a privilege to have worked.

I. Introduction and Synopsis

Hᴇʙʀɪᴅ materials (Fig. 1) are combinations of two or more materials, or of materials and space, assembled in such a way as to have attributes not offered by any one material alone.[1,2] Particulate and fiber composites are examples of one type of hybrid, but there are many others: sandwich structures, foams, lattice structures, segmented structures, zero-expansion material, and more. Here we explore the potential of hybrid materials, emphasizing the choice of the components, their configuration, and their relative volume fraction. The new variables expand design space, allowing the creation of new "materials" with specific property profiles. But how are we to compare a hybrid—a sandwich structure for example—with monolithic materials such as polycarbonate or titanium? To do this we must think of the sandwich (as an example) not only as a hybrid with faces of one material bonded to a core of another, but as a "material" in its own right, with its own set of *effective properties*; it is these that allow the comparison.

The approach adopted here is one of breadth rather than precision. The aim is to assemble methods to allow the properties of alternative hybrids to be scanned and compared with those of monolithic materials, seeking those that best meet a given set of design requirements. Once materials and configuration have been chosen, standard methods—optimization routines, finite-element analyzes—can be used to refine them. But what the standard methods are *not* good at is the quick scan of alternative combinations. That is where the approximate methods developed below, in which material and configuration become the variables, pay off.

In this paper we explore two broad classes of hybrid, each with a number of discrete members.

(1) *Cellular structures* are combinations of material and space giving precise control of density, stiffness, strength, and thermal conductivity.

(2) *Sandwich structures* have outer faces of one material supported by a core of second, low density, material—a configuration that can offer a flexural stiffness per unit weight that is greater than that offered by either component alone.

We use simple continuum and micromechanical models to estimate the *equivalent properties* of each configuration, allowing direct comparison with monolithic engineering materials. The examples of Section IV illustrate how this is done. The methods are developed further in a White Paper.[3]

II. Holes in Material-Property Space

Material properties can be "mapped" as *Material Property Charts* of which Fig. 2 is an example: it is a chart of Young's modulus plotted against density. The small bubbles show the range of these two properties exhibited by a given material type. The larger colored envelopes enclose material families. All the charts have in common that parts of them are populated with materials but other parts are not: there are *holes*. Some parts of the holes are inaccessible for fundamental reasons that relate to the size of atoms and the nature of the forces that bind them together. But others are empty even though, in principle, they could be filled.

Is anything to be gained by developing materials (or material combinations) that lie in these holes? To answer this we need criteria of excellence to assess the merit of any given hybrid. These are provided by the material indices, described fully elsewhere.[4] If a possible hybrid has a value of any one of these that exceed those of existing materials, it has the potential to increase performance.

The axes in Fig. 2 are Young's modulus, E, and density ρ. The material indices E/ρ, $E^{1/2}/\rho$, and $E^{1/3}/\rho$, indicated by the "Guide lines" at the lower right, are criteria of the excellence or material indices for selecting material for light, stiff structures. A

F. Zok—contributing editor

Manuscript No. 29018. Received December 7, 2010; approved March 9, 2010.
A contribution to the celebration of the life of Tony Evans.
[†]Author to whom correspondence should be addressed. e-mail: mfa2@eng.cam.ac.uk

Feature

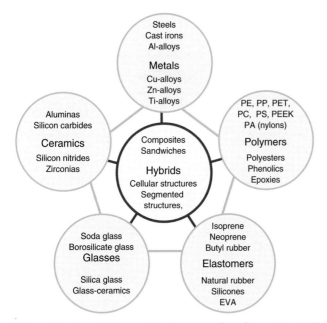

Fig. 1. Hybrid materials combine the properties of two (or more) monolithic materials, or of one material and space. They include fibrous and particulate composites, foams and lattices, sandwiches, and almost all natural materials.

grid of lines of one index—E/ρ—is plotted in the Figure. The orange arrow lies normal to the index lines. If the filled areas can be expanded in the direction of this arrow (i.e. to greater values of E/ρ) the materials so created will enable lighter, stiffer structures. The arrow thus defines a *vector for material development*.

Figure 3 shows a second example. The axes are density ρ and strength σ_f, meaning the yield strength for ductile materials, the modulus of rupture for those that are brittle. Once again small bubbles enclosed the property ranges of individual material types, the larger colored envelopes enclose families. Again some regions are filled and some are empty. The material indices σ_f/ρ, $\sigma_f^{2/3}/\rho$, and $\sigma_f^{1/2}/\rho$, characterizing materials that allow light, strong structures, are indicated by "Guide lines." A grid showing σ_f/ρ with superimposed vector, indicates the desired direction of development for materials for structures that are both light and strong.

So far, so good. But how much of the holes are accessible? Does the underlying physics of stiffness, strength, and density prohibit the existence of materials within the holes? Figure 4 is an attempt to answer this question. The stable element with the highest density is osmium ($E = 560$ GPa, $\rho = 22\,550$ kg/m^3). Vegard's law, though approximate, effectively rules out the creation of an alloy or compound with a larger density this. The bulk material with the highest modulus is diamond ($E = 1,050$ GPa, $\rho = 3.520$ kg/m^3), a consequence of the great stiffness of the covalent carbon–carbon bond and the small size of the carbon atom. It is conceivable that structures with more densely packed, more tightly bound atoms might exist but no examples of bulk materials with such a combination are at present known. Structures with diamond-like bonding but with lower density than diamond could be made by creating lattice structures like those described in the next section of this paper. It is shown there that the modulus of such structures falls linearly with density. Combining these facts allows part of the modulus-density space to be blanked out as inaccessible, as indicated in Fig. 4. That still leaves possible room of expansion by creating materials that lie in the area marked "accessible."

A similar, approximate boundary for the accessible space can be plotted on the strength-density chart (Fig. 5). The idea that

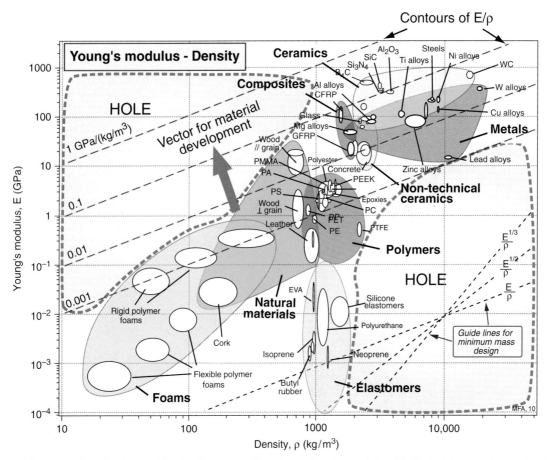

Fig. 2. A material property chart showing modulus-density space, with contours of specific modulus, E/ρ. Part of the space is occupied by materials, part it empty (the "holes"). Material development that extended the occupied territory in the direction of the arrow (the "vector for material development") allows components with greater stiffness-to-weight than is at present possible.

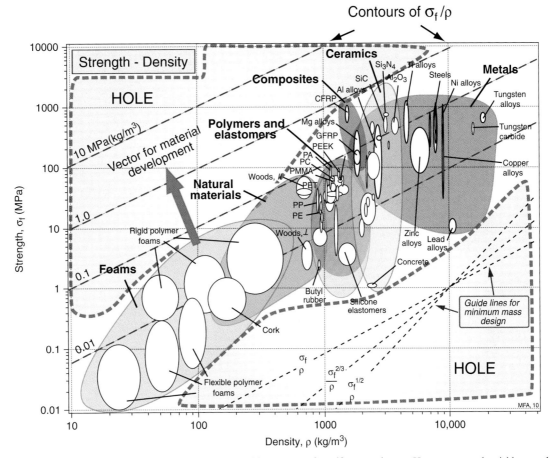

Fig. 3. A material property chart showing strength-density space, with contours of specific strength, $\sigma_{f/\rho}$. Here σ_f means the yield strength metals and polymers, the modulus of rupture of ceramics. Like Fig. 2, it has holes. Material development that extended the occupied territory in the direction of the arrow enables components with greater strength-to-weight than is at present possible.

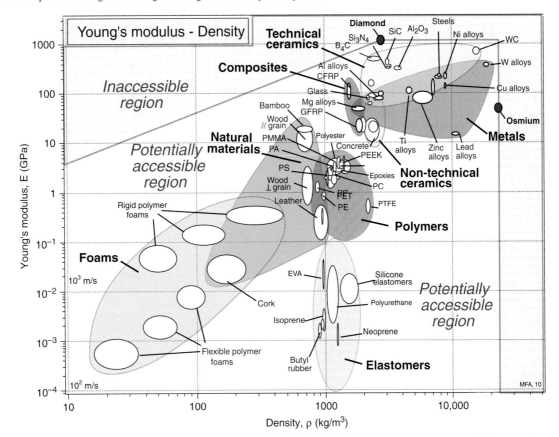

Fig. 4. Some unfilled areas of modulus-density space are inaccessible for fundamental reasons, but other areas are potentially accessible, suggesting that materials might be created to fill them. Two "limiting" solids, osmium and diamond, are indicated.

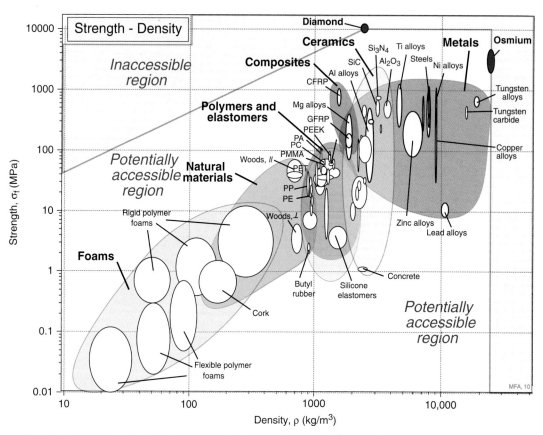

Fig. 5. Like the previous figure, the strength-density space has potentially accessible areas that are not yet filled.

the "ideal" strength of a solid is a fraction of its modulus—$E/30$ is often cited—has a sound physical basis. The upper boundary on the figure is based on the conservative value $\sigma_f = E/100$ Once again it appears that room exists for developing materials with higher strength-to-weight ratios than those we already have.

One approach to filling these holes—the long-established one—is that of developing new metal alloys, new polymer chemistries, and new compositions of glass and ceramic so as to create monolithic materials that expand the populated areas of the property charts. But developing new materials can be expensive and uncertain, and the gains tend to be incremental rather than step like. An alternative is to combine two or more existing materials so as to allow a superposition of their properties—in short, to create hybrids. The great success of carbon and glass–fiber reinforced composites at one extreme, and of foamed materials at another (hybrids of material and space) in filling previously empty areas of the property charts is encouragement to explore ways in which such hybrids can be designed.

When is a hybrid a "material"? There is a certain duality about the way in which hybrids are thought about and discussed. Some, like filled polymers, composites or wood are treated as materials in their own right, each characterized by its own set of macroscopic material properties. Others—like galvanized steel—are seen as one material (steel) to which a coating of a second (zinc) has been applied, even though if could be regarded as a new material with the strength of steel but the surface properties of zinc. Sandwich panels illustrate the duality, sometimes viewed as two sheets of face material separated by a core material, and sometimes—to allow comparison with monolithic materials—as a "material" with their own density, axial and flexural stiffness and strength, thermal conductivity, expansion coefficient, etc. To call any one of these a "material" and characterize it as such is a useful shorthand, allowing designers to use existing methods when designing with them. But if we are to design the hybrid itself, we must deconstruct it, and think of it as a combination of materials (or of material and space) in a chosen configuration.

III. Approximate Models for Hybrid Properties

In this section we assemble approximate equations for the density, moduli, strengths, and toughnesses for two families of hybrid materials. These are used in Section IV to compare hybrids with monolithic engineering materials.

(1) Cellular Structures: Foams and Lattices

Cellular structures—foams and lattices—are hybrids of a solid and a gas. The properties of the gas might at first sight seem irrelevant, but this is not so. The thermal conductivity of low-density foams of the sort used for insulation is determined by the conductivity of the gas contained in its pores; and the dielectric properties, and even the compressibility, can depend on the gas properties.

There are two distinct species of cellular solid. The first, typified by foams, are *bending-dominated structures*; the second, typified by triangulated lattice structures, are *stretch dominated*—a distinction explained more fully below.[5–8]

Foams are cellular solids made by expanding polymers, metals, ceramics, or glasses with a foaming agent. Figure 6, left, shows an idealized cell of a low-density foam. It consists of solid cell walls or edges surrounding a void space containing a gas or fluid. Foams have the characteristic that, when loaded, the cell walls bend. *Lattice-structures* (Fig. 6, right) are configured to suppress bending, so the cell edges have to stretch instead.

(A) Density: Cellular solids are characterized by their *relative density*, which for the structure shown here (with $\ll L$) is

$$\frac{\tilde{\rho}}{\rho_s} = C_1\left(\frac{t}{L}\right)^2 \tag{1}$$

where $\tilde{\rho}$ is the density of the foam, ρ_s is the density of the solid of which it is made, L is the cell size, t is the thickness of the cell edges, and C_1 is a constant, approximately equal to 1.

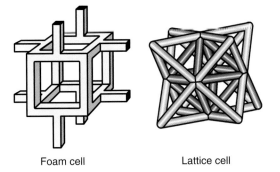

Foam cell   Lattice cell

Fig. 6. Left: an idealized cell in a low density foam. Right: a fully triangulated lattice structure.

(B) Elastic Moduli of Bending-Dominated Foams: A remote compressive stress σ exerts a force $F \propto \sigma L^2$ on the cell edges, causing them to bend and leading to a bending deflection δ, as shown in Fig. 7(a). For the open-celled structure shown in the figure, the bending deflection scales as

$$\delta \propto \frac{FL^3}{E_s I} \qquad (2)$$

where E_s is the modulus of the solid of which the foam is made and $I = \frac{t^4}{12}$ is the second moment of area of the cell edge of square cross section, $t \times t$. The compressive strain suffered by the cell as a whole is then $\varepsilon = 2\delta/L$. Assembling these results gives the modulus $\tilde{E} = \sigma/\varepsilon$ of the foam as

$$\tilde{E} = C_2 \left(\frac{\tilde{\rho}}{\rho_s}\right)^2 E_s \quad \text{(bending-dominated behavior)} \qquad (3)$$

Because $\tilde{E} = E_s$ when $\tilde{\rho} = \rho_s$, we expect the constant of proportionality C_2 to be close to unity a speculation confirmed both by experiment and numerical simulation. The quadratic dependence means that a small decrease in relative density causes a large drop in modulus. When the cells are equiaxed in shape, the foam properties are isotropic with shear modulus, bulk modulus, and Poisson's ratio given

$$\tilde{G} = \frac{3}{8}\tilde{E} \qquad \tilde{K} = \tilde{E} \qquad v = \frac{1}{3} \qquad (4)$$

(C) Elastic Moduli of Stretch-Dominated Lattices: The structure on the right in Fig. 6, is fully triangulated. This means that the cell edges must stretch when the structure is loaded elastically. On average one third of its edges carry tension when the structure is loaded in simple tension, regardless of the loading direction. Thus

$$\tilde{E} = C_3 \left(\frac{\tilde{\rho}}{\rho_s}\right) E_s \quad \text{(strech-dominated behavior)} \qquad (5)$$

with $C_3 = 1/3$. The modulus is linear, not quadratic, in density giving a structure that is stiffer than a foam of the same density. The structure is almost isotropic, so we again approximate the shear modulus, bulk modulus, and Poisson's ratio by Eq. (4).

(D) Strength of Bending-Dominated Foams: When the foam structure in Fig. 6 is loaded beyond the elastic limit, its cell walls may yield, buckle elastically, or fracture as shown in Fig. 7. Consider yielding first (Fig. 7(b)). Cell edges yield when the force exerted on them exceeds their fully plastic moment.

$$M_f = \frac{\sigma_{y,s} t^3}{4} \qquad (6)$$

where $\sigma_{y,s}$ is the yield strength of the solid of which the foam is made. This moment is related to the remote stress by $M \propto FL \propto \sigma L^3$. Assembling these results gives the compressive failure strength when yield dominates, $\tilde{\sigma}_c$

$$\tilde{\sigma}_c = C_4 \left(\frac{\tilde{\rho}}{\rho_s}\right)^{3/2} \sigma_{y,s} \quad \text{(yield of foams)} \qquad (7)$$

where the constant of proportionality, $C_4 \approx 0.3$, has been established both by experiment and by numerical computation.

Elastomeric foams collapse not by yielding but by elastic buking; brittle foams by cell wall fracture (Figs. 7 (c) and (d)). As with plastic collapse, simple scaling laws describe this behavior well. Collapse by buckling occurs when the stress exceeds

$$\tilde{\sigma}_c \approx 0.05 \left(\frac{\tilde{\rho}}{\rho_s}\right)^2 E_s \quad \text{(buckling of foams)} \qquad (8)$$

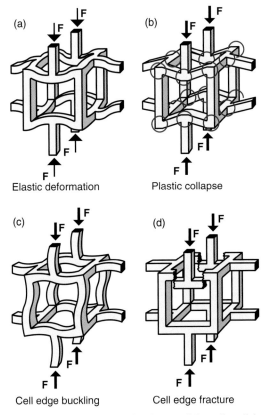

Fig. 7. (a) Elastic deformation of a foam cell by cell wall bending. (b) Plastic collapse: the cell edges bend plastically. (c) An elastomeric foam, by contrast, collapses by the elastic buckling of its cell edges. (d) A brittle foam collapses by the successive fracturing of cell edges.

(a) Elastic deformation   (b) Plastic collapse

(c) Cell edge buckling   (d) Cell edge fracture

We identify the compressive strength $\tilde{\sigma}_c$ with the lesser of Eqs (7) and (8). We further set the yield strength $\tilde{\sigma}_y$ the flexural strength $\tilde{\sigma}_{flex}$ equal to $\tilde{\sigma}_c$.

(E) Strength of Stretch-Dominated Lattices: Collapse occurs when the cell edges yield, giving the collapse stress

$$\tilde{\sigma}_c \approx \frac{1}{3}\left(\frac{\tilde{\rho}}{\rho_s}\right)\sigma_{y,s} \quad \text{(yield of lattices)} \qquad (9)$$

This is an upper bound because it assumes that the struts yield in tension or compression when the structure is loaded. If the struts are slender, they may buckle before they yield. They do so at the stress

$$\tilde{\sigma}_C \approx 0.2 \left(\frac{\tilde{\rho}}{\rho_s}\right)^2 E_s \qquad (10)$$

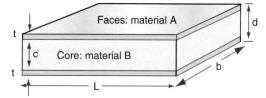

Fig. 8. The sandwich. The face thickness is t, the core thickness c, and the panel thickness d.

We identify the compressive strength $\tilde{\sigma}_c$ with the lesser of Eqs (9) and (10) .We further identify the yield strength $\tilde{\sigma}_y$, the flexural strength $\tilde{\sigma}_{flex}$ with $\tilde{\sigma}_c$.

(F) Fracture Toughness of Bending-Dominated Foams: Foams that contain crack-like flaws that are long compared with the cell size ℓ fail by fast fracture, meaning that the crack propagates unstably if the stress intensity factor exceeds the critical value $\tilde{K}_{Ic}$, which is the lesser of

$$\tilde{K}_{Ic} = 0.5\left(\frac{\ell}{a}\right)^{1/2}\left(\frac{\tilde{\rho}}{\rho_s}\right)^{3/2} K_{Ic,s} \quad \text{and} \quad K_{Ic,s} \tag{11}$$

where a is the intrinsic flaw size of the material of the cell edges, $K_{Ic,s}$ is its fracture toughness.[9]

(G) Fracture Toughness of Stretch-Dominated Lattices: Lattices that contain crack-like flaws long compared with the cell size ℓ fail by fast fracture if the stress intensity factor exceeds the critical value $\tilde{K}_{Ic}$, which is the lesser of

$$\tilde{K}_{Ic} = 0.5\left(\frac{\ell}{a}\right)^{1/2}\left(\frac{\tilde{\rho}}{\rho_s}\right) K_{Ic,s} \quad \text{and} \quad K_{Ic,s} \tag{12}$$

ℓ/a is the ratio of the cell size of the foam to the flaw size in the material.

(2) Sandwich Structures

A sandwich panel epitomizes the concept of a hybrid. It combines two materials in a specified geometry and scale, configured such that one forms the faces, the other the core, to give a structure of high bending stiffness and strength at low weight (Fig. 8). The separation of the faces by the core increases the moment of inertia of the section, I, and its section modulus, Z, producing a structure that resists bending and buckling loads well. The faces, each of thickness t, carry most of the load, so they must be stiff and strong; and they form the exterior surfaces of the panel so they must also tolerate the environment in which it operates. The core, of thickness c, occupies most of the volume, it must be light, and stiff and strong enough to carry the shear stresses necessary to make the whole panel behave as a load-bearing unit.[10–15]

(A) A Sandwich as a "Material": So far we have spoken of the sandwich as a *structure*: faces of material A supported on a core of material B, each with its own density, modulus and strength. But we can also think of it as a *material* with its own set of properties, and this is useful because it allows comparison with more conventional materials. To do this we calculate equivalent material properties for the sandwich, identifying them, as with foams, by a tilde (e.g. $\tilde{\rho}$, $\tilde{E}$). The quantities $\tilde{\rho}$ and $\tilde{E}$ can be plotted on the modulus-density chart, allowing a direct comparison all the other materials on the chart. All the constructions using material indices apply unchanged. We base the analysis on the symmetric sandwich with the dimensions defined in Fig. 8. The symbols that appear in this section are defined in Table I.

(B) Density: The equivalent density of the sandwich (its mass divided by its volume) is

$$\tilde{\rho} = f\rho_f + (1-f)\rho_c \tag{13}$$

here f is the volume fraction occupied by the faces: $f = 2t/d$.

In-plane and through-thickness moduli are simply the arithmetic and harmonic means of the components.

$$\tilde{E} = fE_f + (1-f)E_c \quad \text{(in-plane)} \tag{14}$$

$$E_{tt} = \frac{1}{f/E_f + (1-f)/E_c} \quad \text{(through thickness)} \tag{15}$$

(C) Equivalent Flexural Modulus: The flexural properties are quite different. The flexural compliance (the reciprocal of the stiffness) has two contributions, one from the bending of the panel as a whole, the other from the shear of the core (Fig. 9). They add. The bending stiffness is

$$EI = \frac{b}{12}\left(d^3 - c^3\right)E_f + \frac{bc^3}{12}E_c$$

The shear stiffness is

$$AG = \frac{bd^2}{c}G_c$$

Summing deflections gives

$$\delta = \frac{12PL^3}{B_1 b\{(d^3 - c^3)E_f + c^3 E_c\}} + \frac{PLc}{B_2 d^2 bG_c} \tag{16}$$

Table I. The Symbols Used in Describing Sandwich Structures

Symbol	Meaning and usual units
t, c, d	Face thickness, core thickness, and overall panel thickness (m)
L, b	Panel length and width (m)
m_a	Mass per unit area of the panel (kg/m^2)
$f = 2t/d$	Relative volumes occupied by the faces
$(1-f) = c/d$	Relative volume occupied by the core
I	Second moment of area (m^4)
ρ_f, ρ_c	Densities of face and core material (kg/m^3)
$\tilde{\rho}$	Equivalent density of panel (kg/m^3)
E_f	Young's modulus of the faces (GN/m^2)
E_c, G_c	Young's modulus and shear modulus of the core (GN/m^2)
$\tilde{E}, \tilde{E}_{tt}, \tilde{E}_{flex}$	Equivalent in-plane, through-thickness and flexural modulus of panel (GN/m^2)
$\sigma_{y,f}$	Yield strength of faces (MN/m^2)
$\sigma_{y,c}, \sigma_{ts,c}, \sigma_{c,c}, \sigma_{flex,c}$	Yield strength, tensile strength, compressive strength, and flexural strength of core (MN/m^2)
$\tilde{\sigma}_{in\text{-}plane}$	Equivalent in-plane strength of panel (MN/m^2)
$\tilde{\sigma}_{flex1}, \tilde{\sigma}_{flex2}, \tilde{\sigma}_{flex3}$	Equivalent flexural strength of panel, depending on mechanism of failure (MN/m^2)

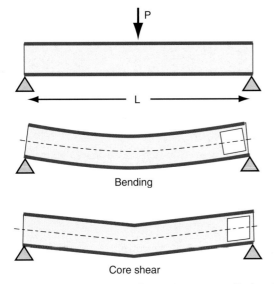

Fig. 9. Sandwich panel flexural stiffness. There are contributions from bending and from core shear.

where the dimensions, d, c, t, and L are identified in Fig. 8, E_f is Young's modulus of the face sheets and G_C is the shear modulus of the core. The load configuration determines the constant values (B_1, B_2, B_3, and B_4) as summarized in Table II. Comparison with $\delta = \frac{12PL^3}{B_1 \tilde{E}_{flex} \tilde{I}}$ (where $\tilde{I}$ is the second moment of area, $\tilde{I} = bd^3/12$) for the "equivalent" homogeneous material gives the equivalent flexural modulus[6,10–14]:

$$\frac{1}{\tilde{E}_{flex}} = \frac{1}{E_f \left\{ \left(1 - (1-f)^3 \right) + \frac{E_c}{E_f}(1-f)^3 \right\}} + \frac{B_1}{B_2} \left(\frac{d}{L} \right)^2 \frac{(1-f)}{G_c} \tag{17}$$

The bending stiffness for the sandwich is recovered by forming the quantity $\tilde{E}_{flex} \tilde{I}$.

(D) In-plane and Through-Thickness Yield Strength: We equate the in-plane strength to the sum of the contributions of face and core, assuming they both must yield for the panel as a whole to yield:

$$\tilde{\sigma}_{y,ip} = f\sigma_{y,f} + (1-f)\sigma_{y,c} \tag{18a}$$

The through-thickness yield strength is the lesser of those of the faces and core:

$$\tilde{\sigma}_{y,tt} = Lesser\, of\, (\sigma_{y,f},\ \sigma_{y,c}) \tag{18b}$$

(E) Flexural Strength: Sandwich panels can fail in flexure in many different ways (Fig. 10). The failure mechanisms compete, meaning that the one that happens at the lowest load dominates. We calculate an *equivalent strength* for each mode, then seek the lowest.
(a) Face Yield: The fully plastic moment of the sandwich is

$$M_f = \frac{b}{4} \left\{ (d^2 - c^2)\sigma_{y,f} + c^2\sigma_{flex,c} \right\}$$

The equivalent flexural strength when face yielding is the dominant failure mode is then

$$\tilde{\sigma}_{flex1} = \frac{4M_f}{bd^2} = \left(1 - (1-f)^2 \right)\sigma_{y,f} + (1-f)^2\sigma_{flex,c} \tag{19}$$

This is the flexural strength the equivalent "material" of a homogenous beam with the same width b, depth d and bending strength as the sandwich panel.
(b) Face Buckling: In flexure, one face of the sandwich is in compression. If it buckles, the sandwich fails. The face-stress at which this happens[10] is

$$\sigma_b = 0.57 \left(E_f E_c^2 \right)^{1/3} \tag{20}$$

The failure moment M_f is then well approximated by

$$M_f = \sigma_b \frac{b}{6d} \left(d^3 - (d - 2t)^3 \right)$$

Table II.   Constants to Describe Modes of Loading

Mode of loading	Description	B_1	B_2	B_3	B_4
	Cantilever, end load	3	1	1	1
	Cantilever, uniformly distributed load	8	2	2	1
	Three point bend, central load	48	4	4	2
	Three point bend, uniformly distributed load	384/5	8	8	2
	Ends built in, central load	192	4	8	2
	Ends built in, uniformly distributed load	384	8	12	2

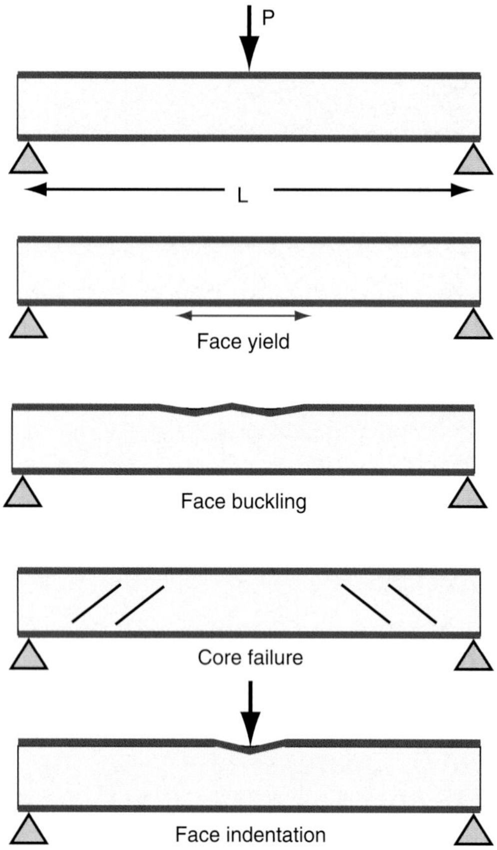

Fig. 10. Failure modes of sandwich panels in flexure.

Equating this to

$$M_f = \tilde{\sigma}_{flex2}\frac{bd^2}{6}$$

gives

$$\sigma_{flex2} = 0.57(1 - (1-f)^3)(E_f E_c^2)^{1/3} \tag{21}$$

(c) **Core Shear:** Failure by core shear occurs at the load

$$P_f = B_4 bc(\tau_{y,c} + \frac{t^2}{cL}\sigma_{y,f})$$

Here the first term results from shear in the core (shear strength $\tau_{y,c}$), the second from the formation of plastic hinges in the faces. Equating to

$$P_f = \frac{B_3 bd^2}{4L}\tilde{\sigma}_3$$

gives the equivalent strength when failure is by shear:

$$\tilde{\sigma}_{flex3} = \frac{B_4}{B_3}\left\{4\frac{L}{d}(1-f)\tau_{y,c} + f^2\sigma_{y,f}\right\} \tag{22}$$

(The load configuration determines the constant values B_3 and B_4, as summarized in Table II) When the core material is roughly isotropic (as foams are) $\tau_{y,c}$ can be replaced by $\sigma_{c,c}/2$. When it is not (an example is that of a honeycomb core), $\tau_{y,c}$ must be retained.

(d) **Core Bending:** In the extreme case that face sheets have very low strength, or that the face-sheet thickness is set to zero, the "sandwich" still exhibits the bending strength of the core itself:

$$\tilde{\sigma}_{flex4} = \sigma_{flex,c} \tag{23}$$

(e) **Indentation:** A load P distributed over a length a and width b of the panel exerts a local indentation pressure of

$$\frac{P}{ab} = p_{ind} = \frac{2t}{a}(\sigma_{y,f}\sigma_{y,f})^{1/2} + \sigma_{y,c}$$

(Ashby *et al.*,[6] eq. [10.13]) from which we find the minimum face thickness t^* to avoid indentation

$$t^* \geq \frac{a}{2}\left(\frac{P}{ab} - \sigma_{y,c}\right)\cdot\left(\frac{1}{(\sigma_{y,c}\cdot\sigma_{y,c})^{1/2}}\right)$$

The flexural strength is set equal to
Greater of σ_{flex4} and (Least of σ_{flex1}, σ_{flex2}, and σ_{flex3}). Indentation has to be treated separately as a constraint on the face-sheet thickness.

IV. Can hybrids expand the Filled Part of Material-Property Space?

The figures that follow compare the properties of hybrids with those of monolithic materials using the models of Section III. Many have the ability to expand the occupied region of property space.

(1) Foams and Lattices Based on Aluminum–SiC Composites (Figs. 11 and 12)

This is an example of the modeling of modulus and density of cellular structures. The starting point is aluminum 20% SiC(p),[‡] a mix that is the basis of one of the Cymat[15] range of metal foams. The equations describing the modulus and density of foams and lattices of Section III*(1)* have been evaluated for a range of relative densities. They appear on the Modulus—Density chart in Fig. 11 as lines of yellow and green ellipses. The starting material, Al 20% SiC(p), is identified at the upper right. The modeled foams, plotted yellow, are labeled in red with their relative densities in brackets. Measured data for real aluminum–SiC(p) foams, also with relative densities in brackets, are labeled in black, allowing a comparison. Note that the lattices, plotted in green, lie at higher values of modulus for the same density. At a relative density of 3.5% (0.035) the lattice is predicted to be 10 times stiffer than a foam of that density.

Figure 12 extends the comparison to a chart on which 3000 materials from all classes are plotted, using $E^{1/2}/\rho$ as a criterion of excellence. The lattice structures out-perform all of them, expanding the populated area of the chart.

(2) Stiff Sandwich Structures: Aluminum-Faced PVC Foam-Cored Sandwiches (Fig. 13)

In designing sandwich panels and shells it makes sense to choose stiff, strong materials for the thin faces and light weight materials for the core, the role of which is to separate the faces and carry shear stresses. The model equations will, however, work for other choices.

Figure 13 shows the flexural moduli and densities of a set of sandwich panels with faces of 6061 T4 aluminum alloy and cores of a PVC foam with a density of 100 kg/m³ (0.1 Mg/m³). The face-sheet thickness is varied from 0.1 to 3 mm in three (logarithmic) steps. For each of these there are five thicknesses of core, ranging from 10 to 50 mm, distinguished by color. The choice of the parameters B_1/B_2 and of B_3/B_4 are determined by the way the panel is loaded, as explained in Section III*(2)*.

[†](p) means "particulate."

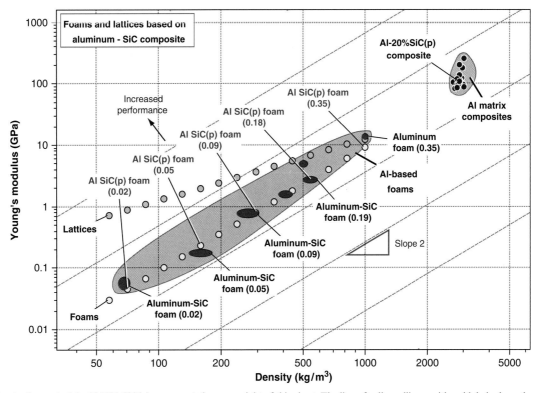

Fig. 11. The starting material,, Al 20% SiC(p) appears at the upper right of this chart. The line of yellow ellipses with red labels show the modulus and density of foams made from Al–SiC(p) They should be compared with the measured values for real aluminum SiC(p) foams, shown in red with black labels. The relative densities are listed in brackets. The modulus and density of lattices made of the same material are shown in green for comparison.

Here they are set at those for a panel centrally loaded in 3-point bending. The face-sheet material is identified at the top right, the core material at the lower left, both labeled in black. The sandwich panels, labeled in red, lie on an arc linking the two.

How good are they? Here we need a criterion of excellence. The index M for selecting materials for a light, stiff panel[4] is

$$M = \frac{E_{\text{flex}}^{1/3}}{\rho}$$

that plots as a set of lines of slope 3 in Fig. 13 (here E_{flex} is the flexural strength and ρ is the density). A contour that is tangent

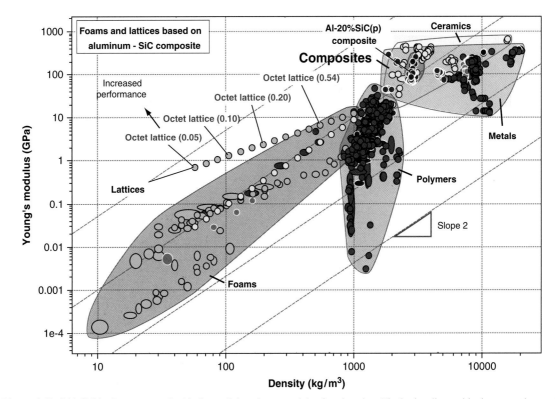

Fig. 12. The moduli of Al–SiC lattices compared with those of the other materials of engineering. The lattices lie outside the currently populated areas of the chart. Numbers in brackets are relative densities.

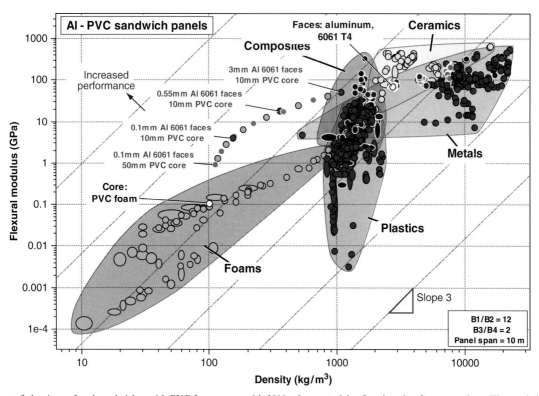

Fig. 13. A set of aluminum-faced sandwiches with PVC foam cores, with 2800 other materials of engineering for comparison. The sandwiches expand the filled part of the space.

to the arc of sandwiches identifies the optimum choice of face and core thicknesses.

The best of the sandwiches ranks more highly than any other material by this criterion. The chart illustrates how effectively sandwich structures can populate holes in material-property space.

(3) Strong Sandwich Structures: Aluminum-Faced PVC Foam-Cored Sandwiches (Fig. 14)

Figure 14 is a chart in which strength, rather than stiffness, is explored. The core is PVC foam with a density of 100 kg/m^3, as in the previous figure. The face-sheet material, in this case, is a high strength aluminum alloy: 7075 in the T6 condition.

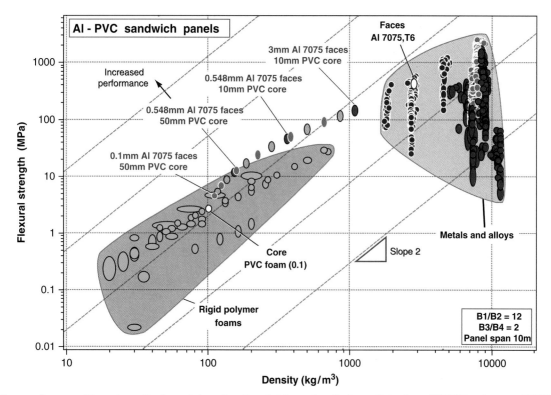

Fig. 14. The arc of orange ellipses shows the flexural strengths of sandwich panels with faces of aluminum 7075 T6 and cores of PVC foam with a density of 100 kg/m^3.

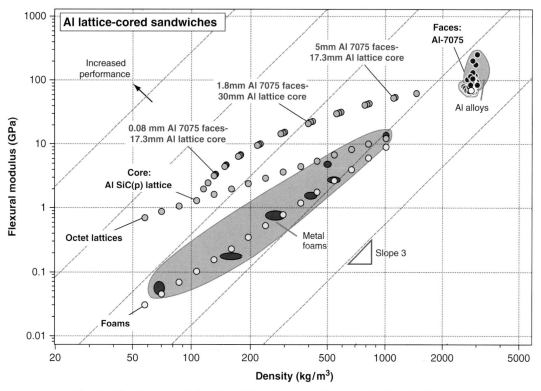

Fig. 15. The flexural moduli and densities lattice-cored aluminum-faced sandwich panels.

The other parameters are set at the same values as in Example 4.2.

The chart shows the flexural strength σ_{flex} and density ρ of this set. The face-sheet material is identified at the top right, the core material at the lower left, both labeled in black. The modeled panels, labeled in red, lie on an arc linking the two.

The criterion for selecting materials for a light, strong panel is the index

$$M = \frac{\sigma_{flex}^{1/2}}{\rho}$$

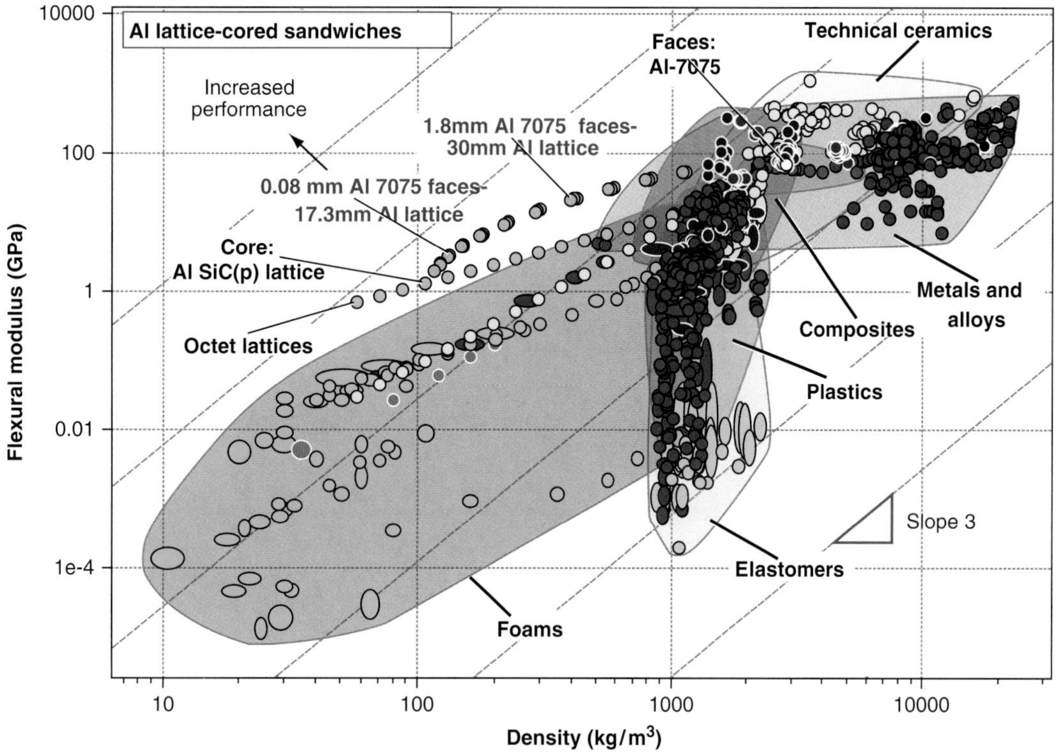

Fig. 16. The same data and axes as Fig. 15, with 2800 other engineering materials added for comparison. The lattice-cored sandwiches offer significant gain in performance.

which plots as a set of lines of slope 2 in Fig. 14. In this example, too, sandwiches perform exceptionally well expanding the filled part of the property space.

(4) Advanced sandwich structures: Aluminum-Faced Lattice-Cored Sandwiches (Figs. 15 and 16)

Sandwich panels are often made with foam cores. Figs. 11 and 12 demonstrated, however, that lattice structures offered considerably better performance than foams at the same density. Here we first calculate properties of lattice structures, then use the one of these as input for the core for a sandwich.

Figure 15 shows the result. The figure has axes of flexural modulus and density. It shows the predicted properties of a string of Al–SiC octet lattices as green circles. One of these, identified on the figure, was selected as the core material and combined with face sheets of aluminum 7075 T6 to create sandwiches. The result is the arc of sandwiches that appear as orange ellipses.

The index M for selecting materials for a light, stiff panel is

$$M = \frac{E_{\text{flex}}^{1/3}}{\rho}$$

which plots as a set of lines of slope 3 in Fig. 15 (here E_{flex} is the flexural strength and ρ is the density). A contour that is tangent to the arc of sandwiches identifies the optimum choice of face and core thicknesses.

In Fig. 16 some 2800 representative engineering materials have been added in Fig. 15. They map out the part of the space that is occupied by conventional monolithic engineering materials. The lattices already expand the occupied area. The lattice-cored sandwiches expand it further.

V. Summary and Conclusions

Hybrid materials have the capacity to expand the ranges of stiffness and strengths of structural materials, spreading the occupied areas of material-property space. This is demonstrated here by comparing the properties of lattice and sandwich structures with those of some 2800 conventional materials. The comparison is made possible by evaluating the "effective" properties of the hybrids: the properties that would be attributed to a monolithic material with the same density, axial and flexural stiffness and strength as the hybrid. The performance of the hybrids, when measured by an appropriate criterion of excellence, surpasses those of any conventional material.

Acknowledgments

I would like to recognize the insights and suggestions that have been provided by discussions with Professors John Hutchinson, Lorna Gibson, Norman Fleck, Vikram Deshpande and David Cebon. In particular I wish to acknowledge the profound influence of a close collaboration with Tony Evans.

References

[1]F. X. Kromm, J. M. Quenisset, R. Harry, and T. Lorriot, "An Example of Multimaterial Design"; *Proceedings Euromat'01*, Rimini, Italy, 2001.

[2]M. F. Ashby and Y. Brechet, "Designing Hybrid Materials," *Acta Mater.*, **51** [19] 5801–2 (2003).

[3]M. F. Ashby, D. Cebon, C. Bream, C. Cesaretto, and N. Ball, "The CES Hybrids Synthesizer—a White Paper. Granta Design, Cambridge, U.K., 2010.

[4]M. F. Ashby, *Materials Selection in Mechanical Design. ISBN 978-1-85617-663-7*, 4th edition, Butterworth Heinemann, Oxford, U.K., 2011.

[5]L. J. Gibson and M. F. Ashby, *Cellular Solids, Structure and Properties. ISBN 0-521-49560-1*, 2nd edition, Cambridge University Press, Cambridge, U.K., 1997.

[6]M. F. Ashby, A. G. Evans, N. A. Fleck, L. J. Gibson, J. W. Hutchinson, and H. N. G. Wadley, *Metal Foams: a Design Guide. ISBN 0-7506-7219-6*. Butterworth Heinemann, Oxford, 2000.

[7]H. N. G. Wadley, N. A. Fleck, and A. G. Evans, "Fabrication and structural performance of periodic metal sandwich structures," *Compos. Sci. Technol.*, **63**, 2331–43 (2003).

[8]V. S. Deshpande, M. F. Ashby, and N. A. Fleck, "Foam Topology: Bending Versus Stretching Dominated Architectures," *Acta Mater.*, **49**, 1035–40 (2001).

[9]R. G. Hutchinson and N. A. Fleck, "The structural performance of the periodic truss," *J. Mech. Phys. Solids.*, 756–82 (2006).

[10]H. G. Allen, *Analysis and Design of Structural Sandwich Panels*. Pergamon Press, Oxford, U.K., 1969.

[11]D. Zenkert, *An Introduction to Sandwich Construction. ISBN 0 947817778*. Engineering Advisory Services Ltd., Solihull, U.K., 1995 published by Chameleon Press Ltd., London, U.K.

[12]J. Pflug, B. Vangrimde, and I. Verpoest, "Material Efficiency and Cost Effectiveness of Sandwich Materials"; *Sampe US Proceedings*, Orlando, FL, 2003.

[13]J. Pflug, I. Verpoest, and D. Vandepitte. (2004) SAND.CORE Workshop, Brussels, December, 2004.

[14]J. Pflug, "Sandwich Materials Selection Charts," *J. Sandwich Struct. Mater.*, **8** [5] 407–21 (2006).

[15]Cymat Technologies Ltd. Available at http://www.cymat.com, 2010 (accessed December 2, 2010). ☐

J. Am. Ceram. Soc., **94** [S1] S15–S34 (2011)
DOI: 10.1111/j.1551-2916.2011.04599.x
© 2011 The American Ceramic Society

journal

Protocols for the Optimal Design of Multi-Functional Cellular Structures: From Hypersonics to Micro-Architected Materials

Lorenzo Valdevit,[‡,†] Alan J. Jacobsen,[§] Julia R. Greer,[¶] and William B. Carter[§]

‡ Mechanical and Aerospace Engineering Department and Chemical Engineering and Materials Science Department,
University of California, Irvine, California, 92697

§ HRL Laboratories, Malibu, California, 90265

¶ Materials Science Department, California Institute of Technology, Pasadena, California, 91125

Cellular materials with periodic architectures have been extensively investigated over the past decade for their potential to provide multifunctional solutions for a variety of applications, including lightweight thermo-structural panels, blast resistant structures, and high-authority morphing components. Stiffer and stronger than stochastic foams, *periodic* cellular materials lend themselves well to geometry optimization, enabling a high degree of tailorability and superior performance benefits. This article reviews a commonly established optimal design protocol, extensively adopted at the macro-scale for both single and multifunctional structures. Two prototypical examples are discussed: the design of strong and lightweight sandwich beams subject to mechanical loads and the combined material/geometry optimization of actively cooled combustors for hypersonic vehicles. With this body of literature in mind, we present a motivation for the development of *micro-architected materials*, namely periodic multiscale cellular materials with overall macroscopic dimensions yet with features (such as the unit cell or subunit cell constituents) at the micro- or nano-scale. We review a suite of viable manufacturing approaches and discuss the need for advanced experimental tools, numerical models, and optimization strategies. In analyzing challenges and opportunities, we conclude that the technology is approaching maturity for the development of micro-architected materials with unprecedented combinations of properties (e.g., specific stiffness and strength), with tremendous potential impact on a number of fields.

I. Introduction

STOCHASTIC cellular materials (i.e., foamed materials that contain significant amounts of porosity) have long been used for their low weight, high sound absorption, crashworthiness, and thermal properties.[1] Approximately 15 years ago, advances in manufacturing technologies spearheaded a large academic and industrial interest in metallic foams,[2] which combine all the properties listed above with increased specific strength and stiffness and high-temperature capabilities. An important feature of open-cell foams is the interconnected open space, which can be employed to enable additional capabilities, such as active cooling[2,3] or energy storage,[2,4–6] thus enabling multifunctionality. More recently, detailed mechanical experiments on metallic foam-based sandwich panels under bending and compressive loads revealed that all foams are bending-dominated, i.e., they deform by compliant and weak bending modes of the cell walls and ligaments, inefficiently using the base constituent material in the foam by leaving much of it out of the load path.[7–9] In addition, their stochastic nature inevitably introduces imperfections that further depress their mechanical properties.[10] Vastly increased specific stiffness and strength (i.e., stiffness and strength per unit weight) can be obtained in *periodic cellular architectures* (such as those depicted in Fig. 1); if designed properly, under global bending and compressive loadings these architectures will deform by stretching of the ligaments, a stiff and strong local deformation mode that makes maximal use of the base constituent and maximizes load carrying capacity.[4,11] As an additional benefit over "semi-engineered" open-cell foams, periodic cellular topologies have many more geometrical features that can be engineered and optimized. A large body of research has been published in the past decade on optimally designed metallic periodic cellular systems, with emphasis on specific strength,[11–17] active cooling,[18] and combinations thereof,[19–21] combined strength and thermal conductivity (through a heat pipe design),[22] high-velocity impact absorption,[23–26] and high-authority shape morphing potential.[27–31]

In spite of the variety of applications (each imposing different objective functions and constraints), the same protocol for optimal design has been consistently (and successfully) adopted in nearly all cases. This protocol consists of a combination of analytical, numerical, and experimental techniques, and is reviewed in Section II of this article, with emphasis to mechanical and thermo-mechanical optimization. In Section III of this article, we pose three questions: (i) Are there any mechanical benefits in designing *micro-architected materials* (namely, macro-scale periodic cellular materials with unit cells at the micro-scale)? (ii) Are there suitable and cost-effective manufacturing processes for micro-architected materials? (iii) Is the optimal design protocol (including analytical, numerical and experimental techniques) which has been successfully adopted for large-scale structures appropriate to harness the full potential of micro-architected materials? By answering these questions, we conclude that the technology is approaching maturity for the development, characterization, and optimal design of a novel class of multifunctional materials with the potential to achieve unprecedented combination of properties.

II. Micro-Architected Materials

This section briefly reviews the well-established optimal design protocol for cellular periodic structures. Manufacturing approaches are described first, to offer a flavor of the topologies and materials combination that are readily available. The optimal design protocol (consisting of a combination of

T. M. Pollock—contributing editor

Manuscript No. 28794. Received October 16, 2010; approved April 02, 2011.
†Author to whom correspondence should be addressed. e-mail: valdevit@uci.edu

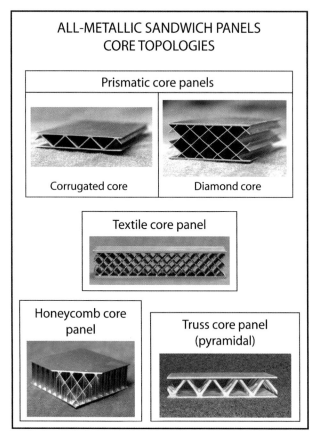

ALL-METALLIC SANDWICH PANELS CORE TOPOLOGIES

Prismatic core panels

Corrugated core

Diamond core

Textile core panel

Honeycomb core panel

Truss core panel (pyramidal)

Fig. 1. Examples of all-metallic sandwich panels manufactured by Transient Liquid Phase (TLP) bonding. Modified from Lu *et al.*[3]

analytical, numerical, and experimental techniques) is presented for two archetypal structures: a simple single-function structure (a lightweight periodic sandwich beam with a prismatic corrugated core, designed for resistance to mechanical loads—bending and transverse shear), and a more complex multifunctional structure (a lightweight actively cooled sandwich plate, designed for resistance to mechanical and thermal loads—and additional design constraints). The latter is presented in the context of materials development for hypersonic vehicles. In both cases, materials selection is addressed.

(1) Manufacturing Approaches

Reliable manufacturing techniques have been developed for metallic sandwich structures with a number of core topologies. Cores are generally assembled by folding a plate (as in case of corrugated cores or truss cores) or slotting and assembling a large number of beam and/or plate elements (honeycomb cores, diamond prismatic cores, textile cores); in the latter case, the constituents need to be metallurgically bonded to impart strength and stiffness to the structure.[6] In both approaches, face sheets are bonded to the core. As bonded nodes are inevitably subjected to substantial in-service loads (in tension, compression and shear), a manufacturing process that results in strong nodes is essential. For a number of materials such as steels, copper, and aluminum alloys, Transient Liquid Phase Bonding (TLP), a high-temperature brazing process involving significant inter-diffusion at the joints, is the ideal technique.[32] TLP enables much stronger structures than conventional, lower-temperature brazing because the resulting nodes have nearly the same chemical composition as the base metal. TLP is also better than welding, both for simplicity and scalability (all the nodes are formed at once, without need for line-of-sight access) and because solidification at the brazing temperature ensures much lower residual stresses in the bonded region

compared to welding. Examples of sandwich panels with various core topologies obtained with TLP are provided in Fig. 1. Most of the experimental work published to date on all metallic sandwich panels pertains to steel panels, for which TLP brazing agents are readily available. TLP recipes for aluminum and copper alloys also exist.[33]

Recently, alternative manufacturing approaches were developed for aerospace-relevant high-temperature alloys. Titanium (Ti6Al4V) panels were manufactured with diffusion bonding, resulting in good nodal microstructure and strengths.[34] The extension to even higher temperatures requires nickel superalloys. Unfortunately, high-strength, γ'-rich nickel superalloys are not formable at room temperature. A clever solution for thin-gage panels was recently proposed, whereby a formable, single-phase γ-Ni superalloy is assembled in the right shape, all the components are assembled via TLP bonding, and subsequently the finished structure is aluminized and precipitation hardened, resulting in a high-strength, γ'-rich alloy.[35]

For lower temperature applications, polymer-matrix composites (e.g., Carbon-epoxy) are available with significantly higher weight efficiency than other metals.[36] The primary manufacturing issue is ensuring sufficient nodal strength. Carbon-epoxy honeycomb core panels obtained by a slotting procedure were recently demonstrated and optimized for compressive loads.[37] For corrugated core panels, 3D weaving is a natural option, albeit at an increase in cost and manufacturing complexity.[38] A number of simpler, prepreg-based approaches are currently under consideration.[36]

(2) Design Protocol for Maximum Specific Strength

Periodic cellular materials have the prominent feature of being naturally suitable to optimization. In addition to selecting the ideal base material (or combinations thereof in the case of a hybrid), the architecture can be optimized for a specific objective (or multiple objectives) subject to a number of constraints. The general multi-step procedure can be summarized as follows:

1. Fundamental properties of the base material(s) are obtained, either from data sheets or through appropriate experimental characterization, e.g., dog-bone tensile testing resulting in a stress-strain curve—possibly including temperature and time effects: visco-elasticity, visco-plasticity, fatigue, etc.

2. The evolution of the variables of interest (stress, strain, temperature, electric potential, etc.) is modeled analytically as a function of the structure geometry and the applied loads (mechanical, thermal, electrical, etc.). Constraints are formulated for the specific application under consideration (e.g., no yielding or buckling anywhere in the structure, no melting of the material).

3. Numerical analyses are performed, typically employing commercial Finite Elements packages, to verify the validity of the simplifying assumptions underlying step (2).

4. A combination of steps (2) and (3) is coupled with an optimization routine (quadratic optimizers for convex problems, discrete algorithms for problems featuring many local minima) and the structure geometry and/or material are optimized subject to all the prescribed constraints. The objective function strongly depends on the specific problem.

5. A prototype (possibly to scale) of the *entire* optimal or near-optimal structure (or at least substructure) is manufactured and its performance is verified *experimentally* to verify all the modeling assumptions (underlying both (2) and (3)).

As an example, we examine optimization of a metallic corrugated-core sandwich panel loaded by any combination of bending and transverse shear loads (Fig. 2(a)).[12,13] The objective is maximum strength at a prescribed weight. Nondimensional load intensity and weight are defined as:

$$
\begin{aligned}
\Pi &= \frac{V^2}{EM} \\
\psi &= \frac{W}{\rho \ell^2} = 2\frac{d_f}{\ell} + \frac{1}{\cos\theta}\frac{d_c}{\ell}
\end{aligned}
\tag{1}
$$

where V and M are the maximum shear force and bending moment per unit width of the panel, respectively, $\ell = M/V$ is the governing length-scale in the problem, E and ρ are the Young's modulus and density of the base material, respectively, and the geometric variables d_f, d_c, θ are defined in Fig. 2(a). The length-scale ℓ defines the actual loading condition (e.g., $\ell = L/2$ for three-point bending, L being the span of the panel between the loading points); normalizing all the dimensions with ℓ renders generality. Four possible failure mechanisms are identified: face (FY) and core (CY) yielding, and face (FB) and core (CB) buckling. For transverse loadings (bending about an axis parallel to the corrugation—Fig. 2(a)), analytical expressions are readily derived:

$$
\begin{aligned}
\left(\frac{V^2}{EM}\right)_{FY} &= \frac{\varepsilon_Y d_f}{\ell}\left(\frac{H_c}{\ell} + \frac{d_f}{\ell}\right) \\
\left(\frac{V^2}{EM}\right)_{CY} &= \frac{\varepsilon_Y d_c \sin\theta}{\ell} \\
\left(\frac{V^2}{EM}\right)_{FB} &= \frac{k_f \pi^2 \tan^2\theta}{48}\left(\frac{H_c}{\ell} + \frac{d_f}{\ell}\right)^{-1}\left(\frac{d_f}{\ell}\right)^3 \\
\left(\frac{V^2}{EM}\right)_{CB} &= \frac{k_c \pi^2 \sin^3\theta}{12}\left(\frac{H_c}{\ell} + \frac{d_f}{\ell}\right)^{-2}\left(\frac{d_c}{\ell}\right)^3
\end{aligned}
\tag{2}
$$

where ε_Y is the yield strain of the constituent material. Similar equations can be derived for longitudinal loadings. Note that ε_Y is the *only* material property governing the problem. The implication is that the optimal material for a corrugated-core sandwich panel subject to any combination of bending and transverse shear loads is simply the material with the largest yield strain. The same conclusion applies to any other core topology and a number of mechanical loading conditions. This important result allows separation of materials selection and optimal topological design. Multifunctional problems involving more complex physics often lack this feature, requiring material and topology to be concurrently optimized (see Section II(3)).

The interplay of failure mechanisms is best illustrated with failure mechanisms maps (Fig. 2(b)). If the corrugation angle, θ, and the panel weight, Ψ, are fixed, panel geometry is entirely defined by the thickness of the core, H_c and face sheet, d_f. Hence, each point on the map represents a possible panel design, with all designs having the same weight. The various regions denote design spaces where panel strength is governed by each failure mechanism (core yielding is never active under these loading conditions and weight). Strength contours (expressed in nondimensional form) clearly identify that the best design occurs at the intersection of three failure mechanisms (Incidentally, the confluence of three failure mechanisms at the optimal design point is a recurring feature for many core topologies,[11,12,15–17,39] but this condition is not universal (C. A. Steeves, Personal Communication).

Before it can be used with confidence, this model needs to be verified with a combination of numerical (FE) analyses and validated with a selected set of experiments. Figure 2(c) shows excellent agreement between analytical (white dot) and numerical predictions with experimental results for one particular design loaded in three-point bending (black dot in Fig. 2(b)). The inset in Fig. 2(c) compares the deformed shape of the panel at the end of the experiment. Note that both face and core buckling are evident (face yielding was also

verified with a strain gage during the experiment), consistent with the analytical predictions of Fig. 2(b).

Numerical and experimental validation of the analytical model allows computationally efficient design optimization for a wide range of applied load intensities. With reference to Eqs. (1–2), the problem can be stated as follows: for any given load intensity, Π, minimize the panel weight, Ψ, subject to four constraints ($\Pi < \Pi_{FY}$, $\Pi < \Pi_{FB}$, $\Pi < \Pi_{CY}$, $\Pi < \Pi_{CB}$). As all functions are convex, a simple quadratic optimizer was successfully used. Results for aluminum panels are presented in Fig. 2(d). This master figure compares the weight efficiency of a number of optimally designed core topologies; the corrugated core panel loaded transversely (discussed herein) is much lighter than solid plates, but more efficient topologies can be devised (hexagonal honeycombs are optimally efficient in this loading condition, and have often been used as benchmarks).

(3) Design Protocol for Multifunctional Structures: An Example From Hypersonics

(A) Preliminaries: For multifunctional applications, the challenge is choosing a cellular material with the best combination of constituent material and architecture to optimize all the desired objective functions under a series of design constraints. The multi-step protocol of Section II(2) can be adapted to this more challenging scenario, although the computational intensity quickly grows as the physics of the problem becomes more complex. Herein, we discuss the optimization of a simple architected material (a prismatic sandwich panel with hollow rectangular channels) for minimum weight under the simultaneous application of mechanical and thermal loads (subject to a number of design constraints). The motivation is a feasibility study for metallic actively cooled combustors for hypersonic vehicles (Fig. 3(a)). Details beyond this concise treatment are provided in a number of references.[19–21,40]

(B) Thermo-Mechanical Loads on Combustor Wall of a Hypersonic Vehicle: Whereas acreage (and to some extent, leading edges) of thermally balanced hypersonic vehicles can be engineered to passively dissipate heat by radiation, combustor walls inevitably require active cooling strategies to contend with the large heat fluxes arising from the combustion process. The prototypical structure, a sandwich panel with prismatic channels that provide active cooling by the fuel before injection (Fig. 3(b)), is subjected to significant thermo-mechanical loads. The thermal loads are represented by a heat transfer coefficient and a hot gas temperature on one side of the panel. The mechanical loads are (i) pressure in the combustion chamber (which, depending on the boundary conditions, can induce panel-level bending), and (ii) pressure in the cooling channels (dictated by fuel injection requirements). As a result, significant thermo-mechanical stresses arise. Withstanding these stresses at operating temperatures necessitates careful design. A viable solution must resist several failure modes: yielding or rupture due to (a) thermal stresses, (b) pressure or inertial stresses, (c) combined thermo-mechanical stresses, as well as (d) softening of the material, (e) coking of the coolant, and (f) excessive pressure drop in the cooling ducts. The objective is twofold: (i) identify the optimal material and (ii) optimize the structure for minimum weight. The challenge is to assure that none of the failure modes is active over the pertinent ranges of coolant flow rate, V_{eff} (often nondimensionally expressed in terms of the air/fuel mixture richness, ϕ, relative to stoichiometric combustion) and thermal loads (expressed by the heat-transfer coefficient between the combustion gas and the solid surface, h_G). The intensity of the mechanical loads is assumed constant for simplicity. Geometry and loads are depicted in Fig. 3(b).

(C) Optimal Design Protocol: The multi-step approach of Section II(2) is applicable, albeit with the complication that even a simplistic analytical model precludes the extraction of a

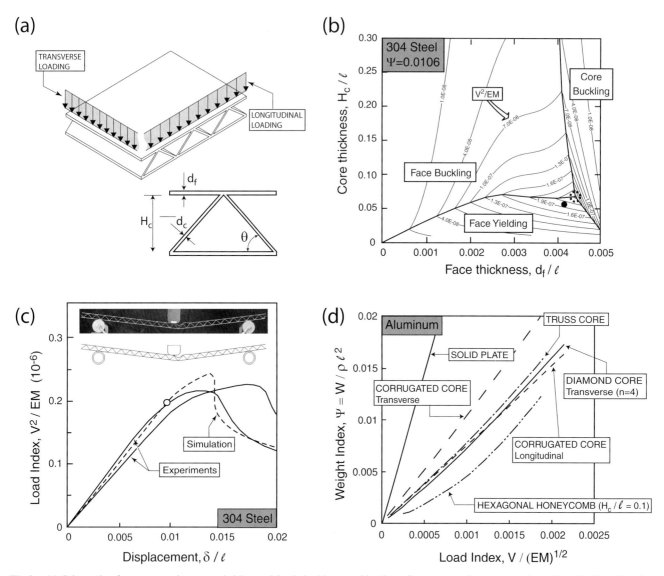

Fig. 2. (a) Schematic of a corrugated-core sandwich panel loaded with a combination of moment and transverse shear. Two loading directions (referred to as transverse and longitudinal) are indicated. (b) Failure map for steel corrugated core panels loaded in the transverse direction. Each point represents a different design (all at the same weight). In this case, the confluence of the three regions denotes the maximum strength design. (c) Comparison of analytical and numerical predictions and experimental results for a 3-point bending experiment on the steel corrugated core panel indicated by a black dot in (b). (d) Weight-efficiency of optimized aluminum panels for a number of core topologies. Adapted from Valdevit et al.[12,13]

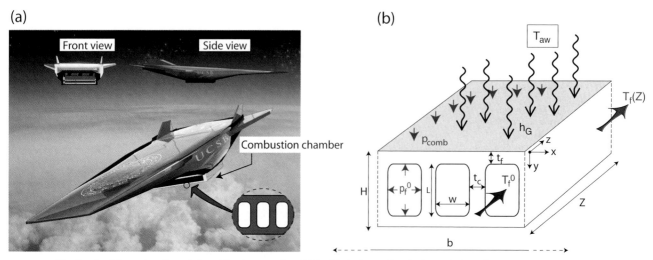

Fig. 3. (a) Schematic of an air-breathing hypersonic vehicle. (b) Archetypal actively cooled combustor with applied loads.

simple materials performance index. The implication is that materials and structural designs must be tackled concurrently. A flowchart of the optimization protocol is presented in Fig. 4. To explore the feasibility of a number of materials over a series of operating conditions, ranges of thermal load (i.e., heat transfer coefficient from the combustor side), and cooling efficiency (i.e., coolant flow rate) are explored. A suite of high-temperature metallic materials were investigated (with and without thermal barrier coatings[19]), all benchmarked with the state-of-the-art ceramic matrix composite C-SiC. Integration of the actively cooled panel with the rest of the vehicle largely affects the thermo-mechanical stresses. Although several conditions were investigated, herein we will focus on a flat panel supported in discrete locations, separated by a span, L. Once a material is chosen, and specific values of thermal loads and cooling efficiency are selected, the thermo-mechanical problem is fully defined. Analytical models based on a thermal network and plate/beam theory provide the temperature and stress distributions. Please see Valdevit *et al.*[19] for details. The accuracy of these models was checked against selected Computational Fluid Dynamics (CFD) and Finite Elements calculations. The model/FE agreement for the Von Mises stress distribution is illustrated in Fig. 5. The graphs track stress variations along four paths in the unit cell. Notice that the agreement for thermal, mechanical, and thermo-mechanical stresses is excellent throughout, except for two cases. (a) At the internal nodes (points 2 and 3 in Fig. 5), significant stress intensifications (naturally not predicted by the analytical model) arise. This discrepancy is disregarded for three reasons[19]: (i) For this particular simulation, the temperatures at the corners are relatively low (and hence the yield strength relatively high), so that the corners remain elastic. We speculate that this concept generalizes to all metallic systems of interest, although a formal proof requires further analysis. (ii) The fillet radius can be increased in actual designs, ameliorating the stress intensification. (iii) Local plasticity at the nodes upon a few cycles can be accepted, provided that it is followed by shakedown. (b) Thermal stresses are underpredicted by ~20% at the cold face (points 3, 4, 7, and 8 in Fig. 5). This discrepancy can be related to modeling assumptions. Simple expressions for the thermal stresses were obtained assuming that the entire core is at the same temperature as the cold face sheet. FE analyses confirmed that this assumption results in accurate stress predictions on the hot face, whereas it underestimates the thermal stress in the cold face. As the yield strength of the materials decreases with increasing temperature, the cold face is never prone to failure, rendering this inaccuracy inconsequential.

Once the analytical model is validated by numerical analyses, it can be successfully used for efficient optimization studies (Experimental investigation is ultimately necessary to close the design loop, but this requires substantial dedicated test facilities and is beyond the scope of this work). A simple quadratic optimizer (FMINCON, available in the MATLAB suite) is used to minimize the panel mass subject to the constraints defined above. As this thermo-mechanical problem is more complex than the simple mechanical optimization discussed in Section II(2) (in that a number of local optima arise), a set of randomly generated initial guesses is introduced to seek the global optimum. An alternative would be the use of discrete (i.e., non gradient-based) optimization algorithms, inherently more robust against local optima (see Section III(5)(C)). When the optimizer finds a solution, the set of geometric parameters yielding the minimum weight design is stored. Conversely, when the optimizer fails to find a solution within the standard number of iterations defined in FMINCON, the material under consideration is deemed infeasible for the specific set of thermal loads and cooling efficiency. The procedure is repeated for a set of points scanning the thermal loads/cooling efficiency space, and for a suite of high-temperature materials.

(D)   Results: Optimal Designs and Ideal Materials: The ensuing information can be presented in two complementary

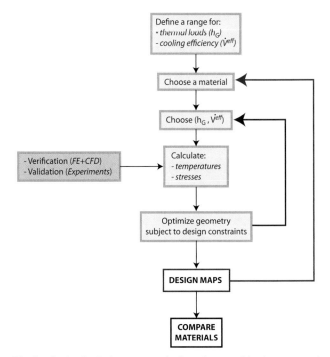

Fig. 4. Optimal design protocol for the combined geometry/materials selection of actively cooled panels for combustor liners in hypersonic vehicles. Reprinted from Valdevit *et al.*[19]

ways: (i) materials robustness maps (Fig. 6(a)) and (ii) weight-efficiency plots (Fig. 7). Materials robustness maps depict the region in the thermal load/cooling efficiency space where a given material provides a feasible solution, irrespective of its weight (orange area in the maps of Fig. 6(a)). The gray area in the maps extends the feasible region to higher thermal loads and/or lower cooling efficiency by allowing a thermal barrier coating to be interposed between the panel hot side and the combustion chamber (A conventional YSZ columnar TBC is assumed, with through-thickness thermal conductivity of 1 W/m K, in-plane conductivity of 0 W/m K, and mass density of 3000 kg/m³). The TBC thickness (not to exceed 300 μm and 25% of the face sheet thickness) is chosen by the optimizer, as a compromise between added weight and reduced temperatures in the underlying metallic structure (See Valdevit *et al.*[19] and Vermaak *et al.*[21] for details). Four different materials are illustrated in Fig. 6(a): a Niobium alloy (C-103), the ceramic matrix composite C-SiC (benchmark material), a high-temperature Copper alloy (GrCop-84), and a Nickel superalloy (Inconel X-750). Materials properties are provided in Valdevit *et al.*[19] and Vermaak *et al.*[21] For the particular set of boundary conditions adopted here, the four materials show similar robustness (loosely defined as the area of design feasibility), but this conclusion can change greatly as the span between panel supports is shortened.[19,21]

Importantly, the optimal design tool described in Section II(3)(C) (and depicted schematically in Fig. 4) can be used as a preliminary screening tool for new materials development. The mechanical properties of Inconel X-750 can be improved by alloying or heat treatment, generally resulting in increased flow stress or increased softening temperature, but not both in the same material (Fig. 6(b)). The question is which of the two property improvements would be most beneficial to the application being considered. The answer is provided in Fig. 6(c); for the boundary conditions used in this study, a 20% increase in the flow stress (without extending the softening temperature) has a much larger impact on the robustness of the material than a similar increase in softening temperature (without elevating the flow stress). Again, the conclusion changes if the panel span is shortened. This information is very important for the materials developer, and it

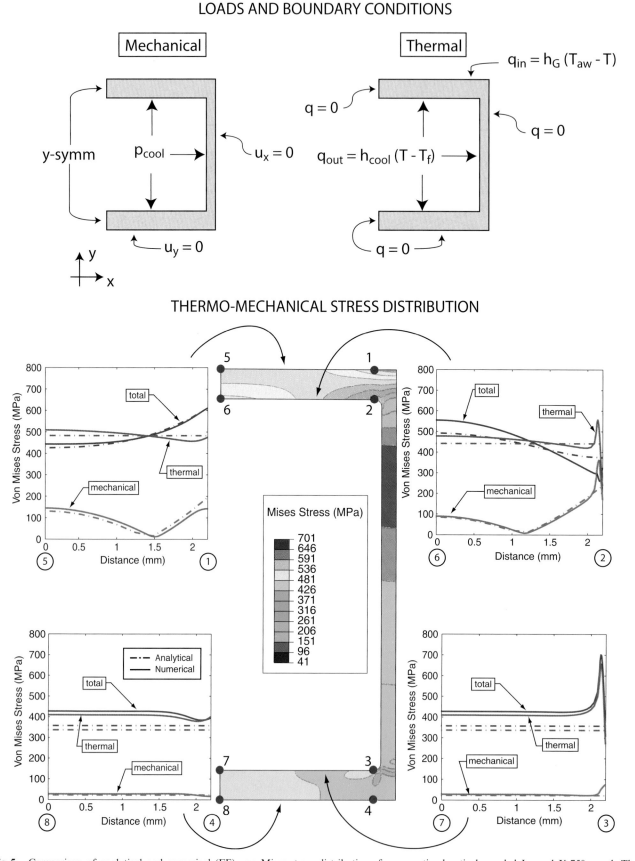

Fig. 5. Comparison of analytical and numerical (FE) von Mises stress distributions for an optimal actively cooled Inconel X-750 panel. The insets show the results for thermal, mechanical, and combined thermomechanical stresses along the four paths depicted. With the exception of Points 2 and 3, clearly affected by stress intensification, the agreement is very satisfactory, validating the optimization results presented in Fig. 6. Modified from Valdevit et al.[19]

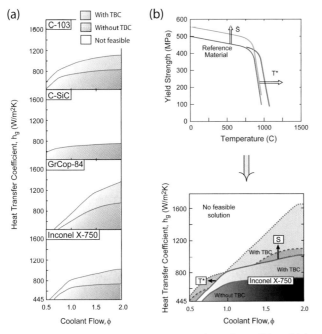

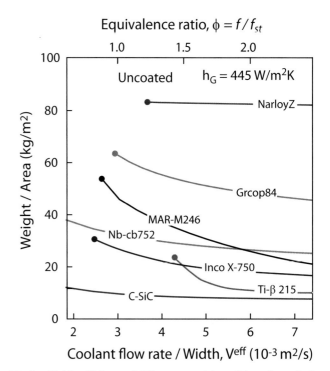

Fig. 6. (a) Performance maps for a number of candidate high-temperature materials for actively cooled walls of hypersonic combustors. The normalizing parameter for the coolant flow equivalence ratio ($\phi = f/f_{st}$) is that expected for steady-state Mach 7 flight conditions. Each map is the result of two independent optimizations. One (yellow) is performed absent a coating. When solutions exist only with a coating (light gray), the optimization is conducted using the TBC thickness as a variable. Areas without a feasible solution are in white. (b) Reclaimed feasibility space provided by two notional materials based on INCONEL X-750. The first notional material (designated S), examines the effect of elevating the yield strength at intermediate temperatures by 25%, whereas the second (designated T*) probes the effect of extending the maximum use temperature by 25%; analyses are shown both with (light) and without (dark) a TBC. For comparison, the notional results are superimposed on the original performance maps for INCONEL X-750 (Modified from Vermaak *et al.*[21])

Fig. 7. Weight efficiency of different material candidates for actively cooled walls of hypersonic combustors. The equivalence ratio, ϕ, is defined as for Fig. 6. From Valdevit *et al.*[19]

is not readily accessible in any other forms (i.e., it does not transparently appear from the equations for temperature and stress distributions). This is a perfect example of the need for efficient optimal design tools to tackle these inherently multifunctional problems.

In Fig. 7, we show weight efficiency for a specific thermal load (realistic for a Mach 7 hydrocarbon-powered vehicle). Each curve tracks the minimum weight of an optimized structure for a given material. Notice that different materials have vastly different weight efficiencies, even if their robustness (the area of the curve in Fig. 6(a)) is somewhat similar. The benchmark material, C-SiC, offer by far the lowest weight. High-temperature titanium alloys (Ti-β 215) offer the lightest metallic systems, but their feasibility is limited to low thermal loads. Among the metals that offer robust solutions over a range of thermal loads, Niobium alloys (e.g., Nb-cb752, C-103) and Nickel superalloys (e.g., Inconel X-750, MAR-M246) are the most promising materials. Allowing the metals to shakedown upon thermo-mechanical cycling results in lighter systems (by as much as 30%–40%, depending on the boundary conditions), further increasing the competitiveness of metallic solutions.[40]

In closing, notice that all these conclusions are not transparently available from the equations, as thermal and mechanical properties are deeply intertwined in this heavily constrained thermo-mechanical problem.

III. Optimal Design of Micro-Architected Cellular Materials

In this section, we present the case for the development of a new class of multifunctional materials, characterized by a periodic cellular architecture with unit cell at the micro-scale and a characteristic dimension for the constituent material in the sub-micrometer region. If manufactured and designed correctly, *these micro-architected materials enable exploitation of potentially useful nano-scale mechanical effects (e.g., size effects in plasticity and fracture) that enhance mechanical properties relative to bulk macro-scale structures.* After reviewing viable manufacturing schemes to exploit this vision, we present the technical rationale for the expected performance and assess both the applicability of the multi-step optimal design approach described in Section II and the availability of suitable experimental and computational tools. In the interest of brevity we focus on mechanical design, although similar concepts can be extended to other functionalities.

(1) Manufacturing Approaches

A viable manufacturing approach for lattice-based micro-architected materials must possess the following key features: (i) dimensional control down to the 0.1–10 μm range; (ii) scalability to macroscopic part dimensions; and (iii) acceptable throughput to enable cost-effective manufacturing. Architectural flexibility (i.e., the capability to generate different unit cell topologies) and a wide suite of base materials are additional desirable attributes. To the best of the authors' knowledge, three families of manufacturing approaches exist today for the fabrication of lattice-based micro-architected materials: scaled-down versions of wire layup[41] and other modular assembly methods[42] (discussed in Section II(1)), stereolithography[43] (including the most advanced 2-photon approach[44]), and a new self-propagating photopolymer waveguide (SPPW) process,[45] recently developed at HRL Laboratories. Key attributes of each method are assessed in Table 1. Modular assembly methods can be useful for a wide range of end-materials, but are currently limited by the achievable resolution (minimum unit cell sizes ~100 μm and minimum feature sizes ~10 μm) and scalability (as the number of unit cells becomes very large, the assembly procedure becomes more and more cumbersome). Stereolithography allows incredible resolution (sub-micron feature sizes and unit cells of the order of a few microns for the most recent two-photon process) and nearly infinite architectural freedom (virtually anything that can be CAD drawn can be made). As a serial process, stereolithography is extremely slow: for a given sample size, the total processing time roughly scales with the inverse of the minimum feature size, implying that macroscopic quantities of micro-architected materials could take days to make. For proof-of-concept and basic research, stereolithography is a very powerful technique, but its difficult scalability makes it currently inadequate for industrial processing.

For a wide range of desirable end-geometries, the best compromise among resolution, architectural freedom, and scalability may be provided by the SPPW process.[45] Polymeric lattices are formed by UV exposure of a two-dimensional photolithographic mask with a pattern of circular apertures that is covering a reservoir containing an appropriate photomonomer (Fig. 8). Within the photomonomer, self-propagating polymer waveguides originate at each aperture in the direction of each collimated UV beam, forming a three-dimensional array of polymer fibers that polymerize together at all points of intersection. After removing the uncured monomer, three-dimensional lattice-based open-cell polymeric materials can be rapidly fabricated. Although this method does not allow for *arbitrary* shapes to be formed within the starting resin bath, it has the potential to form a wide range of free-standing 3D polymer structures based on linear mechanically efficient truss-type elements. In striking contrast with stereolithography, the optical waveguide process can form all truss-type elements in the structure in parallel with a single exposure step, typically lasting less than 1 min. With current UV exposure capabilities, cellular materials with truss member diameters ranging from ~10 μm to >1 mm and a relative density <5% up to 30% have been demonstrated.[46] The overall material thickness, H, can range from 100 μm to over 25 mm (although generally $H < 100 \cdot d$, where d is the truss diameter). Examples of ordered unit cell architectures with different symmetries are shown in Fig. 9; however, this process is not limited to such architectures. Nonsymmetric architectures, functionally graded materials, and hierarchical microlattice structures are all easily obtained.

For manufacturing techniques that result in a polymer template, such as stereolithography and the SPPW process, a number of postprocessing techniques are available to replicate the micro-architectural features with a metal or a ceramic[47,48] (Fig. 10). Continuous metallic film, such as nickel, can be deposited by electroplating or electroless process on the surface of the polymer micro-lattice structure and the polymer template can be subsequently removed with a chemical etch.[49] Controlled thickness coatings are obtained by varying the plating time. Ceramic films can be deposited with chemical vapor deposition (CVD) techniques. To withstand the high temperatures required for CVD of refractory metals or ceramics, the polymeric template must be pyrolyzed with minimal geometric distortion as recently demonstrated.[50] After the CVD process, the carbon micro-lattice template can be removed by oxidation (>600°C in air), leaving a hollow tube ceramic micro-lattice structure, such as the SiC sample shown in Fig. 10.

One key advantage of the polymer→metal or polymer→ceramic conversion process is to capture the strengthening effects associated with a constituent material in thin-film form factor in a bulk form (Section III(2)). These "film form" properties generally require film thicknesses in the micro- (or even nano-) scale, dictating truss diameters ~10–100 μm. This makes the fabrication approach described above ideally suited for fabricating optimal open-cell periodic architectures with exceptionally strong metallic or ceramic constituent materials.

(2) Challenges and Opportunities

The lattice materials manufactured with the SPPW process described in Section III(1) possess two distinct features, not readily available with competing concepts: (i) *hollow truss configurations* and (ii) *multi-scale architectures*, with global sample size on the order of several inches and sub-millimeter unit cell dimensions. These two features provide unique opportunities to expand the current bounds of material properties spaces and achieve combinations of properties currently unavailable in any existing material (including the macro-scale architected materials described in Section II). The target regions for specific strength and stiffness are depicted in Fig. 11. Importantly, micro-architected materials fabricated as described herein maintain an open core architecture, enabling multifunctionality: with reference to Section II(3), strong and stiff structures amenable to efficient active cooling are obviously an attractive possibility. Herein, we briefly review the rationale for these opportunities (limiting our attention to mechanical properties), and summarize the outstanding challenges that must be overcome in order to exploit the full potential of micro-architected materials.

(A) Advantages of a Hollow Truss Configuration: Under any mechanical loadings, the strength of metallic lattice materials designed to operate in the elastic regime is limited by the onset of either yielding or elastic buckling. A simple analysis reveals the benefits of a hollow truss configuration; uniform compressive loading is assumed for simplicity, but the same conclusions qualitatively apply to other loading conditions. Consider a lattice material with a solid truss pyramidal unit cell, defined by truss member length, l, truss diameter, $2a$, and truss angle, ω. The relative density can be expressed as[51]:

$$\bar{\rho} = \frac{2\pi}{\cos^2 \omega \sin \omega} \left(\frac{a}{l}\right)^2 \tag{3}$$

and the compressive strength is:

$$\frac{\sigma_{\text{comp}}}{\sigma_{\text{bar}}} = \bar{\rho} \sin^2 \omega \tag{4}$$

where $\sigma_{bar} = \min\{\sigma_Y, \sigma_b\}$ represents the strength of the individual truss member, with σ_Y the yield strength of the base material and $\sigma_b = k^2 \pi^2 E a^2 / 4 l^2$ the elastic buckling strength. For conservativeness, it is customary to idealize each bar as pin-jointed, resulting in $k = 1$. Solid truss lattice materials are buckling-limited at low relative density, and transition to the yielding-limited regime at $\bar{\rho}_{trans} = 8\varepsilon_Y/(\pi \sin \omega \cos^2 \omega)$ (Fig. 12). For most metals, assuming a truss angle of 45–70°, the yield strain $\varepsilon_Y \sim 10^{-3}$, indicating a transition at $\bar{\rho} \sim 1 - 2\%$. As trusses with relative densities <<1% can be manufactured with the approach described in Section III(1), the implication is that the lightest lattice materials based on solid trusses will be inevitably buckling-dominated. The situation improves when hollow trusses are employed. Invoking a thin-wall approximation, the relative density of hollow truss structures scales as:

$$\bar{\rho} = \frac{4\pi}{\cos^2 \omega \sin \omega} \left(\frac{a}{l}\right) \left(\frac{t}{l}\right) \tag{5}$$

with t the truss wall thickness. The compressive strength scales as before, but the strength of the bar, σ_{bar}, is now:

$$\sigma_{\text{bar}} = \min \begin{cases} \sigma_Y & \text{yielding} \\ \sigma_{\text{gb}} = \dfrac{k^2 \pi^2 E a^2}{2 l^2} & \text{global (Euler) buckling} \\ \sigma_{\text{lb}} = \dfrac{E}{\sqrt{3(1 - v^2)}} \dfrac{t}{a} & \text{local buckling} \end{cases} \tag{6}$$

The local buckling load corresponds to the chessboard mode.[52] Again, we assume $k = 1$ in the global buckling

Table I.   Comparison of Known Fabrication Methods for Open-Cellular Lattice-Type Materials. Reported Values Have Been Extrapolated Based on Published Techniques. Modified From Jacobsen et al.[46]

Approach to lattice-type cellular structures	Wire or textile layup; modular assembly[41,42]	Stereolithography[43]	2-photon stereo-lithography[44]	Self-propagating polymer waveguide (SPPW) process[45]
Unit cell type	3D Periodic	3D Periodic or aperiodic	3D Periodic or aperiodic	3D, Periodic or aperiodic
Multiple sizes of trusses in unit cell	Difficult	Yes (maximum flexibility)	Yes	Yes (Intrinsic to process)
Control of solid member diameter	Limited	Independent control	Independent control	Independent control
Member angle control*	0–90°	0–90°	0–90°	~50–90°
Fabrication area	0.1 m²	0.25 m²	0.005 m²	~0.4 m²
Thickness range	1–20 cm	1 mm–0.1 m	10–100 µm	100 µm–5 cm
Nominal max. volume	0.02 m³	1 m³	10^{-6} m³	0.02 m³
Max no. unit cells through thickness.	~10	10's	~10	~10
Min. unit cell size	~100 µm	<10 µm	<2 µm	<50 µm
Min. feature size	~10 µm	<1 µm	<<0.1 µm	<5 µm
Max. no. truss elements in a cubic volume	~4000	>10 000	~1000	~4000
Potential to grade properties	Limited	Yes (highest flexibility)	Yes	Yes
Possible base materials	Metals, polymers	Polymers	Polymers	Polymers
Post-processing material options	Annealing	Template for CVD/CVI, electroplating, casting, slurry coating, carbonization	Limited for very small structures: electroplating or CVD are possible	Template for CVD/CVI, electroplating, investment casting, slurry coating, carbonization
Rate of manufacture	Hours	Hours to days	Minutes to hours	Minutes
Potential for scalable manufacturing	Medium (single parts only)	Low (single parts only)	Low (single parts only)	High (>1 m²/min for continuous process)

*(relative to horizontal).

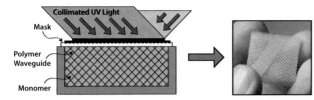

Fig. 8. Schematic representation of the process used to form micro-truss structures from self-propagating polymer waveguides (SPPW) and a prototypical structure formed by this process.[45]

load for conservativeness. For optimal structures in the buckling-dominated regime, $\sigma_{lb} = \sigma_{gb}$, resulting in

$$(t/l) = \frac{\pi^2 \sqrt{3(1 - v^2)}}{2} (a/l)^3.$$

The transition between buckling- and yielding-dominated regimes now occurs at:

$$\bar{\rho}_{trans} = \frac{8 \sqrt{3(1 - v^2)}}{\pi \sin \omega \cos^2 \omega} \varepsilon_Y^2.$$

For a metal, $\bar{\rho}_{trans} \sim 0.002\%$. Hence, metallic hollow trusses are yielding-dominated in the entire range of feasible relative densities, with substantial benefits on the strength (Fig. 12). For applications exploiting the local plastic buckling modes of hollow trusses, such as energy absorption, the advantage is even more significant. The amount of energy dissipated in crushing a bar by global (Euler) buckling is insignificant compared to the amount of energy absorbed in local modes. The implication is that hollow truss lattice materials will exhibit unique properties as cores of impact resistant sandwich structures.[49]

The situation is qualitatively identical for ceramic materials, whereby the yield strength is replaced by a defect-sensitive fracture strength. See Section III(2)(B) for more details.

(B) Advantages of Small-Scale Architecture: The strength-density relation derived in Section III(2)(A) for both solid and hollow trusses is *length-scale independent:* proportional scaling of all dimensional geometric variables (truss bar length, l, radius, a, and wall thickness, t) has no effect on either relative density or specific strength. The fundamental assumption is that constituent material properties are themselves scale-independent. Although reasonable for wall thicknesses as small as a few microns, powerful strengthening effects will

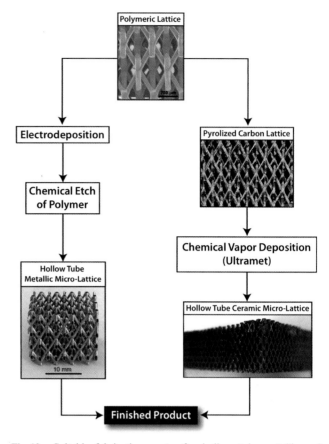

Fig. 10. Suitable fabrication routes for hollow tube metallic and ceramic micro-lattice structures (Images from Jacobsen *et al.*[45,50]).

emerge as sub-micron dimensions are approached. These recently documented effects arise from three phenomena: (a) yield strength elevation in metals due to strain gradient effects and/or (b) dislocation/surface interactions, and (c) fracture strength elevation in ceramics due to reduced average flaw size. Herein, we briefly review all the three mechanisms.

Scale Effects in Plasticity in the Presence of Strain Gradients: A large body of experimental investigations reveals the presence of size effects in plastic response that become more pronounced as the size of the sample (or the relevant length scale) approaches μm or sub-μm dimensions. Notable examples are the increase of indentation strength at shallower indentation depths,[53] increase in flow stress and

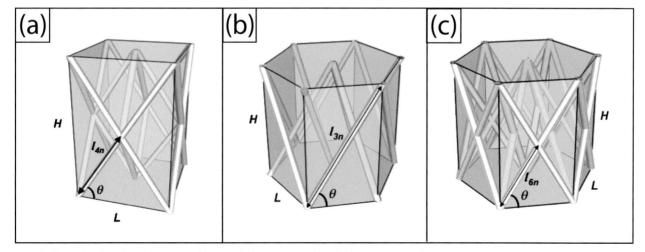

Fig. 9. Archetypal unit cell architectures with (a) 4-fold symmetry, (b) 3-fold symmetry, and (c) six-fold symmetry, as examples of structures that can be manufactured with the SPPW process depicted in Fig. 8 (from Jacobsen *et al.*[156]).

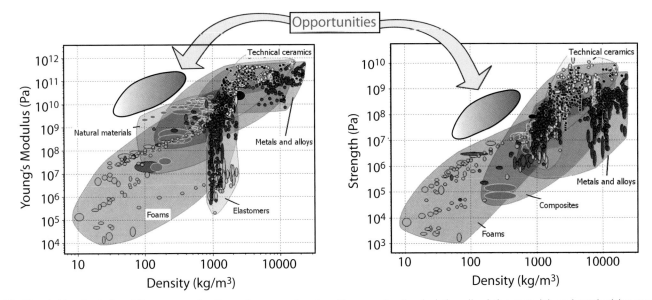

Fig. 11. Ashby charts for stiffness versus density, and compressive strength versus density, depicting all existing materials and emphasizing two target areas for micro-architected materials.

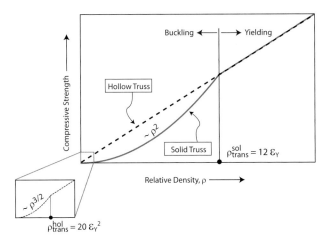

Fig. 12. Maximum compressive strength of solid and hollow metallic trusses. Notice that hollow trusses are yielding-dominated for any reasonable relative density.

hardening rate for thinner wires in torsion[54] and bending,[55] the classic Hall-Petch effect on the grain size dependence of flow stress,[56] and the increase in particle strengthening as the reinforcement size is reduced.[57] While these problems are vastly different, they all require a natural length scale for interpretation. A fundamental commonality in all the afore-mentioned situations is the presence of substantial plastic strain gradients during deformation. The development of constitutive laws that capture stress dependence on both strain and strain gradients is a natural modeling strategy. A number of strain gradient plasticity theories that reduce to the classic J_2 theory as the strain gradients are progressively reduced have been proposed over the past 25 years, most prominently by Fleck and Hutchinson,[58,59] and Nix and Gao.[60–62] The fundamental differences between the two theories in predicting experimental results were recently reviewed by Evans and Hutchinson.[63] Regardless of the differences, central to both theories is the concept of geometrically necessary dislocations (GND), initially introduced by Ashby.[64] Geometric considerations demonstrate that plastic strain gradients often require the storage of GNDs to maintain displacement compatibility. The GND density, ρ_G (total GND line length per unit volume) can be calculated once the active slip systems are identified. In an averaged sense, ρ_G can be related to the strain gradient, ε_p^*, through a natural length scale: $\rho_G \sim \varepsilon_p^*/\ell$, where ℓ is generally extracted from experimental results. Both theories predict hardening effects (and in the case of Fleck/Hutchinson, initial yield strength elevation) increasing with ρ_G.

Although a comprehensive strain gradient plasticity theory capable of capturing all the experimental phenomena while reducing to J_2 theory at large scale is still incomplete, there exists a general agreement on the marked effect of strain gradients on flow stress. These effects appear even at relative large sample sizes (~10 μm) and have the potential to substantially elevate the performance of micro-architected materials relative to their macro-scale counterparts. Although the lattice structures manufactured as described in Section III(1) (Fig. 10) will initially experience nearly zero strain gradients when loaded in compression and/or bending (all the truss members will uniformly compress or stretch), as the deformation progresses beyond first yield and the hollow truss members plastically buckle, substantial plastic strain gradients will arise. Although no initial yield strength elevation due to strain gradients is anticipated, both the collapse strength and the crushing energy (plastic dissipation) of micro-architected materials may be significantly higher than for conventional macro-scale materials. Recently, significant size effects have also been observed even in the absence of strain gradients, with further potential benefits to micro-architected materials. These effects are reviewed in the following subsection.

Size Effects in Plasticity in the Absence of Strain Gradients: Over the past 5 years, a multitude of room-temperature uniaxial compression and tension experiments have been performed on a wide range of single-crystalline metallic nano-pillars and nano-dogbones, including fcc metals (Ni and Ni-based superalloys,[65–67] Au,[68–71] Cu,[72–80] and Al[81,82]), bcc metals (W, Nb, Ta, and Mo[83–89]), hcp metals (Mg[90,91] and Ti[92]), tetragonal low-temperature metals (In[93]), Gum metal[94,95], nanocrystalline metals (Ni[96,97]), shape memory alloys (NiTi[98–102] and Cu-Al-Ni[103,104]), and a variety of metallic glasses[105–107]. For samples with nonzero initial dislocation densities (i.e., excluding whiskers and nano-fibers), a strong size effect on the flow strength was ubiquitously demonstrated as the sample size approached μm and sub-μm dimensions.[108] The compressive strength data for all single crystalline face-centered cubic (fcc) metals (Au, Al, Ni, and Cu) show a unique trend, suggesting the existence of a universal law of the form:

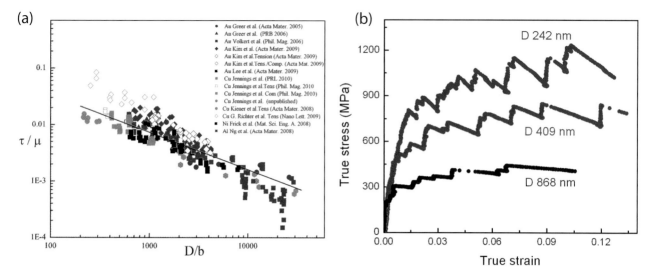

Fig. 13. Sample size effects on the flow stress in nanoscale experiments. (a) Shear flow stress (normalized with the shear modulus) VS sample diameter for a number of small-scale experiments on FCC metals reported in the literature. Resolving the stress on the relevant slip system and normalizing the sample diameter with the Burger's vector allow comparison of different metals on the same chart.[157] Reprinted from Greer and De Hosson (2011), with permission. (b) Compressive stress versus strain for uniaxial compression of single crystal Nb nano-pillars of different diameters (above each curve).[86]

$$\sigma_{res}/\mu = A(d/b)^{-m} \qquad (7)$$

where μ is the shear modulus, σ_{res} is the resolved shear stress onto the $\{111\}/\langle 110 \rangle$ slip system, d is the pillar diameter, b is the Burgers vector, and A and m are constants (Fig. 13(a)). A similar observation was reported by Dou and Derby.[109] Based on the existing data for Au, Al, and Ni, $A \sim 0.71$, and $m \sim 0.66$. This exponent is nearly identical to those reported for nearly all other fcc micro- and nano-pillars where the samples contain initial dislocations.[108]

Figure 13(b) depicts some representative stress-strain curves for single crystalline Nb nano-pillars subjected to uni-axial compression.[86] The size-dependent strengthening effect cannot be explained through well-known thin-film mechanisms, such as grain size hardening,[110] the confinement of dislocations within a thin film by the substrate,[111] or the presence of strong strain gradients.[112] Intriguingly, unlike Taylor hardening, the flow strength does not appear to scale with the evolving density of mobile dislocations. Several models attempting to explain the causality between the declining dislocation density and attained strengths have been put forth. For example, the *dislocation starvation* model, first proposed by Greer and Nix[113] hypothesizes that the mobile dislocations inside a small nano-pillar have a greater probability of annihilating at a free surface than of interacting with one another, thereby shifting plasticity into nucleation-controlled regime.[66,68,69,114] Other models include source exhaustion hardening,[115,116] source truncation,[117,118] and weakest link theory.[116,118] The general commonality in all these theories is the representation of dislocation source operations in a discrete fashion, enabling an evaluation of the effect of sample size on the source lengths, and therefore on their operation strengths. Some of these models also capture the ubiquitously observed stochastic signature of the experimental results, showing either marginal dislocation storage[114,116,119] or no storage at all.[66,69,80]

The vast majority of samples for the nano-mechanical characterization described above have been produced by Focused Ion Beam (FIB). Figure 14 shows a number of examples. Unfortunately, the effect of the ion implantation introduced as a result of the FIB processing is not well characterized, and hinders an accurate interpretation of the experimentally observed size effects in plasticity. Although several investigators have reported Ga+ ion bombardment damage resulting in altered microstructural features (e.g., dislocation sources,

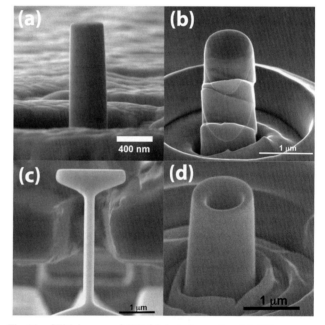

Fig. 14. SEM images of FIB-fabricated samples. (a) 400 nm nano-crystalline Ni-W nano-pillar. (b) Compressed 600 nm Nb pillar with significant slip offsets. (c) Typical dog-bone shaped tensile Au sample,[139,140] and (d) Mo anti-pillar with a hollow center.[158]

lattice rotations), attempts to attribute the observed strain hardening to these features have been inconclusive. Recent evidence (both experimental and computational, within and outside of the authors' groups) convincingly demonstrates that the size effect is a strong function of the initial dislocation density rather than the fabrication technique.[80,120] The size effects observed in the FIB-prepared specimens, for example, are identical to those obtained for the samples fabricated by a completely FIB-less technique, which contain similar initial dislocation densities.[121] Further, it has been reported that introducing dislocations into an initially pristine structure actually weakens rather that strengthens the sample.[122,123] In a recent computational study, it was reported that FIB-induced damage could contribute up to ~10% of the observed flow stress increase only for a particular size range between 500 nm and 1 μm, whereas for the larger and smaller specimens, the effects of the FIB on strength are marginal.[124]

Based on these arguments and the now ubiquitously reported presence of power-law size effects for all non-pristine pillars produced with or without the use of FIB, the authors are confident that the size effects are real and are *not* a function of the fabrication technique.

The ability to manufacture a macro-scale micro-architected material with a hollow truss topology characterized by truss wall thickness in the μm (and sub-μm) scale has the potential to exploit these beneficial size effects to achieve exceptional constituent materials properties. When combined with optimal design of the truss architecture, this approach should result in a macro-scale material with unprecedented specific strength. Admittedly, the metallic films deposited on the trusses will be polycrystalline, likely with a nanoscale grain size. The strengthening effects described above for fcc single crystals are much less understood in multi-grain surface-dominated small-scale systems. In fact, both homogeneous (grain boundaries, twin boundaries, etc.) and heterogeneous (i.e., phase boundaries, precipitate-matrix boundaries, free surfaces, and passivated surfaces) interfaces in size-limited features are crucial elements in the structural reliability of most modern materials. Yet very little work has been done on characterizing the combined effects of interfaces and surfaces—extrinsic (sample size in a surface-dominated structure) and intrinsic (microstructural features like grain boundaries, twin boundaries, phase boundaries, etc.)—on the mechanical response of materials. Furthermore, a vast majority of the above-mentioned experiments was conducted at room temperature, limiting our understanding of athermal versus thermal contribution to size-dependent strength. Significant efforts must be focused on investigating mechanical properties and identifying particular deformation mechanisms operating in boundary-containing metallic material systems with reduced dimensions (for example, nano-pillars containing two or three grains, twin boundaries, and homo- and heterogeneous nano-laminates). The knowledge of the specific deformation mechanisms as a function of feature size and initial microstructure will be essential for the design, manufacturing, and property control of new, revolutionary lightweight metallic micro-architected materials with unprecedented combinations of properties.

Fracture Strength Elevation in Ceramics at Small Scales: Ceramic thin films (e.g., carbon, silicon carbide, silicon nitride) possess yield strengths >10 GPa. Unless the constraining environment is such that crack growth is impeded (as would be the case for a uniform film adhered to a substrate and loaded in compression normal to the plane of the wafer), failure will generally occur by fracture. Linear elastic fracture mechanics predicts a fracture strength $\sigma_f \sim K_c^{solid}/\sqrt{a}$, with K_c^{solid} the fracture toughness of the constituent material and a the size of the largest crack. Assuming a statistical distribution of crack directions, mode I conditions will generally dominate the strength, whereby $K_c^{solid} = K_{Ic}^{solid}$ and a is the size of the largest crack oriented favorably to mode I propagation. The smaller the sample dimension, the smaller its largest crack. The implication is that the strength of a ceramic material will substantially increase as the sample length scale is reduced. As an example, a 5–10 μm thick polycrystalline diamond film deposited on a hollow truss might have[125] $a \sim 1 \mu m$, $K_c \sim 4.6 MPa\sqrt{m}$, resulting in $\sigma_f \sim 4 GPa$. The relationship between the fracture toughness of a lattice and that of its constituent material has been recently elucidated by Fleck *et al.* for the case of planar lattices[126]: $K_{Ic}^{lattice}/K_{Ic}^{solid} \sim \rho^d\sqrt{\ell/a}$, with ρ the relative density of the lattice, ℓ the unit cell size, a a typical crack size in the constituent material, and the exponent d is a strong function of the lattice topology *(0.5 < d < 2)*. Similar relationships can be derived for 3D lattices. If the architecture is properly chosen to minimize d, and assuming that the constituent crack size can be reduced together with the unit cell size, micro-architected lattice materials can have substantial fracture toughness, at a fraction of

the weight of solid materials. At the same time, ceramics are exceptionally stiff (E~1TPa, for polycrystalline diamond[125]). Such strengths and stiffnesses are unattainable with any metallic system, offering ceramic micro-architected materials the potential to leap into currently unclaimed areas in a number of materials property charts. Importantly, the manufacturing technology described in Section III(1) is a key enabler for this vision: only an approach capable of manufacturing *large-scale* materials with *micron-level control* of the lattice architecture allows the base material to be deposited in the form of a sub-μm thin film. This has two enormous benefits: (i) it allows use of materials not available in the bulk (e.g., polycrystalline diamond), and (ii) it allows accurate control of the maximum flaw size, with enormous increases in the fracture strength relative to bulk values.

(3) Optimal Design Protocol for Micro-Architected Materials

The multi-step optimal design protocol presented in Section II in the context of large-scale periodic cellular structures is generally applicable to micro-architected materials. Although the general methodology is unchanged, a fundamental complexity emerges. *The bulk properties of micro-architected materials are a strong function of phenomena occurring across three length scales* (Fig. 15): a macroscopic level (bulk), a mesoscopic level (unit-cell) and a micro/nanoscopic level (the characteristic length scale of the constituent materials). Unique critical phenomena occur at each length scale, requiring experimental investigation. The size effects on plastic flow stress, strain hardening, and fracture strength (discussed above) are clearly micro/nano-scale phenomena, as are microstructure (and properties) anisotropy possibly arising from the film deposition process. Film thickness/microstructure variation along the truss members and details of the node topology—and their effects on the mechanical properties of the material—occur at the unit-cell level. The same length-scale can also affect the fracture strength, as variation in flaw distributions along members and around nodes can play a substantial role. Finally, the vast number of unit cells composing the bulk material may introduce large-scale effects previously unnoticed in macro-scale lattice materials: geometric imperfections (i.e., deviation from a perfect lattice) and the possible occurrence of buckling modes with characteristic length scale much larger than the unit cell level (and hence not captured with the type of analysis presented in Section II(2)) might play a significant role on the overall stiffness and strength of the bulk material. Novel characterization techniques and numerical strategies must be implemented to enable multi-scale studies. These are discussed below.

(4) Experimental Characterization of Micro-Architected Materials

(A) Macro-Scale Mechanical Characterization: At the macro-scale, conventional techniques traditionally employed for the characterization of large-scale lattice materials (and in fact, any other material) are perfectly adequate to analyze micro-architected materials. Traditional universal test frames (e.g., INSTRON, MTS) equipped with tensile, bending and shear fixtures can be used to measure stress-strain response in different loading scenarios and validate/calibrate analytical and numerical models for failure prediction.

(B) Micro-Scale Mechanical Characterization: As the unit cells of micro-architected materials can take many shapes and sizes, and several base materials can be used, the ideal device for mechanical characterization at this scale should have the following features: (i) be adaptable to samples of vastly different sizes and shapes; (ii) allow controlled displacement actuation and independent load measurement;

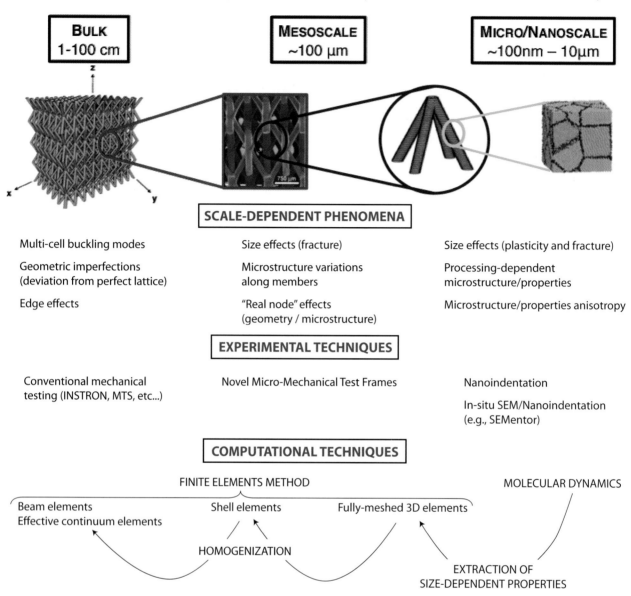

Fig. 15. Expected phenomena, mechanical characterization tools and computational techniques for the three length-scales of interest in micro-architected materials. (SEM images from Jacobsen *et al.*[45,159])

(iii) be capable of extreme force resolution (~1–100 nN) and range (~1 N), displacement resolution (~10–100 nm) and range (1–10 mm); and (iv) allow optical (or SEM) access to the test coupon with potential for strain mapping (via Digital Image Correlation). Hybrid micro-test frames, encompassing a MEMS force sensor and an off-chip displacement actuator, are necessary to meet the requirements listed above. A number of such devices have been developed in the past two decades.[127,128] An economical and versatile device, capable of covering the entire range of force and displacement described above, has been recently introduced by one of the authors (Fig. 16).[129] In the proposed design, the sensor is a micro-fabricated Silicon double-ended tuning fork (DETF), whose working principle is the change in natural frequency of vibration in a pair of parallel and connected beams upon application of an external axial force. DETF sensors deflect axially rather than laterally, hence exhibiting essentially infinite stiffness relative to the sample being tested and exceptional force range (in the Newton range). This is in stark contrast with the more commonly employed capacitive or visual force detection schemes, in which load cell and sample have comparable compliance. At the same time, the strong dependence of the natural frequency of a beam on the axial load and the extreme

precision available in frequency measurement (a change of a fraction of a Hertz is easily detected in a 100 kHz signal) grant the device nN resolution. Samples of different geometry can be handled with micro-fabricated custom fixtures, enabling a variety of testing conditions (bending, compression, tension, etc....).

(C) Nanoscale Mechanical Characterization: Uniaxial Mechanical Testing at the Nanoscale: The two best-established techniques for nanoscale mechanical characterization are Atomic Force Microscopy (AFM) and Nanoindentation. The former controls the displacement through a piezo-actuator and senses the force through the deflection of a micro-cantilever, typically measured optically. Although extreme force and deflection resolutions are possible,[1,130–132] the force range is small, on the order of a few pico-Newtons, with a vertical distance resolution smaller than ~0.1 nm. Such a small load range limits the applicability of AFM to the characterization of stiff materials (e.g., metals and ceramics). The mechanical properties of stiff materials can be well characterized by nanoindentation, whose premise involves forcing a generally sharp diamond indenter tip into the surface of a material, while measuring the imposed force, the corresponding displacement of the indenter, and in some cases, the contact stiffness.[62,133–138] Nanoindenters are inherently

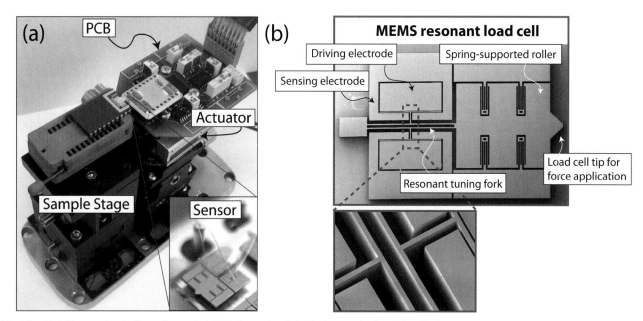

Fig. 16. (a) Micro-mechanical test frame (μ-MTF) for unit-cell level characterization. The displacement actuator is a commercially available nano-stage, while the load cell (inset) is microfabricated. (b) Detail of the microfabricated load cell (From Torrents *et al.*[129] and Azgin *et al.*[160])

load-controlled instruments, where the load is applied through an electromagnetic coil assembly, and the displacement is measured by capacitive gages. Typical modern-day nanoindenters have load resolution of several nano-Newtons and sub-nanometer displacement resolution. From a measurement of the indentation depth, the local hardness, H, of the material is readily accessed, and the yield strength is generally estimated as $\sigma_Y \approx H/3$. Obtaining more detailed mechanical information is unfortunately very difficult: the stress and strain fields induced by the indenter are complex and tri-axial, rendering the interpretation of experimental results challenging. Carving nanoscale samples in the shape of pillars or dog-bones addresses this difficulty, by enabling the introduction of nearly uniaxial stress and strian fields (see Section III(2)(B)). To allow *in-situ* observation of the sample deformation, nanoindenters have been coupled with electron microscopes. A unique such instrument (called the "SEMentor") was recently developed by one of the authors.[139,140] The SEMentor is comprised of a nanomechanical module, similar to the DCM assembly of a commercial Agilent nanoindenter, inside of a SEM. The former offers a precise control and high resolution of load (~1 nN) and displacement (<1 nm) and their rates, as well as contact stiffness during the experiment, while the latter allows for visualization of the process. Custom-made grips were fabricated to conduct nanoscale *in-situ* experiments in uniaxial compression and tension. Uniaxial tensile investigations will be essential in determining nanoscale yield and ultimate tensile strength and fracture toughness of nanoscale materials, as well as in elucidating the origins of tension/compression asymmetry likely to be observed in nanoscale polycrystalline samples. All these features are critically important to the development of micro-architected materials. Furthermore, this *in-situ* testing technique will allow correlation of the macroscopic stress-strain behavior with some microstructural activity by direct observation of, for example, the dislocation glide "avalanches" manifested by multiple slip lines, shear offsets, and phase delamination (if any).

Microstructural Characterization: A key analytical technique allowing direct observation of dislocations and their interactions with various boundaries and surfaces is High Resolution Transmission Electron Microscopy (HR-TEM). While, of course, the post-mortem TEM analysis is not capable of providing any information about the mobile defect activity, it is powerful in revealing the post-deformation micro-

structure, i.e., the evolved dislocation networks, the final grain configurations, and most importantly it sheds light on the particular interactions of dislocations with the individual interfaces and surfaces in the deformed samples. This information is useful in uncovering some of the fundamental mechanisms that might have operated during deformation of these nano-volumes with specified interfaces.

(5) Numerical Modeling of Micro-Architected Materials

(A) Continuum-Based Approaches (Finite Elements Analysis): Once the properties of the thin film base materials are known (including their size effects), traditional contin-

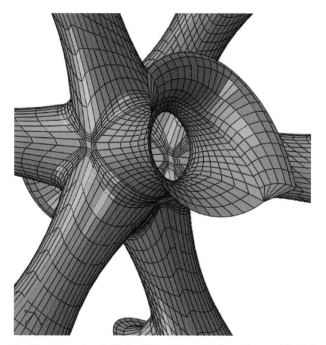

Fig. 17. Example of Finite Elements mesh for micro-architected hollow truss structures. Node fillet radii, nonuniform wall thicknesses, noncircular cross-sections and bar tapers are parametrically defined with geometric modeling tools, for efficient integration with optimization algorithms.

uum finite elements approaches can be used to extract both the unit cell properties as well as the averaged bulk properties of micro-architected materials. Although the numerical tools are identical to those used for large-scale lattice materials (see Section II), some additional steps must be taken to merge the unit cell and the bulk length scales (Fig. 15).

Finite Elements Modeling of Individual Trusses and Unit Cells: Individual hollow truss members and unit cells can be modeled with shell elements and/or solid elements. Besides validating the analytical predictions for an ideal structure, FE modeling must quantify the effect of three potentially critical factors on stiffness, strength, ductility, and collapse mechanisms. *(i) Geometric non-uniformities* (e.g., curviness in nominally straight truss members, non-ideal node size and shape, and wall thickness variations along members and within nodes). *(ii) Heterogeneity and anisotropy in materials properties* (arising from variations in grain size and texture along the trusses and around the nodes). *(iii) Interface strength and toughness in multi-materials systems* (e.g., a ceramic film deposited on a metallic wall). These phenomena are in principle present in large-scale lattice structures as well, but their effect on the overall properties is typically negligible. Geometric algorithms for automatic meshing are essential to capture point (i) above (Fig. 17).

Large-Scale (Multi-Cells) Finite Elements Modeling: The vast difference between the unit cell and the bulk scales requires numerical strategies for efficient modeling. Generally, the plethora of Finite Elements results extracted from the fully meshed unit-cell models described above must be condensed into a lower-order model (homogenization). Two sequential approaches can be envisioned, in increasing order of complexity and computational efficiency: (a) beam-elements unit-cell model; (b) effective 3D solid elements model. For approach (a), each unit cell is modeled with a number of beam elements and the effects of nodal size/shape and local buckling that are lost in transitioning from shell to beam elements can be incorporated by introducing fictitious (non-uniform) stress-strain response for the material. Approach (b) follows homogenization procedures similar to those implemented for large-scale period core sandwich panels.[26,141] The challenge is to define the constitutive behavior of the material in a way that a single solid element (possibly with size spanning several unit cells) elastically and plastically deforms consistently with buckling phenomena that are dominated by length scales of the order of the unit cell size, truss radius and wall thickness.

(B) Atomistic Approaches (Molecular Statics and Dynamics): Standard molecular dynamics (MD) and molecular statics (MS) techniques can be used to develop quantitative predictions of the elastic behavior of brittle ceramics and the elasto-plastic behavior of ductile metals and nanostructured composites of ductile and brittle materials. The Embedded Atom Method (EAM) potential[142] and the Modified EAM (MEAM) potential[143] are appropriate for metals, whereas materials with significant angular bonding (such as some ceramics) require dedicated MEAM potentials.[144] Molecular statics (MS) approaches are adequate for stiffness calculations, both in single crystal and nanocrystalline samples, and enable investigation of the role of all strain components on grain size and orientation at T~0K (this is acceptable as stiffness is not expected to be a strong function of temperature). If necessary, temperature effect can be quantified with MD simulations at non-zero temperature. Prediction of the plastic behavior of micro-architected materials at the nanoscale is significantly more challenging, as significant size and orientation effects might dominate strength and ductility. Controlled defects can be introduced to examine trade-offs between homogeneous and heterogeneous dislocation and void nucleation. As plastic flow is highly temperature dependent, MS is inadequate for elasto-plastic investigations, and MD simulations must be performed at nonzero temperature (typically at T~273K). A substantial challenge is bridging the simulation and experimental time scales:

computational resources demand strain rates $>10^6 \, s^{-1}$ for all MD simulations, whereas experiments typically take place at rates of $1 \, s^{-1}$ or less. To avoid recoding fictitious strain rate effects in the data, high strain rate MD simulations must be extrapolated to lower strain rates, possibly using mechanism-based modeling. Validation of these calculations with nano-scale experiments (Section III(4)(C)) is critical to ensure that deformation mechanisms unveiled by the MD simulations persist even at much lower strain rates. MD capabilities to accurately predict the initial yield point have been well established. Quantitatively capturing phenomena subsequent to failure initiation (e.g., hardening and ultimate failure) is less straightforward. Once again, nanomechanical experiments will be essential to validate and supplement MD models. MD and FEM analyses can be interfaced *off-line* with phenomenological constitutive laws: such laws are informed by MD (e.g., yield strength VS sample size) and can be easily imported in commercial FE packages (either directly or via user-defined subroutines). Naturally, combining MD numerical schemes at the nanoscale with FE calculations at the micro and macro scale in *real-time* in a truly multi-scale algorithm is the Holy Grail of computational mechanics. After decades of research in multi-scale mechanics, this is still a daunting task. Nonetheless, *off-line* combinations of atomistic simulations at the nanoscale and Finite Elements and analytical continuum mechanics at the micro/macro-scale have been capable of macro-scale properties predictions.[145] Experimental results as those described in Section III(4) are often necessary for model calibration.

In addition to MD and MS simulations, Discrete Dislocations Simulations (DDS) have been introduced to compute the effect of dislocation motion and interaction on plastic flow in crystals.[146–148] These models generally utilize a FE continuum framework to solve for stresses, strains and displacements, and treat dislocations as singularities, which both affect and are affected by the global strain and stress fields. Dislocations motion is governed by the Peach-Koehler equation and standard dislocation-dislocation interaction laws. Dislocation sources like Frank-Read and single-arm sources are introduced at statistically random locations. These models (both in 2D and 3D) are computationally more efficient than full-scale MD simulations, but have not yet fully succeeded in duplicating key aspects of the experimentally measured size effects in plasticity.[149]

(C) Optimization Algorithms: The large number of variables in the optimal design of micro-architected materials, coupled with the possible difficulty in obtaining closed-

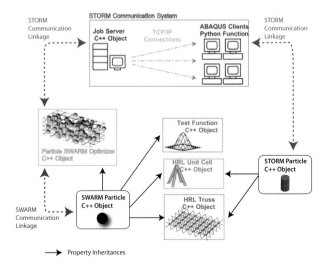

Fig. 18. Schematic of a possible efficient optimization code for the optimal design of complex micro-architected materials.[36] The optimizer can be coupled with commercial Finite Elements packages for on-line fully numerical optimization.

form solutions for some of the objectives and constraints, require an optimization procedure more sophisticated than that typically used for the optimization of macro-scale structures (Section II). Importantly, most of the structural optimization work performed to date assumes that the *constituent material(s)* and the *architecture* of the cellular structure is decided *a priori,* and the optimizer must simply select the best values for all the *geometric parameters* that define that architecture (e.g., unit-cell size, truss angle, wall thickness). A large number of architectures were independently studied at the macroscopic scales.[11,12,15,16,150] Standard quadratic optimizers were typically used, and relied on close-form expressions for the objective function and all the constraints. Although easy and convenient, this approach might be unsuitable for the optimization of micro-architected multifunctional materials, for two reasons: (i) complex geometric features (e.g., details of the node architecture) might strongly affect stress and strain distributions, and hence impede close-form expressions of objective and constraints, and (ii) a large number of local optima might appear, which could confuse quadratic optimizers. Discrete optimizers (e.g., genetic or particle swarm optimizers) do not require close-form expressions for objectives and constraints and are rather insensitive to local minima. Unfortunately, their robustness and efficiency are not as good as for gradient-based algorithms. A discrete optimization protocol that allows on-line interaction with a commercial Finite Elements program for objective function evaluation is schematically described in Fig. 18.[36] The on-line integration of FE analyses within the optimization loops is extremely resource-intensive. Although novel software technology (e.g., Microsoft HPC) will help, this remains a key bottleneck in the optimization process.

Ultimately, to take full advantage of the nearly infinite freedom in designing with micro-architected materials, the topological architecture of the material should be itself a variable in the optimization process. Topology optimization algorithms were developed over the past couple of decades,[151–154] with the scope of defining the ideal arrangement of two or more phases of matter (one might be air to define a cellular solid) to achieve extremal values in one or more macroscopic properties. These algorithms are naturally more complex and resource-intensive than the simple procedure described in Section II. In addition, although exceptionally interesting and unexpected results might occur (e.g., when the Kagome structure was identified by Torquato's group as an ideal stiff and statically determinate lattice,[155] spearheading dozens of studies on its mechanical properties and technological importance,[28,30,150]) in many cases the resulting topologies are nearly impossible to manufacture. Nonetheless, incorporation of topology optimization concepts within the computational framework described in Section II is a promising way to explore a massive design space.

IV. Conclusions

Periodic cellular materials and structures have been extensively investigated over the past decade for a large number of multifunctional applications. A common optimal design protocol was successfully adopted in several studies to select the geometry, architecture, and base material that maximizes the relevant objective function(s). The protocol employs a combination of analytical, numerical, and experimental techniques, and was reviewed herein in the context of mechanical and thermo-mechanical structures. We extended this concept to the multi-scale design of a micro-architected material. Enabling this vision are newly available manufacturing approaches for suitable periodic cellular architectures with unit cell sizes in the sub-millimeter scale and constituent material characteristic length in the micro/nano scale. The resulting micro-architected material is characterized by three different length scales: a bulk scale (~1–100 cm), a unit-cell scale (~100 μm–1 mm) and a constituent material scale

(~100 nm–10 μm). We surmise that micro-architected materials will possess unique superior mechanical properties (primarily specific stiffness and strength), by exploiting recently observed size effects at the micro and nano scales. Extending optimal design protocols for large-scale periodic cellular structures to the case of micro-architected materials entails the added difficulty that bulk mechanical properties are now a strong function of physical phenomena occurring at three different length scales. Existing modeling tools must be adapted to handle such difficulty. We briefly reviewed the state-of-the-art in small-scale mechanical characterization and numerical modeling and concluded that the necessary techniques are sufficiently mature to underpin the development of micro-architected materials. As the length scale ℓ of a periodic cellular structure shrinks to sub-millimeter dimensions, and the number of unit cells in the bulk material increases as ℓ^{-3}, the difference between a *structure* and a *material* gets progressively blurred: in this sense, micro-architected materials represent a new class of materials, characterized by a unique variable amenable to optimization: the *topological architecture*. We hope that this overview will stimulate intense research efforts aimed at developing new micro-architected materials with properties outside of the boundaries of current materials properties spaces and into new, previously unclaimed regions.

Acknowledgments

The authors are grateful to DARPA for financial support through grant no. W91CRB-10-C-0305 on Materials with Controlled Microstructural Architecture (Judah Goldwasser, program manager). LV acknowledges partial funding from the California-Catalonia Engineering Program. JRG acknowledges the financial support from NSF CAREER Award (DMR-0748267) and ONR Grant no. N000140910883. WBC and AJJ also acknowledge prior DARPA support through contract no. W911NF-08-C-0038 and internal support funds from HRL. The authors dedicate this article to the memory of Anthony G. Evans, who inspired and directed much of the work on periodic cellular materials. His leadership, enthusiasm, and mentorship are dearly missed.

References

[1]L. J. Gibson and M. F. Ashby, *Cellular Solids: Structure and Properties.* Cambridge University Press, Cambridge, UK, 1999.
[2]M. F. Ashby, A. G. Evans, N. A. Fleck, L. J. Gibson, J. W. Hutchinson, and H. N. G. Wadley, Metal Foams: A Design Guide. Butterworth-Heinemann, Oxford, UK, 2000.
[3]T. J. Lu, L. Valdevit, and A. G. Evans, "Active Cooling by Metallic Sandwich Structures With Periodic Cores," *Prog. Mater. Sci.,* **50** [7] 789–815 (2005).
[4]A. G. Evans, J. W. Hutchinson, N. A. Fleck, M. F. Ashby, and H. N. G. Wadley, "The Topological Design of Multifunctional Cellular Metals," *Prog. Mater. Sci.,* **46** [3-4] 309–27 (2001).
[5]H. N. G. Wadley, "Cellular Metals Manufacturing," *Adv. Eng. Mater.,* **4** [10] 726–33 (2002).
[6]H. N. G. Wadley, N. A. Fleck, and A. G. Evans, "Fabrication and Structural Performance of Periodic Cellular Metal Sandwich Structures," *Compos. Sci. Technol.,* **63** [16] 2331–43 (2003).
[7]H. Bart-Smith, A. F. Bastawros, D. R. Mumm, A. G. Evans, D. J. Sypeck, and H. N. G. Wadley, "Compressive Deformation and Yielding Mechanisms in Cellular Al Alloys Determined Using X-ray Tomography and Surface Strain Mapping," *Acta Mater.,* **46** [10] 3583–92 (1998).
[8]H. Bart-Smith, J. W. Hutchinson, and A. G. Evans, "Measurement and Analysis of the Structural Performance of Cellular Metal Sandwich Construction," *Int. J. Mech. Sci.,* **43** [8] 1945–63 (2001).
[9]A. F. Bastawros, H. Bart-Smith, and A. G. Evans, "Experimental Analysis of Deformation Mechanisms in a Closed-Cell Aluminum Alloy Foam," *J. Mech. Phys. Solids,* **48** [2] 301–22 (2000).
[10]H. Bart-Smith, J. W. Hutchinson, N. A. Fleck, and A. G. Evans, "Influence of Imperfections on the Performance of Metal Foam Core Sandwich Panels," *Int. J. Solids Struct.,* **39** [19] 4999–5012 (2002).
[11]N. Wicks and J. W. Hutchinson, "Optimal Truss Plates," *Int. J. Solids Struct.,* **38** [30–31] 5165–83 (2001).
[12]L. Valdevit, J. W. Hutchinson, and A. G. Evans, "Structurally Optimized Sandwich Panels With Prismatic Cores," *Int. J. Solids Struct.,* **41** [18–19] 5105–24 (2004).

[13]L. Valdevit, Z. Wei, C. Mercer, F. W. Zok, and A. G. Evans, "Structural Performance of Near-Optimal Sandwich Panels With Corrugated Cores," *Int. J. Solids Struct.*, **43** [16] 4888–905 (2006).

[14]N. Wicks and J. W. Hutchinson, "Performance of Sandwich Plates With Truss Cores," *Mech. Mater.*, **36** [8] 739–51 (2004).

[15]F. W. Zok, H. Rathbun, M. He, E. Ferri, C. Mercer, R. M. McMeeking, and A. G. Evans, "Structural Performance of Metallic Sandwich Panels With Square Honeycomb Cores," *Philos. Mag.*, **85** [26–27] 3207–34 (2005).

[16]F. W. Zok, H. J. Rathbun, Z. Wei, and A. G. Evans, "Design of Metallic Textile Core Sandwich Panels," *Int. J. Solids Struct.*, **40** [21] 5707–22 (2003).

[17]F. W. Zok, S. A. Waltner, Z. Wei, H. J. Rathbun, R. M. McMeeking, and A. G. Evans, "A protocol for Characterizing the Structural Performance of Metallic Sandwich Panels: Application to Pyramidal Truss Cores," *Int. J. Solids Struct.*, **41** [22–23] 6249–71 (2004).

[18]L. Valdevit, A. Pantano, H. A. Stone, and A. G. Evans, "Optimal Active Cooling Performance of Metallic Sandwich Panels With Prismatic Cores," *Int. J. Heat Mass Transfer*, **49** [21–22] 3819–30 (2006).

[19]L. Valdevit, N. Vermaak, F. W. Zok, and A. G. Evans, "A Materials Selection Protocol for Lightweight Actively Cooled Panels," *J. Appl. Mech.*, **75**, 061022 (2008).

[20]N. Vermaak, L. Valdevit, and A. G. Evans, "Influence of Configuration on Materials Selection for Actively Cooled Combustors," *J. Propul. Power*, **26** [2] 295–302 (2010).

[21]N. Vermaak, L. Valdevit, and A. G. Evans, "Materials Property Profiles for Actively Cooled Panels: An Illustration for Scramjet Applications," *Metall. Mater. Trans. A.*, **40A** [4] 877–90 (2009).

[22]C. S. Roper, "Multiobjective Optimization for Design of Multifunctional Sandwich Panel Heat Pipes With Micro-Architected Truss Cores," *Int. J. Heat Fluid Flow*, **32**, 239–48 (2011).

[23]H. J. Rathbun, D. D. Radford, Z. Xue, M. Y. He, J. Yang, V. Deshpande, N. A. Fleck, J. W. Hutchinson, F. W. Zok, and A. G. Evans, "Performance of Metallic Honeycomb-Core Sandwich Beams Under Shock Loading," *Int. J. Solids Struct.*, **43** [6] 1746–63 (2006).

[24]M. T. Tilbrook, V. S. Deshpande, and N. A. Fleck, "The Impulsive Response of Sandwich Beams: Analytical and Numerical Investigation of Regimes of Behaviour," *J. Mech. Phys. Solids*, **54** [11] 2242–80 (2006).

[25]H. N. G. Wadley, K. P. Dharmasena, M. Y. He, R. M. McMeeking, A. G. Evans, T. Bui-Thanh, and R. Radovitzky, "An active Concept for Limiting Injuries Caused by Air Blasts," *Int. J. Impact Eng.*, **37** [3] 317–23 (2010).

[26]Z. Y. Xue and J. W. Hutchinson, "Crush Dynamics of Square Honeycomb Sandwich Cores," *Int. J. Numer. Meth. Eng.*, **65** [13] 2221–45 (2006).

[27]L. H. Han, T. J. Lu, and A. G. Evans, "Optimal Design of a Novel High Authority SMA Actuator," *Mech. Adv. Mater. Struct.*, **12** [3] 217–27 (2005).

[28]R. G. Hutchinson, N. Wicks, A. G. Evans, N. A. Fleck, and J. W. Hutchinson, "Kagome Plate Structures for Actuation," *Int. J. Solids Struct.*, **40** [25] 6969–80 (2003).

[29]T. J. Lu, J. W. Hutchinson, and A. G. Evans, "Optimal Design of a Flexural Actuator," *J. Mech. Phys. Solids*, **49** [9] 2071–93 (2001).

[30]S. Lucato, J. Wang, P. Maxwell, R. M. McMeeking, and A. G. Evans, "Design and Demonstration of a High Authority Shape Morphing Structure," *Int. J. Solids Struct.*, **41** [13] 3521–43 (2004).

[31]N. Wicks and J. W. Hutchinson, "Sandwich Plates Actuated by a Kagome Planar Truss," *J. Appl. Mech.-Trans. Asme*, **71** [5] 652–62 (2004).

[32]W. D. Macdonald and T. W. Eagar, "Transient Liquid-Phase Bonding," *Annu. Rev. Mater. Sci*, **22**, 23–46 (1992).

[33]Wall-Colmonoy Corporation. Available at http://www.wallcolmonoy.com

[34]D. T. Queheillalt and H. N. G. Wadley, "Titanium alloy Lattice Truss Structures," *Mater. Des.*, **30** [6] 1966–75 (2009).

[35]S. J. Johnson, B. Tryon, and T. M. Pollock, "Post-Fabrication Vapor Phase Strengthening of Nickel-Based Sheet Alloys for Thermostructural Panels," *Acta Mater.*, **56** [17] 4577–84 (2008).

[36]S. W. Godfrey, "Optimal Design of Orthotropic Fiber-Composite Corrugated-Core Sandwich Panels Under Axial Compression"; MS Thesis, Mechanical and Aerospace Engineering, University of California, Irvine, (2010).

[37]B. P. Russell, V. S. Deshpande, and H. N. G. Wadley, "Quasi-Static Deformation and Failure Modes of Composite Square Honeycombs," *J. Mech. Mater. Struct.*, **3** [7] 1315–40 (2008).

[38]A. P. Mouritz and B. N. Cox, "A Mechanistic Interpretation of the Comparative In-Plane Mechanical Properties of 3D Woven, Stitched and Pinned Composites," *Compos. Part A - Appl. S.*, **41** [6] 709–28 (2001).

[39]V. S. Deshpande, N. A. Fleck, and M. F. Ashby, "Effective Properties of the Octet-Truss Lattice Material," *J. Mech. Phys. Solids*, **49** [8] 1747–69 (2001).

[40]N. Vermaak, "Thermostructural Design Tools for Hypersonic Vehicles," Ph.D. Thesis, Materials Department, University of California, Santa Barbara, 2010.

[41]D. Queheillalt and H. Wadley, "Cellular Metal Lattices With Hollow Trusses," *Acta Mater.*, **53** [2] 303–13 (2005).

[42]Q. Li, E. Y. Chen, D. R. Bice, and D. C. Dunand, "Mechanical Properties of Cast Ti-6Al-2Sn-4Zr-2Mo Lattice Block Structures," *Adv. Eng. Mater.*, **10** [10] 939–42 (2008).

[43]F. P. W. Melchels, J. Feijen, and D. W. Grijpma, "A Review on Stereolithography and its Applications in Biomedical Engineering," *Biomaterials*, **31** [24] 6121–30 (2010).

[44]H.-B. Sun and S. Kawata, "Two-Photon Photopolymerization and 3D Lithographic Microfabrication," pp. 169–273 in *NMR, 3D Analysis, Photopolymerization*, Vol. 190. Edited by N. Fatkullin. Springer, Berlin, Germany, 2004.

[45]A. J. Jacobsen, W. Barvosa-Carter, and S. Nutt, "Micro-Scale Truss Structures Formed From Self-Propagating Photopolymer Waveguides," *Adv. Mater.*, **19** [22] 3892–6 (2007).

[46]A. J. Jacobsen, J. A. Kolodziejska, R. Doty, K. D. Fink, C. Zhou, C. S. Roper, and W. B. Carter, "Interconnected Self-Propagating Photopolymer Waveguides: An Alternative to Stereolithography for Rapid Formation of Lattice-Based Open-Cellular Materials" in *Twenty First Annual International Solid Freeform Fabrication Symposium - An Additive Manufacturing Conference*, University of Texas, Austin (2010).

[47]J. Banhart, "Manufacture, Characterisation and Application of Cellular Metals and Metal Foams," *Prog. Mater. Sci.*, **46**, 559–632 (2001).

[48]M. Scheffler and P. Colombo, *"Cellular Ceramics: Structure, Manufacturing, Properties and Applications*. WILEY, Weinheim, 2005.

[49]A. G. Evans, M. Y. He, V. S. Deshpande, J. W. Hutchinson, A. J. Jacobsen, and W. B. Carter, "Concepts for Enhanced Energy Absorption Using Hollow Micro-Lattices," *Int. J. Impact Eng.*, **37** [9] 947–59 (2010).

[50]A. J. Jacobsen, S. Mahoney, W. B. Carter, and S. Nutt, "Vitreous Carbon Micro-Lattice Structures," *Carbon*, **49** [3] 1025–32 (2011).

[51]V. S. Deshpande and N. A. Fleck, "Collapse of Truss Core Sandwich Beams in 3-Point Bending," *Int. J. Solids Struct.*, **38** [36–37] 6275–305 (2001).

[52]H. G. Allen and P. S. Bulson, *"Background to Buckling."* McGraw-Hill, London, 1980.

[53]M. R. Begley and J. W. Hutchinson, "The Mechanics of Size-Dependent Indentation," *J. Mech. Phys. Solids*, **46** [10] 2049–68 (1998).

[54]N. A. Fleck, G. M. Muller, M. F. Ashby, and J. W. Hutchinson, "Strain Gradient Plasticity – Theory and Experiment," *Acta Metall. Mater.*, **42** [2] 475–87 (1994).

[55]J. S. Stolken and A. G. Evans, "A microbend Test Method for Measuring the Plasticity Length Scale," *Acta Mater.*, **46** [14] 5109–15 (1998).

[56]N. J. Petch, "The Cleavage Strength of Polycrystals," *J. Iron Steel I.*, **174** [1] 25–8 (1953).

[57]D. J. Lloyd, "Particle Reinforced Aluminum and Magnesium Matrix Composites," *Int. Mater. Rev*, **39**, 1–23 (1994).

[58]N. A. Fleck and J. W. Hutchinson, "Strain Gradient Plasticity," *Adv. Appl. Mech.*, **33**, 295–361 (1997).

[59]N. A. Fleck and J. W. Hutchinson, "A Reformulation of Strain Gradient Plasticity," *J. Mech. Phys. Solids*, **49** [10] 2245–71 (2001).

[60]H. Gao, Y. Huang, W. D. Nix, and J. W. Hutchinson, "Mechanism-Based Strain Gradient Plasticity – I. Theory," *J. Mech. Phys. Solids*, **47** [6] 1239–63 (1999).

[61]Y. Huang, H. Gao, W. D. Nix, and J. W. Hutchinson, "Mechanism-Based Strain Gradient Plasticity – II. Analysis," *J. Mech. Phys. Solids*, **48** [1] 99–128 (2000).

[62]W. D. Nix and H. J. Gao, "Indentation Size Effects in Crystalline Materials: A Law for Strain Gradient Plasticity," *J. Mech. Phys. Solids*, **46** [3] 411–25 (1998).

[63]A. G. Evans and J. W. Hutchinson, "A Critical Assessment of Theories of Strain Gradient Plasticity," *Acta Mater.*, **57** [5] 1675–88 (2009).

[64]M. F. Ashby, "Deformation of Plastically Non-Homogeneous Materials," *Philos. Mag.*, **21** [170] 399 (1970).

[65]D. M. Dimiduk, M. D. Uchic, and T. A. Parthasarathy, "Size-Affected Single-Slip Behavior of Pure Nickel Microcrystals," *Acta Mater.*, **53**, 4065–77 (2005).

[66]Z. W. Shan, R. Mishra, S. A. Syed, O. L. Warren, and A. M. Minor, "Mechanical Annealing and Source-limited Deformation in Submicron-diameter Ni Crystals," *Nat. Mater.*, **7** [116] 115–9 (2008).

[67]M. D. Uchic, D. M. Dimiduk, J. N. Florando, and W. D. Nix, "Sample Dimensions Influence Strength and Crystal Plasticity," *Science*, **305** [5686] 986–9 (2004).

[68]J. R. Greer, W. C. Oliver, and W. D. Nix, "Size Dependence of Mechanical Properties of Gold at the Micron Scale in the Absence of Strain Gradients," *Acta Mater.*, **53**, 1821–30 (2005).

[69]J. R. Greer and W. D. Nix, "Nanoscale Gold Pillars Strengthened through Dislocation Starvation," *Phys. Rev. B*, **73**, 245410–6 (2006).

[70]C. A. Volkert and E. T. Lilleodden, "Size Effects in the Deformation of Sub-Micron Au Columns," *Philos. Mag.*, **86**, 5567–79 (2006).

[71]A. Budiman, S. Han, J. R. Greer, N. Tamura, J. Patel, and W. D. Nix, "A Search for Evidence of Strain Gradient Hardening in Au Submicron Pillars Under Uniaxial Compression Using Synchrotron X-ray Microdiffraction," *Acta Mater.*, **56** [3] 602–8 (2007).

[72]D. Kiener, C. Motz, M. Rester, M. Jenko, and G. Dehm, "FIB Damage of Cu and Possible Consequences for Miniaturized Mechanical Tests," *Mater. Sci. Eng. A*, **459** [1–2] 262–72 (2006).

[73]G. Dehm, "Miniaturized Single-Crystalline fcc Metals Deformed in Tension: New Insights in Size-Dependent Plasticity," *Prog. Mater. Sci.*, **54** [6] 664–88 (2009).

[74]D. Kiener, C. Motz, T. Schoberl, M. Jenko, and G. Dehm, "Determination of Mechanical Properties of Copper at the Micron Scale," *Adv. Eng. Mater.*, **8** 1119–25 (2006).

[75]D. Kiener, W. Grosinger, G. Dehm, and R. Pippan, "A Further Step Towards an Understanding of Size-Dependent Crystal Plasticity: In Situ Tension Experiments of Miniaturized Single-Crystal Copper Samples," *Acta Mater.*, **56**, 580–92 (2008).

[76]D. Kiener, W. Grosinger, and G. Dehm, "On the Importance of Sample Compliance in Uniaxial Microtesting," *Scr. Mater.*, **60** [3] 148–51 (2009).

[77]D. Kiener, C. Motz, and G. Dehm, "Micro-Compression Testing: A Critical Discussion of Experimental Constraints," *Mater. Sci. Eng. A*, **505**, 79–87 (2009).

[78]R. Maass, S. Van Petegem, D. Grolimund, H. Van Swygenhoven, D. Kiener, and G. Dehm, "Crystal Rotation in Cu Single Crystal Micropillars:

In Situ Laue and Electron Backscatter Diffraction," *Appl. Phys. Lett.*, 017905 (2008).

[79]A. T. Jennings and J. R. Greer, "Tensile Deformation of Electroplated Copper Nanopillars," *Phil. Mag. A.*, **91** [7–9] 1108–20 (2010).

[80]A. T. Jennings, M. J. Burek, and J. R. Greer, "Size Effects in Single Crystalline cu Nano-Pillars Fabricated Without the Use of Fib," *Phys. Rev. Lett.*, **104**, 135503 (2010).

[81]K. S. Ng and A. H. W. Ngan, "Breakdown of Schmid's Law in Micropillars," *Scr. Mater.*, **59** [7] 796–9 (2008).

[82]K. S. Ng and A. H. W. Ngan, "Effects of Trapping Dislocations Within Small Crystals on Their Deformation Behavior," *Acta Mater.*, **57**, 4902–10 (2009).

[83]H. Bei, E. P. George, and G. M. Pharr, "Small-Scale Mechanical Behavior of Intermetallics and Their Composites," *Mater. Sci. Eng. A*, **483–484**, 218–22 (2008).

[84]H. Bei, S. Shim, G. M. Pharr, and E. P. George, "Effects of Pre-Strain on the Compressive Stress–Strain Response of Mo-Alloy Single-Crystal Micropillars," *Acta Mater.*, **56** [17] 4762–70 (2008).

[85]H. Bei and S. Shim, "Effects of Focused Ion Beam Milling on the Mechanical Behavior of a Molybdenum-Alloy Single Crystal," *Appl. Phys. Lett.*, **91** [11] 11915 (2007).

[86]J.-Y. Kim, D. Jang, and J. R. Greer, "Insights into Deformation Behavior and Microstructure Evolution in Nb Single Crystalline Nano-Pillars Under Uniaxial Tension and Compression," *Scr. Mater.*, **61** [3] 300–3 (2009).

[87]J. Y. Kim and J. R. Greer, "Tensile and Compressive Behavior of Gold and Molybdenum Single Crystals at the Nano-Scale," *Acta Mater.*, **57** [17] 5245–53 (2009).

[88]J. Y. Kim, D. C. Jong, and J. R. Greer, "Tensile and Compressive Behavior of Tungsten, Molybdenum, Tantalum and Niobium at the Nanoscale," *Acta Mater.*, **58** [7] 2355–63 (2010).

[89]A. S. Schneider, D. Kaufmann, B. G. Clark, C. P. Frick, P. A. Gruber, R. Monig, O. Kraft, and E. Arzt, "Correlation Between Critical Temperature and Strength of Small-Scale bcc Pillars," *Phys. Rev. Lett.*, **103**, 105501 (2009).

[90]E. Lilleodden, "Microcompression Study of Mg (0001) Single Crystal," *Scr. Mater.*, **62** [8] 532–5 (2010).

[91]C. M. Byer, B. Li, B. Cao, and K. T. Ramesh, "Microcompression of Single-Crystal Magnesium," *Scr. Mater.*, **62** [8] 536–9 (2010).

[92]Q. Yu, Z.-W. Shan, J. Li, X. Huang, L. Xiao, J. Sun, and E. Ma, "Strong Crystal Size Effect on Deformation Twinning," *Nature*, **463**, 335–8 (2010).

[93]G. Lee, J.-Y. Kim, A. S. Budiman, N. Tamura, M. Kunz, K. Chen, M. J. Burek, J. R. Greer, and T. Y. Tsui, "Fabrication, Structure, and Mechanical Properties of Indium Nanopillars," *Acta Mater.*, **58** [4] 1361–8 (2010).

[94]E. A. Withey, A. M. Minor, Jr, D. C. Morris, J. W. Morris, and S. Kuramot , "The Deformation of Gum Metal Through In Situ Compression of Nanopillars," *Acta Mater.*, **58** [7] 2652–65 (2010).

[95]E. A. Withey, J. Ye, A. M. Minor, S. Kuramoto, D. C. Chrzan, and J. W. Morris Jr, "Nanomechanical Testing of Gum Metal," *Exp. Mech.*, **50** 37–45 (2010).

[96]A. Rinaldi, P. Peralta, C. Friesen, and K. Sieradzki, "Sample-Size Effects in the Yield Behavior of Nanocrystalline Nickel," *Acta Mater.*, **56** 511–7 (2008).

[97]D. Dang and J. R. Greer, "Size-Induced Weakening and Grain Boundary-Assisted Deformation in 60nm Grained Ni Nanopillars," *Scripta Mater.*, **64**, 77–80 (2011).

[98]B. G. Clark, D. S. Gianola, O. Kraft, and C. P. Frick, "Size Independent Shape Memory Behavior of Nickel-Titanium," *Adv. Eng. Mater.*, **12** [8] 805–15 (2010).

[99]C. P. Frick, S. Orso, and E. Arzt, "Loss of Pseudoelasticity in Nickel-Titanium Submicron Compression Pillars," *Acta Mater.*, **55**, 3845–55 (2007).

[100]C. P. Frick, B. G. Clark, S. Orso, A. S. Schneider, and E. Arzt, "Size Effect on Strength an Strain Hardening of Small-Scale (111), Nickel Compression Pillars," *Mater. Sci. Eng. A*, **489** [1–2] 319–29 (2008).

[101]C. P. Frick, B. G. Clark, S. Orso, P. Sonnweber-Ribic, and E. Arzt, "Orientation-Independent Pseudoelasticity in Small-Scale NiTi Compression Pillars," *Scr. Mater.*, **59**, 7–10 (2008).

[102]C. P. Frick, B. G. Clark, A. S. Schneider, R. Maaß, S. Van Petegem, and H. Van Swygenhoven, "On the Plasticity of Small-Scale Nickel-Titanium Shape Memory Alloys," *Scr. Mater.*, **62**, 492–5 (2010).

[103]J. M. San Juan, M. L. Nó, and C. A. Schuh, "Nanoscale Shape-Memory Alloys for Ultrahigh Mechanical Damping," *Nat. Nanotechnol.*, **4**, 415–9 (2009).

[104]J. M. San Juan, M. L. Nó, and C. A. Schuh, "Superelasticity and Shape Memory in Micro- and Nanometer-Scale Pillars," *Adv. Mater.*, **20**, 272–8 (2008).

[105]D. Jang and J. R. Greer, "Transition From a Strong-Yet-Brittle to a Stronger-and-Ductile State by Size Reduction of Metallic Glasses," *Nat. Mater.*, **9**, 215–9 (2010).

[106]C. A. Volkert, A. Donohue, and F. Spaepen, "Effect of Sample Size on Deformation in Amorphous Metals," *J. Appl. Phys.*, **103** [8] 083539 (2008).

[107]C. Q. Chen, Y. T. Pei, and J. T. M. De Hosson, "Strength of Submicrometer Diameter Pillars of Metallic Glasses Investigated with in Situ Transmission Electron Microscopy," *Philos. Mag. Lett.*, **89** [10] 633–40 (2009).

[108]M. D. Uchic, P. A. Shade, and D. M. Dimiduk, "Plasticity of Micrometer-Scale Single Crystals in Compression," *Annu. Rev. Mater. Res.*, **39** [1] 361–86 (2009).

[109]R. Dou and B. Derby, "A Universal Scaling Law for the Strength of Metal Micropillars and Nanowires," *Scr. Mater.*, **61** [5] 524–7 (2009).

[110]C. V. Thompson, "The Yield Stress of Polycrystalline Thin Films," *J. Mater. Res.*, **8**, 237 (1993).

[111]B. von Blanckenhagen, P. Gumbsch, and E. Arzt, "Dislocation Sources and the Flow Stress of Polycrystalline Thin Metal Films," *Philos. Mag. Lett.*, **83** [1] 1–8 (2003).

[112]M. S. De Guzman, G. Neubauer, P. Flinn, and W. D. Nix, "The Role of Indentation Depth on the Measured Hardness of Materials," *Mater. Res. Soc. Symp. Proc.*, **308** 613 (1993).

[113]J. R. Greer, W. C. Oliver, and W. D. Nix, "Size Dependence of Mechanical Properties of Gold at the Micron Scale in the Absence of Strain Gradients," *Acta Mater.*, **53** [6] 1821–30 (2005).

[114]K. S. Ng and A. H. W. Ngan, "Stochastic Nature of Plasticity of Aluminum Micro-Pillars," *Acta Mater.*, **56** [8] 1712–20 (2008).

[115]S. I. Rao, D. M. Dimiduk, T. A. Parthasarathy, M. D. Uchic, M. Tang, and C. Woodward, "Athermal Mechanisms of Size-Dependent Crystal Flow Gleaned From Three-Dimensional Discrete Dislocation Simulations," *Acta Mater.*, **56** [13] 3245–59 (2008).

[116]D. M. Norfleet, D. M. Dimiduk, S. J. Polasik, M. D. Uchic, and M. J. Mills, "Dislocation Structures and Their Relationship to Strength in Deformed Nickel Microcrystals," *Acta Mater.*, **56** [13] 2988–3001 (2008).

[117]T. A. Parthasarathy, S. I. Rao, D. M. Dimiduk, M. D. Uchic, and D. R. Trinkle, "Contribution to size effect of yield strength from the stochastics of dislocation source lengths in finite samples," *Scr. Mater.*, **56** [4] 313–6 (2007).

[118]S. I. Rao, D. M. Dimiduk, M. Tang, T. A. Parthasarathy, M. D. Uchic, and C. Woodward, "Estimating the Strength of Single-Ended Dislocation Sources in Micron-Sized Single Crystals," *Philos. Mag. A*, **87** [30], 4777–94 (2007).

[119]D. M. Dimiduk, C. Woodward, R. LeSar, and M. D. Uchic, "Scale-Free Intermittent Flow in Crystal Plasticity," *Science*, **312**, 1188–90 (2006).

[120]O. Kraft, P. A. Gruber, R. Monig, and D. Weygand, "Plasticity in Confined Dimensions," *Annu. Rev. Mater. Res.*, **40**, 293–317 (2010).

[121]A. T. Jennings, M. J. Burek, and J. R. Greer, "Size Effects in Single Crystalline cu Nano-Pillars Fabricated Without the Use of Fib," *Phys. Rev. Lett.*, **104**, 135503 (2010).

[122]S. Lee, S. Han, and W. D. Nix, "Uniaxial Compression of fcc Au Nanopillars on an MgO Substrate: The Effects of Prestraining and Annealing," *Acta Mater.*, **57** [15] 4404–15 (2009).

[123]S. Shim, H. Bei, M. K. Miller, G. M. Pharr, and E. P. George, "Effects of Focused Ion Beam Milling on the Compressive Behavior of Directionally Solidified Micropillars and the Nanoindentation Response of an Electropolished Surface," *Acta Mater.*, **57**, 503–10 (2009).

[124]J. A. El-Awady, C. Woodward, D. M. Dimiduk, and N. M. Ghoniem, "Effects of Focused Ion Beam Induced Damage on the Plasticity of Micropillars," *Phys. Rev. B*, **80**, 104104 (2009).

[125]R. Ikeda, H. Ogi, T. Ogawa, and M. Takemoto, "Mechanical Properties of CVD Synthesized Polycrystalline Diamond," *Ind. Diamond Quat.*, **69**, 35–8 (2009).

[126]N. A. Fleck, V. Deshpande, and M. F. Ashby, "Micro-Architected Materials: Past, Present and Future," *Proc. Royal Soc. A*, **466**, 2495–516 (2010).

[127]D. J. Bell, T. J. Lu, N. A. Fleck, and S. M. Spearing, "MEMS Actuators and Sensors: Observations on Their Performance and Selection for Purpose," *J. Micromech. Microeng.*, **15** S153–64 (2005).

[128]M. A. Haque and T. Saif, "Mechanical Testing at the Micro/Nano Scale." in *Springer Handbook of Experimental Solid Mechanics*. Edited by W. N. Sharpe. Springer, New York, 2008.

[129]A. Torrents, K. Azgin, S. W. Godfrey, E. S. Topalli, T. Akin, and L. Valdevit, "MEMS Resonant Load Cells for Micro-Mechanical Test Frames: Feasibility Study and Optimal Design," *J. Micromech. Microeng.*, **20** 125004 (17pp) (2010).

[130]F. Giessibl, "Advances in Atomic Force Microscopy," *Rev. Mod. Phys.*, **75** [3] 949–83 (2003).

[131]P. Hinterdorfer and Y. Dufrene, "Detection and Localization of Single Molecular Recognition Events Using Atomic Force Microscopy," *Nat. Methods*, **3** [5] 347–55 (2006).

[132]G. Schitter and M. J. Rost, "Scanning Probe Microscopy at Video-Rate," *Mater. Today*, **11**, 40–8 (2008).

[133]M. D. Kriese, W. W. Gerberich, and N. R. Moody, "Quantitative Adhesion Measures of Multilayer Films: Part I. Indentation Mechanics," *J. Mater. Res.*, **14** [7] 3007–18 (1999).

[134]M. D. Kriese, W. W. Gerberich, and N. R. Moody, "Quantitative Adhesion Measures of Multilayer Films: Part II. Indentation of W/Cu, W/W, Cr/W," *J. Mater. Res.*, **14** [7] 3019–26 (1999).

[135]E. T. Lilleodden, W. Bonin, J. Nelson, J. T. Wyrobek, and W. W. Gerberich, "In-Situ Imaging of MU-N Load Indents into GAAS," *J. Mater. Res.*, **10** [9] 2162–5 (1995).

[136]W. D. Nix, "Elastic and Plastic Properties of Thin Films on Substrates: Nanoindentation Techniques," *Mat. Sci. Eng. A - Struct.*, **234**, 37–44 (1997).

[137]W. C. Oliver and G. M. Pharr, "An Improved Technique for Determining Hardness and Elastic Modulus Using Load and Displacement Sensing Indentation Experiments," *J. Mater. Res.*, **7** [6] 1564–83 (1992).

[138]W. C. Oliver and G. M. Pharr, "Measurement of Hardness and Elastic Modulus by Instrumented Indentation: Advances in Understanding and Refinements to Methodology," *J. Mater. Res.*, **19** [1] 3–20 (2004).

[139]J. R. Greer, J.-Y. Kim, and M. J. Burek, "The In-Situ Mechanical Testing of Nanoscale Single-Crystalline Nanopillars," *J. Mater.*, **61** [12] 19–25 (2009).

[140]J.-Y. Kim and J. R. Greer, "Tensile and Compressive Behavior of Gold and Molybdenum Single Crystals at the Nano-Scale," *Acta Mater.*, **57**, 5245–53 (2009).

[141]Z. Y. Xue and J. W. Hutchinson, "Constitutive Model for Quasi-Static Deformation of Metallic Sandwich Cores," *Int. J. Numer. Meth. Eng.*, **61** [13] 2205–38 (2004).

[142]J. E. Angelo, N. R. Moody, and M. I. Baskes, "Trapping of Hydrogen to Lattice-Defects in Nickel," *Modelling Simul. Mater. Sci. Eng.*, **3** [3] 289–307 (1995).

[143]M. I. Baskes, "Modified Embedded-Atom Potentials for Cubic Materials," *Phys. Rev. B*, **46** [5] 2727–42 (1992).

[144]W. Xiao, M. I. Baskes, and K. Cho, "MEAM Study of Carbon Atom Interaction With Ni Nano Particle," *Surf. Sci.*, **603** [13] 1985–98 (2009).

[145]Y. Jiang, Y. G. Wei, J. R. Smith, J. W. Hutchinson, and A. G. Evans, "First Principles Based Predictions of the Toughness of a Metal/Oxide Interface," *Int. J. Mater. Res.*, **101** [1] 8–15 (2010).

[146]V. S. Deshpande, A. Needleman, and E. Van der Giessen, "Plasticity Size Effects in Tension and Compression of Single Crystals," *J. Mech. Phys. Solids*, **53** [12] 2661–91 (2005).

[147]H. Tang, K. W. Schwarz, and H. D. Espinosa, "Dislocation-Source Shutdown and the Plastic Behavior of Single-Crystal Micropillars," *Phys. Rev. Lett.*, **100**, 185503 (2008).

[148]E. Van der Giessen and A. Needleman, "Discrete Dislocation Plasticity – A Simple Planar Model," *Modelling Simul. Mater. Sci. Eng.*, **3** [5] 689–735 (1995).

[149]M. D. Uchic, P. A. Shade, and D. M. Dimiduk, "Plasticity of Micrometer-Scale Single Crystals in Compression," *Annu. Rev. Mater. Res.*, **39**, 361–86 (2009).

[150]J. Wang, A. G. Evans, K. Dharmasena, and H. N. G. Wadley, "On the Performance of Truss Panels With Kagome Cores," *Int. J. Solids Struct.*, **40** [25] 6981–8 (2003).

[151]M. P. Bendsoe and O. Sigmund, "*Topology Optimization*," pp. 370. Springer, Berlin, Germany, (2002).

[152]P. W. Christensen and A. Klabring, "*An Introduction to Structural Optimization*," pp. 214. Springer, Berlin, Germany, (2008).

[153]O. Sigmund and S. Torquato, "Design of Materials With Extreme Thermal Expansion Using a Three-Phase Topology Optimization Method," *J. Mech. Phys. Solids*, **45** [6] 1037–67 (1997).

[154]O. Sigmund and S. Torquato, "Design of Smart Composite Materials Using Topology Optimization," *Smart Mater. Struct.*, **8** [3] 365–79 (1999).

[155]S. Hyun and S. Torquato, "Optimal and Manufacturable Two-Dimensional, Kagome-Like Cellular Solids," *J. Mater. Res.*, **17** [1] 137–44 (2002).

[156]A. J. Jacobsen, W. Barvosa-Carter, and S. Nutt, "Micro-Scale Truss Structures With Three-Fold And Six-Fold Symmetry Formed From Self-Propagating Polymer Waveguides," *Acta Mater.*, **56** [11] 2540–8 (2008).

[157]J. R. Greer and J. T. M. De Hosson, "Plasticity in Small-Sized Metallic Systems: Intrinsic Versus Extrinsic Size Effects," *Prog. Mater. Sci.*, **56**, 654–724 (2011).

[158]J.-Y. Kim and J. R. Greer, "Size-Dependent Mechanical Properties of Mo Nanopillars," *Appl. Phys. Lett.*, **93**, 101916 (2008).

[159]A. J. Jacobsen, W. Barvosa-Carter, and S. Nutt, "Compression Behavior of Micro-Scale Truss Structures Formed From Self-Propagating Polymer Waveguides," *Acta Mater.*, **55** [20] 6724–33 (2007).

[160]K. Azgin, C. Ro, A. Torrents, T. Akin, and L. Valdevit, "A Resonant Tuning Fork Sensor with Unprecedented Combination of Resolution and Range." in *IEEE MEMS Conference*, Paper # 0700. Cancun, Mexico, 2011.

J. Am. Ceram. Soc., **94** [S1] S35–S41 (2011)
DOI: 10.1111/j.1551-2916.2011.04561.x
© 2011 The American Ceramic Society

journal

A Shape-Morphing Ceramic Composite for Variable Geometry Scramjet Inlets

Richard Miles,[†] Phillip Howard, Christopher Limbach, and Syed Zaidi

Department of Mechanical and Aerospace Engineering, Princeton University, Princeton, New Jersey 08544

Sergio Lucato, Brian Cox, and David Marshall

Teledyne Scientific Company, Thousand Oaks, California 91360

Angel M. Espinosa, and Dan Driemeyer

Boeing Phantom Works, Saint Louis, Missouri 63166

The development of ceramic composites with three-dimensional fiber reinforcement architectures formed by textile methods has led to the potential for active shape-morphing surfaces that can operate in high temperature and variable pressure environments. This technology is of particular interest for hypersonic applications, where SCRAM jet engines require variable inlet geometry to achieve efficient flight over realistic flight profiles and variable flight conditions. The experiments reported here show that significant shape morphing can be achieved and good control of the shape sustained even in the presence of large temperature and pressure gradients. Experiments were carried out using a subscale morphing hypersonic inlet with rectangular cross-section in a Mach 8 wind tunnel facility with a total temperature of 800 K. The upper surface of the inlet consisted of a C–SiC composite plate (0.7 mm thick, 37.5 cm long, and 11 cm wide) connected to five actuators through a triangular truss support structure. The lower surface was a flat plate instrumented with an array of pressure taps along the flow centerline. As the shape varied, the surface contour was reliably controlled for high efficiency, low loss compression. A factor of six inlet area ratio variation was achieved and good agreement with model predictions was observed.

I. Introduction

HIGH-TEMPERATURE shape-morphing structural materials could potentially revolutionize the design of air breathing hypersonic vehicles. The performance of the RAM/SCRAM engine is strongly affected by shapes and contours of the compression ramp and inner surfaces, including the inlet, the isolator, the combustor and the nozzle.[1,2] A contour that is designed for operation at one Mach number can be very inefficient at off-design Mach numbers. Even weak shock waves can substantially reduce engine performance, while instabilities in the flow can lead to upstream shock motion and engine unstart. High-temperature shape-morphing materials could (i) enable continuous variation of the inlet area ratio to accommodate changing Mach number and air density associated with

climb and descent; (ii) improve engine and isolator performance by accommodating area and shape changes associated with engine mode transition from subsonic to supersonic combustion (RAM to SCRAM); (iii) optimize the compression ramp and inlet contour to minimize shock formation and reduce pressure losses; (iv) permit dynamic adaptation to compensate for flight maneuvers and correct for distortion due to thermal expansion, as well as suppress upstream shock motion to reduce the probability of engine unstart, and if unstart occurs, reduce the recovery time; and (v) improve nozzle performance by controlling the nozzle area ratio and contour for efficient thrust generation and minimum pressure loss. On both internal and external surfaces of the vehicle, high-temperature morphing materials could replace hinged control surfaces with smoothly varying surfaces, thereby eliminating flow disturbances associated with discontinuous changes in slope.[3]

High-temperature shape-morphing materials are also of interest for hypersonic and supersonic ground test facilities.[4] In this case, the shape-morphing capability would allow for dynamic optimization of the expansion contour of the facility, improving the flow quality, reducing shock and expansion waves and thus minimizing losses and improving performance. Shape morphing would also provide the opportunity for real time control of the facility Mach number by changing the expansion area ratio continuously. Low-temperature variable Mach capability has been previously demonstrated for a Mach 2.5–3.2 in-draft wind tunnel nozzle.[5] This leads to the possibility of simulating portions of the flight mission including climb and descent. Asymmetric shape morphing can be used to simulate asymmetric inlet flows for direct connect engine test applications.[6] Optimizing the performance of flow deflectors and other aero appliances and the diffuser is also important for ground test facilities because they affect the flow quality and the pressure recovery and thus the facility operating envelope and run time. High-temperature shape-morphing aero appliance surfaces and diffuser geometries could accommodate configuration variation of test articles and allow for dynamic changes to occur during the ground test, such as changes in angle of attack.

To operate in these applications, a high-temperature shape-morphing material must be both flexible and structurally stiff, with strength sufficient to withstand the stresses associated with high thermal gradients and pressure gradients. The material must be capable of withstanding temperatures in excess of 1500 K for even modest Mach numbers (Mach 8), resistant to oxidation, and resistant to thermomechanical fatigue. The purpose of this paper is to describe the design and fabrication of a shape-morphing ceramic composite (CMC) structure that satisfies these requirements and to present results of preliminary testing of its performance under hypersonic conditions, as the wall of a subscale inlet in a Mach 8 wind tunnel.

B. McMeeking—contributing editor

Manuscript No. 29043. Received December 13, 2010; approved March 9, 2011.
Funding at Princeton was provided by the Office of Naval Research Multi University Research Initiative through the University of California, Santa Barbara. Funding at TSC was provided by Boeing Phantom Works for development of the morphing CMC structures and by the Air Force Office of Scientific Research and NASA through the National Hypersonic Science Center for Materials and Structures for collaboration during testing at Princeton.
[†]Author to whom correspondence should be addressed. e-mail: miles@princeton.edu

II. CMC Structure

(1) General Design Considerations for Morphing Ceramic Composite

The challenge of designing morphing surfaces that are sufficiently flexible to change shape by bending or twisting yet capable of supporting large pressure loads has been addressed by combining thin face sheets with stiff, statically determinate truss structures.[7–10] A class of those high authority shape-morphing structures has been initially described by Lu *et al.*[11] and further generalized by Hutchinson *et al.*[10] Previous studies have also suggested that these face sheet truss structures are a lighter-weight alternative to metallic foam core structures.[12] A simple example building on Lu's work is shown in Fig. 1, where a thin face sheet attached to a triangular truss structure is actuated by heating selected truss elements with an electric current. The active elements consist of a shape memory alloy, which undergoes a reversible strain on heating and cooling. This class of morphing structures can be actuated by internal active elements, as in Fig. 1, or by forces applied externally. Previous studies have been directed at optimizing morphing structures with other truss structures capable of more intricate shape changes with metallic or polymer face sheets.[7–10]

With recent advances in forming textile-based CMCs,[10] hybrid systems consisting of a CMC face sheet attached to metallic truss structure are also feasible. In this case, with the high-temperature CMC face sheet facing the hot gas flow, the temperature of the metallic truss structure may be controlled if necessary by active or passive cooling to maintain sufficient temperature gradient across the hot face sheet. Structures of this type are enabled by several characteristics of textile-based composites. One is the ability to form mechanically robust thin skins with through-thickness fiber reinforcements that eliminate delamination as a failure mode. Thin C–SiC and SiC–SiC composite skins (thickness <1 mm) with woven reinforcement architectures have been fabricated and shown to survive combustion environments with surface temperatures of 1800 K and temperature gradients >1000 K/mm across the hot skin.[13] The face sheet of the structure in Fig. 1 consists of such a composite (SiC–SiC), with two-layer, angle-interlock reinforcement architecture.[14]

The other enabling characteristic is the ability to form arrays of connections between the backside of the hot skin and the truss structure capable of transmitting both tensile and compressive loads. A simple example of a smoothly pinned joint, from the structure in Fig. 1, is shown in Figs. 2(a) and (b). The connectors consist of fiber tows drawn out from the weave structure of the composite skin to form loops held in shape by carbon rods during weaving of the fiber preform. After using chemical vapor infiltration (CVI) to deposit a thin coating of BN on all the fibers (which serves as a debond layer to enable toughening), the preform was rigid and the carbon rods were removed. The SiC matrix was then infiltrated into the preform, a

Fig. 2. Smoothly pinned joints connecting CMC skin and truss structure: (a) schematic of weave structure; (b) configuration used in the morphing structure in Fig. 1 with superalloy struts attached; (c) configuration used in the morphing structure in Fig. 3.

superalloy wire was inserted into the loops, and the truss elements were attached to the wire by laser welding.

The maximum load capability (or lifetime in oxidizing environments) of the attachment feature shown in Fig. 2(b) is limited by damage generated by local stress concentrations at the contact between the superalloy wire and the inside of the CMC loops. A more robust design is shown in Fig. 2(c). In this case, the fiber tows forming the connectors were drawn over molybdenum tubes (outer diameter 2 mm, inner diameter 1 mm) during weaving. The tubes remained in place during subsequent processing of the composite, whereupon selected sections of the tubes between the loops were cut and removed, leaving the sections of the tubes beneath the loops as an integral part of the attachment structure. The insides of these tube sections provide a smoother, better aligned, and more damage resistant bearing surface for the superalloy rods.

The achievable deflections of a structure such as Fig. 1 are dictated by the face sheet thickness and strain limit. The critical radius of curvature ρ_c for bending of an initially flat face sheet is given by

$$\rho_c = \frac{t}{2\varepsilon_c} \tag{1}$$

where t is the face sheet thickness and ε_c the critical strain. CMCs generally show relatively high fatigue limits in cyclic loading.[15,16] If we take a conservative limit $\varepsilon_c \sim 0.25\%$, less than half of the typical failure strain for SiC–SiC and C–SiC composites, the minimum radius of curvature for a face sheet of thickness of 1 mm is ~200 mm. The radius of curvature in Fig. 1(b), with center deflection of 19 mm, is ~600 mm.

(2) Fabrication of morphing duct

A morphing CMC structure that would serve as the top wall of a model inlet duct of length 37.5 cm and width 11 cm was constructed as shown in Fig. 3. The CMC skin was fabricated using a woven preform of carbon fibers (T300-3K) with a two-layer angle-interlock structure and attachment sites formed with the method shown in Fig. 2(c). The composite was processed by CVI in two steps, the first to deposit a coating of pyrolytic carbon (~0.5 μm thickness) on individual fibers, and the second to form a matrix of SiC.

The supporting metal truss structure, consisting of a triangular array of solid superalloy plates, was connected to the CMC skin with superalloy rods passing through the molybdenum tubes of the attachment sites and matching holes in lugs that had been cut at the edges of the plates in a hinge-like configuration. This structure is easily flexed in the longitudinal direction but has high stiffness in the lateral direction.

Fig. 1. Shape-morphing structure with SiC–SiC face sheet, superalloy support structure, and shape memory actuating struts.

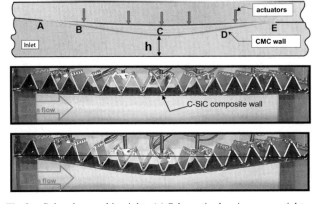

Fig. 3. Subscale morphing inlet. (a) Schematic showing sequential target shapes of top CMC wall during morphing; (b) and (c) side view of inlet showing CMC wall in positions for maximum and minimum inlet area ratio.

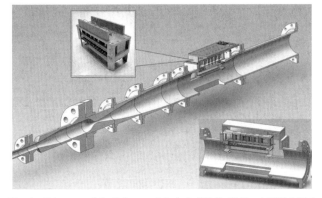

Fig. 4. Diagram of the Princeton Mach 8, 800 K, 1050 psi (7.24 MPa) facility showing details of the morphing inlet insertion and the morphing inlet geometry. A copper plate is placed between the surface and the five actuators to isolate them from the high temperature. Inset shows morphing inlet seen through the schlieren side windows.

The target family of deflected shapes for the wall, illustrated in Fig. 3(a), was achieved using five actuators connected to the support structure through pivot plates visible in Fig. 3. The actuator rods were driven by stepper motors controlled through a Labview program, so that all actuators moved in synchronization and the shape transitioned smoothly between successive contours. The position control relied on simple counting of steps in 3.05 μm/step increments starting from an initial calibrated reference state. The family of shape contours satisfied the constraints that the morphing CMC section of the duct must transition smoothly at the ends to a mating fixed metallic structure (i.e., the slope at positions A and E remained fixed during all actuation movements) and that the deflections in the central throat region (C) extend over a range of 25 mm. At the extreme position Fig. 3(c), the radius of curvature of the CMC wall is approximately 250 mm in the throat region (C) and approximately 500 mm at positions B and C. Note that while the actuator displacements are in the same direction at positions B, C and D, the forces applied at the CMC skin are compressive at position C and tensile at positions B and D.

III. Wind Tunnel Testing

A Mach 8 wind tunnel at Princeton University was used to assess the potential of the composite structure in Fig. 3 for hypersonic morphing applications. Issues to be examined include the performance of the material under thermal stress, the flow quality achievable, the performance of the actuators, the impact of pressure gradients on the material, including start up and shut down transients, the stiffness and dynamic stability of the material in the presence of high frequency buffeting from a hypersonic flow, and the ability of the material to deflect sufficiently far and sustain a specified contour sufficiently well to control the flow without leakage or mechanical damage. The wind tunnel operates at a total temperature of 790 K and at a stagnation pressure of 1050 psi (7.24 MPa). The tunnel is operated with an air ejector system and can be run for several minutes at a time, limited by the capacity of the electrically driven storage heater. At Mach 8, the static temperature is ∼60 K and the static pressure is ∼5 Torr (0.28 kPa). The tunnel has a 9 in. (23 cm) diameter circular test section.

The morphing CMC skin of Fig. 3 was mounted in the fixture shown in the inset of Fig. 4, which formed a rectangular duct with fixed side walls and base plate beneath the morphing CMC top wall. The upstream end of the CMC skin was held in a fixed position against the mating inlet ramp on the tunnel wall, while at the downstream edge sliding occurred between the CMC face sheet and the mating metal duct surface during actuation. The side walls were quartz to allow optical observations during testing. The fixture was placed at the top of the tunnel, with bypass air flowing below it to avoid choking of the tunnel itself. Figure 4 is a schematic of the wind tunnel showing the location of the test fixture.

As the flow is compressed in the morphing inlet, the pressure increases significantly above the free stream static pressure. The pressure at the back of the morphing surface is determined by the downstream static pressure, which remains well below the pressure of the air in the inlet duct, leading to an outward compressive force on the actuators. This force increases nonlinearly as the morphing surface approaches the opposite wall. The pressure in the duct ranged from about 0.5 psi to ∼14 psi (2.5–100 kPa). The edges of the morphing wall were not sealed against the side walls (the clearance was <0.5 mm), so some air leakage occurred from the duct to the backside cavity. However, the leak rate was sufficiently small that there was very little perturbation to the flow inside the duct.

Because the actuator driver rods could not be conveniently vacuum sealed from the morphing surface in this particular setup, the entire structure including the actuators and driving electronics was enclosed within the wind tunnel test chamber, as shown in the Fig. 5. A 7° inlet ramp was installed from the top of the curved inlet wall to divert the flow into the morphing section. The leading edge of the morphing plate was mounted at a 5° angle to minimize discontinuity and shock formation at the joint with the inlet ramp. The rear of the morphing section was captured into a sliding retainer. The back end of the actuator section was left open to the tunnel diffuser so that large pressure gradients would not occur across the morphing panel during

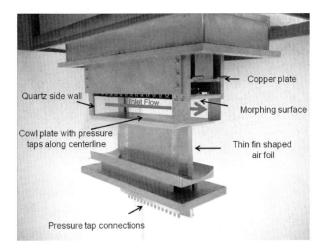

Fig. 5. Morphing inlet tunnel model including the 13 pressure taps that are routed to the bottom of the wind tunnel through a thin fin. The water-cooled copper plate can be seen near the top. The curved boundary beneath the air-foil mates flush with the wind tunnel's inner perimeter.

start up and shut down. Even though the static temperature of the air in the test section is low, the boundary layer and other stagnant air rises to a temperature close to the total temperature of the plenum (800 K). Because the stagnant hot air permeates the chamber behind the morphing panel, a water-cooled copper thermal insulation plate was installed between the CMC wall and the actuators to protect the actuators from the high temperatures.

In order to quantify the static pressure profile, a series of 13 pressure tap holes were drilled along the centerline of the flat bottom plate at 16 mm intervals. The pressure tap tubes were routed across the hypersonic flow below the flat plate to the bottom of the wind tunnel through a thin fin-shaped air-foil to minimize disturbance of the wind tunnel below the inlet flow. The fin structure, with a bottom curved plate that fits the contour of the lower section of the wind tunnel, is shown in Fig. 5. The diagnostics included the static pressure profile from the 13 pressure taps as well as schlieren images of the shock structure taken through the side windows and an overall video of the motion of the surface.

IV. Flow Modeling

Initial modeling of the inlet was undertaken using the proprietary Boeing 3D BCFD flow solver, assuming laminar adiabatic flow. This modeling was undertaken to assure that the insertion of the morphing inlet did not choke the wind tunnel and to determine the impact of the tunnel boundary layer on the inlet flow field. The flow solver computed the flow along the entire wind tunnel, following the expansion contour from the throat of the tunnel through the test region. The flow was solved for half of the tunnel using 6.3 million cells assuming a symmetry plane. The computation was performed for the flow from the nozzle throat through and below the morphing inlet and it included the boundary layer evolution. Because of the conservation of enthalpy and entropy in an ideal expansion, the plenum has to be operated at high temperature and pressure to provide the kinetic energy and dynamic pressure required for testing. For the tests conducted here, the objective was to create a Mach 8 simulation so that aerodynamic properties of the flow including shock and expansion wave structure could be determined. Even though the proper Mach number is reached, the facility is not capable of producing the true flight air speed and static temperature because that requires plenum temperatures on the order of 2500 K. For Mach number duplication what is necessary is to have high enough plenum temperature to avoid condensation of the air as it cools by a factor of almost 14 in the expansion. The pressure drops by a factor of ~10 000 (assuming ideal gas with a ratio of heat capacities of 1.4). Modeled conditions at

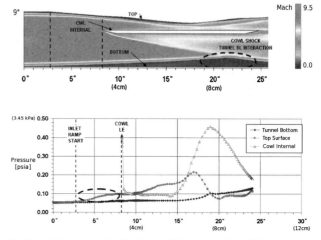

Fig. 7. (a) Mach number field around the test section. (b) Surface pressures computed for Mach 8 flow. All simulations were done using the Boeing 3D BCFD code with laminar flow.

the throat of the tunnel were 800 K (1440 R) and 450 psia (3.1 MPa), corresponding to plenum conditions of 960 K (somewhat higher than actual operating conditions of the facility) and 860 psi (5.9 MPa) (somewhat lower than the actual operating conditions of the facility). The geometry used for the modeling is shown in Fig. 6. This inlet geometry was an earlier configuration with the cowl plate located 2.31 s (59 mm) from the flat morphing surface. For the run reported here, the location of the cowl plate was 1.2 s (30 mm) below the flat morphing surface, significantly closer to that surface so a more substantial compression ratio could be achieved.

Figure 7 shows the predicted centerline Mach numbers (7a) and the wall pressure profiles (7b) of the upper (morphing) and low surfaces of the tunnel and the top of the cowl plate. In this realization the surface is morphed as shown in Fig. 6. The model predicts that the insertion of the inlet into the top of the wind tunnel test section will not choke the wind tunnel, although there is a significant subsonic region at the bottom of the tunnel below the inlet where the cowl shock interacts with the lower boundary layer.

The boundary layer at the top ramp of the inlet is thick due to the fact that it has grown along the wind tunnel wall throughout the expansion. This thick boundary layer and the elliptical intersection of the inlet ramp with the round surface of the tunnel

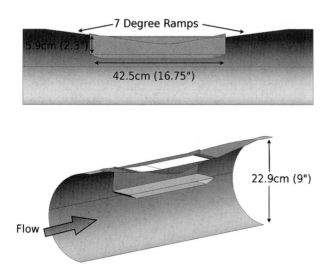

Fig. 6. Wind tunnel geometry used for the Boeing 3D BCFD code computation.

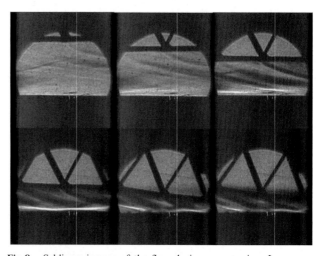

Fig. 8. Schlieren images of the flow during compression. Images are taken at 1.6-s intervals starting at 34.1 s into the run and show a 1.85 in. (4.7 cm) long section of the flow path. The weak shock structure from the upper (morphing) surface arises from the slight bumps in the woven ceramic associated with the connector pin locations.

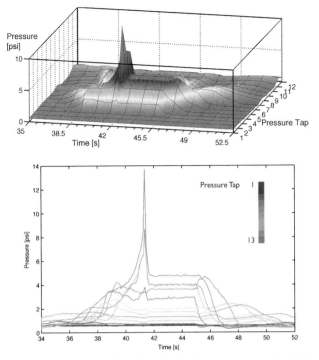

Fig. 9. Pressure versus time plots for the 13 pressure taps. Tap #1 is farthest upstream. The rapid pressure drop at 41.6 s corresponds to the slippage of an actuator due to the high surface pressure loading at maximum compression.

cause the shock from the 7° inlet ramp to be curved and unsteady, and thus not seen with schlieren imaging across the flow. The dashed ellipse in Fig. 7(b) highlights the slow pressure rise along the upper compression ramp associated with this three-dimensional (3D) interaction. Even though the shock is not apparent from the modeled schlieren images, it does lead to a reduction of the centerline Mach number and an increase in the static pressure along the morphing surface and along the upper surface of the cowl plate, as noted in Fig. 7(b). The high-pressure peak at the leading edge of the cowl plate results from the small stagnation point at that location.

A 2D method of characteristics model of the centerline shock structure and pressure contour inside the morphing inlet was developed as an on-site working tool for direct comparisons with experimental data. By implementing the rotational method of characteristics due to Ferri,[17] the expected pressure profile resulting from the morphing surface contour could be rapidly calculated. The simulations were performed using an inverse-

marching scheme with 60 characteristics and 10^3 stream-wise steps. As input to the model the pressure and Mach number were specified at the inlet using experimental data and isentropic flow relationships.

V. Results

Results are shown in Figs. 8 and 9 from a series of tests in which the CMC wall was morphed as shown in Fig. 3 between positions corresponding to $h = 30$ and 5 mm, reducing the area ratio between the inlet and throat by a factor of 6. Figure 8 shows a sequence of schlieren images of the morphing inlet. These images were taken at 1.6-s interval starting at 34.1 s after the start of the run. Figures 9(a) and (b) are renderings of the time evolution of the pressure tap measurements during the shape morphing, with tap #1 the farthest upstream and tap #13 the farthest downstream. It is apparent that the morphing continued until the maximum pressure reached 13.7 psi (94.5 kPa), at which point the force on the actuators exceeded their load capacity and slippage occurred, immediately reducing the compression. The force on the central actuator at this point was estimated by multiplying the maximum pressure with the area supported by the central actuator (approximately 10 cm × 5 cm, or 4 in. × 2 in.), which results in a load of over 100 lbf (445 N). Subsequent static load tests indicated the actuators slip with a force of approximately 108 lbf (480 N).

The static pressure of 0.5 psi (3.4 kPa) in the unmorphed condition indicates that the Mach number of the flow through the inlet is 6.3. This is below the Mach 8 free stream condition due to the shock and compression waves from the front of the compression ramp and the leading edge of the cowl. Other weak shock structure in the facility upstream of the test section may also contribute to the reduction of the inlet Mach number. At the maximum displacement, the area ratio is approximately a factor of 6. The thickness of the boundary layer along the morphing surface can be estimated from the curvature of weak shocks visible in the in the schlieren images. These shocks were generated from small distortions of the woven ceramic plate beneath the actuator attachment points. The slope of these weak shocks can be used to approximate the local Mach number. These weak shocks show the Mach number increasing away from the wall and approaching the free stream angle. In the unmorphed condition, they reach the free stream angle at about 10 mm from the morphing surface. This corresponds to about 30% of the cross sectional area. In the highly deflected cases, the boundary layer shrinks due to compression, but occupies a larger fraction of the flow cross section, leading to an increase in the effective compression area ratio between the undeflected and fully deflected configurations. There is only a thin laminar

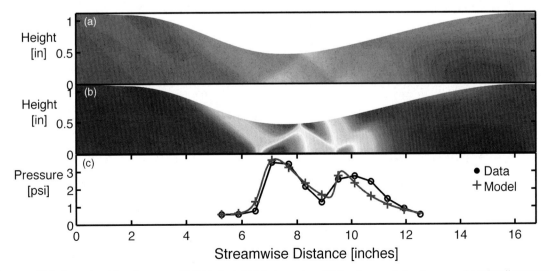

Fig. 10. Computed Mach number (a) and pressure (b) fields for inlet Mach number of 6.3 and computed and measured cowl wall pressure profiles (c).

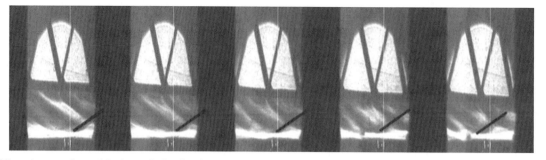

Fig. 11. Schlieren images taken at 0.2-s intervals showing the upstream advance of the reflected shock. The middle image corresponds to the 38.9 s point corresponding to the computed result in Fig. 10. The images show a 1.9 in. (4.7 cm) long length of the test section and are stretched to match the geometry of Fig. 10.

boundary layer on the cowl surface (lower plate) because the cowl lip is in the tunnel free stream and thus does not ingest the tunnel wall boundary layer.

The results of the analysis by the 2D method of characteristics for one snapshot during the experiment at $t = 38.9$ s are shown in Fig. 10. This time step was chosen about 2 s before the maximum compression point to avoid the region with multiple reflected shocks and associated increased unsteadiness. Figures 10(a)–(c) show calculated contours of Mach number and static pressure (a and b) as well as the computed pressure profile along the bottom surface (c). Also shown in Fig. 10(c) is the measured pressure profile from Fig. 9 at the 38.9 s point. The best fit to the data corresponds to an entrance Mach number of 6.3, as expected from the unmorphed pressure measurements. Five schlieren images taken at 0.2 s intervals over from 39.4 to 40.2 s are shown in Fig. 11, with the images distorted to match Fig. 10 and the contrast enhanced. The weak reflected shock is indicated with a red line in each frame and can be seen advancing upstream as the compression increases. The middle image is closest to the 39.8 s point corresponding to the computed result.

Figure 12 shows nine sequential schlieren images taken during the steady operation plateau from 42 to 44 s in Fig. 9. Note here that the morphing structure stays firmly in place with no observable shape fluctuations in the plate even between the pin joints. The weak shocks generated at the pin locations are evident and their local slopes provide a measure of local Mach number, which is about Mach 2.4 in the center of the flow for this configuration.

These results show that, other than the actuator slippage that occurred at the highest compression, the shape-morphing material provided reliable inlet shape control and had sufficient strength and stiffness to maintain the designated profile even in

the presence of significant temperature gradients and pressure variations. Even with the small irregularities in the surface the overall performance of the morphing surface is in good agreement with predicted values.

VI. Conclusions

The development of CMCs with integrally woven fiber reinforcements has led to the potential for active shape morphing of surfaces in high temperature and variable pressure environments. Large temperature and pressure gradients can be sustained without significant distortion of the shape. For the experiments reported here a 0.7 mm thick, 37.5 cm long, and 11 cm wide plate of C–SiC composite was connected to an actuated truss structure and mounted as the upper surface of a shape-morphing hypersonic inlet. The lower surface was a flat steel plate instrumented with an array of pressure taps. The operation of this morphing inlet in an 800 K, 9 in. (23 cm) diameter Mach 8 wind tunnel facility demonstrated that the actuator-controlled CMC surface could be shaped to a desired contour and maintained at that contour in the presence of significant temperature gradients and pressure loads. The measured pressure profile matched the computed profile to good accuracy. The integrally woven pin-joint attachment method proved robust, although there were weak shock waves observed emanating from the slight distortions of the front surface at the attachment locations. These distortions can be easily eliminated in future developments. The actuators accurately controlled the surface contour until the back-pressure force rose above 100 lbf (445 N) per actuator, the limit of the actuator operation envelope, and at that point some slippage occurred. These results demonstrate that textile-based CMCs are excellent candidates for applications in hypersonic vehicles and hypersonic ground test facilities, where large variations in shape and accurate shape control are required in the presence of severe temperature and pressure loads.

References

[1]E. T. Curran, and S. N. B. Murthy (eds.), *Scramjet Propulsion, Progress in Astronautics and Aeronautics*, Vol. 189. Paul Zarchan, Editor in Chief AIAA, Reston, VA, 2000.

[2]C. R. McClinton, "High Speed/Hypersonic Aircraft Propulsion Technology Development"; pp. 1-1 to 1-32 in *Advances on Propulsion Technology for High-Speed Aircraft. Educational Notes RTO-EN-AVT-150, Paper 1*. RTO, Neuilly-sur-Seine, France, 2008. Available at http://www.rto.nato.int

[3]P. Balakumar, H. Zhao, and H. Atkins, "Stability of Hypersonic Boundary-Layers over a Compression Corner"; AIAA Paper No. 2002-2848, June 24–26, 2002.

[4]R. J. Engers, D. R. Rubin, F. Marconi, and A. F. Bartlett, "Missile Radome Development Testing at ATK GASL"; AIAA 2007-1649, 2007.

[5]K. H. Timpano, S. Zaidi, L. Martinelli, and R. B. Miles, "Design and Test of a Morphing Supersonic Nozzle"; AIAA Paper No. 2008-851, 48th AIAA Aerospace Sciences Meeting and Exhibit. Reno, NV, 2008.

[6]M. Hagenmaier, D. Eklund, and R. Milligan, "Improved Simulation of Inlet Distortion Effects in a Direct-Connect Scramjet Combustor"; AIAA-2011-233.

[7]S. L. dos Santos e Lucato, J. Wang, P. Maxwell, R. M. McMeeking, and A. G. Evans, "Design and Demonstration of a High Authority Shape Morphing Structure," *Int. J. Solids Struct.*, **41** [13] 3521–43 (2004).

Fig. 12. Sequential 0.2 s images of constant geometry morphed surface showing steady plate contour. The visible section is 1.9 in. (4.7 cm) long in the horizontal direction.

[8]S. L. dos Santos e Lucato and A. G. Evans, "The Load Capacity of a Kagome Based High Authority Shape Morphing Structure," *Trans ASME*, **73**, 128–33 (2006).

[9]J. Wang, A. Nausieda, S. L. dos Santos e Lucato, and A. G. Evans, "Twisting of a High Authority Morphing Structure," *Int. J. Solids Struct.*, **44** [9] 3076–99 (2007).

[10]R. G. Hutchinson, N. Wicks, A. G. Evans, N. A. Fleck, and J. W. Hutchinson, "Kagome Plate Structures for Actuation," *Int. J. Solids Struct.*, **40**, 6969–80 (2003).

[11]T. J. Lu, J. W. Hutchinson, and A. G. Evans, "Optimal Design of a Flexural Actuator," *J. Mech. Phys. Solids*, **49**, 2071–93 (2001).

[12]V. S. Deshpande and N. A. Fleck, "Collapse of Truss Core Sandwich Beams in 3-Point Bending," *Int. J. Solids Struct.*, **38**, 6275–305 (2001).

[13]D. B. Marshall and B. N. Cox, "Integral Textile Ceramic Structures," *Annu. Rev. Mater. Res.*, **38**, 425–43 (2008).

[14]S. Flores, A. G. Evans, F. W. Zok, M. Genet, B. N. Cox, D. B. Marshall, O. Sudre, and Q. Yang, "Treating Matrix Nonlinearity in the Binary Model Formulation for 3D Ceramic Composite Structures," *Compos.: Part A*, **41**, 222–9 (2010).

[15]S. F. Shuler, J. W. Holmes, X. Wu, and D. Roach, "Influence of Loading Frequency on the Room Temperature Fatigue of a Carbon-Fiber/SiC-Matrix Composite," *J. Am. Ceram. Soc.*, **76** [9] 2327–36 (1993).

[16]J. W. Holmes, "Influence of Stress Ratio on the Elevated-Temperature Fatigue of a Silicon Carbide Fiber-Reinforced Silicon Nitride Composite," *J. Am. Ceram. Soc.*, **74** [7] 1639–45 (1991).

[17]A. Ferri, "The Method of Characteristics for the Determination of Supersonic Flow Over Bodies of Revolution at Small Angles of Attack"; NACA Report No. 1044, 1951.   □

J. Am. Ceram. Soc., **94** [S1] S42–S54 (2011)
DOI: 10.1111/j.1551-2916.2011.04503.x
© 2011 The American Ceramic Society

journal

The Design of Bonded Bimaterial Lattices that Combine Low Thermal Expansion with High Stiffness

Jonathan Berger,[†,‡] Chris Mercer,[§] Robert M. McMeeking,[‡,¶] and Anthony G. Evans[‡,¶]

[‡]Materials Department, University of California, Santa Barbara, California 93106

[§]Hybrid Materials Center, NIMS 1-2-1 Sengen, Tsukuba, Ibaraki 305 0047, Japan

[¶]Mechanical Engineering Department, University of California, Santa Barbara, California 93106

In engineered systems where thermal strains and stresses are limiting, the ability to tailor the thermal expansion of the constituent materials independently from other properties is desirable. It is possible to combine two materials and space in such a way that the net coefficient of thermal expansion (CTE) of the structure is significantly different from the constituents, including the possibility of zero and negative thermal expansion. Bimaterial lattices that combine low, negative, or an otherwise tailored CTE with high stiffness, when carefully designed, have theoretical properties that are unmatched by other known material systems. Of known lattice configurations with tailorable CTE, only one geometry, a pin-jointed lattice, has been shown to be stretch dominated and thus capable of having stiffness that approaches its theoretical upper bound. A related lattice with bonded joints, more amenable to fabrication, is developed that has a stiffness and CTE similar to the pinned structure. Analytical models for this rigid-jointed lattice's CTE and stiffness are developed and compared successfully with numerical results. A near space-filling, negative thermal expansion version of this lattice is devised and fabricated from titanium and aluminum. CTE measurements on this lattice are made and are well predicted by the analytical and numerical models. These insights guide the design of a family of bonded lattices with low areal density, low or negative CTE, and high stiffness to density ratio. Such lattices are shown to have a thermomechanical response that converges on pin-jointed behavior when the lattice elements are long and slender.

I. Introduction

R ECENT assessments have elucidated bimaterial, planar lattice concepts that attain zero (or low) thermal expansion coefficients (Fig. 1).[1–5] Among these, only the configuration depicted in Fig. 1(d) is known to combine low thermal expansion with high stiffness and strength. The lattice in Fig. 1(b) is the result of topology optimization and has biaxial stiffness near theoretical bounds, but has poor uniaxial stiffness, suffers from edge effects in lattices with limited periodicity, and has a complex geometry.[2] Furthermore, the lattice in Fig. 1(d), hereby known as the UCSB lattice, has properties that are transversely isotropic. Other stiff, strong, planar lattices have been identified that have zero, negative, or low thermal expansions in specific directions within its plane,[6] but are anisotropic, with significant thermal expansions in other in-plane orientations.

T. M. Pollock—contributing editor

Manuscript No. 28994. Received December 02, 2010; approved February 12, 2011.
This work was financially supported by the Office of Naval Research through the MURI program "Revolutionary Materials for Hypersonic Flight" (Contract N00014-05-1-0439).
[†]Author to whom correspondence should be addressed. e-mail: berger@engineering.ucsb.edu

In the lattice of Fig. 1(d), the members that govern its response are defined within the unit cell depicted in Fig. 2. In this lattice, the outer, hexagonal (type I) members (length L_1 and width w_1) have the lower coefficient of thermal expansion (CTE) α_1, while the triangular, inner (type II) members (length L_2 and width w_2) have relatively higher CTE, α_2. At the nodal points A, J, F in the lattice (Fig. 2), the ratio of the effective thermal expansion $\bar{\alpha}$ to the CTE of the type I material α_1 is dictated by the constituent CTE ratio, $\lambda = \alpha_2/\alpha_1$, and by the skewness angle, θ, depicted in Fig. 2. For a pin-jointed lattice, the expansion coefficient, $\bar{\alpha}$ has been derived as[5]:

$$\frac{\bar{\alpha}}{\alpha_1} = \frac{1 - (1/2)\lambda\sin(2\theta)(1/\sqrt{3} + \tan\theta)}{1 - (1/2)\sin(2\theta)(1/\sqrt{3} + \tan\theta)} \quad (1)$$

For a representative material combination, Al alloy and Ti alloy, with $\lambda \approx 2.6$, zero expansion prevails for a pin-loaded lattice at skewness $\theta \approx 25°$.[5] When the lattices are bonded, or otherwise mechanically attached at the nodes to make them rigid joints, bending moments are introduced into the type I members and the thermal expansions are larger.[5]

To assess these predictions, lattices based on Ti and Al alloys have been made, and their thermal expansion characterized.[7] Measurements obtained for a pin-jointed unit cell are in close agreement with the prediction of Eq. (1). Those measured for lattices assembled using mechanical (dovetail) attachments, i.e. with rigid joints, give larger thermal expansions. A variety of features adversely affect the reliability and the repeatability of the thermal expansion of such rigid-jointed lattices: (i) local plastic strains induced by the thermal expansion difference between constituents at the bimaterial attachment. (ii) Bending moments associated with the reduction in effective member lengths due to member overlap at the nodal points where six type I members from three neighboring unit cells converge (Type J Node, Fig. 2). (iii) The reduction in effective length of type I members due to the intersection of struts at Type J and D nodes. (iv) The reduction in effective length of type I members due to material added at the dovetail bimaterial interface. A previous assessment has demonstrated that the detrimental effect of plasticity at the bimaterial attachment is minimal, because the plastic strains are highly localized, facilitating shakedown after the first few cycles.[7] Consequently, while the CTE mismatch at the attachment generates a nonlinear, hysteretic contribution to the thermal strain during the first cycle, the thermal expansion remains repeatable during all subsequent cycles. A preconditioning treatment is sufficient to initialize the system and stabilize the thermal expansion.

The detrimental influence of the nodal geometry on the bending moments is more substantive. The elevation in the bending stiffness of the type I members associated with nodes of Type J significantly increases the overall thermal expansion coefficient, as determined both experimentally and by finite-element (FE) analysis.[7] This detriment motivates a systematic

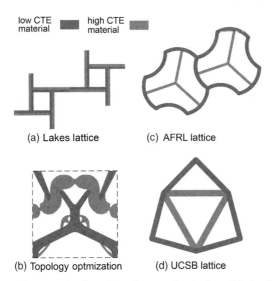

Fig. 1. Concepts for low thermal expansion lattices; (a) the Lakes lattice[1]; (b) the lattice obtained by topology optimization[2,3]; (c) the AFRL design[4]; and (d) the UCSB lattice.[5]

study that seeks geometrically straightforward configurations, amenable to manufacture, that impart thermal expansion closer to the pin-jointed prediction of Eq. (1), while retaining high stiffness and strength. The objective of this article is to seek such configurations, but with a focus, for the time-being, on stiffness.

The thermal expansion characteristics to be pursued emphasize designs that reduce the bending moments in the slender type I members. Analytic results are presented in Section II for the original design and extended to an alternative, offset design (Fig. 3). The alternative design is comprehensively analyzed in Section III by the FE method and specific designs discussed. It will be demonstrated that bonded, offset configurations can be conceived that have thermal expansions essentially the same as the pin-jointed lattice. Given the minimal-member bending stiffness for these new lattices, basic elasticity results are derived for pin-jointed systems in Section IV. These specify the salient trends in elastic response. Thereafter, a series of FE results for rigid-jointed systems are generated to ascertain deviations in stiffness from pin-joint predictions.

II. Design Principles for Low Thermal Expansion

Predicated on the foregoing assessment that the thermal expansions in excess of the pin-jointed lattice are primarily affected by the bending moments in the type I members, a beam theory methodology has been devised that identifies geometrically straightforward designs that converge to pin-jointed behavior. Two design layouts are considered: (i) One conforms to Fig. 2,

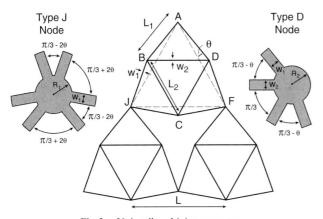

Fig. 2. Unit cell and joint geometry.

with the effective rigidity of the material around the joints taken into account. (ii) The other allows the unit cells to be separated by spacers (Fig. 3), both relaxing the joint rigidity limitation and permitting greater motions of type I members. The latter feature allows the implementation of a unit cell with a negative thermal expansion, so that, when averaged with the positive expansion of the spacer, the lattice has a zero or extremely low CTE. The joints are represented as circular flanges (Fig. 2), having radius R_1 (for Type A, F, and J) and radius R_2 (for Type B, C, and D). The radius R_1 is defined as the convergence of struts of type AD and CJ, and R_2 by the convergence of struts of type AD with type BD and CD. The rigidity of the flanges reduces the effective length of the adjacent struts.

By considering the incremental changes in member length during thermal expansion, and eliminating the increment of θ, we find for the original design of Fig. 2

$$\frac{\Delta L}{L} = \frac{\sqrt{3}\frac{\Delta L_1}{L_1} - \sin\theta(\cos\theta + \sqrt{3}\sin\theta)\frac{\Delta L_2}{L_2}}{\cos\theta(\sqrt{3}\cos\theta - \sin\theta)} \tag{2}$$

where

$$L_1 = \frac{L}{2\cos\theta} \tag{3}$$

$$L_2 = \frac{L}{2}(1 + \sqrt{3}\tan\theta) \tag{4}$$

and the symbol Δ followed by the letter L, with or without subscripts, indicates change of length. Because type II struts do not bend, but sustain only an axial load T_2, their strain is

$$\frac{\Delta L_2}{L_2} = \left(\frac{T_2}{E_2 A_2} + \varepsilon_2^t\right)\left(1 - 2\frac{R_2}{L_2}\right) + 2\hat{\varepsilon}_2^t\frac{R_2}{L_2} \tag{5}$$

where E_2 is Young's modulus for type II members, A_2 is their cross-sectional area, ε_2^t is the thermal strain in the struts and $\hat{\varepsilon}_2^t$ is the effective thermal strain of joints of type D. Type I struts bend and stretch, and their axial strain is given by

$$\frac{\Delta L_1}{L_1} = \left(\frac{T_1}{E_1 A_1} + \varepsilon_1^t\right)\left(1 - \frac{R_1 + R_2}{L_1}\right) + \hat{\varepsilon}_1^t\frac{R_1}{L_1} + \hat{\varepsilon}_2^t\frac{R_2}{L_1} \tag{6}$$

where T_1 is the tension in these members, E_1 is Young's modulus for type I members, A_1 is their cross-sectional area, ε_1^t their thermal strain and $\hat{\varepsilon}_1^t$ is the effective thermal strain of joints of type J.

Type I struts sustain a uniform shear force V and a nonuniform bending moment M (Fig. 4). Equilibrium at joints A and D requires that

$$T_1 = \frac{\cos\theta + \sqrt{3}\sin\theta}{\sqrt{3}\cos\theta - \sin\theta}V \tag{7}$$

and

$$T_2 = -\frac{2}{\sqrt{3}\cos\theta - \sin\theta}V \tag{8}$$

Inspection of Fig. 2 reveals that, by geometry, the transverse bending deflection of AD is given by

$$\delta = L_1\Delta\theta = \frac{\Delta L_2 - (\cos\theta + \sqrt{3}\sin\theta)\Delta L_1}{\sqrt{3}\cos\theta - \sin\theta} \tag{9}$$

while Euler–Bernoulli beam theory gives

$$\delta = \frac{V(L_1 - R_1 - R_2)^3}{12E_1 I_1} \tag{10}$$

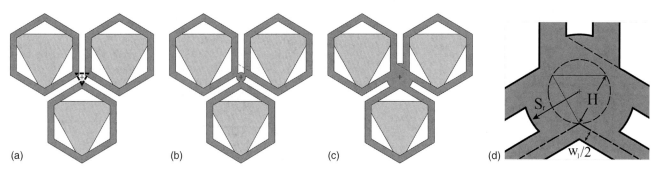

Fig. 3. Spaced lattices with $\theta = \pi/6$, with no additional material (a), undersized stiffener (b), and appropriately sized spacer (c and d) where axial member loads can be transmitted directly to adjacent unit cells. Spacer has characteristic dimension H while the stiffener has radius S_r.

where I_1, for type I members, is the second moment of area of the cross section about the neutral axis, so that $E_1 I_1$ is the bending stiffness of type I members, specified for bending in the

$$\frac{\Delta L}{L} = \frac{\sqrt{3}\left(\varepsilon_1^t \frac{\hat{L}_1}{L_1} + \hat{\varepsilon}_1^t \frac{R_1}{L_1} = \hat{\varepsilon}_2^t \frac{R_2}{L_1}\right) - (\cos\theta + \sqrt{3}\sin\theta)\sin\theta\left(\varepsilon_2^t \frac{\hat{L}_2}{L_2} + 2\hat{\varepsilon}_2^t \frac{R_2}{L_2}\right)}{(\sqrt{3}\cos\theta - \sin\theta)\cos\theta}$$
$$+ \frac{12 I_1 (\cos\theta + \sqrt{3}\sin\theta)\left[\sqrt{3}\cos\theta + \left(3 + \frac{2E_1 A_1 \hat{L}_2}{E_2 A_2 L_1}\right)\sin\theta\right]\left(\varepsilon_2^t \frac{\hat{L}_2}{L_2} + 2\hat{\varepsilon}_2^t \frac{R_2}{L_2} - \varepsilon_1^t \frac{\hat{L}_1}{L_1} - \hat{\varepsilon}_1^t \frac{R_1}{L_1} - \hat{\varepsilon}_2^t \frac{R_2}{L_1}\right)}{A_1 \hat{L}_1^2 (\sqrt{3}\cos\theta - \sin\theta)\cos\theta\left\{(\sqrt{3}\cos\theta - \sin\theta)^2 + \frac{12 I_1}{A_1 \hat{L}_1^2}\left[\frac{2E_1 A_1 \hat{L}_2}{E_2 A_2 L_1} + (\cos\theta + \sqrt{3}\sin\theta)^2\right]\right\}} \quad (17)$$

plane of the lattice. Elimination of δ and V among Eqs. (7)–(10) provides

$$T_1 = \frac{12 E_1 I_1 L_1 (\cos\theta + \sqrt{3}\sin\theta)^2 \left(\frac{\Delta L_2}{L_2} - \frac{\Delta L_1}{L_1}\right)}{(L_1 - R_1 - R_2)^3 (\sqrt{3}\cos\theta - \sin\theta)^2} \quad (11)$$

and

$$T_2 = \frac{24 E_1 I_1 L_1 (\cos\theta + \sqrt{3}\sin\theta)\left(\frac{\Delta L_2}{L_2} - \frac{\Delta L_1}{L_1}\right)}{(L_1 - R_1 - R_2)^3 (\sqrt{3}\cos\theta - \sin\theta)^2} \quad (12)$$

Use of Eqs. (5) and (6) allows a solution for the strut tensions as

$$T_1 =$$
$$\frac{12 E_1 I_1 L_1 (\cos\theta + \sqrt{3}\sin\theta)^2 \left(\varepsilon_2^t \frac{\hat{L}_2}{L_2} + 2\hat{\varepsilon}_2^t \frac{R_2}{L_2} - \varepsilon_1^t \frac{\hat{L}_1}{L_1} - \hat{\varepsilon}_1^t \frac{R_1}{L_1} - \hat{\varepsilon}_2^t \frac{R_2}{L_1}\right)}{\hat{L}_1^3 \left\{(\sqrt{3}\cos\theta - \sin\theta)^2 + \frac{12 I_1}{A_1 \hat{L}_1^2}\left[\frac{2E_1 A_1 \hat{L}_2}{E_2 A_2 L_1} + (\cos\theta + \sqrt{3}\sin\theta)^2\right]\right\}}$$
$$(13)$$

and

$$T_1 =$$
$$\frac{24 E_1 I_1 L_1 (\cos\theta + \sqrt{3}\sin\theta)\left(\varepsilon_2^t \frac{\hat{L}_2}{L_2} + 2\hat{\varepsilon}_2^t \frac{R_2}{L_2} - \varepsilon_1^t \frac{\hat{L}_1}{L_1} - \hat{\varepsilon}_1^t \frac{R_1}{L_1} - \hat{\varepsilon}_2^t \frac{R_2}{L_1}\right)}{\hat{L}_1^3 \left\{(\sqrt{3}\cos\theta - \sin\theta)^2 + \frac{12 I_1}{A_1 \hat{L}_1^2}\left[\frac{2E_1 A_1 \hat{L}_2}{E_2 A_2 L_1} + (\cos\theta + \sqrt{3}\sin\theta)^2\right]\right\}}$$
$$(14)$$

where the notation

$$\hat{L}_1 = L_1 - R_1 - R_2 \quad (15)$$

and

$$\hat{L}_2 = L_2 - 2R_2 \quad (16)$$

has been introduced. When these results are inserted into Eqs. (5) and (6) and the outcome used in Eq. (4), the thermal expansion of the lattice becomes

When the bending stiffness is negligible ($I_1 = 0$), and the joint radii R_1 and R_2 are neglected, this result simplifies to Eq. (1).

The result in Eq. (17) highlights the two effects of bonded joints. The first involves the finite extent of the joints (R_1 and R_2 are nonzero), implied by the first term on the right hand side. The second arises because the bending of type I struts adds strain to the thermal response, signified by the second term on the right hand side.

Appropriate choices for the joint radii (Fig. 2) are

$$R_1 = \frac{3w_1}{\pi - 6\theta} \quad (18)$$

and

$$R_2 = \max\left(\frac{3(w_1 + w_2)}{2\pi - 6\theta}, \frac{3w_2}{\pi}\right) \quad (19)$$

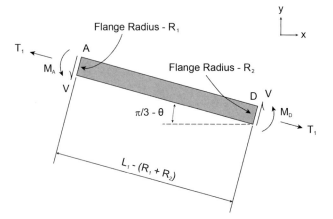

Fig. 4. Type I members sustain an axial load, T, shear, V, and bending moment, M.

The thermal expansion for nodes A, J, F is

$$\hat{\varepsilon}_1^t = \varepsilon_1^t \tag{20}$$

An appropriate choice for nodes B, C, D is a weighted average of the expansion of type I and II materials,

$$\hat{\varepsilon}_2^t = \frac{1}{3}(2\varepsilon_1^t + \varepsilon_2^t) \tag{21}$$

Results for the linear thermal expansion for various material combinations are plotted against θ in Fig. 5, with a range of values for the bending stiffness parameter $I_1/A_1 L_1^2$ used for illustration.

The above results affirm that the design of joints of Type A, J, F is critical to the realization of low thermal expansion behavior, because the low angle included between type I members from adjacent unit cells leads to excessively large R_1. This limitation is obviated by using an offset design wherein the centers of neighboring unit cells are displaced, allowing R_1 to become small even when $\theta \to \pi/6$. Such offset designs are achieved by insertion of a spacer made from type I material, as shown in Fig. 3. The spacer removes the coupling of $\hat{L}_1$ and θ, for a fixed L, enabling the use of larger values of θ. Offset designs can have negligible bending, whereupon, the thermal expansion can be closely approximated by the first term on the right hand side of Eq. (17), with R_1 and R_2 neglected. The CTE for the offset lattice is thus

$$\frac{\bar{\alpha}}{\alpha_1} = \frac{\sqrt{3} - \lambda F(\theta)}{\sqrt{3} - F(\theta)}\left(\frac{L}{L+H}\right) + \frac{H}{L+H} \tag{22}$$

where H is the size of the spacer and $F(\theta) = (\cos\theta + \sqrt{3}\sin\theta)\sin\theta$. When H is zero, Eq. (1) is recovered once more. The formula in Eq. (22) can be used to guide low thermal expansion lattice designs; specifically, Eq. (22) predicts that zero thermal expansion of the lattice occurs when

$$\frac{H}{L} = \frac{\lambda F(\theta) - \sqrt{3}}{\sqrt{3} - F(\theta)} \tag{23}$$

This design has unit cells that contract upon heating, compensating for the expansion of the spacers. Such solutions are feasible when $\sqrt{3}/\lambda < F(\theta) < \sqrt{3}$ (note that, within this range, the unit cell has negative thermal expansion). The upper limit of this range coincides with $\theta = \pi/3$. For illustration, when $\lambda = 2$, the lower limit is $\theta = \pi/6$. Beyond these limits is the requirement that joints A, J, F remain physically small, requiring that $\frac{\geq H\sqrt{3}w_1\sin\theta}{F(\theta)}$, whereupon the type I elements from

neighboring unit cells have no intersection. From Eq. (23), this requirement provides

$$\frac{w_1}{L} \leq \frac{[\lambda F(\theta) - \sqrt{3}]F(\theta)}{[\sqrt{3} - F(\theta)]\sqrt{3}\sin\theta} \tag{24}$$

revealing that the slenderness of type I elements has to be respected. Consequently, when $\theta = \pi/6$, Eq. (24) gives $w_1/L \leq (\lambda - 2)$, indicating that the design can be satisfied provided that $\lambda = \alpha_2/\alpha_1 > 2$, ensuring that w_1 is positive. Moreover, any material combination with λ slightly in excess of 2 (say more than 2.1) is acceptable, because w_1/L must be small to ensure low bending stiffness.

Alternatively, Eq. (24) can be recast as a condition on θ with λ and w_1/L already selected. The outcome is not transparent because it involves a combination of trigonometric functions of θ without obvious simplification. Nevertheless, inspection indicates that any θ slightly above the lower limit $F(\theta) = \sqrt{3}/\lambda$ will satisfy Eq. (24). Specific cases should be assessed numerically to ensure a satisfactory design, as elaborated below.

A chosen design is limited in its range of temperature operation by the requirement that the triangle of type II elements has space into which material can expand when the lattice is heated. The critical condition occurs when joint C (Fig. 2) touches its counterparts from the two adjacent unit cells. It is straightforward to ascertain that the strain increment in type II bars that causes this critical condition is

$$\varepsilon_2^t = \frac{\cos\theta - \sqrt{3}\sin\theta + \frac{2H}{L}\cos\theta}{\cos\theta + \sqrt{3}\sin\theta} \tag{25}$$

Consequently, the operating temperature range must be chosen to ensure that thermal straining of type II elements is smaller. For $\theta \sim \pi/6$ and greater a careful choice of $\frac{H}{L}$ is needed, because $\cos\theta - \sqrt{3}\sin\theta \leq 0$ for $\theta \geq \pi/6$.

III. Specific, Low Expansion, Offset Designs

(1) FE Method

Lattices are modeled using a representative volume element (RVE) FE technique. Three-dimensional models are subject to periodic boundary conditions for the two in-plane dimensions in the form of uniform macroscopic strains. The relative displacements between pairs of boundary nodes are controlled by tying their displacements together consistent with the strains we wish to impose.[8,9] The third, z-direction, is left free. A single node on the interior of the model is held fixed in space to prevent rigid-body translations. The commercial FE code ABAQUS[10] is used for mesh generation and to perform analysis. MATLAB[11] code is used extensively to manipulate meshes, apply boundary conditions, and for postprocessing. Periodic boundary conditions were implemented in Cartesian coordinates. A typical RVE is pictured in Fig. 6. Typical meshes consist of $\sim20\,000$ to $\sim70\,000$ eight-noded linear hexahedral elements (type C3D8R). The number of elements varies greatly with the relative density of the lattice geometry being analyzed, which ranged from 8% to 98% for slender to space-filling designs, respectively. Mesh sensitivity is studied to ascertain model resolution at which solutions converge. Because the average strains are small, there is no distinction between the macroscopic Cauchy stress and the macroscopic nominal stress, so that the macroscopic stress can be computed by simply dividing force resultants by section areas for the undeformed volume of the RVE.

(2) Offset Lattices

In stretch-dominated structures, where loads are well distributed, stress contours are rather uniform, and consequently the structure is relatively efficient in its utilization of material. On the other hand, FE calculated thermomechanical stress distributions in the original lattice design reveal large nonuniform

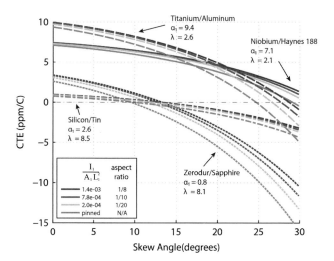

Fig. 5. Lattice coefficient of thermal expansion as a function of skew angle θ and type I member aspect ratios.

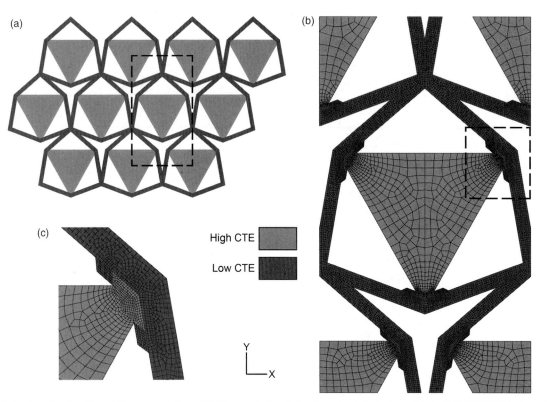

Fig. 6.   (a) Perimeter of unit cell used for computations. (b) Representative finite-element mesh that consisted of 50 000–70 000 three-dimensional elements. (c) Detail of a bimaterial joint region.

bending stresses associated with rigidity and member intersection at the joints (Fig. 7(a)). The lattices analyzed to obtain these results are modeled with the temperature-dependent properties of Ti and Al alloys listed in Table Ib and subject to a temperature excursion from 40° to 250°C. The bimaterial interface, in reality press-fit, is considered to be perfectly bonded for simplicity. Simulations that model the interface between sublattices with contact in compression, and a frictional coefficient of $\mu = 1$, give results that are negligibly different from welded models.

The maximum tensile equivalent stress (also known as the von Mises stress) is located in the bimaterial joint region and results from CTE mismatch between constituents. In the temperature range analyzed, the maximum tensile equivalent stress is lower than the tensile yield stress in both materials.

Constraints on the motion of the lattice's components drive the macroscopic CTE toward that of the constituents. By placing a spacer (Fig. 3), with characteristic dimension H, between unit cells and eliminating excess material around the bimaterial joint, the lattice geometry can be designed so that it behaves according to the original concept for low CTE, i.e. with negligible stress (Fig. 7(b)). Contours for the revised design show greatly reduced bending stresses and the overall lattice behavior is in good agreement with new analytical predictions (Fig. 8).

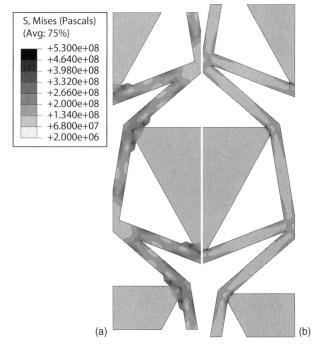

Fig. 7.   Finite-element thermomechanical stress distribution in Ti and Al lattices given the material properties listed in Table I and subject to a 175°C temperature excursion—(a) original design and (b) new offset design.

Table Ia.   Temperature Average Values of Material Properties—For the Range 40°–215°C

		Coefficient of thermal expansion (ppm/°C)	Young's modulus (GPa)	σ_y (MPa)
Type-1 material	Ti–6Al–4V	9.4	110	800
Type-2 material	7075-T6 aluminum	24.3	70	400

Table Ib.   Temperature-Dependent Material Properties

		Coefficient of thermal expansion (ppm/°C)	Young's modulus (GPa)	σ_y (MPa)	Temperature (°C)
Type-1 material	Ti–6Al–4V	9.2	110	1100	20
		9.6		858	300
Type-2 material	7075-T6	22.9	70	434	20
		26.3		391	200
				339	300

(3) Lattice Design

For any pair of constituent materials and their associated properties, there is a range of skew angles, θ, over which a desired lattice CTE can be achieved. Properties such as stiffness, the Poisson ratio, and strength vary over this range so that any one solution may be superior to others depending on design requirements. For a unit temperature change, $\Delta T = 1$ so that $\varepsilon_i^t = \alpha_i$, Eq. (17) gives the thermal expansion coefficient of the lattice and can be inserted into Eq. (26),

$$\bar{\alpha} = \frac{\Delta L + H\alpha_1}{L + H} \tag{26}$$

to calculate the expansion of a spaced lattice. The macroscopic thermal strain in a spaced lattice is the weighted sum of the lattice and spacer strain, the latter having thermal expansion coefficient α_1. Equation (26) can be used to identify regions of design parameter space where lattices with the desired thermomechanical response exist. Designs in this neighborhood can then be investigated through FE to address specific geometries and to investigate their thermomechanical response in comparison with the results in Section II.

The thermomechanical strain response of the bimaterial joints in Eq. (21) is assumed to be an average of the constituents. While this approximation is sufficient for exploratory investigations, specific mechanical bimaterial interface geometries must be considered when designing real structures, as their behavior may differ substantially from this idealization. The dovetail joints used by Steeves *et al.*,[5,7] and in the current effort, are examples of bimaterial attachments. Mechanical connections capable of carrying tensile loads are necessary for transmitting all applied macroscopic loading situations except biaxial tension. The size of these joints, given by their characteristic dimension, R_2, reduce the effective length of members ($\hat{L}_1$ and $\hat{L}_2$, Eqs. (15) and (16)).

For fixed values of w_1 and w_2, the thermomechanical response of the system is a strong function of the member effective lengths, and therefore of R_1 and R_2. In many designs, a single radius R_2 does not exist by which both type I and II members are reduced equally at the Type D joints (Eqs. (15) and (16)). The actual reduction in effective length of members at these nodes is a function of member width and the relative angle at which they are incident to the joint. To identify designs with a tailored and well-predicted CTE, which is amenable to fabrication, more detailed modeling is necessary.

The addition of the spacer (Fig. 3) allows practical access to previously unachievable skew angles and corresponding higher relative densities. In previous designs without the spacer, at skew angles approaching 30°, members from adjacent unit cells

intersect to an extent, and prevent the desired thermomechanical response, and the desired CTE cannot be realized. A skew angle of 30°, without a spacer present, results in type I members in adjacent cells being parallel and a hexagonal unit cell appearance. If only the area bounded by the unit cell is considered, these designs can achieve near-maximum areal density as an assembly of hexagonal cells with small gaps between them. The size of the gaps between unit cells is directly related to the size of the spacer H. Upon temperature excursion from the reference state, type I members expand, distort, and rotate, causing Type D nodes to translate away from the center of the unit cell. The size of the spacer and the corresponding gap is dictated by the maximum outward deflection of these nodes in the specified temperature range while considering the need to avoid adjacent unit cells impinging upon each other after thermal straining. For some designs with skew angles near or above 30°, an upper use temperature exists at which initially nearly parallel type I members in adjacent cells deform to contact each other causing the lattice to densify. For these designs, the upper use temperature can be increased by expanding the size of the spacer at the cost of driving the CTE of the system toward that of the type I material (Eq. (26)). Densification may also significantly influence the stiffness and strength of these lattices and may be a beneficial feature in some applications. If material continuity is beneficial, such as for aerodynamic surfaces, densified lattices can be useful.

In some spaced lattice designs where the dimension of the spacer is on the order of w_1, no additional material is needed to achieve this offset. In others, designs with skew angles in the neighborhood of 30°, unit cells may intersect minimally or not at all (Fig. 3) and the strength and stiffness joint will suffer. A disk of material centered on the spacer, a stiffener with radius S_r can be added to the lattice to maintain continuity and transmit stresses between unit cells. If the disk of material is too small, stress concentrations in these regions can be design limiting. An appropriate radius for this disk is one where axial loads in type I members can be directly transmitted to adjacent cells (Figs. 3(c) and (d)). This stiffener has a dimension independent of the spacer and may be useful in facilitating a connection between the lattice and a substructure.

As type II members are not subject to thermally induced bending stresses, members with geometries other than truss or beam-like forms can be considered without altering the mechanics of the system.[5] In Fig. 10, the lattices pictured in the middle and on the right (b and c) are variations of the lattice on the left (a) incorporating type II members that are not simple prismatic beams. Stiffening of type II members drives the thermal expansion of the system to lower values. The increased stiffness of these members can be modeled by using an effective modulus E_2^* for E_2 in Eqs. (13) and (14). The only restriction in geometry is that the type II members not impinge on, and reduce the effective length of the slender type I members when thermomechanically strained. Previous fabricated designs have used truss and solid triangular inner type II members (Figs. 9(a) and (b)).[5,7] To explore the potential space-filling properties of this lattice, a hexagonally shaped type II sublattice geometry was chosen for fabrication in this work.

If in-plane geometries of members are specified to have a finite width centered on the lines shown in Fig. 2, the width of the bimaterial joint is limited by the width of type II members, w_2 (for the geometry considered in this work). Selecting a small bimaterial joint resulted in a small value of w_2 in Eq. (17). To account for the much larger cross section of the hexagonal geometry used, an effective modulus of $E_2^* = 10E_2$ is utilized in the analytical model. The order of magnitude increase in stiffness is an estimate; further increases do not significantly influence results. By choosing a slightly skewed, but nearly hexagonal type II element, an upper use temperature densification event can be engineered between high and low CTE sublattices. When both adjacent unit cells and sublattices contact each other at the same ΔT, a nearly or completely densified and continuous structure can be formed.

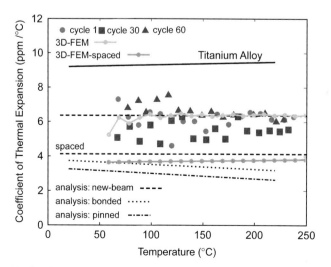

Fig. 8. Comparison of results from finite-element analysis for coefficient of thermal expansion and experiment.

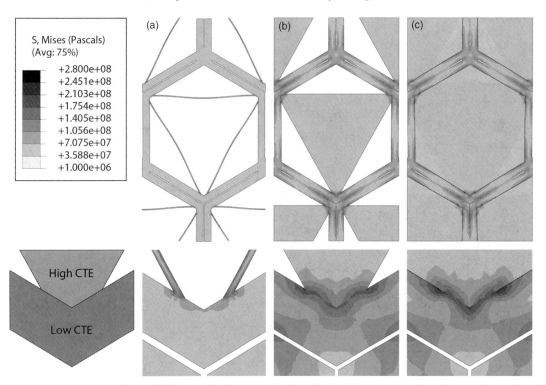

Fig. 9. Thermal stress distributions in Ti- and Al-fabricated lattice geometry with thin type I members (a), triangular type II member (b), and space-filling design (c).

The dovetail joint dimensions were chosen to maximize the effective length of type I members. Joint dimensions were minimized to reduce R_2 with practical consideration for the interface size and the need to maintain a robust mechanical connection between sublattices. Lattices previously investigated[5,7] used dovetail joints where the high CTE type II material composed the inner, male side of the connection. Additional material was added to the lattice at these locations to help facilitate a robust connection. This added material served to reduce the effective length of members resulting in a poorly predicted thermal response (Fig. 8). In the current design, the inner male component is now composed of the low CTE type I material, and the bimaterial joint is relocated to be enclosed in the inner type II member. Switching the low CTE material to the male side reduces thermal stresses resulting from the expansion of the high CTE material, which was previously confined by the low CTE constituent. It is hoped that plasticity can be avoided altogether with this configuration, in contrast to the previous dovetail joint design.

The test specimen designed and fabricated in this work was chosen to have a skew angle of 30° and nearly space-filling inner type II member. By specifying the geometric parameters $w_1 = 3.0$ mm, $w_2 = 0.4$ mm, $\theta = 30°$, H = 2.0 mm, $S_r = 3.80$ mm, and unit cell length L = 50.0 mm, the designs in Fig. 10 are achieved. FE and analytical predictions for the geometries shown are listed in Table II.

(4) Lattice Fabrication

A metallic lattice was fabricated for CTE measurement purposes. The type I lattice was electro discharge machined from 3-mm-thick sheets of Ti–6Al–4V. The type II members were machined from 7075-T6 aluminum alloy. Sublattices were press-fit together. Tolerances were such that assembly with hand pressure was possible; however, a mechanical press was used to ensure proper assembly. The structure consists of 10 unit cells arranged so that two cells in the interior are separated from the edge by another unit cell to minimize edge effects (Fig. 10).

(5) CTE Measurement Methodology

Thermal expansion measurements on the Ti and Al lattice were performed using a two-dimensional (2D) digital image correlation (DIC) system. A high contrast black and white speckle pattern was applied using spray paint. Lattices were heated on a laboratory hot plate (Wensco, model #H1818RA4000) at a rate of 60°C/h from room temperature to 220°C. A frame of

Fig. 10. Lattices of Ti alloy (struts) and Al alloy (hexagonal units) fabricated for measurement of the coefficient of thermal expansion.

Table II.　Analytical and FE CTE Predictions for Thin, Triangular, and Hexagonal Type-2 Member Lattices Pictured in Fig. 9

Coefficient of thermal expansion (ppm/°C)	Thin	Triangular	Hexagonal	Fabricated
Analytical prediction	2.6	−1.4	−1.4	−1.4
Finite element	1.4	−1.6	−1.6	−1.1

Constituent properties and FE results are temperature average values from Table 1b from 20° to 70°C. Thin Type-2 members elastically buckled under compression for larger temperature excursions.

common silica insulation, approximately 1.5 in thick, was placed around the lattice with a glass plate on top. The temperature was recorded by four self-adhering K-type thermocouples (Omega, model SA1XL-K-SRTC, Omega Engineering Inc., Stamford, CT) located on the upper face of the lattice, two each on Ti and Al. Acetone was used to remove the applied paint at the location of the thermocouple attachments to increase heat transfer. The temperature of the lattice was taken to be the average of the four. Digital images were captured by CCD camera (AVT Dolphin F-201B, Allied Vision Technologies, Newburyport, MA) with a zoom lens (Tamron AF 70-300 1:4-5. 6, Tamron USA Inc., Commack, NY) positioned approximately 2 m from the specimen. The focal length was maximized subject in the confines of the laboratory space to minimize the effect of out of plane deformations on in-plane measurements. Images were taken every 5 s to record deformations. Two 300 W incandescent lights in hoods were positioned close above the glass plate for imaging purposes. Using the Vic-2D (Correlated Solutions, Correlated Solutions Inc., Columbia, SC) software, virtual extensometers placed on the reference image, and tracked through the images, measured the displacement between pairs of pixel subsets. A typical area of interest consisted of one unit cell with three Type D nodes visible. Strains were calculated from the relative displacement between pairs of subsets that were typically 23^2 pixels. Strains were measured in the Al and Ti and were calculated as the average of six virtual extensometers. Lattice strains were calculated as the average of three virtual extensometers placed between the three Type D nodes surrounding an interior unit cell. Temperature average CTE was measured by linear fitting to strain–temperature plots, where the average CTE over the temperature range is the slope of the resulting straight line. Extensometers were placed upon relatively unstressed material regions to measure the CTE of the constituents.

Image distortion from convective currents emanating from the specimen is a common problem when using DIC to record thermomechanical strains. A frame of silica fiber insulation was placed around the lattice and a glass plate on top to help thermally isolate the specimen from the camera. A fan is used to mix the air directly above the glass plate and carry hot air away from the lights.

(6)　CTE Results

A typical experimental strain versus temperature plot is shown in Fig. 11 along with FE prediction. Temperature average CTE for the lattice and the constituent materials is reported in Table III. Although the DIC software is capable of measuring microstrains, scatter on the order of 1000 microstrain was present in the measurements due to image distortion. Convective currents emanating from the lighting were sufficient to cause visible inhomogeneous lensing in successive images. Measurements on the Ti were limited to the Type D joints resulting in short gauge lengths and larger scatter.

The average value of CTE for the constituent materials is measured to be ∼4% higher than reported values (24.3 ppm/C for Al, and 9.4 ppm/C for Ti) (Table Ia), but are within experimental error. The lattice has an average measured CTE

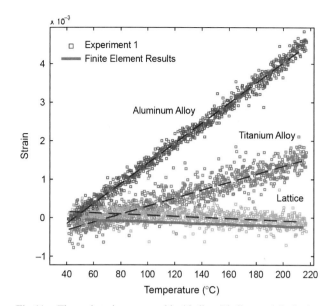

Fig. 11.　Thermal strains measured in Al alloy, Ti alloy, and the lattice. Significant scatter in Ti measurements is due to the relatively short gauge lengths that had to be used.

of −0.9 ppm/C that is well predicted by FE of −1.1 ppm/C, and close to the analytical prediction of −1.4 ppm/C.

3.7　Discussion—Thermal Expansion

Measured thermal expansions of the lattice agree well with analytical and FE predictions. Differences between measured and FE prediction are believed to be due to the reduced effective length of the type I members resulting from fabrication imperfections causing premature contact between sublattices. The hexagonally shaped aluminum members had visible machining imperfections on some edges where they were cut from stock. The small wing-shaped gap between high and low CTE members results in a structure that is more imperfection sensitive than a lattice with triangular or prismatic beam-type members. Geometric imperfections in this gap can cause sublattices to impact each other before the predicted upper use temperature, reducing the effective length of type I members and driving up the CTE. Space-filling lattices with these features, such as aerodynamic surfaces, may be adversely affected by debris in these small gaps.

No plasticity is predicted in the revised dovetail geometry. The limited plasticity present in other designs[7] was a function of the bimaterial joint geometry used and is not inherent to the functional mechanics of the lattice itself.

Scatter in CTE measurements results from several sources. Convective currents in the air column between the specimen and the camera distort images through their associated density gradients and lensing effects.[12] Efforts were made to mitigate convective currents coming from the specimen and hotplate, but the incandescent lighting used for imaging proved a sufficient source of interference. A cold light source or a camera orientation that minimizes the effect of convective currents, by positioning it outside the affected area, would reduce the effect. The large scatter in the CTE measurements for Ti is due to the lattice geometry studied, the size of the unit cells, and the short lengths of relatively unstressed material available for measurements taken in the Type D node region.

The analytical model developed in Section II does a good job of predicting lattice thermal expansion. The model uses the simple assumption that the thermal expansion of the type II joint region is a weighted average of the constituent materials (Eq. (21)). In reality, a distinct material interface exists between the constituents, and the thermal response is much more complicated. The analytical CTE prediction of −1.4 ppm/C for the lattice is close to the FE predicted value of −1.1 ppm/C and

Table III. Measured CTE of Lattices and Constituent Materials (95% Confidence Bounds), and FE Prediction for Lattice using Measured Values (ppm/°C)

	Aluminum (measured)	Titanium (measured)	Lattice (measured)	FE prediction
Exp. 1	26.1 (25.9–26.3)	10.4 (10.0–10.7)	−1.5 (−1.7 to −1.3)	−0.8
Exp. 2	24.9 (24.7–25.1)	10.0 (9.6–10.7)	−0.8 (−0.9 to −0.6)	−0.6
Exp. 3	25.5 (25.4–25.6)	9.1 (8.6–9.6)	−0.6 (−0.7 to −0.5)	−2.6
Exp. 4	24.9 (24.8–25.0)	9.1 (8.7–9.5)	−0.8 (−0.8 to −0.6)	−2.2
Average	25.4	9.7	−0.9	−1.3

sufficiently accurate to help characterize the design space to locate geometries of interest.

CTE measurements performed on a titanium and aluminum lattice using 2D DIC were able to validate the analytical and FE predictive models used in its design. The metal lattice behaved elastically over a temperature range of 175°C, exhibiting consistent negative thermal expansion. The design space identified through the analytical model suggests the ability to realize a family of structures with a wide range of stiffness and thermal expansion properties.

IV. Stiffness

(1) Basic Stiffness Results for Pin-Jointed Lattices

(A) Elastic Properties: The unit cell of the design without spacers, shown in Fig. 12(a), is loaded by a set of forces, parameterized by P, Q, and S. It is presumed that the lattice has been designed in the manner described above, such that the bending stiffness of struts of type I are very low, whereupon their behavior is stretch dominated. In such a situation, the lattice can be analyzed as if it were pin-jointed.

The resulting behavior is isotropic in the plane of the lattice, and stated as

$$\varepsilon_{xx} = \frac{\sigma_{xx}}{E} - \frac{v\sigma_{yy}}{E} + \alpha\Delta T \tag{27a}$$

$$\varepsilon_{yy} = \frac{\sigma_{yy}}{E} - \frac{v\sigma_{xx}}{E} + \alpha\Delta T \tag{27b}$$

$$\varepsilon_{xy} = \frac{\sigma_{xy}}{2G} \tag{27c}$$

where E, v, G, and α are the in-plane Young's modulus, the Poisson ratio, shear modulus, and CTE, respectively, of the lattice.

The biaxial stiffness is the ratio of the biaxial stress to the in-plane strain under equibiaxial loading,[5] and can be deduced as

$$S_b = \frac{(\sqrt{3}\cos\theta - \sin\theta)^2}{2\sqrt{3}L_1\left[\frac{1}{E_1 w_1} + \frac{2\sin^2\theta}{3E_2 w_2}(\cos\theta + \sqrt{3}\sin\theta)\right]} \tag{28}$$

Note that this corrects a misprint in the previously published formula in Steeves et al.,[5] which is missing the leading 2 in the denominator. The shear modulus is

$$G = \frac{(\sqrt{3}\cos 2\theta + \sin 2\theta)^2}{8\sqrt{3}L_1\left[\frac{1}{E_1 w_1} + \frac{2\sin^2 2\theta}{3E_2 w_2}(\cos\theta + \sqrt{3}\sin\theta)\right]} \tag{29}$$

In these expressions for S_b and G, L_1 is used rather than L because L_1/w_1 is the aspect ratio of type I elements.

The Poisson ratio for the lattice may be computed from

$$v = \frac{S_b - 2G}{S_b + 2G} \tag{30}$$

and the Young's modulus

$$E = 2G(1 + v) \tag{31}$$

Note that, in certain circumstances, the Poisson ratio will be zero or negative. For example, when $E_2 \gg E_1$, the Poisson ratio reduces to

$$v = \frac{\cos 2\theta - \sqrt{3}\sin 2\theta}{3 + 2(\sqrt{3}\cos\theta + \sin\theta)\sin\theta} \tag{32}$$

As a consequence, it is zero at $\theta = \pi/12$, and negative for $\theta > \pi/12$.

The introduction of a stiff spacer, as in Fig. 3, has no effect on the elastic properties of the lattice. If the spacer is a stiff component (e.g., composed of a solid plate rather than a set of truss or beam elements), the elastic properties of the lattice are still given by the values in Eqs. (28) and (29). This situation arises because for a stiff spacer, when we neglect its deformation and treat it as rigid, both stress and elastic strain scale in the same way with the size of the spacer. To show this for the case of equibiaxial stress, the loads in Fig. 12(b) are

$$P = (L + H)\sigma_B \tag{33a}$$

$$Q = \frac{\sqrt{3}}{2}(L + H)\sigma_B \tag{33b}$$

where σ_B is the applied biaxial stress. Because the spacers are rigid, the forces experienced by the adjacent nodes are the same. The stress are

$$\sigma_{11} = \frac{2Q}{\sqrt{3}L} \tag{34a}$$

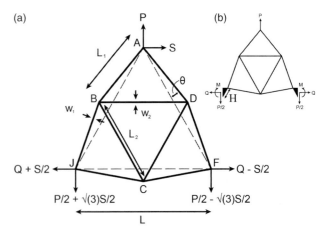

Fig. 12. Pin-jointed unit cell subject to loads P, Q, and S used in stiffness calculation (a). Spaced lattice unit cell with applied loads P and Q (b).

$$\sigma_{22} = \frac{P}{L} \tag{34b}$$

And the changes in dimensions of the lattice are

$$\Delta u_1 = \varepsilon_{11} L = \frac{1}{E}\left[\frac{2Q}{\sqrt{3}} - \nu P\right] \tag{35a}$$

$$\Delta u_2 = \varepsilon_{22} \frac{\sqrt{3}L}{2} = \frac{\sqrt{3}}{2E}[P - \nu Q] \tag{35b}$$

The strains are

$$\bar{\varepsilon}_{11} = \frac{\Delta u_1}{L + H} \tag{36a}$$

$$\bar{\varepsilon}_{22} = \frac{2\Delta u_2}{\sqrt{3}(L + H)} \tag{36b}$$

Using Eqs. (30), (31), and (32), with (36), the strains become

$$\varepsilon_{11} = \varepsilon_{22} = \frac{1}{E}[\sigma_b - \nu\sigma_b] \tag{37}$$

showing that the strains are not a function of the size of the spacer.

For the case of shear loading for a lattice with spacer with an applied shear stress τ, the forces in Fig. 12(b) are

$$P = (L + H)\tau \tag{38a}$$

$$Q = \frac{-\sqrt{3}(L + H)\tau}{2} \tag{38b}$$

Because the spacer is rigid, the resulting macroscopic strains are

$$\bar{\varepsilon}_{11} = \frac{1}{E}\left[\frac{Q}{\frac{\sqrt{3}(L+H)}{2}} - \nu\frac{P}{L+H}\right] = \frac{1}{E}[-\tau - \nu\tau] \tag{39a}$$

$$\bar{\varepsilon}_{22} = \frac{1}{E}\left[\frac{P}{(L+H)} - \nu\frac{Q}{\frac{\sqrt{3}(L+H)\tau}{2}}\right] = \frac{1}{E}[\tau + \nu\tau] \tag{39b}$$

This indicates that there is no stiffness penalty upon introduction of a stiff spacer.

(2) In-Plane Compression Measurements

In-plane compression experiments have been used to generate stress/strain measurements. The objectives are two-fold: (i) allow calibration of the mechanical robustness and stiffness of representative lattices, and (ii) provide validation data for the ensuing FE calculations. For these purposes, it suffices to fabricate monolithic lattices from 1-mm-thick plates by laser cutting. To probe the yielding and strain-hardening characteristics, one set of lattices has been generated from 304 stainless steel. During testing, out of plane buckling was prevented by constraining the lattice between two 12.7-mm-thick tempered glass plates, bolted to an aluminum frame. The experiments were performed in an MTS™ 810 servohydraulic testing system under displacement control, using a displacement rate of 0.5 mm/min. Images of the lattice were recording every 15 s during the tests using a CCD camera connected to an image correlation system.

Two different type II member geometries were used. The stress–strain behaviors, shown in Fig. 14, reveal robust behavior, characterized by yielding followed by strain hardening. In all cases, yielding occurs in the type I members at critical stress levels in the range, $15 \le \sigma_c \le 18$ MPa. The bright regions in the images (Fig. 13) indicate the occurrence of out-of-plane plastic buckling. Unload–reload measurements reveal hysteresis and decreasing stiffness with increase in plastic strain.

(3) Experimental and FE Results

Stress strain curves generated from FE modeling (FEM) are plotted alongside experimental results in Fig. (14). The 304 stainless steel was modeled by linear elastic response followed by yielding with isotropic hardening. Twenty node biquadratic elements (C3D20R) were used in these calculations. Homogeneous strains were applied to the RVE in two directions. The influence of the glass plates used to confine specimens out of plane was not modeled.

Notable features of the experimental stress–strain curves include a reduction in elastic stiffness with increasing strain past initial yield as evident in the unload–reload regions, and significant hysteresis in these regions. No reduction in stiffness is seen in models restricted to in-plane deformations. Models seeded with imperfections to initiate buckling show the same reduction in stiffness with strain as experiments. Frictional interactions between the lattices and the glass plates used to constrain out-of-plane motion are attributed as the source of hysteresis observed in the experiment. The collapse modes, plastic buckling of type I members oriented most obliquely to the loading direction, and out-of-plane plastic hinging in the same yielded members are accurately captured by the FE results (Fig. 15).

(4) Comparison of Pinned and Bonded Stiffness

To compare pin-jointed analytical and FE-bonded predictions for stiffness, the equations in Section II were used to identify lattice geometries with zero thermal expansion and maximum stiffness. For a given skew angle θ and lattice unit cell length L, Eq. (17) can be used to identify values of w_1 and w_2 that produce maximum stiffness for a given pair of constituent materials. Using the temperature-average properties of Ti and Al (Table Ia), a variety of zero CTE lattices as predicted by Eq. (17) are identified (Fig. 16). The associated aspect ratio of type I members, $\hat{L}_1/w_1$, decreases with increasing skew angle from 20.4 to 5.0 (Figs. 16(a–d)). The stiffness of the analogous pin-jointed structures, those having the same member dimensions, is plotted along with FEM results in Fig. 17, where $V_{f,i}$ is the fraction of solid material of type i.

Fig. 13. The 304 stainless steel lattice compressed along the vertical axis. Bright areas indicate out-of-plane deformation associated with member buckling.

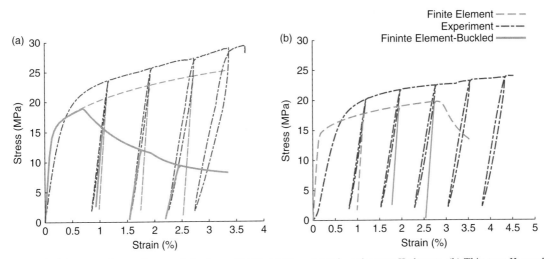

Fig. 14. Uniaxial compression strain–stress behavior or 304 Steel lattices. (a) Triangular type II element. (b) Thin type II members.

(4) Discussion—Stiffness

Experimental and FE results for 304 steel lattices in axial compression show that the geometries tested are prone to out-of-plane deformations, suggesting relatively good in-plane properties. FE analysis of an RVE shows good agreement with experiments conducted on steel lattices. To avoid the complex loading interactions between glass plates used to confine test specimens, thicker lattices not prone to out-of-plane plane deformations should be tested. Lattices incorporated into other structures and attached at nodal locations will have reduced out-of-plane degrees of freedom, such as when they are used as the face sheet in a sandwich panel, and will also be less prone to buckling in this manner. Additional boundary conditions or structural elements within the RVE technique can be used to model the behavior of more confined lattices.

Analytical models for pin-jointed biaxial stiffness and CTE give results that are similar to those from FE analysis of bonded lattices with slender members. Bonded lattices have increased CTE and stiffness over analytical predictions at higher relative densities. Decreasing member slenderness causes more overlap of members near joints leading to potentially larger deviations from the assumptions in Section II regarding the thermomecha-nical response of the joint region. Beam theory cannot accurately predict the deformation of members with aspect ratios less than about 10; in this case, Ti and Al lattices with $\theta > \sim 26°$. However, the associated geometries still have a CTE very close to the value predicted by Eq. (26). Stress distributions in lattices subjected to equibiaxial tensile straining show greater uniformity at lower relative densities suggesting more efficient and stretch-dominated behavior. The stiffness of the pin-jointed lattice is clearly recovered in bonded lattices with slender members as predicted in Section II.

IV. Concluding Remarks

Modifications to the geometry and modeling assumptions of previous bonded lattice designs of the UCSB lattice have resulted in a design scheme capable of rapidly identifying geometries that inherit the CTE and stiffness properties of the parent pin-jointed structure. The pin-jointed structure has been shown to be near optimal in stiffness over a wide range of densities.[4] Similar bonded lattices have obvious advantages in terms of fabricability. The behavior of these lattices is elastic and

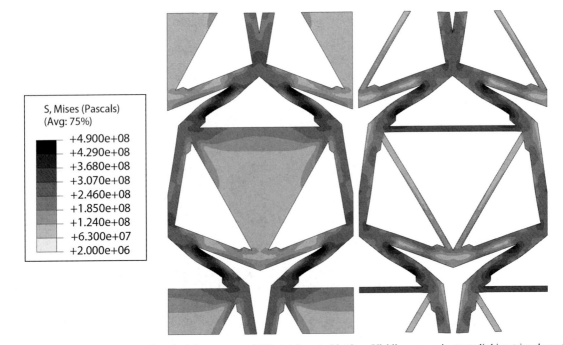

S, Mises (Pascals)
(Avg: 75%)

| +4.900e+08 |
| +4.290e+08 |
| +3.680e+08 |
| +3.070e+08 |
| +2.460e+08 |
| +1.850e+08 |
| +1.240e+08 |
| +6.300e+07 |
| +2.000e+06 |

Fig. 15. Finite-element stress distributions in uniaxially compressed 303 stainless steel lattices. Yielding occurs in struts linking triangle vertices to six-member joints and involves out-of-plane displacements. This behavior agrees well with experimental results.

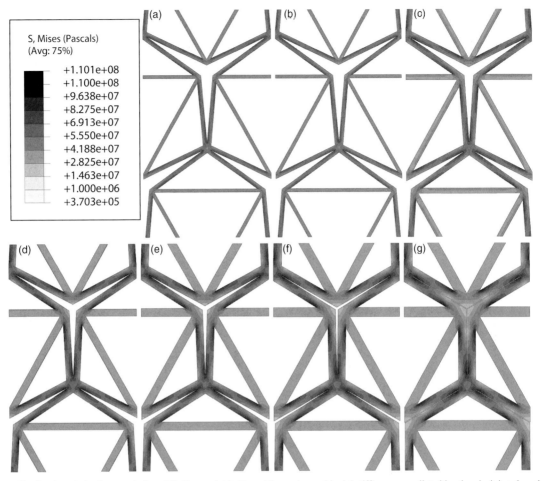

Fig. 16. Stress distributions in lattices made from Ti alloy and Al alloy with maximum biaxial stiffness as predicted by the pin-jointed analytical model, subject to 0.1% biaxial tensile strain. Skew angles range from 24.5° to 30° and volume fractions of solid from 14% to 46% (a–g). Stress distributions become more uniform with increasing slenderness moving toward behavior similar to the pin-jointed response.

amenable to fabrication on length scales ranging from aerospace structures to those relevant to nanotechnology.

The behavior of the bonded structure tends toward that of the pin-jointed lattice at lower relative densities in maximum stiff-

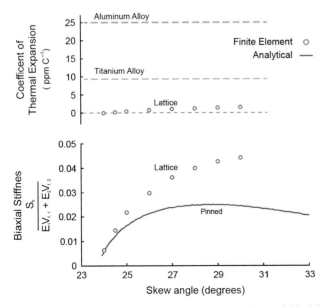

Fig. 17. Results for the coefficient of thermal expansion and biaxial stiffness for lattices shown in Fig. 16. Analytical results derived from a pin-jointed model are shown as are results from finite-element analysis of lattices having bonded joints.

ness lattices having slender members. At higher densities, lattice members have aspect ratios too low for beam-theory-based models to accurately predict their behavior. CTEs in these structures are still significantly different from the mean of the constituents. Three-dimensional computer-aided design and FE can be used to identify and fabricate rigid-jointed lattices with a tailored thermal expansion coefficient that is well predicted by modeling.

A lattice composed of Ti–6Al–4V and 7075-T6 Al was designed, fabricated, and measured to have a negative thermal expansion coefficient. Design space illustrated by the analytical model developed in this work shows the possibility for realizing material systems with a wide range of CTE including significantly negative thermal expansions. Such negative CTE materials can be used in a limited capacity in systems composed mostly of more conventional positive CTE materials, so that the system average is zero or low. The demonstrated ability of this design approach opens the door for the investigation and application of a wide family of materials with novel properties including high stiffness and low thermal expansion.

Further extension of these analytical and numerical techniques can be used to investigate the introduction of anisotropy by allowing geometry to vary among members of the same type in a unit cell allowing properties to be tailored in two directions. Rapid prototyping and other direct fabrication techniques can be used to fabricate volumetric lattices with properties tailored in three dimensions. The analytical and FE techniques can be extended to consider these variations.

Possible combinations lattice constituent materials include ceramics, glasses, and glass ceramics. In high-temperature applications, the thermal strains in these materials can be tailored to match the thermal strains experienced in significantly cooler supporting substructures. A low CTE glass, such as Zerodur, used as the type I material will offer a large CTE ratio, λ, when paired with a wide range of other higher expansion materials making available a wide range of achievable CTE (Fig. 5).

Other considerations exist in the design of these systems. These include (i) transient heating effects resulting from mismatches in thermal conductivity between constituents and non-uniform heating, (ii) the net CTE of the lattice is a function of the relative thermal strain between sublattices that may be varying due to transients and inhomogeneous thermal loading, which can affect the response, and (iii) aerodynamic surfaces are often curved surfaces so that nonflat shapes with a tailored thermal expansion coefficient may be desired. The techniques developed and exercised in this paper can be extended to address these issues, and the strength and failure modes of systems can also be addressed.

References

[1] R. S. Lakes, "Cellular Solid Structures with Unbounded Thermal Expansion," *J. Mater. Sci. Lett.*, **15**, 475–7 (1996).

[2] O. Sigmund and S. Torquato, "Composites with Extremal Thermal Expansion Coefficients," *Appl. Phys. Lett.*, **69**, 3203–5 (1996).

[3] L. V. Gibiansky and S. Torquato, "Thermal Expansion of Isotropic Multiphase Composites and Polycrystals," *J. Mech. Phys. Solids*, **45**, 1223–52 (1997).

[4] G. Jefferson, T. A. Parthasarathy, and R. J. Kerans, "Tailorable Thermal Expansion Hybrid Structures," *Int. J. Solids Struct.*, **46**, 2372–87 (2009).

[5] C. A. Steeves, S. L. Lucato, M. Y. He, E. Antinucci, J. W. Hutchinson, and A. G. Evans, "Concepts for Structurally Robust Materials that Combine Low Thermal Expansion with High Stiffness," *J. Mech. Phys. Solids*, **55**, 1803–22 (2007).

[6] J. N. Grima, P. S. Farrugia, R. Gatt, and V. Zammit, "A System with Adjustable Positive or Negative Thermal Expansion," *Proc. R. Soc. A*, **463**, 1585–96 (2007).

[7] C. A. Steeves, C. Mercer, E. Antinucci, M. Y. He, and A. G. Evans, "Experimental Investigation of the Thermal Properties of Tailored Expansion Lattices," *Int. J. Mech. Mater. Des.*, **5**, 195–202 (2009).

[8] M. Danielsson, D. M. Parks, and M. C. Boyce, "Three-Dimensional Micromechanical Modeling of Voided Polymeric Materials," *J. Mech. Phys. Solids*, **50**, 351–79 (2002).

[9] K. Bertoldi, Harvard University, School of Engineering and Applied Sciences (discussion).

[10] Simulia. *ABAQUS 6.9-EF*. Simulia, Providence, RI, 2009.

[11] Mathworks. *MATLAB 2008b*. Mathworks, Natick, MA, 2008.

[12] M. A. Sutton, J. J. Orteu, and H. W. Schreier, *Image Correlation for Shape, Motion and Deformation Measurements*. Springer, Berlin, 2009.   □

J. Am. Ceram. Soc., **94** [S1] S55–S61 (2011)
DOI: 10.1111/j.1551-2916.2011.04447.x
© 2011 The American Ceramic Society

journal _____

Optimization of Thermal Protection Systems Utilizing Sandwich Structures with Low Coefficient of Thermal Expansion Lattice Hot Faces

Craig A. Steeves[†,‡] and Anthony G. Evans[*,§]

[‡]Institute for Aerospace Studies, University of Toronto, Toronto, ON, Canada M3H 5T6

[§]Department of Materials, University of California, Santa Barbara, California 93106

Atmospheric cruise hypersonic vehicles are subject to high viscous heating over large surface areas. Acreage thermal protection systems (TPSs) must be stiff, strong, and light while withstanding large thermal gradients and protecting the cool interior of the vehicle. It is a challenge to design thermal protection to minimize the thermal stresses caused by thermal expansion mismatch. This paper uses a recent concept for low-thermal-expansion periodic lattices to propose a sandwich configuration for acreage TPSs. A key aspect of these concepts is that they can be attached to coll structures without inducing thermal stresses during heating. Sandwich TPSs are analyzed and optimized for minimum mass for required performance characteristics, and compared with an optimized baseline system. For performance requirements relevant to atmospheric hypersonic flight, the sandwich TPSs using low-thermal-expansion periodic lattices are superior to the baseline system for a large range of operating conditions.

I. Background

THERMAL protection systems (TPSs) for the acreage surfaces of hypersonic vehicles are required to protect the vehicle interior from high viscous heating, as well as to provide structural strength and stiffness in order that the vehicle retains its aerodynamic shape.[1,2] For hypersonic cruise vehicles, the acreage surface temperature may be well above 1000°C,[3] which makes thermal expansion mismatch and the resulting thermal stresses a critical issue when a hot structure is connected to a cold underlying vehicle. Furthermore, TPSs must be designed to minimize mass and volume for required thermal and structural performance.

Because of the problems associated with thermal expansion mismatch, it is attractive to use materials with very low thermal expansion. Periodic lattices that exhibit this characteristic have been proposed, among others Lakes,[4] Sigmund and Torquato,[5] Grima *et al.*,[6] and Jefferson *et al.*[7] A recent concept for *relatively stiff* periodic lattices with low or zero net thermal expansion coefficient[8,9] will be utilized to provide a zero thermal expansion stiff layer, which operates at high temperature and can be incorporated as part of a sandwich-type TPS. This lattice geometry is shown in Fig. 1. The behavior and properties of low-thermal-expansion bimaterial lattices are described in Steeves *et al.*[8,9] To summarize, using standard high-temperature aerospace alloys it is possible to build bimaterial lattices that have a low, tailorable net thermal expansion coefficient, and

nearly optimal stiffness for an elastically isotropic lattice attaining a given coefficient of thermal expansion (CTE). This is achieved because the expansion of the individual members is accommodated by bending and rotation. Two parameters govern the thermal expansion behavior: the ratio Σ of the CTE of the two materials and the skewness θ as depicted in the figure. Increasing either Σ or θ decreases the effective CTE of the lattice. Critical to the functioning of this concept is the existence of "stationary nodes" which, if the lattice is designed to have zero thermal expansion, do not move relative to each other upon heating. These nodes can be connected to a cold structure without the generation of thermal stresses. As indicated in Fig. 1 there are type 1 stationary nodes at the intersections of six type 1 low-CTE struts in adjacent unit cells, as well as type 2 stationary nodes at the centers of the high-CTE type 2 material. While there is always solid material at the type 1 unit cell junctions, the space at the center of the type 2 material need not be filled. In addition, there are type 3 stationary nodes in the gap between unit cells (that is, at the center of an equilateral triangle formed by three type 2 stationary nodes), but this node is never filled by solid material and hence cannot be used as a connection point.

The configuration of the lattice unit cells is flexible; three configurations different from that shown in Fig. 1 are shown in Fig. 2. All of the unit cells have the same thermal expansion properties because the locations of the connections between the two materials are identical. Cell (a) is the most structurally efficient. Cell (b) fills the space within the continuous lattice but the two materials are connected at the same location and hence the thermal behavior is unchanged. Cell (c), described by Berger *et al.*,[10] incorporates spacers at the joints between adjacent cells. The spacers slightly increase the net CTE of the lattice for a fixed skewness angle, but permit much higher lattice skewness. By increasing the skewness, the net CTE can be significantly reduced, which greatly exceeds the effect of the uniform thermal expansion of the spacers. This enables an almost fully flat surface with only small slots between the materials. Moreover, the unit cell geometry can be such that at a design temperature, the gaps are fully closed (Berger *et al.*[10] for details).

The design of a TPS with minimum mass for a required thermal resistance and structural strength is discussed herein. Two configurations of systems are optimized and compared: a baseline TPS comprising a thick layer of low-density insulating material attached to a metallic plate, which provides stiffness and strength; and a sandwich-type TPS with low-density insulation sandwiched between two metallic faces, where the hot exterior face of the sandwich is a lattice-type structure with low or zero thermal expansion, shown in Fig. 1.[8] The two TPS configurations are sketched in elevation in Fig. 3; a three-dimensional schematic of the lattice-based TPS with the low-density ceramic foam removed is given in Fig. 4. For the case of the TPS, it is sensible for aerodynamic reasons to use a unit cell configuration shown as Fig. 2(c) to minimize the surface roughness.

A photograph of a prototype sandwich (with the ceramic foam insulation removed for visual access) is shown in Fig. 5. The cold face and the stiffeners will typically be composed of

R. McMeeking—contributing editor

Manuscript No. 28681. Received September 28, 2010; approved January 10, 2011.
This work was supported by the Office of Naval Research through the MURI program Revolutionary Materials for Hypersonic Flight (Contract No. N00014-05-1-0439).
[*]Member, The American Ceramic Society.
[†]Author to whom correspondence should be addressed. e-mail: csteeves@utias.utoronto.ca

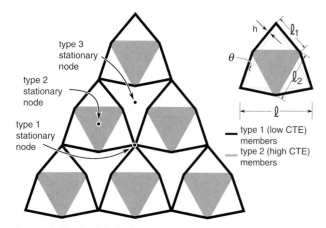

Fig. 1. A sketch of the lattice structure showing the key geometric parameters. The low-CTE type 1 members are black while the high-CTE type 2 members are gray. The key parameters are the lattice skewness θ, the ratio of thermal expansions $\Sigma = \alpha_2/\alpha_1$, the unit cell size ℓ, and the strut width h. "Stationary nodes" exist at the junction of six type 1 members and at the center of the type 2 region.

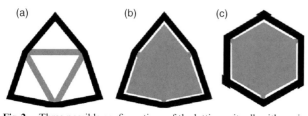

Fig. 2. Three possible configurations of the lattice unit cell, with equivalent thermal expansion properties. (a) Structure with maximum stiffness per mass. (b) Structure with the interior region filled with the high thermal expansion material. (c) Structure with spacers at the unit cell connections accommodated by increased lattice skewness.

one of the lattice materials in order to eliminate local thermal expansion mismatch at the connections. The remainder of the sandwich core is filled with lightweight insulating material, which serves to provide sufficient compressive strength that the metallic core stiffeners need not have high buckling resistance (and hence can be very slender). This configuration has the added benefit that the lightweight insulation, which is usually porous and friable, is protected from impacts by the robust metallic lattice.

One key design challenge with this sandwich-based TPS is that there is a direct metallic connection from the hot lattice to the cool inside face through the shear stiffeners. This provides a path of high thermal conductivity and hence encourages the

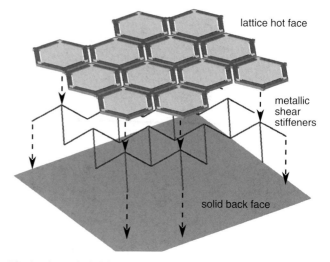

Fig. 4. An exploded sketch of the lattice-based thermal protection system, showing the lattice hot face, the lightweight tetrahedral truss metallic shear stiffeners and the solid cold back face. The truss core would be immersed in low-density ceramic foam; this has been removed for visual ease.

lowest possible metallic content in the core. A second challenge is that the lattices have relatively low stiffness and yield strength compared with solid plate materials, but this is partly mitigated by arranging the lattices in a sandwich configuration. The goal is to provide sufficient stiffness, strength, and thermal resistance for a total mass less than would be expected from the baseline system shown in Fig. 3(a). This is to be accomplished by using very low-density metallic cores to reduce through-thickness thermal conduction, and relatively robust lattice faces to improve stiffness and strength. An analysis of the properties of the baseline system and the sandwich TPS follows, which can be used to generate optimal designs and compare the two systems directly. This paper will focus on the stiffness of the sandwich TPS, but will also provide the corresponding results for strength-governed designs.

II. Analysis of Thermal Resistance, Stiffness, Strength, and Mass

The approach used here is to devise methods that can be used to optimize both the baseline system and the lattice-based sandwich TPS. An approach comparable to that undertaken in Weaver and Ashby,[11] Chen *et al.*,[12] or Steeves and Fleck[13] is used to generate designs that have minimum mass for a required thermal resistance and stiffness or strength. The resulting

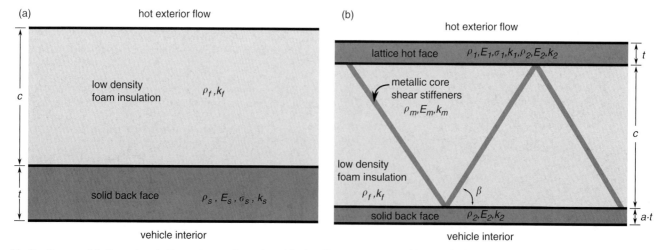

Fig. 3. Two possible thermal protection system configurations: (a) a baseline system comparable to typical shuttle insulation systems with an insulating ceramic foam layer atop a solid metallic plate; (b) a lattice-based sandwich TPS which is comprised of a low-CTE lattice hot face of two high-temperature alloys connected to a solid cold face with a core composed of metallic shear stiffeners and low-density ceramic foam insulation.

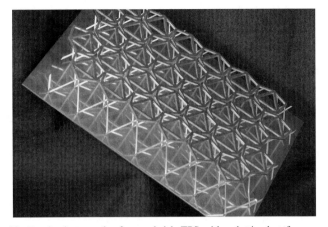

Fig. 5. A photograph of a sandwich TPS with a lattice hot face, as sketched in Fig. 3. The ceramic foam insulation has been removed for illustrative purposes and the lattice is not bimaterial: this prototype is intended only to illustrate the geometry. The fraction of metallic material in the core space is $v_m = 0.0034$.

designs are then normalized for a direct comparison between the baseline system and the sandwich system. The required analysis for predicting thermal resistance, stiffness, and strength follows first for the baseline system and then for the sandwich TPS.

(1) Baseline System

Both TPS configurations are assumed to act as simply supported beams with span L and width b, applicable to panels on evenly spaced unidirectional stringers, supporting a uniformly distributed pressure load w. The baseline TPS, in Fig. 3(a), has a porous ceramic layer of thickness c with density ρ_f and thermal conductivity k_f; and a solid back plate with thickness t, density ρ_s, thermal conductivity k_s, and yield strength σ_s. The thermal and mechanical properties of this TPS are effectively decoupled: the porous ceramic provides all of the thermal resistance and the metallic backing provides all of the stiffness and strength. The thermal resistance R of the baseline system is given by:

$$R = \frac{c}{k_f} + \frac{t}{k_s} \tag{1}$$

Assuming that the stiffness of the ceramic foam material can be neglected, the stiffness is:

$$EI_{eff} = \frac{bt^3 E_s}{12} \tag{2}$$

The solid back face of the baseline system behaves as a beam in bending with negligible strength provided by the ceramic foam insulation, and fails due to yielding under the pressure load, w:

$$w = 2\sigma_s \frac{t^2}{L^2} \tag{3}$$

The overall mass of the baseline system is:

$$M = bLt\rho_s + bLc\rho_f \tag{4}$$

(2) Sandwich TPS

The sandwich TPS shown in Fig. 3(b) is composed of a zero CTE lattice hot face of thickness t and a composite core with thickness c comprising low-density ceramic foam combined with shear stiffeners connecting the lattice to a solid back face of thickness $a \cdot t$, where a is a constant, typically < 1. Although the lattice materials are nearly optimally stiff, they are much less stiffer than solid materials, so that even when $a < 1$ the solid face is much stiffer than the lattice. The lattice has two constituents: a

low-CTE material 1 which comprises the continuous skewed triangular lattice and a high-CTE material 2, which constitutes the discontinuous lattice component (see Fig. 1). The two materials have, respectively, stiffness E_1 and E_2, yield strengths σ_1 and σ_2, thermal expansions α_1 and α_2, densities ρ_1 and ρ_2, and thermal conductivities k_1 and k_2. The lattice strength and stiffness are dominated by the properties of the continuous lattice comprising the low-CTE material 1. Metallic shear stiffeners in a tetrahedral configuration to match the symmetries of the lattice are connected to the sandwich faces at angle β, and have density ρ_m, thermal conductivity k_m, and yield strength σ_m. Typically the stiffeners will be composed of one of the lattice materials to eliminate CTE mismatch at the connections, but this is not necessary. Here, it is initially assumed that the shear stiffeners and the cold face are composed of high CTE material 2 and are connected to the lattice at the type 2 stationary nodes depicted in Fig. 1, though this will be reexamined later. One advantage of connecting the lattice to the core stiffeners at the type 2 stationary nodes is that this region has relatively low stress, while the type 1 stationary nodes are at the highly stressed junction of six type 1 members. The lattice unit cells have length ℓ and the two members have lengths ℓ_1 and ℓ_2. The type 1 members have width h.

The thermal resistance of the sandwich TPS is given by:

$$R = \frac{t}{k_1} + \frac{c}{k_c} + \frac{at}{k_1} \tag{5}$$

where k_1 and k_c are the thermal conductivities of the composite lattice and composite core, respectively. The lattice is porous and permits hot fluid to penetrate through it; k_1 is approximately infinite and provides no contribution to the sandwich thermal resistance. The conductivity of the core is approximated by a rule of mixtures, whereby:

$$k_c = v_m k_m + (1 - v_m)k_f \tag{6}$$

with v_m the volume fraction of the metallic shear stiffeners in the core, typically ranging from 0.1% to 1%.

Both the lattice stiffness and strength are functions of the lattice geometry. The slenderness of the lattice members $\tilde{r} \equiv h/\ell$ is chosen to ensure that the lattice behaves as approximately pin-jointed, which minimizes the net lattice CTE. The lattice skewness θ is determined by solving the implicit equation:

$$\frac{1 - \frac{1}{2}\frac{\alpha_2}{\alpha_1}\sin(2\theta)\left(\frac{1}{\sqrt{3}} + \tan\theta\right)}{1 - \frac{1}{2}\sin(2\theta)\left(\frac{1}{\sqrt{3}} + \tan\theta\right)} = \bar{\alpha} \tag{7}$$

which ensures that the net thermal expansion of the lattice is equal to a required $\bar{\alpha}$. Low-CTE lattices are much less stiffer than solid plates of the same material, and hence the neutral axis of bending of the sandwich beam is very close the solid face. Here, the conservative approximation that the neutral axis is coincident with the solid cold face is used. The overall stiffness of the beam is:

$$EI_{eff} = btc^2 E_1 \tag{8}$$

where E_1 is the uniaxial stiffness of the lattice, which is given by:

$$E_1 = 2\sqrt{3}\tilde{r}E_1\Theta \tag{9}$$

Because the material 2 component experiences much smaller loads than the material 1 component, it is assumed that the material 2 component is effectively rigid. Approximating the joints as pinned, which underestimates the stiffness of the lattice, the lattice stiffness–skewness parameter Θ is:

$$\Theta = \frac{\cos\theta\cos^2(\theta+\pi/6)\cos^2(2\theta-\pi/6)}{\cos^2(2\theta-\pi/6)+\cos^2(\theta)\cos^2(\theta+\pi/6)+\cos^2(\theta+\pi/3)\cos^2(\theta+\pi/6)} \tag{10}$$

For a continuous beam loaded by pressure w on stringers spaced at L, the maximum stress induced in the lattice face is:

$$\sigma_{max} = \frac{wL^2}{12tc} \tag{11}$$

This expression implies that the neutral axis of bending is coincident with the cold solid face of the sandwich as described above. The most likely mode of failure for the lattice is for the low-CTE material 1 members to yield, which occurs at an applied axial stress on the lattice of:

$$\sigma = 2\sigma_1\tilde{r}\sin\left(\frac{\pi}{3}-\theta\right) \tag{12}$$

The axial stress in the lattice is due to the bending moment induced by the external pressure; the sandwich lattice face is assumed to carry all the axial loads from bending. The maximum pressure load the sandwich beam can carry before the lattice yields is, therefore:

$$w_{LY} = \frac{24ct\tilde{r}\cos(\pi/6+\theta)\sigma_1}{L^2} \tag{13}$$

It can be ensured that the lattice fails due to yielding by setting the aspect ratio $\tilde{r}$ to be sufficiently large to prevent buckling and by choosing an overall lattice unit cell size such that the thickness of the lattice, determined later, is large enough to prevent buckling for the chosen lattice strut length. The sandwich beam may alternately fail due to shearing of the core; the core shear strength is provided exclusively by the metallic component, which has strength:

$$\tau_c = \frac{v_m\sin^2\beta\sigma_m}{3} \tag{14}$$

where σ_m is the yield strength of the metallic component of the core and β the angle of the core reinforcing struts with respect to the lattice face. This estimate is very conservative because it assumes that the core members that are in compression, buckle at very low loads and hence make no contribution to the core shear strength. Consequently, the maximum pressure load that may be supported by the beam is:

$$w_{CS} = \frac{2v_m\sin^2\beta c\sigma_m}{3L} \tag{15}$$

For any choice of material and geometry, the sandwich will fail in the mode with minimum w for that configuration.

The mass of the sandwich TPS is:

$$M = bLt(a\rho_2+\rho_1)+bLc\rho_c \tag{16}$$

where ρ_1 and ρ_c are the lattice and core densities, respectively. For a lattice that has minimal gaps, the high-CTE material 2 is the dominant component, and hence:

$$\rho_1 \approx \rho_2 \tag{17}$$

The density of the composite core is given by a rule of mixtures:

$$\rho_c = v_m\rho_m + (1-v_m)\rho_f \tag{18}$$

with ρ_m the density of the shear stiffener material, which is typically the high-CTE material 2.

III. Optimization of TPSs

The goal of this analysis is to generate TPS designs that provide minimum mass for given thermal resistance and stiffness or strength. The optimal designs for both the sandwich TPS and the baseline system will be compared with determine the overall minimum mass design. The following nondimensional expressions will be used in the optimizations. Nondimensional performance indices will be denoted by $(\hat{\cdot})$, while nondimensional material parameters by $(\tilde{\cdot})$ and the geometric design parameters will be identified by $(\bar{\cdot})$. Analogous parameters and indices from the sandwich and baseline systems will use the same labels in order to make the comparison clearer.

For the baseline system, the mass index is:

$$\hat{M} = \frac{M}{bL^2\rho_s} \equiv \bar{c}(\bar{t}+\tilde{\rho}) \tag{19}$$

with the nondimensional parameters:

$$\bar{c}=c/L; \quad \bar{t}=t/c; \quad \tilde{\rho}=\rho_f/\rho_s$$

The stiffness index is:

$$\hat{EI} = \frac{EI_{eff}}{bL^3E_s} \equiv \frac{\bar{c}^3\bar{t}^3}{12} \tag{20}$$

The strength index is:

$$\hat{w} = \frac{w}{\sigma_s} \equiv 2\bar{t}^2\bar{c}^2 \tag{21}$$

The thermal resistance index is:

$$\hat{R} = \frac{R}{L/k_s} \equiv \bar{c}\left(\bar{t}+\frac{1}{\tilde{k}}\right) \tag{22}$$

with $\tilde{k}=k_f/k_s$.

The corresponding indices for the sandwich TPS are:

$$\hat{M} = \frac{M}{bL^2\rho_1} \equiv \bar{c}((\tilde{\rho}_1+a)\bar{t}+\tilde{\rho}_c) \tag{23}$$

again $\bar{t}=t/c$ and $\bar{c}=c/L$, while $\tilde{\rho}_1=\rho_1/\rho_1$ and $\tilde{\rho}_c=\rho_c/\rho_1$. The stiffness index is:

$$\hat{EI} = \frac{EI_{eff}}{bL^3E_1} \equiv 2\sqrt{3}\tilde{r}\Theta\bar{c}^3\bar{t} \tag{24}$$

The two strength indices are, for lattice yielding:

$$\hat{w}_{LY} = \frac{w_{LY}}{\sigma_1} \equiv 24\cos(\pi/6+\theta)\tilde{r}\bar{c}^2\bar{t} \tag{25}$$

and for core shearing:

$$\hat{w}_{CS} = \frac{w_{CS}}{\sigma_1} \equiv 2v_m\sin^2\beta\bar{c}\tilde{\sigma} \tag{26}$$

where $\tilde{\sigma}=\sigma_m/\sigma_1$. The thermal resistance index is:

$$\hat{R} = \frac{R}{L/k_1} \equiv \bar{c}\left((1/\tilde{k}_1+a)\bar{t}+\frac{1}{\tilde{k}_c}\right) \tag{27}$$

where $\tilde{k}_1=k_1/k_1$ and $\tilde{k}_c=k_c/k_1$.

The design requirements for the TPS, which are parameters of the problem, are $\hat{R}_{req}$, $\hat{EI}_{req}$ and $\hat{w}_{req}$.

(1) Optimization for Stiffness

Examining first the behavior of the baseline TPS, it is instructive to draw contours of the performance indices in geometric $(\bar{c},\bar{t})$ design space, in order to illustrate the relationships between the behavior of the stiffness, resistance, and mass. Figure 6(a) shows

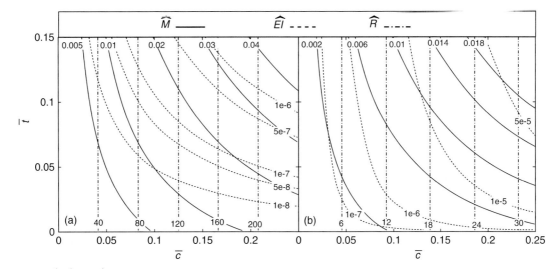

Fig. 6. Contours of $\hat{M}$, $\hat{R}$, and $\hat{EI}$ in $(\bar{c}, \bar{t})$ space (a) for the baseline TPS geometry using aluminum alloy 7075-T6 and ceramic insulating foam LI-900 and (b) for the sandwich TPS geometry using niobium alloy Cb-752, nickel–cobalt alloy NIMONIC PK33 and ceramic insulating foam LI-900 with $v_m = 0.02$, $\tilde{r} = 0.06$ and $a = 0.25$.

an example of the performance of the baseline TPS comprising a solid face of aluminum 7075-T6 and the ceramic foam insulator LI-900; properties for both materials are found in Table I. In Fig. 6(a), the stiffness indices and the mass indices are strongly dependent upon both $\bar{c}$ and $\bar{t}$, while the resistance index is almost independent of $\bar{t}$. This is because the aluminum has a much higher conductivity (by three orders of magnitude) then the ceramic foam and hence does not contribute to the insulating capability of the baseline TPS.

To minimize mass for required minimum stiffness and resistance indices, it is necessary to find the location of the design point that has the smallest $\hat{M}$ for the required $\hat{EI}_{req}$ and $\hat{R}_{req}$. In this example, the feasible design region lies above the contour of $\hat{EI}_{req}$ and to the right the contour of $\hat{R}_{req}$. There are several possibilities for an optimal location. First, if the gradients of the performance indices are parallel to the gradient of the mass index, that is, $\nabla \hat{M} = \lambda \nabla \hat{R}$ or $\nabla \hat{M} = \lambda \nabla \hat{EI}$ (with λ a constant multiplier), optimal design may occur at one of these points provided the other performance requirement is also satisfied. This is not the case in this example: neither of the performance indices have gradients parallel to the gradient of the mass index. Second, if the contours for the desired performance indices, $\hat{w}_{req}$ and $\hat{R}_{req}$, intersect, the minimum mass design occurs at the intersection point. For the baseline system, this condition is met by:

$$\bar{t} = \frac{\left(12\hat{EI}_{req}\right)^{\frac{1}{3}}}{\tilde{k}\left(\hat{R}_{req} - \left(12\hat{EI}_{req}\right)^{\frac{1}{3}}\right)} \tag{28}$$

with the corresponding

$$\bar{c} = \tilde{k}\left(\hat{R}_{req} - \left(12\hat{EI}_{req}\right)^{\frac{1}{3}}\right) \tag{29}$$

Recall that this is the baseline TPS, and the appropriate non-dimensionalizations must be used. If the contours of $\hat{R}_{req}$ and

$\hat{EI}_{req}$ do not intersect (this occurs when the stiffness requirement dominates the design), then the optimal design is found at a boundary of the design region; in this case, the maximum $\bar{t}$ permissible will be chosen and the minimum $\bar{c}$ will be found which satisfies both $\hat{R}_{req}$ and $\hat{EI}_{req}$. Once the optimal pair $(\bar{c}, \bar{t})$ is found, the minimum mass index $\hat{M}_{min}$ can be calculated. Trajectories of optimal design will follow either resistance index contours (for fixed required resistance) or stiffness index contours (for fixed required stiffness).

For the sandwich TPS, the optimization process is similar. Figure 6(b) shows the analogous contour plot of the performance indices and the mass index for the sandwich TPS, utilizing a lattice composed of the low-CTE niobium alloy Cb-752 and the high-CTE nickel–cobalt alloy NIMONIC PK33, with LI-900 as the core foam material and NIMONIC PK-33 as the metallic core component. (Note that the two contour plots are not directly comparable because of the different parameters used in the material property normalizations.) Several other geometric parameters must be specified; the net CTE of the lattice is 0 ppm/C, the volume fraction of metal in the core is $v_m = 0.005$, the slenderness ratio of the lattice struts is $\tilde{r} = 0.05$, and the thickness ratio of the solid face to the lattice is $a = 0.25$. For the sandwich TPS, the gradient of the mass index is parallel to the gradient of the stiffness index at

$$\bar{t} = \frac{\tilde{\rho}_c}{2(\tilde{\rho}_l + a)} \tag{30}$$

for all values of $\bar{c}$. For this set of example materials and geometry, the optimal is found at $\bar{t} \approx 0.020$. The value of $\bar{c}$ corresponding to the particular $\hat{EI}_{req}$ is:

$$\bar{c} = \left(\frac{\hat{EI}_{req}(\tilde{\rho}_l + a)}{\sqrt{3}\tilde{r}\Theta\tilde{\rho}_c}\right)^{\frac{1}{3}} \tag{31}$$

Table I.   Relevant Material Properties for the Materials Used in the Demonstration Optimization (Unused Data are Omitted)

	Density (kg/m³)	Young's modulus (GPa)	Yield strength (MPa)	Thermal conductivity (W·(m·K)⁻¹)	Thermal expansion (ppm/K)
LI-900	144			0.135	
Al 7075-T6	2810	72	505	130	
Cb-752	9030	110	257	48.7	7.4
NIMONIC PK33	8210	221		25.7	17.2
Toray T300	1760		3500	8.5	

Sources: Cb-752: ATI Wah Chang,[14] PK-33: Special metals,[15] LI-900: Banas.[16]

which satisfies the stiffness constraint. If this combination of $(\bar{c}, \bar{t})$ also satisfies the resistance requirement, then this is the optimal solution. If it does not, the minimum mass solution will occur at the junction of the contours of $\hat{R}_{\text{req}}$ and $\hat{E}I_{\text{req}}$, as above for the baseline case. Here, the optimal value of $\bar{c}$ is found by solving the implicit equation:

$$\hat{E}I_{\text{req}} - 2\sqrt{3}\tilde{r}\Theta\bar{c}^3 \left(\frac{\tilde{k}_1(\tilde{k}_c\hat{R}_{\text{req}} - \bar{c})}{\tilde{k}_c(\tilde{k}_1 a + 1)} \right) = 0 \tag{32}$$

The corresponding value of $\bar{t}$ is:

$$\bar{t} = \frac{\tilde{k}_1\left(\hat{R}_{\text{req}}\tilde{k}_c - \bar{c}\right)}{\bar{c}\tilde{k}_c\left(\tilde{k}_1 a + 1\right)} \tag{33}$$

Finding optimal designs in $(\bar{c}, \bar{t})$ space creates loci of combinations of $\hat{E}I_{\text{req}}$, $\hat{R}_{\text{req}}$ and $\hat{M}_{\text{min}}$, and these can be graphed with $\hat{M}_{\text{min}}$ as a function of one performance index for various values of the other performance index. For comparison, trajectories of minimum mass design can be plotted together for different material combinations or geometric arrangements, but the indices must be first normalized to provide a consistent basis for comparison. In the case described here, the sandwich TPS will be normalized to be consistent with the baseline TPS using the normalization factors $\tilde{\rho}^N = \rho_1/\rho_s$, $\tilde{k}^N = k_s/k_1$ and $\tilde{E}^N = E_1/E_s$. Figure 7 shows normalized trajectories of minimum normalized mass index $\hat{M}_{\text{min}}^N$ as a function of the required normalized stiffness index $\hat{E}I_{\text{req}}^N$ for a value of required thermal resistance $\hat{R}_{\text{req}}^N = 50$.

(2) Optimization for Strength

A similar procedure is followed to generate optimal designs for simultaneous strength and thermal resistance requirements, with the strength requirement given by $\hat{w}_{\text{req}}$. For the baseline case, there are no locations where the gradient of the mass index is parallel to the gradient of either the strength or thermal resistance index; consequently, optimal design occurs at locations where the contour of required strength intersects the contour of required thermal resistance, provided the contours intersect. For the baseline system, such a condition is met by:

$$\bar{t} = \left(\frac{\hat{w}_{\text{req}}}{2}\right)^{\frac{1}{2}} \frac{1}{\tilde{k}\left(\hat{R}_{\text{req}} - \left(\frac{\hat{w}_{\text{req}}}{2}\right)^{\frac{1}{2}} \right)} \tag{34}$$

with the corresponding

$$\bar{c} = \tilde{k}\left(\hat{R}_{\text{req}} - \left(\frac{\hat{w}_{\text{req}}}{2}\right)^{\frac{1}{2}} \right) \tag{35}$$

Otherwise, optimal design will occur at the boundaries of the design region.

As a consequence of the competition between the lattice yield and core shear failure mechanisms, the strength index is composed of two parts: curved contours in the lattice yield region and straight contours, independent of $\bar{t}$, in the core shear region. While the gradient of the mass index is never parallel to that of the strength index in the core shearing region, optimal design can occur at locations where the gradient of the strength index in the lattice yield region is parallel to the gradient of the mass index, which occurs at:

$$\bar{t} = \frac{\bar{\rho}_c}{\bar{\rho}_1 + a} \tag{36}$$

for all values of $\bar{c}$. For this set of example materials with $\tilde{r} = 0.05$ and $v_m = 0.01$, the optimum is found at $\bar{t} \approx 0.0096$, which implies that the sandwich core is approximately 100 times as thick as the lattice face for optimal designs in these conditions. The value of $\bar{c}$ corresponding to the particular $\hat{w}_{\text{req}}$ is:

$$\bar{c} = \left(\frac{\hat{w}_{\text{req}}(\tilde{\rho}_1 + a)}{24\cos(\pi/6 + \theta)\tilde{r}\tilde{\rho}_c} \right)^{\frac{1}{2}} \tag{37}$$

If this combination $(\bar{c}, \bar{t})$ satisfies the resistance criterion $\hat{R}_{\text{req}}$ and is within the lattice yield-dominated region, then this is the minimum mass design. Alternately, the optimal design may occur at the junction between the contours of $\hat{R}_{\text{req}}$ and $\hat{w}_{\text{req}}$, or at the boundary between the core shear and lattice yield regions.

For optimized design for strength and thermal resistance requirements, there are again loci of combinations of $\hat{w}_{\text{req}}$, $\hat{R}_{\text{req}}$ and $\hat{M}_{\text{min}}$ in $(\bar{c}, \bar{t})$ space. Again, to compare the results for the sandwich and baseline TPS, further normalization is required; here, the results for the sandwich TPS will be normalized to be consistent with the baseline TPS by using the properties of the solid material of the baseline system. Figure 8 shows normalized trajectories of minimum normalized mass index $\hat{M}_{\text{min}}^N$ as a function of the required normalized strength index $\hat{w}_{\text{req}}^N$ for a value of required thermal resistance $\hat{R}_{\text{req}}^N = 100$. For very low strength

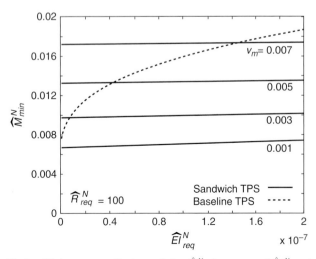

Fig. 7. Minimum normalized mass index $\hat{M}_{\text{min}}^N$ for a range of $\hat{E}I_{\text{req}}^N$ and $\hat{R}_{\text{req}}^N = 100$ for a selection of core relative densities for the sandwich TPS (solid) and the baseline TPS (dashed).

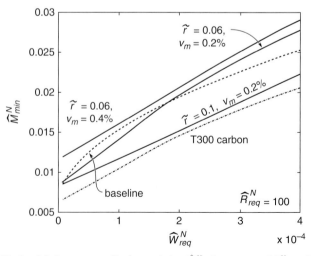

Fig. 8. Minimum normalized mass index $\hat{M}_{\text{min}}^N$ for a range of $\hat{w}_{\text{req}}^N$ and $\hat{R}_{\text{req}}^N = 100$ for a selection of core relative densities for the sandwich TPS and the baseline TPS. Two sandwich configurations are chosen: for the solid lines, the core struts are metallic, of material 2 (NIMONIC PK33), and in dash–dot lines, the core struts are of T300 carbon fiber. Several different combinations of parameters are graphed for the metallic core to illustrate the effect of these parameters on the overall mass of the system.

requirements, the baseline system is preferable, because the sandwich TPS will always have a thicker core due to the high-conductivity struts within it and will consequently have greater mass. For many practical strengths, the sandwich TPS provides a lower mass design, while, due to the dominance of the core shear failure mechanism at high loads, the baseline system again becomes more efficient, at least for all-metal designs.

IV. Discussion

The principal design challenge in using the low-CTE lattice sandwich TPS is minimizing the amount of the metallic component in the core. Because this metallic component provides a thermal path from the hot surface, the sandwich core will always be thicker than the ceramic foam layer on the baseline system. However, because of the stiffness and strength properties of sandwich structures, the amount of material in the sandwich faces is much less than amount of material in the solid metallic layer of the baseline system.

The results here show that the sandwich TPS has lower mass than a baseline TPS when there is a large stiffness requirement: for small stiffness, the baseline system is preferable. As the stiffness requirement increases, the sandwich system clearly produces designs of lower mass, provided the volume fraction of the metallic component of the core can be made small. This is because, to increase stiffness or strength of the sandwich a thicker core can be used, adding lightweight foam material, while for the baseline system increasing stiffness or strength requires increasing the thickness of the metal component. Thus, as the stiffness requirement increases, the sandwich TPS can provide the same thermal resistance and stiffness as the baseline system for one half to two thirds the total mass. An alternate interpretation is that, for the same mass, the bending stiffness of the TPS can be increased by more than an order of magnitude; the principal consequence of this is that the outer aerodynamic shell can be made much stiffer, *reducing the need for stiff interior structural members to support the aerosurface.*

When strength rather than stiffness is the governing design requirement, the sandwich TPS is competitive with, but not clearly superior to, the baseline TPS. The sandwich TPS is also sensitive to an additional parameter: $\bar{r}$, the slenderness of the lattice members. As these become slender, the lattice has a higher propensity to yield or buckle. Making the lattice more robust does not come at a weight penalty, but does increase the net CTE of the lattice; this has the effect of requiring higher lattice skewness. At very low strength requirements, the baseline system is once again superior, because the sandwich core is considerably thicker. As the strength requirement increases, the sandwich is competitive with, and often superior to the baseline system for a range of design variables. Again, low volume fractions of the metallic component in the core are desirable. At higher strength requirements, the baseline system is again superior to many sandwich geometries, because the large loads tend to cause the lattice to yield;

An additional case is plotted in Fig. 8. Here, the metallic core members are replaced by Toray T300 carbon fibers (material properties in Table I). The result for $v_m = 0.03$ is plotted, although for other values of v_m the total mass changes only marginally. The carbon fibers have two significant advantages over the metallic core struts: density approximately one-fifth and thermal conductivity approximately one third as great as the PK-33 metallic members. Consequently, thinner cores can be used for the same thermal protection. Furthermore, increasing the core thickness while decreasing the face thickness will produce equally stiff and strong sandwiches; because of the

lower density carbon fiber strut material, this is an efficient trade.

V. Concluding Remarks

The reason that the sandwich TPS is effective is that to increase its strength or stiffness requires adding lightweight core material; to increase the strength or stiffness of the baseline TPS requires adding heavy metallic material. As a consequence, for TPS designs with high strength or stiffness requirements, the sandwich TPS can generally be designed to be significantly lighter than the baseline TPS, *provided the amount of the metallic component in the core can be kept small.* This has been shown for a large range of possible design requirements: to situate the examples within the practical design space, a baseline TPS with a 1 m span, 120 mm of LI-900 insulation on 8 mm of Al 7075-T6 would have $\hat{EI} = 4.3 \times 10^{-8}$, $\hat{w} = 8.6 \times 10^{-5}$ and $\hat{R} = 115$. (Atmospheric pressure is approximately $\hat{w} = 2 \times 10^{-4}$.) This indicates that for conditions relevant to atmospheric hypersonic flight the sandwich TPS with a low-CTE lattice hot face is a feasible design option.

Acknowledgments

Peter Maxwell, Kirk Field, and Kyle Stehly, all of the University of California, Santa Barbara Materials Department, were very helpful in supporting an experimental component of this research. This paper is dedicated to the memory of Professor Tony Evans, a peerless mentor, colleague, and gentleman.

References

[1]D. Manor, K. Y. Lau, and D. B. Johnson, "Aerothermodynamic Environments and Thermal Protection for a Wave-Rider Second Stage," *J. Spacecraft Rockets,* **420** [2] 208–12 (2005).
[2]Y. Yang, J.-L. Yang, and D.-N. Fang, "Research Progress on Thermal Protection Materials and Structures of Hypersonic Vehicles," *Appl. Math. Mech.,* **290** [1] 51–60 (2008).
[3]D. E. Glass, R. Dirling, H. Croop, T. J. Fry, and G. J. Frank, "Materials Development for Hypersonic Flight Vehicles"; In *14th AIAA/AHI International Space Planes and Hypersonics Systems and Technologies Conference,* Canberra, Australia, AIAA 2006-8122, November 6–9, 2006.
[4]R. S. Lakes, "Cellular Solid Structures with Unbounded Thermal Expansion," *J. Mater. Sci. Lett.,* **150** [6] 475–7 (1996).
[5]O. Sigmund and S. Torquato, "Composites with Extremal Thermal Expansion Coefficients," *Appl. Phys. Lett.,* **690** [21] 3203–5 (1996).
[6]J. N. Grima, P. S. Farrugia, R. Gatt, and V. Zammitt, "A System with Adjustable Positive or Negative Thermal Expansion," *Proc. R. Soc. A,* **4630** [2082] 1585–96 (2007).
[7]G. Jefferson, T. A. Parthasarathy, and R. J. Kerans, "Tailorable Thermal Expansion Hybrid Structures," *Int. J. Solids Struct.,* **460** [11–12] 2372–87 (2009).
[8]C. A. Steeves, S. L. Lucato, M. Y. He, E. Antinucci, J. W. Hutchinson, and A. G. Evans, "Concepts for Structurally Robust Materials that Combine Low Thermal Expansion with High Stiffness," *J. Mech. Phys. Solids,* **550** [9] 1803–22 (2007).
[9]C. A. Steeves, C. Mercer, E. Antinucci, M. Y. He, and A. G. Evans, "Experimental Investigation of the Thermal Properties of Low Expansion Lattices," *Int. J. Mech. Mater. Des.,* **50** [2] 195–202 (2009).
[10]J. Berger, C. Mercer, R. M. McMeeking, and A. G. Evans, "The Design Of Bonded Bimaterial Lattices that Combine Low Thermal Expansion with High Stiffness," *J. Am. Ceram. Soc.,* (2010).
[11]P. M. Weaver and M. F. Ashby, "Material Limits for Shape Efficiency," *Prog. Mater. Sci.,* **410** [1–2] 61–128 (1997).
[12]C. Chen, A.-M. Harte, and N. A. Fleck, "The Plastic Collapse of Sandwich Beams with a Metallic Foam Core," *Int. J. Mech. Sci.,* **430** [6] 1483–506 (2001).
[13]C. A. Steeves and N. A. Fleck, "Collapse Mechanisms of Sandwich Beams with Composite Faces and a Foam Core, Loaded in Three-Point Bending. Part I: Analytical Models and Minimum Weight Design," *Int. J. Mech. Sci.,* **460** [4] 561–83 (2004).
[14]ATI Wah Chang. *Niobium Technical Data Sheet.* ATI Wah Chang, Albany, OR, 2003.
[15]Special Metals. NIMONIC Alloy PK33 Technical Data Sheet, Special Metals, New Hartford, NY, 2004.
[16]R. P. Banas, "Rigid Fibrous Ceramics for Entry Systems"; Technical Report, Lockheed Missiles and Space Company, 1993. □

J. Am. Ceram. Soc., **94** [S1] S62–S75 (2011)
DOI: 10.1111/j.1551-2916.2011.04501.x
© 2011 The American Ceramic Society

journal

Explorations of Hybrid Sandwich Panel Concepts for Projectile Impact Mitigation

Christian J. Yungwirth,[‡] John O'Connor,[§] Alan Zakraysek,[§] Vikram S. Deshpande,[¶] and Haydn N. G. Wadley[†,‡]

[‡]Department of Material Science & Engineering, School of Engineering and Applied Science, University of Virginia, Charlottesville, Virginia 22903

[§]Naval Surface Warfare Center, Indian Head, Maryland 20640

[¶]Engineering Department, Cambridge University, Cambridge CB2 1PZ, U.K.

Previous studies have shown that while stainless-steel sandwich panels with pyramidal truss cores have a superior blast resistance to monolithic plates of equal mass per unit area, their ballistic performance is similar to their monolithic counterparts. Here, we explore concepts to enhance the ballistic resistance without changing the volumetric efficiency of the panels by filling the spaces within the core with combinations of polyurethane, alumina prisms, and aramid fiber textiles. The addition of the polyurethane does not enhance the ballistic limit compared with the equivalent monolithic steel plate, even when aramids are added. This poor performance occurs because the polymer is penetrated by a hole enlargement mechanism which does not result in significant projectile deformation or load spreading and engagement of the steel face sheets. By contrast, ceramic inserts deform and erode the projectile and also comminute the ceramic. The ceramic communition (and resultant dilation) results in stretching of both steel face sheets and leads to significant energy dissipation. The ballistic limit of this hybrid is about twice that of the equivalent monolithic steel plate. The addition of a Kevlar fabric to the ceramic hybrid is shown to not significantly change the ballistic limit but does reduce the residual velocities of the debris.

I. Introduction

IT is well known that sandwich panels, with appropriate distribution of mass between the front and back faces and the core, exhibit superior bending stiffness and strength compared with monolithic (solid) plates of the same mass per unit area. Theoretical studies by Fleck and Deshpande[1] and Hutchinson and Xue[2] also predict that sandwich beams with porous cores will have superior shock resistance to monolithic beams. Subsequent experimental studies[3–5] in which edge clamped metallic panels with low-density lattice cores were subjected to high-intensity shock loading (in air and water) have confirmed these predictions and shown significant reductions in sandwich panel deflections compared with monolithic counterparts.

F. Zok—contributing editor

Manuscript No. 28899. Received November 12 2010; approved February 11 2010.
The ballistic measurements were supported by the Defense Advanced Research Projects Agency and the Office of Naval Research under Grant number N00014-04-1-0299 (Dr. Leo Christodoulou was its program manager). The analysis work has been performed as part of the Ultralight Metallic Panels with Textile Cores Designed for Blast Mitigation and Load Retention program conducted by the University of Virginia and Cambridge University and funded by the Office of Naval Research (ONR) under Grant number N00014-01-1-1051 (Dr. David Shifler was the program manager).
[†]Author to whom correspondence should be addressed. e-mail: haydn@virginia.edu

Explosive events in air are often accompanied by high-velocity fragments. Multifunctional protection systems that combine efficient structural load support with air shock and ballistic impact resistance are therefore of considerable interest. Recent experimental studies indicate that metallic sandwich panels with low relative density cellular cores (optimized for structural load support and shock resistance) have approximately the same ballistic performance as monolithic structures of equal areal mass.[6] The study reported here initiates an exploration of concepts that might be used to enhance the ballistic resistance of these structurally efficient sandwich panels.

The ballistic impact resistance of a simple metallic plate of fixed thickness depends upon its density, strength, ductility, and its strain and strain rate hardening characteristics. It can be significantly enhanced by replacing some of the plate mass with a hard ceramic tile placed on the impact side of the bilayer.[7–10] Even lighter alternatives have been proposed where the metal is replaced by a ballistic fiber composite laminate,[11–14] or sandwich panel,[15] or multilayer.[10] All seek to exploit the same defeat mechanism; namely, during impact, the hard ceramic plastically deforms or fractures the projectile, dissipating the projectiles kinetic energy by plastic work within it. This process is usually accompanied by comminution of the ceramic tile and inelastic stretching of the back face, which further reduces the projectile kinetic energy. Recently, Sarva *et al.*[16] have shown that that by restraining the impact-face of ceramic tiles with a membrane of suitable tensile strength, the ballistic efficiency can be improved by as much as 25% for a 2.5% increase in areal density. Deshpande and Evans[17] have shown that this remarkable effect is due to an enhancement of the strength of the ceramic by the extra confinement.

Blast resistant sandwich panels with low relative density cellular cores have significant empty space within the core: filling all or a part of this empty space with ballistic fibers, polymers, or ceramics offer potential opportunities to enhance the ballistic resistance of multifunctional sandwich panels. In order to explore the mechanisms that might be invoked to enhance the ballistic performance of sandwich structures, we experimentally investigate the effects of filling the empty space within a model stainless steel, pyramidal lattice core sandwich panel with polyurethane elastomers, aramid fiber fabrics, and alumina prisms (and combinations of the same). The study investigates the mechanisms of projectile arrest for fillings that span the disparate possibilities of soft, very high elastic strain to failure polymers to very hard, but brittle ceramics, and examines the implications of these fillings on both the panels ballistic limit and spatial extent of damage which influences multihit performance. The results are contrasted with the ballistic penetration mechanisms of a monolithic stainless plate spanning the same range of aerial masses.

II. Ballistic Penetration Mechanisms

There exists a large literature on the mechanics (experiments and modeling) of penetration of monolithic plates made from metallic alloys as well as ceramics. Here, we give a brief review of the key mechanisms of penetration in order to motivate structural concepts that might enhance penetration resistance. In the 1960s and early 1970s, numerous experimental studies were utilized to explore the impact processes and penetration mechanisms in plates, most notably by Hopkins and Kolsky[18] and Goldsmith.[19] A compendium on the mechanics of projectile penetration was published by Backman and Goldsmith[20] and a more recent review was written by Corbett *et al.*[21] Figure 1 depicts the most common failure modes encountered in cases where the projectile strength exceeds that of the target (so that the projectile remains more or less intact). In general these modes can be divided into regimes based on the ductility of the target.

(1) Low Ductility Regime

Three modes of failure are most commonly observed for low ductility target materials such as hardened aluminum and steel alloys as well as ceramics:

(1) *Spall fracture or scabbing (Fig. 1(a))*: A compressive elastic or shock wavefront emanates from the impact site and reflects as a tensile wave from the back or distal plate/air interface. A spall fracture occurs near the back surface when the tensile stress in the reflected (tensile) wave exceeds the tensile strength of the material.

(2) *Plugging (Fig. 1(b))*: Plugging results in a cylindrical slug, nearly the size of the projectile, being ejected from the target. In metals this typically occurs due to large shears at the periphery of the moving plug which result in adiabatic softening and shear failure of the metal. By contrast, in ceramics this type of failure is a result of the cone cracking mechanism.[13,22]

(3) *Radial fracture (Fig. 1(c))*: If the tensile strength of the target is lower than its compressive strength, then a radial fracture behind the initial compressive stress wave occurs due to the

tensile radial stresses. This failure mode is often observed in both ceramics and high strength (low ductility) metals.

(2) High Ductility Regime

The two failure modes typically observed here occur during the penetration of high ductility metals (e.g., stainless steels) and polymers.

(1) *Petaling (Figs. 1(d) and (e))*: Petaling, both frontal (Fig. 1(d)) and rear (Fig. 1(e)), is produced by high radial and circumferential tensile stresses after passage of the initial wave near the lip of the penetration. This deformation is the result of bending moments created by the forward motion of the plate material as it is pushed ahead of the projectile and is initiated at inhomogeneities or weaknesses in the target. Petaling is often accompanied by large plastic flows and/or permanent flexure. As the material on the distal side of the plate is further deformed, a star-shaped crack is initiated by the tip of the projectile.[23] Finally, the sectors are rotated back by the ensuing motion of the projectile, often forming three to seven symmetric petals. Rearward petaling commonly occurs from ogival- or conical-shaped noses on projectiles penetrating thin ductile plates while frontal petaling occurs in thicker plates.

(2) *Ductile hole enlargement (Fig. 1(f))*: Ductile hole enlargement is a common mode of failure of thick plates impacted by ogival- or small-angle conical-shaped projectiles.[24] At the beginning of contact, the tip of the projectile begins displacing material radially leading to a radial momentum that continues so that a hole in the target is enlarged along the trajectory of the projectile. Heavily dependent on projectile shape, projectile diameter to target thickness ratio and projectile velocity, ductile hole enlargement is favored over plugging when the thickness of the plate exceeds the projectile diameter.

(1) Concepts to Enhance Ballistic Penetration via Hybrid Constructions

Monolithic systems undergo the penetration mechanisms discussed above and typically offer limited penetration resistance. The ballistic performance is enhanced by combining material systems—so-called hybrid constructions. Some examples of such hybrids and the mechanisms by which they enhance ballistic performance are:

(1) *Ceramic tiles backed by metallic or composite plates*: During penetration, the ceramic comminutes and erodes the projectile, reducing its kinetic energy and importantly resulting in a spreading of the applied load. The remnant energy is dissipated by the deformation of the back plate. The ceramic thus performs two roles as illustrated in Fig. 2(a): (a) it deforms and fragments (erodes) the projectile and (b) it spreads the force of the impacting projectile over a large area of the back plate—this inhibits the local penetration mechanisms of the back plate discussed above.

(2) *Ceramic tiles encased within a metallic or polymer case*: Upon impact the projectile penetrates the front cover easily by one of the modes in Fig. 1. Subsequently, it comminutes the ceramic. This comminuted ceramic dilates and expands, and therefore stretches the outer metallic or polymer fiber casing as illustrated in Fig. 2(b). This mechanism efficiently distributes the load and engages a large fraction of the panel in the energy dissipation process.

Ultrahigh specific strength aramid and polyethylene fibers in the form of ballistic fabrics are also widely used with metal/ceramic protection systems to arrest projectile and ceramic debris.[25] The performance of the above-mentioned systems can typically be enhanced by including a ballistic fabric such as Kevlar on the back face of the hybrid system: this fabric catches any small debris from the impacted target or eroded projectile. Here, we explore concepts for enhancing the ballistic response of sandwich panels by *inserting* various materials within them. The study therefore focuses upon methods for increasing performance without increasing the volume of the structure. Figure 3(a) shows the baseline (empty) sandwich panel

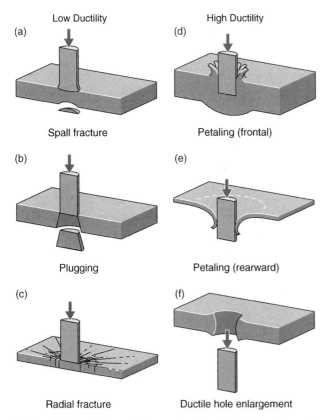

Fig. 1. Sketches of the penetration mode of monolithic plates. The modes are divided into two regimes based on the ductility of the target material.

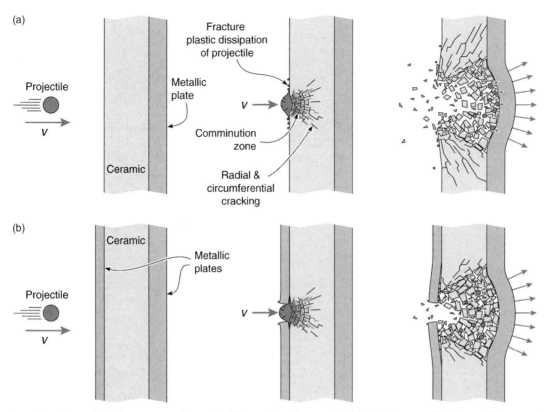

Fig. 2. Sketches of the deformation/failure sequence for a projectile impacting (a) a ceramic tile backed by a metallic plate and (b) a ceramic tile encased within a metallic or polymer case.

and examples of four hybrid panels (Figs. 3(b)–(e)) investigated here. The aim is to discover potential synergies between a hard but brittle ceramic filling and a metallic sandwich panel whereby the metallic casing changes the deformation and failure modes of the ceramic. This will be contrasted with the performance of polymer-filled sandwich panels.

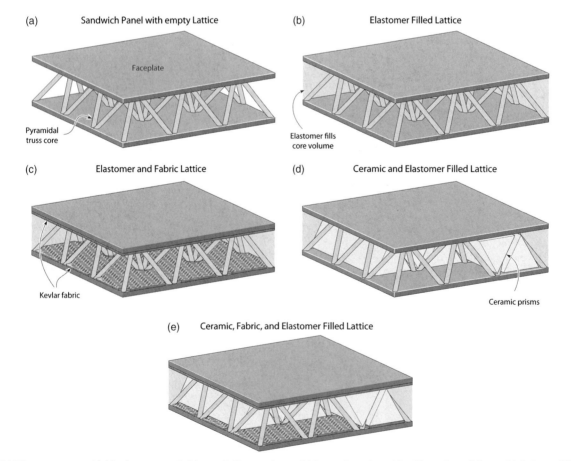

Fig. 3. (a) The empty pyramidal lattice core sandwich panel. Four variants of (a) were investigated in this study as follows. (b) Polymer-filled lattice panel; (c) hybrid polymer-filled lattice panel; (d) ceramic-filled lattice panel; and (e) hybrid ceramic-filled lattice panel.

Table I. Areal Densities of the Sandwich Panels and Monolithic Plates

Panel type	Empty sandwich panel	Polymer-filled sandwich panel	Hybrid polymer-filled sandwich panel	Ceramic-filled sandwich panel	Hybrid ceramic-filled sandwich panel	3-mm-thick monolithic 304 stainless-steel plate	10-mm-thick monolithic 304 stainless-steel plate
Areal density (kg/m^2)	28	55.2	56.8	81.5	78	24	80

III. Panel Configurations and Manufacture

The main aim of this study is to discover concepts that might enhance the ballistic performance of hybrid sandwich panels with cellular cores. We focus on a stainless-steel sandwich panel with a pyramidal truss core (Fig. 3(a)): a previous study[6] has shown that such a panel has a ballistic performance very similar to that of a stainless-steel monolithic plate of equal areal mass.

In this study, the unfilled sandwich panels were identical to those used in Yungwirth *et al.*[6] The air gaps in the cellular core were filled with a polymer and/or a ceramic. The four configurations/fillings investigated are sketched in Fig. 3:

(1) The 304 stainless-steel sandwich panel filled with a polyurethane filling, Fig. 3(b). This will be referred to as the polymer-filled lattice panel.

(2) Both face-sheets of the stainless-steel panel were reinforced with a Kevlar fabric placed on the *inner* surfaces of the face-sheets. This hybrid sandwich panel was then infiltrated with the polyurethane filler. We shall refer to this as the hybrid polymer-filled lattice panel, Fig. 3(c).

(3) Triangular ceramic (alumina) prisms were inserted into 304 stainless-steel sandwich panels and the remaining small gaps plugged with the polyurethane filler. We shall refer to this as the ceramic-filled lattice panel, Fig. 3(d).

(4) Finally, the face-sheets of the stainless-steel sandwich panel were reinforced with a Kevlar fabric (on their inner surfaces) and triangular ceramic prisms inserted into the panel. Polyurethane plugged the remaining gaps in the sandwich core. This panel will be referred to as the hybrid ceramic-filled lattice panel, Fig. 3(e).

We now proceed to summarize the manufacture route for each of these configurations. The measured areal density of each sandwich panel structure is given in Table I.

(1) Construction of the Unfilled Lattice Core Sandwich Plates

Sandwich panels with a pyramidal truss core were manufactured from 304 stainless steel of density, $\rho = 8000$ k/gm^3. The sandwich panels comprised two identical face-sheets of thickness $h = 1.5$ mm and a pyramidal core of thickness $c = 25.4$ mm; see Fig. 4 for de-

tailed dimensions of the sandwich plates. The pyramidal cores had a relative density (ratio of the effective density of the "smeared-out" core to the density of the solid material from which it is made), $\bar{\rho} = 2.6\%$ which implies that the areal mass $m = (2h + c\bar{\rho})\rho$ of the 304 stainless-steel sandwich plates was 29.3 k/gm^2.

The pyramidal lattice cores comprised struts of length 31.75 mm and cross section 1.9 mm × 1.9 mm as shown in Fig. 4(a). The cores were manufactured from 1.9-mm-thick 304 stainless-steel sheets by first punching rhomboidal holes to obtain a perforated sheet, and then folding this sheet node row by node row to obtain regular pyramids as shown in Fig. 5. The sandwich plates were then assembled by laser welding rectangular sheets of dimensions 120.7 mm × 127 mm × 1.5 mm to pyramidal core truss panels comprising 3 × 3 cells (Fig. 4).

(2) Construction of the Polymer-Filled Plates

A low glass transition temperature (T_g) polyurethane elastomer, identical to that described in an earlier study[26] was chosen to infiltrate the pyramidal lattice truss structure. The polyurethane, PMC-780 Dry, was formulated by Smooth-On (Easton, PA) and is a two component, pliable, castable elastomer with an approximate 24 h cure time at room temperature. Part A was composed of the polyurethane prepolymer and a trace amount of toluene diisocyanate while part B was composed of polyol, a proprietary chemical (NJ Trade Secret #221290880-5020P), di(methylthio)toluene diamine, and phenylmercuric neodecanoate. The manufacturers data sheet, Table II, indicates that the polymer has a low elastic modulus (2.8 MPa) and tensile strength (6.2 MPa), but a very high elongation strain to failure of 700%. The assembled pyramidal lattice truss structure was masked on three of the four sides and the polyurethane was poured into the structure and allowed to cure for 48 h.

The hybrid polymer-filled panels had a Kevlar fabric reinforcing the face-sheets. This fabric (made from Kevlar® 29 fiber and designated Kevlar-Flex-Pro) was supplied by Automotive Armor Manufacturing Inc., Miami, FL. Flex-Pro is a laminated composite fabricated from a vulcanized urethane 3000 denier aramid fiber composite. The fabric sheets were 4.8-mm-thick and had an areal mass of approximately 4.88 kg/m^2. The

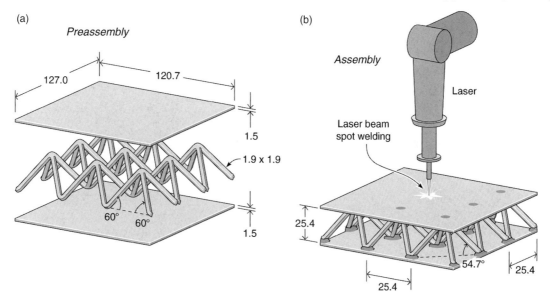

Fig. 4. Illustration of the laser welding process (b) for bonding the pyramidal truss lattice to sandwich plate face-sheets (a). Details of the dimensions of the sandwich plate and the core are included in (a). All dimensions are in mm.

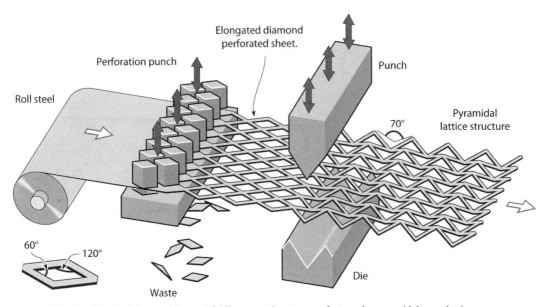

Fig. 5. Sketch of the punching and folding operation to manufacture the pyramidal truss lattice core.

Table II. Manufacturer Reported Properties for the Polyurethane (PU 1)

Property	
Manufacturer	Smooth-On
Product name	PMC-780 dry
Tensile nodulus (MPa)	2.76
Tensile strength (MPa)	6.21
Elongation to break (%)	700
Shore hardness	80 A

Kevlar-Flex-Pro was laser cut into two different designs to accommodate the arrangement of nodal contact points on the front and back face sheets (Figs. 6(a) and (b), respectively). The cut fabric sheets were placed on the inner surfaces of the face sheets before laser welding of the truss cores. Subsequently, the panel was filled with the polyurethane as described above.

(3) Construction of the Ceramic-Filled Panels

Alumina (Al_2O_3) manufactured by CoorsTek (Golden, CO) with the grade designation AD-94 was used to fill the empty truss core sandwich panels. The alumina was diamond saw cut into isosceles triangular cross-section prisms with an apex angle of 70°, base angle of 55°, and base length 3.0 cm. The prisms were 11.4 cm long. The manufacturers' mechanical property data for AD-94 is listed in Table III. After insertion, the remaining small gaps in the sandwich core were filled with the polyurethane described in Section III(2).

The process for manufacturing the hybrid ceramic-filled lattice panels involved first including the Flex-Pro fabric in the sandwich panel as described in Section III(2). Subsequently, ceramic prisms were inserted into the panel and the gaps filled with polyurethane. The main difference here is that the presence of the Flex-Pro meant that the ceramic prisms were necessarily slightly smaller and had a base length of 2.4 cm, a height of 1.8 cm, an apex angle of 70°, and base angles of 55°.

(4) Mechanical Properties of the Steel and Polyurethane

Tensile specimens of dog-bone geometry were cut from each of the as-received steel sheets. The uniaxial tensile responses of the 304 stainless-steel alloy at an applied strain-rate of 10^{-3} s^{-1} is plotted in Fig. 7(a) using axes of true stress and logarithmic strain. The 304 stainless steel displays a linear hardening post-yield response with a tangent modulus $E_t \approx 1$ GPa. This high strain hardening capacity of the stainless steel also stabilizes the tensile specimens against necking, resulting a very high tensile ductility. The polyurethane used here is well above its glass transition at room temperature ($T_g = -56°C$). The compressive stress–strain curves for this material are shown in Fig. 7(b) for a

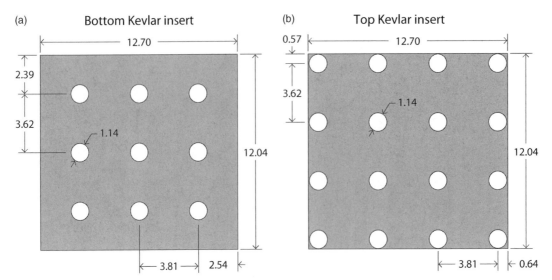

Fig. 6. Sketch of the Kevlar fabric cut to accommodate the nodes of the pyramidal core on the (a) front and (b) back face sheets.

Table III. Manufacturers Data for the Physical Properties of the AD-94 Al_2O_3 Triangular Prisms

Density (g/cm³)	Elastic modulus (GPa)	Flexural strength (MPa)	Tensile strength (MPa)	Compressive strength (MPa)	Fracture toughness (MPa·m^{1/2})	Hardness (GPa)
3.97	303	358	221	2068	4–5	11.5

range of strain rates. These were measured by McShane *et al.*[26] using a split Hopkinson bar set-up. The tangent modulus of the polyurethane increases with strain rate, which is advantageous for the ballistic performance of this polymer.

IV. Dynamic Test Protocol

The ballistic performance of the above-described panel systems was investigated for projectile impact velocities in the range $225 \leq v_p \leq 1700$ m/s. In addition, for comparison purposes, we also investigated the ballistic performance of monolithic 304 stainless-steel plates of thickness 3 and 10 mm with areal masses $m = 24$ and 80 kg/m, respectively. These plates span the areal masses of the sandwich panel systems and provide a useful basis to compare and rank the performance of the different concepts.

Impact experiments were performed on 25 sandwich plates of each configuration. In all cases, the plates were impacted at zero obliquity and midspan by a spherical, 1020 plain carbon steel projectile of diameter 12.5 mm weighing approximately 8.4 g. Ballistic testing was conducted using a powder gun comprising a breech and a gun barrel as sketched in Fig. 8. The gun propelled plastic sabots carrying the 12.5 mm spherical steel projectiles. An electric solenoid activated a firing pin, which initiated 0.38 caliber blank cartridges (Western Cartridge Company, East Alton, IL). The mixture of solid smokeless propellant IMR 3031, manufactured by IMR (Shawnee Mission, KS), and cotton (Fig. 8) in the breech was ignited by this charge and the expanding propellant gas accelerated the sabot through the gun barrel. The purpose of the cotton was to ensure the ensuing pressure wave remained uniform throughout deflagration of the propellant. The sabot was located within a 25.4 mm bore gun barrel: a series of holes placed along the gun barrel were used to dissipate the shock wave and maintain a smooth acceleration of the sabot until it exited the barrel. The velocity at which the sabot exited the gun barrel was adjusted by selecting an appropriate quantity of gunpowder. The plastic sabot comprised four quarters that, upon mating, surrounded the 12.5 mm diameter spherical projectile. The sabot plugs had an inner diameter of 1.8 cm, an outer diameter of 2.7 cm, a height of 3.5 cm, and weighed 18.6 ± 0.12 g. Separation of the sabot from the projectile by air drag was facilitated by a 40° bevel at the sabot opening. The 12.5 mm diameter spherical steel projectiles weighed $M = 8.42 \pm 0.02$ g and were manufactured by National Precision

Ball (Preston, WA) from 1020 plain carbon steel with an ultimate tensile strength of approximately 375 MPa.[6]

The sample test fixture was located within a blast chamber (Fig. 9). A square, 40 cm long, 2.86-cm-thick steel plate was located one meter from the end of the barrel. It had a 3.8 cm diameter hole located in the center through which the projectile entered the test area. Two pairs of brake screens were used to measure the projectile entry and exit velocities (Fig. 9) and provided impact and exit velocity measurements with a precision of ± 2.0 m/s. The test samples were edge clamped along the top and bottom edges so that the effective span of the plate between the clamped edges was approximately 110 mm while the width of the plate was 120 mm; see Fig. 4. The kinetic energy of the projectile dissipated by the sandwich panel structures depends upon the way in which the projectile interacts with the truss cores. To remove this source of variability in the measurements, the sandwich panels were carefully positioned so that the projectile impacted at the center of the square formed by four nodes of the truss on the face-sheet facing the incoming projectile. The projectile usually impacted the back face sheet at the apex of a truss. High-speed photography was used to observe the dynamic transverse deformation and failure of the plates. An Imacon 200 digital framing camera (Intronix Imaging Technologies, Westlake Village, CA) was used for this purpose; this camera is capable of taking up to 16 frames at a maximum rate of 10^8 frames/s. Interframe times of in the range 4.5–50 μs were used and the exposure time was 300 ns. In addition, the plates were examined after each experiment to understand the failure mechanisms.

V. Experimental Observations

We proceed to detail the observations in two steps. First we discuss the ballistic performance of the polymer-filled lattices and then contrast this behavior with that of the ceramic-filled sandwich plates. In order to facilitate comparisons between the different systems, the areal masses of the five sandwich systems (including the unfilled pyramidal core sandwich from Yungwirth *et al.*[6]) and two monolithic plates investigated are listed in Table I.

(1) Polymer-Filled Lattice Plates

The measured projectile exit or residual velocity v_r as a function of the impact velocity v_p is plotted in Fig. 10(a). Full penetration

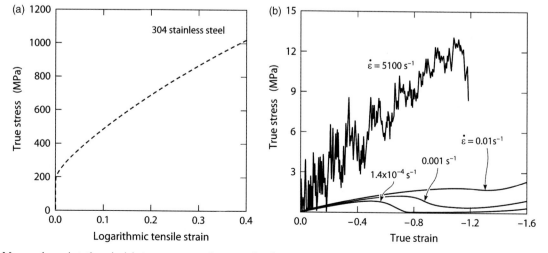

Fig. 7. (a) Measured quasi-static uniaxial stress versus strain curves for the as-received 304 stainless steel and (b) the room temperature compressive response of the polyurethane over a range of strain rates.

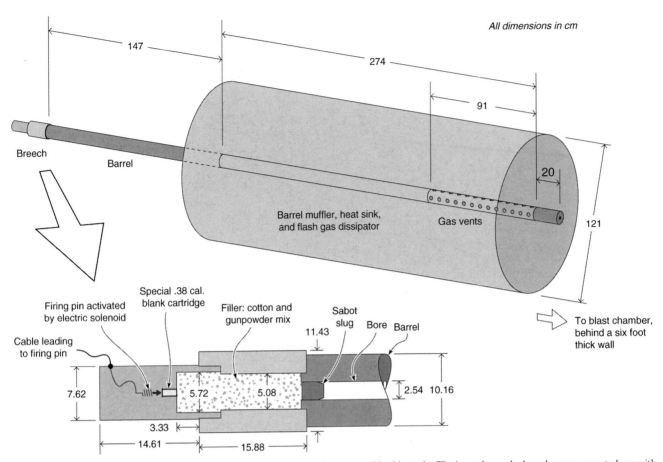

Fig. 8. Sketch showing the principal components of the single-stage powder gun used in this study. The inset shows the breech arrangement along with the initial section of the gun barrel with the sabot slug that carries the 12.5 mm diameter spherical steel projectile.

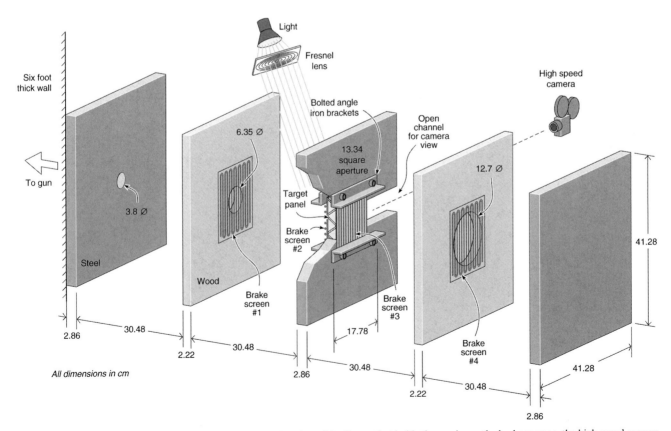

Fig. 9. Schematic illustration of the test set-up including the location of the fixture that holds the specimen, the brake screens, the high-speed camera, the light source, and the projectile catching arrangement.

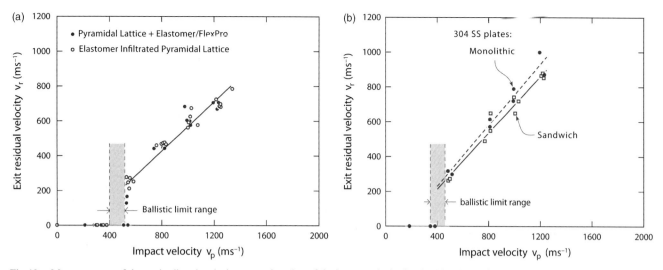

Fig. 10. Measurements of the projectile exit velocity v_r as a function of the impact velocity for the (a) polymer-filled sandwich plates and (b) the 3-mm-thick monolithic steel plate and empty pyramidal truss sandwich plate.

of the panels occurred at impact velocities in the range $v_p = v_{crit} \approx 400$–500 m/s. This is defined as the ballistic limit of these panels for the projectile used in this investigation. Consistent with observations and predictions in Yungwirth et al.,[6] a sharp increase in the projectile residual velocity is observed just above the ballistic limit. The corresponding measurements from Yungwirth et al.[6] of a 3 mm steel plate and the empty pyramidal truss core sandwich plate are included in Fig. 10(b). These plates are also made from 304 stainless steel and impacted with the same projectile and at zero obliquity. Comparing Figs. 10(a) and (b) it is clear that the ballistic limit of the pyramidal core sandwich plates is largely unchanged when the pyramidal core is filled with the PU polymer. This suggests that in terms of the ballistic limit, the monolithic plate is most weight efficient

followed by the empty lattice and then the polymer-filled panel (see areal masses in Table I). It is worth noting here that while adding the elastomer to the sandwich core does not increase the ballistic limit, it seems to reduce the residual velocities of the penetrated projectiles compared with the monolithic and empty sandwich plate structures.

The reasons for the disappointing performance of the polymer-filled sandwich panels can be understood by examining the photographs of the tested specimens included in Fig. 11. These photographs were taken by sectioning the tested specimens along their midplane. It can be seen that the projectile suffered very little deformation during these impact events. As a result, little of its incident kinetic energy was self-absorbed in plastic work and the mechanism of arrest was instead governed by the

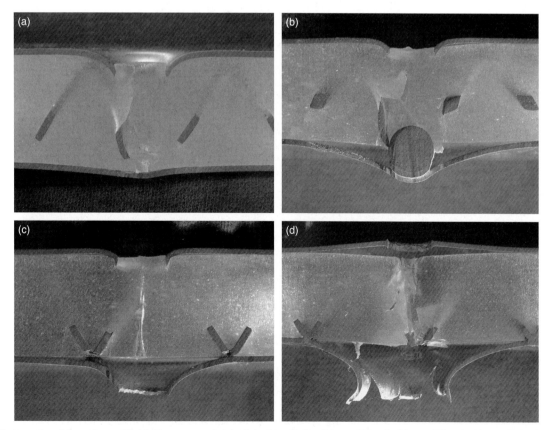

Fig. 11. Photographs of the polymer-filled lattice core sandwich plates impacted at (a) $v_p = 299$ m/s, (b) $v_p = 371$ m/s, (c) $v_p = 516$ m/s, and (d) $v_p = 985$ m/s. The photographs were taken after sectioning the plates along their midplane. Note the front face sheet "plug" attached to the projectile in (b).

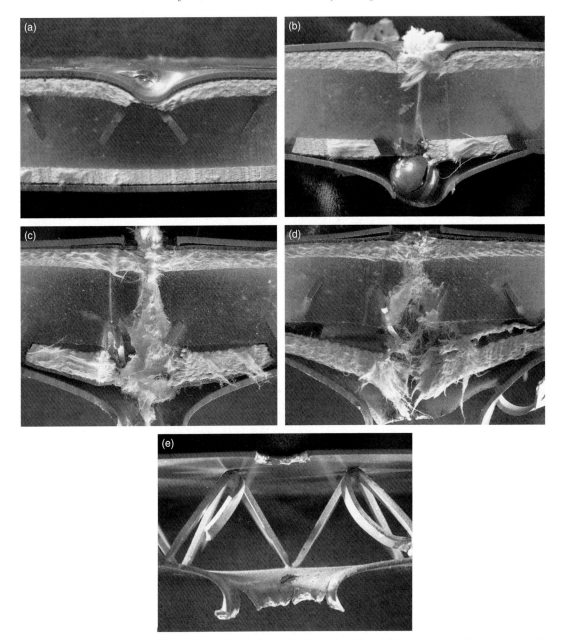

Fig. 12. Photographs of the hybrid polymer-filled lattice core sandwich plates impacted at (a) $v_p = 274$ m/s, (b) $v_p = 499$ m/s, (c) $v_p = 731$ m/s, and (d) $v_p = 1181$ m/s. For comparison purposes, the photograph of the unfilled sandwich panel impacted at $v_p = 1206$ m/s is included in (e). The photographs were taken after sectioning the plates along their midplane.

target response. At impact velocities $v_p = 299$ and 371 m/s (Figs. 11(a) and (b), respectively), the projectile penetrated the front face but was arrested by the rear face of the sandwich panels. In these figures we clearly see (i) a shear-off failure of the front face with nearly no spreading of the deformation; (ii) the tensile failure of the nodes connecting the pyramidal core to the rear face near the impact site; (iii) the stretching and bending of the rear face; and (iv) rehealing of the polymer after penetration. This clearly illustrates that the front face and to some extent the polymer-filled core slow the projectile sufficiently so that the mode of deformation/failure of the rear face switches from shear-off on the front face to stretching on the rear face and this engages a significantly larger fraction of the face sheet material. At the higher impact velocities of $v_p = 516$ and 985 m/s (Figs. 11(c) and (d), respectively), the projectile penetrates the rear face sheet, by a petaling mechanism associated with face sheet stretching. Note that, in all cases, penetration through the polymer core seems to occur by a cavity expansion mechanism but the very high elastic recovery strain of the polymer means that the cavity closes behind the penetrating projectile leaving behind an incipient flaw as seen in Fig. 11(c).

The addition of the FlexPro Kevlar fabric into the sandwich core (i.e., the hybrid polymer-filled lattice plates) does not significantly affect the ballistic limit (Fig. 10(a)) but again seems to reduce the residual velocities of the penetrated projectiles compared with the polymer-filled lattice plates. Given that the areal masses of both these systems is approximately the same (Table I), we can conclude that the FlexPro fabric does serve in increasing the energy absorption capacity of the panel (as the deformation of the projectiles is negligible, the kinetic energy of the incoming projectile is primarily dissipated by the deformation of the panel). A sequence of images of the tested and sectioned panels is included in Fig. 12: these images clearly show that the Kevlar fabric does not change the shear-off mode on the front face but undergoes some stretching on the rear face compared with the polymer-filled lattices shown in Fig. 11. This stretching is the likely reason why the addition of the fabric seems to further reduce the residual velocities of the projectile. The presence of a pattern of circular holes in the fabrics (Fig. 6) may also have contributed to the poor performance because this resulted in a fraction of the Kevlar tows having an effective length of only 30 mm. It is worth emphasizing here that placing the Kevlar

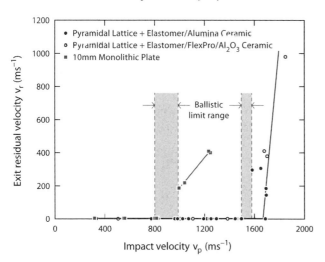

Fig. 13. Measurements of the projectile exit velocity v_r as a function of the impact velocity for the ceramic-filled sandwich plates and hybrid ceramic-filled plates. The corresponding data for the 10-mm-thick steel plate is also included.

fabric within the polymer-filled core highly constrained the transverse deformation of the fabric, preventing it from fully stretching and acting in a "catcher mode." This is the primary reason for the rather disappointing performance of these hybrid panels—placing the Kevlar behind the back face of the sandwich panel is expected to give improved performance.

These polymer-filled lattice plates show minimal performance benefits over their monolithic counterparts. We argue that this is due to two main reasons: (i) the polymer filling is sufficiently soft that it does not plastically deform or erode the steel projectile, and (ii) the cavity expansion deformation mode of the polymer does not efficiently distribute the load to the panel faces so the mode of deformation of the polymer-filled sandwich plate remains unchanged from the empty sandwich plate; compare Figs. 12(d) and (e).

(2) Ceramic-Filled Lattice Plates

The polymer-filled lattice plates show minimal performance benefits over their monolithic counterparts for the reasons outlined above. Filling the sandwich plates with ceramics is expected to address both of these drawbacks.

The measured projectile exit or residual velocity v_r as a function of the impact velocity v_p is plotted in Fig. 13 for both the types of ceramic-filled sandwich panels. The corresponding data for a 10-mm-thick monolithic 304 stainless-steel plate (with areal mass approximately equal to the ceramic-filled lattice panel; see Table I) is included in the figure. The ballistic limit is seen to increase from about 900 m/s for the monolithic plate to about 1600 m/s for the ceramic-filled lattice panels with an equal areal mass.

This increase in the ballistic limit can be rationalized by observing the deformation/failure modes of the sectioned (as-tested) panels. Figures 14 and 15 include photographs of the as-tested 10-mm-thick monolithic steel plate and ceramic-filled lattice panels specimens, respectively. First consider the monolithic plates. At an impact velocity below the ballistic limit ($v_p = 540$ m/s), the projectile is seen to undergo extensive deformation with a small indent forming on the plate (Fig. 14(a)). At just about the ballistic limit ($v_p = 804$ m/s), the projectile again undergoes extensive deformation, but in addition the plate is seen to fail/deform in a plugging type mode wherein a plug is pushed out of the plate directly ahead of the projectile with little deformation away from the impact site (Fig. 14(b)). At higher impact velocities, the projectile penetrates by pushing through a shear plug. These shear plugs leave behind plastic lips at the impact and distal surfaces of the plate.

These deformation modes of the monolithic plate can be contrasted with the corresponding observations for the ceramic-filled lattice plates shown in Fig. 15. At low impact velocities

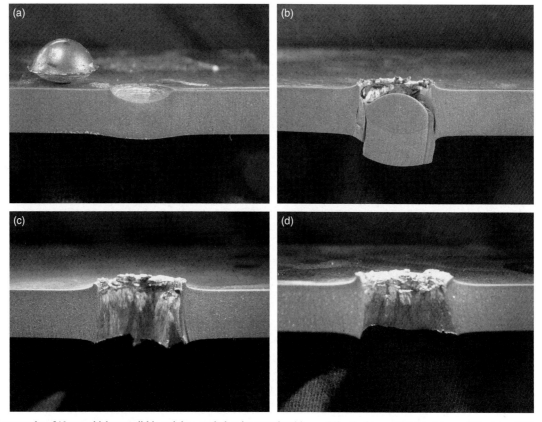

Fig. 14. Photographs of 10-mm-thick monolithic stainless-steel plate impacted at (a) $v_p = 542$ m/s, (b) $v_p = 804$ m/s, (c) $v_p = 979$ m/s, and (d) $v_p = 1218$ m/s. The photographs were taken after sectioning the plates along their midplane. In (b) the heavily deformed projectile and still attached "plug" can be seen.

Fig. 15. Photographs of the ceramic-filled lattice core sandwich plates impacted at (a) $v_p = 326$ m/s, (b) $v_p = 543$ m/s, (c) $v_p = 1060$ m/s, and (d) $v_p = 1233$ m/s. The photographs were taken after sectioning the plates along their midplane.

($v_p = 326$ and 543 m/s), the impact face sheet deforms by a shear plugging mode with cracking observed in the ceramic prism directly underneath. No other significant residual deformation is observed. This is rationalized as follows: the impacting projectile easily forms a shear plug in the steel face sheet but then undergoes extensive plastic deformation as it encounters the hard ceramic. This ceramic prism is constrained by the surrounding prisms, the steel face sheet and the incompressible elastomer. Recall that Deshpande and Evans[17] have shown that confinement significantly enhances the fracture strength of the ceramic and thus confined ceramic prisms here undergoes limited cracking. This limited cracking combined with the high elastic wave speed in the ceramic enables the impacted ceramic prism to very effectively redistribute the load exerted by the projectile over a significant fraction of the back face sheet and thus we observe little or no deformation of the back face sheet of the panel. At higher impact velocities (Figs. 15(c) and (d)), the deformation is more widespread with more extensive cracking of the ceramic.

This cracking results in dilation (or bulking) of the ceramic, which is resisted by the stretching of the lattice trusses until they fail at their nodes and by the stretching of the steel face sheets of the sandwich panel thereafter.

Like the polymer-filled panels, the addition of the Kevlar FlexPro fabric to the panels does not change the ballistic limit significantly (Fig. 13). This is because the FlexPro does not seem to significantly change the deformation mode of the panels compared with the panels without the Kevlar; contrast Figs. 15 and 16. The Kevlar fabric once again was sufficiently confined within the sandwich that stretching of the fabric (and thus energy dissipation) was inhibited. This resulted in the fabric failing by a shear-off (cutting) mechanism rather than the efficient energy absorbing stretching mode.

In order to gain further insight into these deformation mechanisms a high-speed photographic sequence of the deformation of the hybrid ceramic-filled lattice panel impacted at $v_p = 789$ m/s is included in Fig. 17. The timing of each photograph is marked

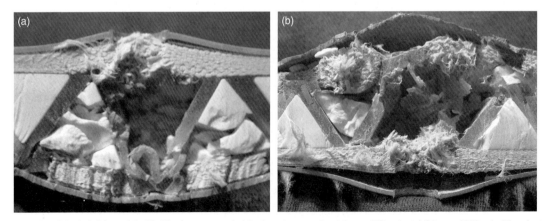

Fig. 16. Photographs of the hybrid ceramic-filled lattice core sandwich plates impacted at (a) $v_p = 790$ m/s and (b) $v_p = 1243$ m/s. The photographs were taken after sectioning the plates along their midplane.

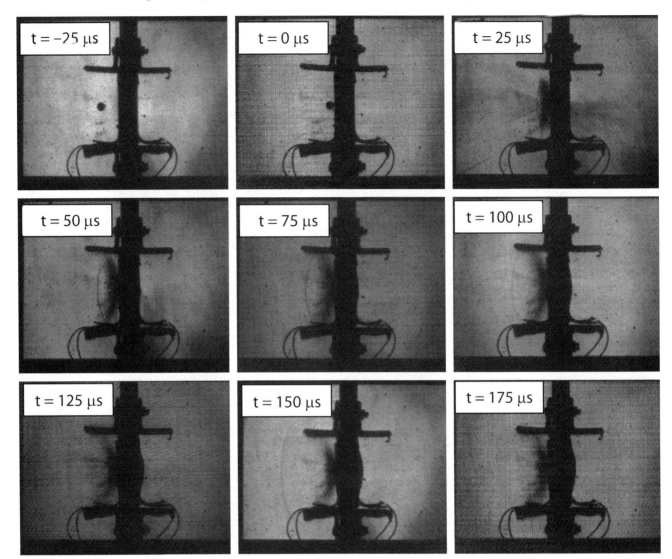

Fig. 17. High-speed photographic sequence (exposure time of 300 ns) of the impact of the hybrid ceramic-filled sandwich plate at $v_p = 789$ m/s. Time after the impact of the projectile against the proximal face-sheet is indicated for each frame.

on the figures with $t = 0$ taken as the instant of the impact against the proximal or front face-sheet. Consistent with the basic supersonic flow theory,[27] at $v_p = 789$ m/s (i.e., Mach number $M_\infty \equiv v_p/c \approx 2.25$, where $c = 340$ m/s is the velocity of sound in the undisturbed air), a detached shock wave is formed in front of sphere with a shock wave angle[||] greater than the Mach angle $\omega = \sin^{-1}(c/v_p)$. Penetration of the front face by the projectile results in a debris cloud being ejected backwards. The projectile then enters the sandwich plate and a significant volumetric expansion of the sandwich plate is observed: consistent with the images in Fig. 16, the front face undergoes minimal deformation while the rear face is seen to undergo significant stretching over nearly its entire span. This deformation mode results in the capture of the projectile within the target.

VI.　Discussion

The ballistic limit v_c of all the panels investigated is summarized in Fig. 18 as a function of the panel areal masses (sometimes referred to as the specific density m_s). We emphasize that these results are valid for only the spherical, 8 g 1020 plain carbon steel projectile used in this study. Figure 18 clearly illustrates that the monolithic, empty and polymer-filled sandwich panels have an equivalent performance, i.e. the ballistic limit of these panels increases approximately linearly with areal mass at the

rate of 11 m/s for an increase in areal mass of 1 kg/m. We thus conclude that there is no performance benefit in terms of the ballistic limit in utilizing either empty or polymer-filled sandwich panels. This is rationalized by noting that the hydrodynamic stresses generated within the polymer (i.e., inertial stresses of order ρv^2, where ρ is the polymer density and v the impact velocity) are significantly larger than the material strength (Fig. 7) for all velocities near and above the ballistic limit. This implies that the only resistance to penetration provided by the polymer is due to its mass, i.e. no performance gains on a mass basis. The incorporation of layers of a Kevlar fabric within the elastomer-filled lattice had surprisingly little effect upon the ballistic resistance of the panel (Fig. 10) because of its confinement by the elastomer and by the panel's rear face sheet.

By contrast, the ceramic-filled sandwich panels have a significantly higher ballistic resistance compared with their monolithic counterparts and lie well above the $11 \text{ m} \cdot (\text{s} \cdot \text{kg} \cdot \text{m}^2)^{-1}$ line that defines the performance of the monolithic, empty and polymer-filled sandwich plates. In fact the ballistic limit v_c of the ceramic-filled sandwich plates is about twice that of their monolithic counterparts. These performance benefits arise because the ceramic inserts completely change the penetration modes of the sandwich plates by switching-off the plugging deformation mode as discussed above.

This study has shown that the ballistic performance of sandwich structures can be significantly enhanced by using ceramic inserts to fill the empty spaces within the truss core sandwich structures. The ceramic used here is an armor grade alumina

---

[||]The shock wave angle is the angle that the shock wave makes with the direction of motion of the projectile.

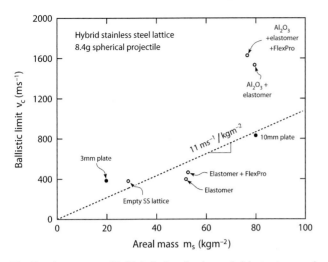

Fig. 18. A summary of ballistic limit v_c for the sandwich structures and solid 304 stainless-steel plates that span the areal density range of the hybrid sandwich panels.

ceramic. We note in passing that using higher performance ceramics such as SiC could further enhance the performances reported here. For example C. J. Yungwirth, and H. N. G. Wadley (private communication) have conducted a study wherein the alumina prisms utilized here in the hybrid ceramic-filled sandwich panels were replaced by SiC prisms. The ballistic limit was approximately equal to the equivalent alumina-filled panel. However, the density of SiC was about 18% less than that of alumina and thus in terms of mass efficiency the SiC panels had a superior ballistic performance. We note that the ballistic performance of ceramics could be significantly enhanced (especially for brittle projectiles) by increasing the ceramic confinement.[28,29] Espinosa *et al.*[30] have shown that this inhibits comminution and penetration through the cracked ceramic.[31] While the focus of this study has explored ways of improving the ballistic performance of sandwich panels by inserting materials *within* the structure, significant improvements are also possible by placing the ceramic on the outside of strike face (H.N.G. Wadley *et al.*) or attaching the Kevlar fabric to the rear face of the structure (both at the expense of a reduced volumetric efficiency). We also note that the study has not attempted to optimize the mass distribution amongst the various components of the system. Using the data and dimensionless parameter approach of Cuniff,[32] it can be shown that a monolithic panel of Kevlar 29 composite with an areal density of 80 kg/m^2 has a ballistic limit several hundred meters per second higher than that of the ceramic-filled system. Allocating a larger fraction of the panels mass to a Kevlar-based textile attached to the rear of the structure is therefore anticipated to result in significant performance improvements.

In this study we have demonstrated that filling empty spaces within sandwich panels with ceramic inserts enhances their bal-

listic performance. It remains to be determined what effect these fillings have on the performance of sandwich panels in terms of their blast resistance. If core compressibility is an essential requirement for the superior blast performance of sandwich panels, multi-layer concepts (e.g., see Fig. 19) that combine a compressible core with a hard core might be required in order to endow the structures with this multifunctional capability. This is a topic for future investigations.

VII. Concluding Remarks

The ballistic performance of stainless-steel pyramidal core sandwich panels with the empty spaces filled with polyurethane or alumina prisms was investigated. The measurements show that addition of the polyurethane does not enhance the ballistic limit compared with the equivalent monolithic steel plate, but it reduces the residual velocity of the penetrated projectiles. This poor performance of the polymer-filled lattice is due to the fact that the polymer is penetrated by a hole enlargement mechanism which does not spread the load and engage the steel sandwich casing adequately, i.e. the failure mode of the polymer-filled sandwich panel remains unchanged from that of the empty sandwich panel. By contrast, the ceramic inserts comminute and dilate while eroding the projectile. This results in the stretching of the steel face sheets and significant energy dissipation. The ballistic limit of this structure is about twice that of the equivalent monolithic steel plate. The addition of a Kevlar fabric within the systems is shown to not significantly change the ballistic limit but help reduce the residual velocities of the penetrated projectile and panel debris. This is attributed to the fact that the Kevlar fabric is significantly constrained within the sandwich and is unable to flex and stretch and thus fails by its weak shear-off (cutting) mechanism.

Acknowledgments

The views, opinions, and/or findings contained in this article are those of the authors and should not be interpreted as representing the official views or policies, either expressed or implied, of the Defense Advanced Research Projects Agency or the Department of Defense. Distribution Statement "A" (Approved for Public Release, Distribution Unlimited).

References

[1]V. S. Deshpande and N. A. Fleck, "One-Dimensional Response of Sandwich Plates to underwater Shock Loading," *J. Mech. Phys. Solids*, **53** [11] 2347–83 (2005).
[2]J. W. Hutchinson and Z. Xue, "Metal Sandwich Plates Optimized for Pressure Impulses," *Int. J. Mech. Sci.*, **47** [4–5] 545–69 (2005).
[3]L. F. Mori, S. Lee, Z. Y. Xue, A. Vaziri, D. T. Queheillalt, K. P. Dharmasena, H. N. G. Wadley, J. W. Hutchinson, and H. D. Espinosa, "Deformation and Fracture Modes of Sandwich Structures Subjected to underwater Impulsive Loads," *J. Mech. Mater. Struct.*, **2** [10] 1981–2006 (2007).
[4]H. N. G. Wadley, K. P. Dharmasena, D. T. Queheillalt, Y. Chen, P. Dudt, D. Knight, K. Kiddy, Z. Xue, and A. Vaziri, "Dynamic Compression of Square Honeycomb Structures during Underwater Impulsive Loading," *J. Mech. Mater. Struct.*, **2** [10] 2025–48 (2007).
[5]K. P. Dharmasena, D. T. Queheillalt, H. N. G. Wadley, P. Dudt, Y. Chen, D. Knight, A. G. Evans, and V. S. Deshpande, "Dynamic Compression of Metallic Sandwich Structures during Planar Impulsive Loading in Water," *Eur. J. Mech.-A/Solids*, **29** [1] 56–67 (2010).
[6]C. J. Yungwirth, H. N. G. Wadley, J. O'Connor, A. Zakraysek, and V. S. Deshpande, "Impact Response of Sandwich Plates with a Pyramidal Lattice Core," *Int. J. Impact Eng.*, **35** [8] 920–36 (2008).
[7]M. L. Wilkins, C. F. Cline, and C. A. Honodel, "Fourth Progress Report of Light Armour Program"; Report UCRL 50694, Lawrence Radiation Laboratory, University of California, 1969.
[8]M. L. Wilkins, "Mechanics of Penetration and Perforation," *Int. J. Eng. Sci.*, **16** [11] 793–807 (1978).
[9]M. Mayseless, W. Goldsmith, S. P. Virostek, and S. A. Finnegan, "Impact on Ceramic Targets," *J. Appl. Mech.*, **54** [2] 373–8 (1987).
[10]B. A. Gama, T. A. Bogetti, B. K. Fink, C. Yu, T. D. Claar, H. H. Eifert, and J. W. Jr. Gillespie, "Aluminum Foam Integral Armor: A New Dimension in Armor Design," *Comput. Struct.*, **52** [3–4] 381–95 (2001).
[11]C. Navarro, M. A. Martinez, R. Cortes, and V. Sanchez-Galvez, "Some Observations on the Normal Impact on Ceramic Faced Armours Backed by Composite Plates," *Int. J. Impact Eng.*, **13** [1] 145–56 (1993).
[12]I. S. C. Benloulo and V. Sanchez-Galvez, "A New Analytical Model to Simulate Impact onto Ceramic/Composite Armours," *Int. J. Impact Eng.*, **21** [6] 461–71 (1998).

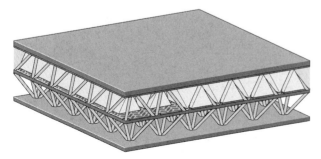

Fig. 19. Sketch of a multilayer sandwich structure comprising a sandwich core with ceramic inserts as the top layer and the empty sandwich panel as the bottom layer. This structure is expected to have multifunctional capabilities in terms of both blast and ballistic performances.

[13]M. J. Normandia, J. C. LaSalvia, W. A. Gooch, and J. W. McCauley, "Protecting the Future Force: Ceramics Research Leads to Improved Armor Performance," *Amptiac Q.*, **8** [4] 21–7 (2004).

[14]P. M. Cunniff, "Dimensionless Parameters for Optimization of Textile Based Body Armor Systems"; pp. 1303–10 *Proceedings of the 18th International Symposium on Ballistics*, San Antonio, TX, 1999.

[15]H. Senf, E. Strassburger, and H. Rothenhausler, "Investigation of Bulging during Impact in Composite Armor," *J. Phys. IV Fr.*, **7** [C3] 301–6 (1997).

[16]S. Sarva, S. Nemat-Nasser, J. McGee, and J. Isaacs, "The Effect of Thin Membrane Restraint on the Ballistic Performance of Armor Grade Ceramic Tiles," *Int. J. Impact Eng.*, **34** [2] 277–302 (2007).

[17]V. S. Deshpande and A. G. Evans, "Inelastic Deformation and Energy Dissipation in Ceramics: A Mechanism-Based Dynamic Constitutive Model," *J. Mech. Phys. Solids*, **56**, 3077–100 (2008).

[18]H. G. Hopkins and H. Kolsky, "Mechanics of Hypervelocity Impact of Solids"; *Proceedings of the 4th Hypervelocity Impact Symposium APGC-TR-60-39*, Eglin Air Force Base 1, 1960.

[19]W. Goldsmith, *Impact*. Arnold, New York, 1960.

[20]M. Backman and W. Goldsmith, "The Mechanics of Penetration of Projectiles into Targets," *Int. J. Eng. Sci.*, **16**, 1–99 (1978).

[21]G. Corbett, S. Reid, and W. Johnson, "Impact Loading of Plates and Shells by Free-Flying Projectiles: A Review," *Int. J. Impact Eng.*, **18**, 131–230 (1996).

[22]F. C. Frank and B. R. Lawn, "On the Theory of Hertzian Fracture," *Proc. R. Soc.*, **A299**, 291–306 (1967).

[23]G. Corbett, S. Reid, and W. Johnson, "Impact Loading of Plates and Shells by Free-Flying Projectiles: A Review," *Int. J. Impact Eng.*, **18**, 131–230 (1996).

[24]M. Backman and W. Goldsmith, "The Mechanics of Penetration of Projectiles into Targets," *Int. J. Eng. Sci.*, **16**, 1–99 (1978).

[25]B. A. Cheeseman and T. A. Bogetti, "Ballistic Impact into Fabric and Compliant Composite Laminates," *Compos. Struct.*, **61** [1–2] 161–73 (2003).

[26]G. J. McShane, C. Stewart, M. T. Aronson, H. N. G. Wadley, N. A. Fleck, and V. S. Deshpande, "Dynamic Rupture of Polymer–Metal Bilayer Plates," *Int. J. Solids Struct.*, **45** [16] 4407–26 (2008).

[27]H. W. Liepmann and A. Roshko, *Elements of Gas-Dynamics*. Dover Publications Inc., New York, 1985.

[28]G. Hauver, P. Netherwood, R. Benck, and L. Kecskes, "Ballistic Performance of Ceramic Targets"; *Proceedings of the Army Symposium on Solid Mechanics*, Plymouth, MA, 1993.

[29]F. Malaise, J. Y. Tranchet, and F. Collombet, "An Experimental Investigation of Ceramic Block Impenetrability to High Velocity Long Rod Impact"; *Proceeding of the 6th International Conference on Mechanical and Physical Behaviour of Materials under Dynamic Loading*, Cracow, Poland, 2000.

[30]H. D. Espinosa, N. S. Brar, G. Yuan, Y. Xu, and V. Arrieta, "Enhanced Ballistic Performance of Confined Multi-Layered Ceramic Targets Against Long Rod Penetrators through Interface Defeat," *Int. J. Solids Struct.*, **37** [36] 4893–914 (2000).

[31]B. A. Gailly and H. D. Espinosa, "Modeling of Failure Mode Transition in Ballistic Penetration with a Continuum Model Describing Microcracking and Flow of Pulverized Media," *Int. J. Numerical Methods Eng.*, **54** [3] 365–98 (2002).

[32]P. M. Cuniff, "Dimensionless Parameters for Optimization of Textile-Based Body Armor Systems"; *18th International Symposium on Ballistics*, San Antonio, TX, November 15–19, 1999. □

J. Am. Ceram. Soc., **94** [S1] S76–S84 (2011)
DOI: 10.1111/j.1551-2916.2011.04601.x
© 2011 The American Ceramic Society

journal

In-Plane Compression Response of Extruded Aluminum 6061-T6 Corrugated Core Sandwich Columns

Russell Biagi, Jae Yong Lim, and Hilary Bart-Smith[†]

Mechanical and Aerospace Engineering Department, University of Virginia, Charlottesville, Virginia, 22904, U.S.A.

The compression response of extruded aluminum 6061-T6 corrugated core sandwich columns is investigated. Analytical equations that predict the collapse load are used to generate failure mechanism maps. From these maps dominant failure mechanisms can be identified as a function of various geometric parameters and material properties. Experimental testing and numerical simulations are performed to test the fidelity of the analytical predictions. Fabrication of the sandwich panels involves extrusion of an aluminum billet through a specially designed die. To create longer columns the extruded panels are joined together using friction stir welding (FSW). Studies of the thermo-mechanically affected zone (TMAZ) show that the hardness within these regions drops by approximately 50%. This significantly influences the observed failure load and failure mechanism and hence heat treatment post welding is required to ensure uniformity of properties. Good agreement is achieved between the predictions and experiment. Lastly, optimal designs are calculated based on the analytical analysis and results compared with hat-stiffened and truss core sandwich columns.

I. Introduction

RECENT studies of the column response of truss core sandwich columns demonstrate that they can compete with current industry standard hat-stiffened panels.[1] With advances in fabrication methods, there are many alternatives for the lightweight core structure, which necessitates the need to characterize the panel response. In this spirit, the following study attempts to characterize the column response of aluminum alloy 6061-T6 corrugated core sandwich columns compressed perpendicular to the corrugations (Fig. 1). This study employs a new fabrication process that has recently been developed to create sandwich panels.[2] This process uses extrusion in conjunction with friction stir welding to create aluminum alloy 6061-T6 corrugated core sandwich structures.

The properties of the core play an integral role in the overall behavior of the sandwich structure, governing the capacity of the whole structure (i.e., sandwich beam bending). For this reason, much of the recent research has focused on the compressive and shear response of various core topologies and of sandwich beams in bending (for example).[3–6] Moreover, several of the core options present opportunities for additional functionality, such as thermal transport and energy absorption.[7–12] However, the in-plane compressive response has received limited attention in the literature. As metallic core sandwich structures are being considered for various new applications, such as energy absorbent ship hulls, in-plane compression is a loading scenario that will inevitably be encountered and must be addressed. This may be the exclusive loading direction, or may occur in conjunction with out-of-plane loading. In either case, understanding the in-plane loading response is essential in order for sandwich structures to achieve widespread use in next generation multifunctional applications. The present work focuses on corrugated core sandwich structures loaded perpendicular to the corrugations. This orientation allows for greater exploitation of multifunctional capabilities. To date, there has not been a comprehensive study of the competing failure modes of a corrugated core sandwich column under this loading direction. This work attempts to address this deficiency.

The outline of the article is as follows. Section II describes the manufacturing process to create the corrugated core sandwich columns and testing done to characterize the base material. Section III presents analytical equations that predict the collapse load for competing failure modes as a function of geometric parameters and material properties. Mechanism maps—where non-dimensional geometrical parameters represent the axes—are generated that highlight the dominant failure mode. Section IV presents finite element simulations to compare with the analytical predictions. Using the mechanism maps presented in Section III, columns are designed to fail by predetermined mechanisms and the results from experiments are compared with the predictions. These results are presented in Section V. Section VI presents minimum mass designs for the corrugated core columns and comparisons are made with hat-stiffened and pyramidal core designs. Implications and conclusions are presented in Section VII.

II. Manufacturing

Aluminum columns were created using an extrusion process, producing the core and faceplates in a single continuous piece. The manufacturing procedure will be presented along with the experimental protocol for parent material characterization and sandwich column compression.

Aluminum alloy 6061 is used to manufacture extruded corrugated core columns in the same fashion described by Queheillalt *et al.*[2] This alloy was chosen based on a variety of considerations including its availability, economy compared to other alloys, ability to be extruded into high tolerance corrugated structures, and it can also be easily machined and heat-treated. A 17.8 cm diameter, 300 ton direct extrusion press was used to force the aluminum billet through a regular prismatic structure die at 482°C. This process creates corrugated structures that are 3.05 m long, parallel to the corrugations. The die was designed by Cellular Materials International Inc. (Charlottesville, VA) and the extrusion process was performed by Mideast Aluminum Inc. (Mountaintop, PA). The extruded panels were then solutionized, water quenched and heat-treated to the T6 condition. A cross section of an extrusion is shown in Fig. 2.

The relative density for this extruded core structure is given as

F. Zok—contributing editor

Manuscript No. 28960. Received November 22, 2010; approved April 02, 2011.
[†]Author to whom correspondence should be addressed. e-mail: hb8h@virginia.edu

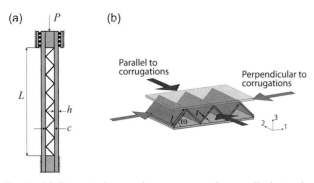

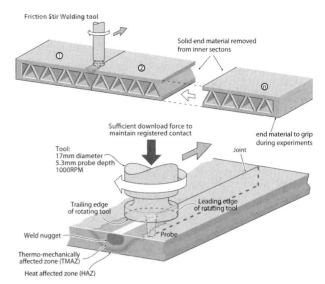

Fig. 1. (a) Corrugated core column compressed perpendicular to the corrugations with fixed end conditions. (b) Illustration of the in-plane loading direction for a corrugated core sandwich structure.

Fig. 3. Illustration of the friction stir weld process. Adapted from figure in[2].

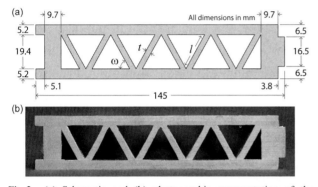

Fig. 2. (a) Schematic and (b) photographic representation of the cross section of an extruded aluminum corrugated sandwich structure.

$$\bar{\rho} = \frac{2}{\sin(2\omega) + (2t/l)} \frac{t}{l} \qquad (1)$$

where l is the ligament length, t is the ligament thickness and ω is the core angle.[2] The measured relative density $\bar{\rho} = 26.2\% \pm 0.1\%$ and core angle $\omega = 61.4° \pm 0.1°$ were very close to the target geometry of $\bar{\rho} = 25\%$ and $\omega = 60°$. The geometry of the core, and therefore $\bar{\rho}$, was constrained by the dimensions of the die available. The specific dimensions were chosen to produce consistent extrusions with high geometric tolerances.

The extrusions were initially 3.05 m long parallel to the corrugations, but only four cells wide. The in-plane column compression experiments carried out in this study load the structures perpendicular to the corrugations (see Fig. 1). To create columns of varying length, 203 mm long prepared sections were joined together using a friction stir welding process (Friction Stir Link Inc., Waukesha, WI). The span length of the designed column determines the total number of sections to be joined. Those sections placed in the interior of the column length were modified by removing the solid vertical material on both outer edges, whereas the first and last sections retain the outer vertical material to enable clamping during the compression experiments (Fig. 3). This resulted in a continuous corrugation along the length of the panel. The face sheets of each whole panel were then machined to a desired thickness, h. Three, 2-inch wide columns were water jet cut from each panel.

Friction stir welding is a process where two metals are joined through frictional heat and mechanical deformation induced by a rotating tool traveling along the line of contact between the pieces.[13–15] The tool heats the metal, but does not melt it, and mechanically mixes the softened material through severe plastic deformation, creating a weld by solid-state plastic flow at an elevated temperature.[14,15] A 17-mm

diameter welding tool rotating at 1000 rpm with a traverse speed of 400 mm/min is used to join the Al 6061-T6 corrugated extrusion segments. This process creates two zones that can affect the material properties of the original alloy. The area of material that undergoes plastic deformation from the rotating tool is referred to as the thermomechanically affected zone (TMAZ).[13] The area adjacent to this region, known as the heat affected zone (HAZ), experiences elevated temperatures that can affect the heat treatment of the material.[15] Friction stir welding is an attractive joining technique because it avoids many of the difficulties and defects associated with traditional liquid-phase welding such as porosity in the weld region, filler metal composition, and compatibility and environmental and safety issues.[13,15] This process is depicted in Fig. 3.

As stated above, the material properties can be affected by the friction stir welding process. It was found that the material properties within the weld region were significantly degraded from those of the bulk material, which can influence the column response. To correct this, the columns were subsequently re-tempered to the T6 condition to ensure uniform material properties throughout the corrugated columns. Comments on the manufacturing process will be discussed in Section VII.

Following ASTM standard E8[16], tension tests were carried out on Al 6061-T6 specimens machined directly from the face of a corrugated extrusion. The Young's modulus, E, and 0.1% offset yield stress, σ_y, were found from the stress–strain response (Fig. 4). The specimen was oriented such that the tensile loading direction was parallel to the direction of extrusion. It is anticipated that there will be a slight difference between the response of the material in the direction of the extrusion and the response of the material perpendicular to the extrusion direction. However, it is assumed that this discrepancy will be minimized as a result of the two heat treatments performed subsequent to the extrusion process. The aluminum 6061-T6 shows very little strain hardening after yield, behaving similarly to an elastic-perfectly plastic solid with Young's modulus $E = 75$ GPa and yield stress $\sigma_y = 290$ MPa.

To quantify the effect of friction stir welding on the joint material, hardness tests were performed in the bulk material and across a weld joint, perpendicular to the weld direction. Measurements were taken both before and after the re-temper stage of the manufacturing process and will be referred to as pre-temper hardness and post-temper hardness herein. Rockwell B hardness tests were carried out in accordance with ASTM standard E18[17] using a Wilson Instrument

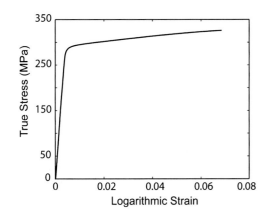

Fig. 4. Uniaxial tensile stress versus strain response for Al 6061-T6.

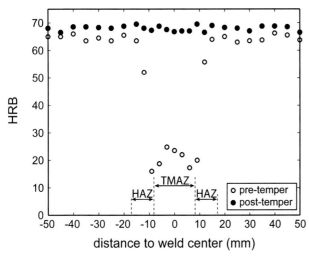

Fig. 5. Hardness distribution across a friction stir weld joint.

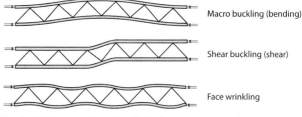

Fig. 6. Failure modes identified for a corrugated core column compressed perpendicular to the corrugations.

ior of sandwich structures subject to in-plane compression is dependent on the overall panel geometry, core topology, and parent material properties. Columns with fixed end conditions are loaded perpendicular to the corrugations (Fig. 1). The corrugated column of span, L, consists of two face sheets of thickness, h, and core with ligament length, l, thickness, t, and core angle, ω (Fig. 1). The core and face sheets are composed of the same isotropic material. The failure mechanisms for a corrugated core column compressed perpendicular to the corrugations, composed of a strain hardening material, have been identified as: (i) macro elastic buckling, (ii) macro plastic buckling, (iii) elastic face wrinkling, (iv) plastic face wrinkling, and (v) elastic core shear failure. These can be seen in Fig. 6. Following the analysis of Côté *et al.*[1] analytical predictions governing the sandwich column failure are presented.

(1) Macro Buckling

Euler buckling, characterized by bending of the structure, and core shear failure are two modes of macroscopic failure that have been identified. Rather than being separate mechanisms, the bending and shear buckling modes are coupled. This can readily be seen in a simplified formula where, assuming thin face sheets with negligible bending stiffness about their own axis, the critical sandwich buckling load is given as[18–20]

$$\frac{1}{P_{cr}} = \frac{1}{P_E} + \frac{1}{P_S}, \tag{2}$$

where P_E and P_S are the Euler buckling load and shear buckling load, respectively. Equation 2 matches Engesser's expression for the critical buckling load of an elastic column taking into account shear deformation[21,22] but is valid only for sandwich structures with thin face sheets. The general formula for the critical macro elastic bucking load of a sandwich column with thick faces is given by Allen[18] and Zenkert[20] to be

$$P_{cr} = \frac{(2k^4\pi^4 D_f D_0/L^4) + (k^2\pi^2 D/L^2)}{(k^2\pi^2 D_0/L^2) + S} \tag{3}$$

with the flexural rigidities for a corrugated core sandwich defined as

$$D_0 = \frac{Ebh(h + l\sin(\omega))^2}{2(1 - v^2)}$$

$$D_f = \frac{Ebh^3}{12(1 - v^2)} \tag{4}$$

$$D = 2D_f + D_0$$

where E is the Young's modulus and v is the Poisson's ratio of the material and the geometric properties are as defined above. Fixed column end conditions are assumed with buckling coefficient $k = 2$. The contribution of the core to

Rockwell hardness tester (Model OUR-a; Milford, CT) with a 1.588-mm (1/16 in) diameter ball indenter and 980N (100 kgf) major test load. A separation of 5 mm was held between sampling points. The pre-temper and post-temper hardness of the bulk aluminum material is reported as the average of 21 measurements, taken in a region far from the weld. The average pre-temper hardness and post-temper hardness were calculated to be 63 ± 0.7 HRB and 67 ± 0.5 HRB, where uncertainties are reported as the standard error of the mean with 95% confidence. Figure 5 shows the measured hardness distribution across the weld. The width of the TMAZ marked is equal to the diameter of the welding tool, $d_t = 17$ mm, that plastically deformed the material. It is clear that the material properties in this region are degraded from the welding process. The material transitions back to the bulk properties through the HAZ, which extends $\sim d_t/2$ from the TMAZ edge. This profile is consistent with the typical hardness of a metallic friction stir weld in which the weld center is slightly harder than the weld edges due to a recrystallized fine-grained microstructure referred to as the weld nugget.[13,15] The post-temper measurements show consistent hardness across the weld region, indicating uniform material properties over the length of the corrugated column.

III. Analytical Predictions

Failure modes are identified for a corrugated core column in compression and analytical predictions are presented. The predictions are used to construct failure mechanism maps that are a function of column geometry, from which specific experimental geometries were chosen. The mechanical behav-

the overall flexural rigidity has been neglected. When loaded in-plane, perpendicular to the corrugations, the deformation of the core is bending dominated and has a negligible contribution to the overall flexural rigidity of the sandwich structure. Note that plane strain conditions are assumed and a factor of $(1-v^2)$ is introduced to the expressions.

The shear rigidity, S, of the sandwich column is assumed to be set by the shear rigidity of the corrugated core, G_{31}, only[1], neglecting the shear stiffness of the faces. Following the analyses of Deshpande and Fleck[4] and Côté *et al.*[23], the shear rigidity in terms of the core geometry is given as

$$S = G_{31} bl \sin(\omega) = \frac{Ebt \sin(\omega) \sin^2(2\omega)}{2(1 - v^2)(\sin(2\omega) + 2(t/l))}. \qquad (5)$$

The preceding equations for macro buckling are valid if the material remains elastic, or if

$$P_{cr} < \frac{4bh\sigma_y}{\sqrt{3}} \qquad (6)$$

where $2\sigma_y/\sqrt{3}$ is the material yield stress assuming plane strain conditions. If this criterion is not satisfied, macro buckling is governed by the plastic response of the material. The macro plastic buckling load is calculated using Eq. 3, with the flexural rigidities modified using the Shanley tangent modulus assumption.[24] The elastic plane-strain modulus, $E/(1-v^2)$, is replaced with the tangent modulus, E_t, where $E_t \equiv d\sigma_s/d\varepsilon_s$ is the tangent modulus of the plane strain true tensile stress versus logarithmic strain curve of the parent material, evaluated at $\sigma_s = P_{cr}/2bh$. In practice, an iterative process is used to obtain the macro plastic buckling failure load. Equation 3 is evaluated at various E_t along the material data curve until $P_{cr}/2bh = \sigma_s$. It is assumed that small transverse deflections take place at the bifurcation load and the core remains elastic with shear rigidity given by Eq. 5. If the material is composed of an elastic-perfectly plastic material, the column is unable to support additional load if the elastic limit is reached and the appropriate failure is face yielding. Face yielding occurs when the stress in the face sheets equals the plane strain yield stress of the material:

$$P_{FY} = \frac{4bh\sigma_y}{\sqrt{3}}. \qquad (7)$$

(2) Face Wrinkling
The face wrinkling response is a short wavelength instability, set by buckling failure of the face between points of attachment to the core. Dictated by the core geometry, the wrinkling span length is $2l \cos(\omega)$. For a column with face sheets of thickness h, the elastic wrinkling load is

$$P_{FW} = \frac{k_1^2 \pi^2 bE}{24 \cos^2(\omega)(1 - v^2)} \frac{h^3}{l^2} \quad \text{if} \quad \frac{h}{l} < \sqrt{\frac{96 \cos^2(\omega)(1 - v^2)\sigma_y}{\sqrt{3} k_1^2 \pi^2 E}} \qquad (8)$$

If the inequality in Eq. 9 is not met, the stress in the faces exceed the elastic limit of the material and the plastic face wrinkling load is given by

$$P_{FW} = \frac{k_1^2 \pi^2 bE_t}{24 \cos^2(\omega)} \frac{h^3}{l^2} \quad \text{if} \quad \frac{h}{l} \geq \sqrt{\frac{96 \cos^2(\omega)(1 - v^2)\sigma_y}{\sqrt{3} k_1^2 \pi^2 E}} \qquad (9)$$

where, as above, E_t is the tangent modulus of the plane strain true stress versus true strain curve of the material evaluated at $\sigma_s = P_{FW}/2bh$. An iterative process is again used to obtain the plastic face wrinkling failure load with Eq. 9

evaluated at various points along the material data curve until the wrinkling stress equals the material stress in the faces; $P_{FW}/2bh = \sigma_s$. Due to the geometry of the core/node interface of the extruded column, the end constraint is chosen to be built-in, with $k_1 = 2$. If the parent material is elastic-perfectly plastic and cannot support additional load beyond yield, the column fails by face yielding given by Eq. 7.

(3) Failure Mechanism Maps
Using the above analytical expressions, failure mechanism maps for the in-plane compression of aluminum 6061-T6 corrugated core columns were constructed. These give a visual representation of failure, making it easier to identify the operative collapse mode (i.e., the mode associated with the lowest critical load P) for a given column geometry. These maps are developed as a function of non-dimensional geometric parameters, h/l and L/l, for a sandwich structure with constant core ligament slenderness ratio, t/l, core angle, ω and given material response. The active failure mode boundaries are generated by evaluating the minimum normalized collapse load $\bar{P} \equiv P/(\sigma_y bl \sin(\omega))$ over a range of h/l and L/l. The failure mechanism map for the fabricated core construct is presented in Fig. 7. Lines of constant normalized critical failure load, $\bar{P}$, and normalized mass, $\bar{M} = M/\rho bl^2$, are indicated on the maps. Although not shown in this map, it is possible to construct columns whose geometries ($\bar{\rho} < \sim 4.6\%$) allow core shear to become an active mechanism in the buckling regime.[25] Experimental geometries were selected to probe the failure regimes along lines of constant force and are compared to analytical predictions and finite element simulations. The measured column geometries tested are indicated on the map. Column geometries that lie in the macro elastic buckling region could not be tested due to specimen length limitations in existing experimental facilities.

(4) Transverse Loading
It is instructive to consider the response of the panel loaded in the transverse direction. For the corrugated core column, this loading scenario is parallel to the corrugations (Fig. 1). For an elastic-perfectly plastic material, the failure modes identified are: (i) macro elastic buckling, (ii) local elastic face buckling, (iii) local elastic core buckling, and (iv) material yielding. The macro elastic buckling failure of a corrugated core column of length, L, and width, b, compressed parallel to the corrugations is taken to be described by the critical

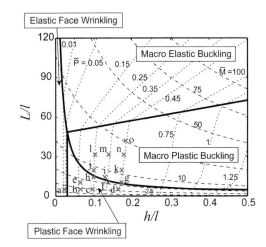

Fig. 7. Failure mode map for the measured Al 6061-T6 extruded corrugated core geometry with $t/l = 0.1488$ and $\omega = 61.4°$. The experimentally tested column geometries are marked on map.

sandwich buckling load given in Eq. 3. When loaded in this orientation, the core deformation is stretch dominated and the core rigidity, D_c, cannot be neglected in the overall flexural rigidity of the sandwich structure as it was when loaded perpendicular to the corrugations. The core flexural rigidity is expressed as

$$D_c = \frac{Ebtl^2\sin^2(\omega)}{12(l\cos(\omega) + t/\sin(\omega))} \qquad (10)$$

and the total flexural rigidity is now

$$D = 2D_f + D_0 + D_c \qquad (11)$$

with D_0 and D_f given in Eq. 4.

Following Côté *et al.*,[23] the shear rigidity is obtained as

$$S = \frac{Ebt\sin^3(\omega)}{(1 + v)(\sin(2\omega) + (2t/l))}. \qquad (12)$$

Local face buckling failure is the short wavelength plate buckling that can occur along the length of the column, between points of attachment of the core and faceplates. The width of the plate is set by the core geometry to be $2l\cos(\omega)$ Following the analysis of Budiansky[26] for the local plate failure in a hat-stiffened panel, the local elastic failure load can be calculated. The critical elastic face buckling load is

$$P_{FB} = \frac{K\pi^2 Eh^2 A}{48(1 - v^2)l^2\cos^2(\omega)} \ \text{if} \ \frac{h}{l} < \sqrt{\frac{48(1 - v^2)\cos^2(\omega)\sigma_y}{K\pi^2 E}}, \qquad (13)$$

where K is the buckling coefficient set by the boundary conditions and the aspect ratio $L/(2l\cos(\omega))$. It is assumed that $L >> 2l\cos(\omega)$ and the face buckles in a pattern of half-sine waves of length equal to plate width $2l\cos(\omega)$ along L.[22,26] Therefore, simply supported conditions are estimated on the loaded edges.[27] Due to the geometry of the extruded construction, the unloaded edges are assumed to be fixed giving $K = 7$.

Similarly, the local elastic core failure is short wavelength buckling along the length of a core web with width, l. The critical local elastic core load is expressed as

$$P_{FB} = \frac{K\pi^2 Et^2 A}{12(1 - v^2)l^2} \ \text{if} \ \frac{t}{l} < \sqrt{\frac{12(1 - v^2)\sigma_y}{K\pi^2 E}} \qquad (14)$$

as above, it is assumed that the unloaded edges are fixed and the loaded edges are simply supported, giving $K = 7$.

The elastic limit is set by the plane strain material yield stress as

$$P_{cr} < \frac{2}{\sqrt{3}}\sigma_y A \qquad (15)$$

where the cross-sectional area is

$$A = \frac{bt}{\cos(\omega) + t/l\sin(\omega)} + 2bh \qquad (16)$$

If the critical load exceeds the elastic limit, the critical plastic failure load is

$$P_y = \frac{2}{\sqrt{3}}\sigma_y A \qquad (17)$$

IV.　Numerical Simulations

Numerical simulations were performed for the in-plane loading of corrugated core sandwich columns, compressed perpendicular to the corrugations. Abaqus-Standard finite element software (Simulia of Dessault Systemes, Providence, RI) was used to for all simulations. Four node, first order continuum plane-strain elements with reduced integration (Abaqus CPE4R) were used with the measured material properties from the uniaxial tensile test results in Fig. 4. These elements prevent shear and volumetric locking that can occur with fully integrated elements in bending.[28] Loading was simulated using a general static analysis with a displacement boundary condition applied to one end of the column. The nodes at the column ends were fixed from all other displacements and all rotations to simulate built-in end conditions. No initial imperfection was introduced into the column. Lack of symmetry about the column centroid was enough to cause a preferential buckling direction. Therefore, the simulations are an upper bound on the column failure, to be compared with the analytical predictions, which are also derived under the assumption of "perfect" column geometry.

V.　Comparison with Experimental Results

Extruded aluminum corrugated core columns, designed to probe the macro plastic buckling and face wrinkling regimes, were fabricated and tested. Column designs a–d (Fig. 7) were constructed from a single extrusion, in which no friction stir welding was necessary. All other geometries required friction stir welding to produce columns of necessary length, L, as described in Section II. Three specimens were tested for each geometry, except point o with two viable samples available for testing. The measured results, compared to the analytical and finite element predictions, are summarized in Table I. Slight variations in column geometry existed due to the fabrication process: specifically with respect to face sheet thickness. As such, the analytical and numerical predictions are calculated using the average measured face thickness of each column.

(1) Plastic Face Wrinkling
Various column geometries were manufactured to probe the plastic face wrinkling regime (b–f, h; Fig. 7). The measured compressive response for a specimen from sample geometry c ($h = 2.12$ mm, $L = 117$ mm) is given in Fig. 8. After initial loading there is a nonlinear material response to a peak load of 65.46 kN, followed by post-buckling softening. Face wrinkling occurs concurrently in both faces as seen in Fig. 8(b). Both the analytical and numerical peak load are within 7% of the measured value, showing very good agreement. Recall geometry c did not require friction stir welding. Figure 9 presents the plastic face wrinkling loading response for a specimen with geometry f, a geometry requiring friction stir welding, which exhibits a similar response to sample c. It should be noted that the face wrinkling failure did not occur at the weld locations.

(2) Macro Plastic Buckling
Macro plastic buckling is the predicted failure mechanism for eight sample geometries manufactured and tested (samples g, i–o). The measured response of a specimen from sample point o ($h = 4.08$ mm, $L = 934$ mm) is given in Fig. 10. The load increases linearly with displacement followed by an increasing nonlinear response up to a peak load of 88.75 kN. After the peak load, post-buckling softening occurs. Photographs of the deformation history show global buckling failure in a shape consistent with a column having clamped end conditions (Fig. 10(b)). The analytical and numerical predictions for

Table I. Summary of the Observed and Predicted Failure Loads and Failure Mechanisms for the Column Geometries Tested. For All Columns, the Predicted Failure Load and Mechanism are Reported Using the Average Measured Face Thickness, h_{avg}. For the Columns that Exhibit Face Wrinkling Failure, the Face Wrinkling Load Using the Minimum Measured Face Thickness, h_{min}, is also reported

Pt	L/l	h/l	$P_{predicted}$ h_{avg}	$FW(h_{min})$	$P_{simulation}$	$P_{measured}$	Predicted	Simulation	Observed
	5.3	0.023	0.157	0.112	0.140	0.090	EFW	EFW	EFW
	5.3	0.024	0.186	0.154	0.165	0.092	EFW	EFW	EFW
a	5.3	0.023	0.165	0.136	0.148	0.102	EFW	EFW	EFW
	5.3	0.065	0.912	0.885	0.897	0.759	PFW	PFW	PFW
	5.3	0.065	0.905	0.879	0.89	0.743	PFW	PFW	PFW
b	5.3	0.064	0.892	0.865	0.869	0.768	PFW	PFW	PFW
	5.3	0.096	1.370	1.353	1.377	1.27	PFW	PFW	PFW
	5.3	0.096	1.372	1.360	1.377	1.29	PFW	PFW	PFW
c	5.3	0.096	1.367	1.360	1.369	1.24	PFW	PFW	PFW
	5.3	0.163	2.402	2.395	2.467	2.27	PFW	PFW	MPB/PFW
	5.3	0.164	2.423	2.381	2.467	2.41	PFW	PFW	MPB/PFW
d	5.3	0.165	2.430	2.353	2.474	2.42	PFW	PFW	MPB/PFW
	10.5	0.064	0.886	0.8449	0.869	0.648	PFW	PFW	PFW
	10.5	0.064	0.886	0.8047	0.869	0.688	PFW	PFW	PFW
e	10.5	0.064	0.892	0.8115	0.876	0.635	PFW	PFW	PFW
	10.6	0.136	1.982	1.877	2.000	1.910	PFW	PFW	PFW
	10.5	0.137	2.005	1.919	2.021	1.860	PFW	PFW	PFW
f	10.5	0.137	1.997	1.919	2.015	1.882	PFW	PFW	PFW
	10.5	0.174	2.556	2.515	2.603	2.482	MPB	MPB	MPB/PFW
	10.6	0.173	2.553	2.500	2.589	2.441	MPB	MPB	MPB/PFW
g	10.6	0.172	2.527	2.437	2.568	2.388	MPB	MPB	MPB/PFW
	14.5	0.098	1.407	1.326	1.406	1.129	PFW	PFW	PFW
	14.5	0.099	1.416	1.312	1.413	1.138	PFW	PFW	PFW
h	14.5	0.097	1.385	1.251	1.385	1.113	PFW	PFW	PFW
	14.5	0.128	1.841	1.732	1.865	1.660	MPB	MPB/PFW	PFW
	14.5	0.127	1.834	1.704	1.851	1.640	MPB	MPB/PFW	PFW
i	14.5	0.128	1.847	1.746	1.865	1.621	MPB	MPB/PFW	PFW
	19.8	0.098	1.387	1.101	1.393	1.119	MPB	MPB/PFW	PFW
	19.8	0.099	1.401	1.142	1.407	1.117	MPB	MPB/PFW	PFW
j	19.8	0.098	1.393	1.128	1.393	1.118	MPB	MPB/PFW	PFW
	19.8	0.173	2.457	2.479	2.488	1.765	MPB	MPB	PFW @ weld
	19.8	0.174	2.469	2.486	2.508	1.754	MPB	MPB	PFW @ weld
k[†]	19.8	0.173	2.461	2.463	2.495	1.743	MPB	MPB	PFW @ weld
	31.6	0.101	1.381	1.251	1.384	1.050	MPB	MPB	PFW
	31.6	0.102	1.399	1.230	1.402	1.082	MPB	MPB	PFW
l	31.6	0.102	1.400	1.203	1.402	1.041	MPB	MPB	MPB/PFW
	31.5	0.138	1.888	1.725	1.894	1.594	MPB	MPB	PFW
	31.6	0.139	1.901	1.711	1.900	1.570	MPB	MPB	MPB/PFW
m	31.6	0.138	1.881	1.704	1.888	1.568	MPB	MPB	MPB/PFW
	31.6	0.179	2.456		2.456	2.261	MPB	MPB	MPB
	31.6	0.179	2.466		2.463	2.230	MPB	MPB	MPB
n	31.6	0.179	2.469		2.469	2.246	MPB	MPB	MPB
	42.1	0.185	2.484		2.452	2.043	MPB	MPB	MPB
o	42.1	0.184	2.467		2.434	1.969	MPB	MPB	MPB

[†]Specimens did not undergo re-tempering process after friction stir welding.

peak load agree within 1.5%. The measured failure load is 20% lower than predicted.

VI. Minimum Weight Design

Corrugated core sandwich structures are optimized such that the geometric parameters minimize the panel mass for a given critical load. Comparisons are made with optimized hat-stiffened panels, which are considered an industry standard for axial load bearing applications and widely used. The corrugated sandwich panels are also compared with pyramidal core sandwich structures, which are a competing sandwich design. Non-dimensional parameters are used in the optimization allowing various designs to be compared.

The non-dimensional loading index is defined as $\bar{P}^* \equiv P_{cr}/\sigma_y bL$ and the non-dimensional panel mass index is given as $\bar{M}^* \equiv M/\rho bL^2$. The panel mass $\bar{M}^*$ is minimized for a given $\bar{P}^*$ subject to the possible critical failure loads as constraints. For all panels, the minimum mass is found numerically using an extensive search method in which the non-dimensional column geometries are varied over the design space with the minimum mass configuration being calculated at each loading index.

The optimization carried out assumes an elastic-perfectly plastic material with the modulus and yield stress of Al 6061-T6, defined from the tension test described earlier. This is a reasonable assumption as the material response of Al 6061-T6 demonstrates very little strain hardening after yield (Fig. 4).

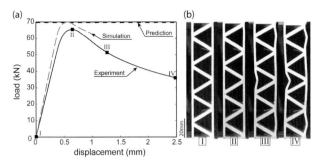

Fig. 8. (a) The measured loading response for a specimen from sample geometry *c* with the finite element and analytical predictions included. (b) Photographs illustrating the deformation history at points marked in (a).

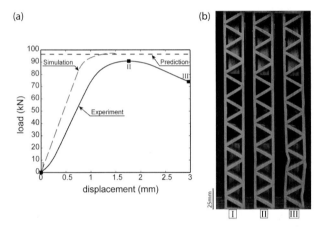

Fig. 9. (a) The measured loading response for a specimen from sample geometry *f* with the finite element and analytical predictions included. (b) Photographs illustrating the deformation history at points marked in (a).

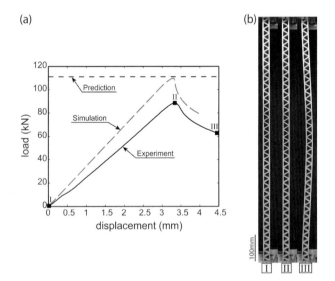

Fig. 10. (a) The measured loading response for a specimen from sample geometry *o* with the finite element and analytical predictions included. (b) Photographs illustrating the deformation history at points marked in (a).

The extruded corrugated core column mass is expressed as

$$M = 2bL\rho\left(h + \frac{t}{2\cos(\omega) + (2t/l\sin(\omega))}\right). \quad (18)$$

The overall length, *L*, of the load-bearing column is typically a design constraint and is not varied in this optimiza-

tion. In addition, the core angle, ω, is held constant ($\omega = 60°$) so as to correspond to the fabricated extrusions. For a corrugated core column compressed perpendicular to the corrugations, the non-dimensional geometric variables, h/L, t/L, and l/L, are sought (with ω held constant) that minimize the mass index $\bar{M}^*$ at a given load $\bar{P}^*$, subject the critical failure load constraints given in Section III for an elastic-perfectly plastic material.

Optimization of a hat-stiffened panel follows the analysis set forth by Budiansky[26] for a simplified geometry that has square stiffeners of length *w* spaced at regular intervals, *w*. Compressed in the transverse direction, perpendicular to the stiffeners, this panel behaves as a monolithic plate. The mass optimization for an extruded pyramidal core column follows the analysis of Côté *et al.*,[1] with only a slight variation to the core relative density due to the extruded geometry (see[2]). The pyramidal core is constrained to a square base, and thus displays transverse isotropy. Therefore, the in-plane loading response of column is the same in the two orthogonal directions. The optimized loading results are presented in Fig. 11. Figure 12 illustrates the corresponding non-dimensional geometric parameters for a corrugated core column loaded perpendicular to the corrugations and a pyramidal core column.

VII. Discussion

(1) Column Compression

The results presented clearly demonstrate that the analytical predictions accurately predict the column failure for both face wrinkling and macro buckling. The peak load and failure mechanism agreement between the predictions and experiments, particularly along the predicted failure boundary, confirm that fixed face wrinkling boundary conditions are an appropriate assumption. Also, for all geometries tested, the analytical predictions are in excellent agreement with the finite element simulations. The discrepancies seen in the failure response highlight the importance of both the precision and sequence of the fabrication process used to construct extruded sandwich panels. It is essential that all stages of the fabrication be rigorously monitored to produce reliable panels, thus allowing analytical and numerical tools to be implemented. Variations in column geometry can cause degradation in failure load and initiate unanticipated failure mechanisms. The friction stir welding process is an attractive and capable manufacturing method to join sandwich panel sections. However, the panels must be manufactured in a way that minimizes imperfections. For example, large panels should be tempered immediately following the friction stir welding, prior to any machining, where all panel geometries (e.g., face thickness and panel width) are greatest, thereby reducing the likelihood of warping. Then the faces should be machined to the desired thickness and columns extracted. It should also be highlighted that, as expected, precise fabrication is vital to the performance of the sandwich panel. The extrusion-friction stir welding process can produce corrugated panels with very few defects if closely monitored. However, small variations in panel geometry can have significant influence on performance.

Looking more closely at some of the results there are some samples designed in the MPB region that exhibit a failure mechanism discrepancy (Table I). Face wrinkling was observed to be the active failure mode or occur in conjunction with global buckling. This is reasonable for the samples with geometries that lie close to failure boundaries (geometries *g*, *i*, and *j*), where the plastic face wrinkling and macro plastic buckling failure loads are approximately equal. Slight imperfections (geometric or experimental) can activate plastic face wrinkling mechanism prior to macro buckling. However, the observed face wrinkling response is not reasonable for samples designed to lie well within the macro plastic buckling region, away from failure boundaries (such as geometries *l*

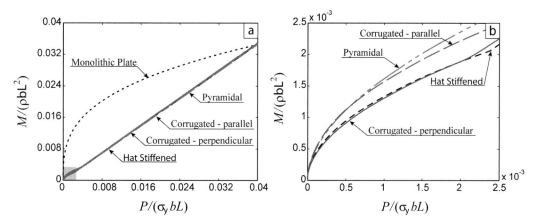

Fig. 11. (a) Failure load comparison for minimum mass optimized corrugated core sandwich columns ($\omega=60°$), pyramidal core sandwich columns ($\omega=60°$) and hat-stiffened panels. The sandwich panels are fabricated using an extrusion process. The corrugated core sandwich columns are optimized for loading perpendicular to the corrugations and parallel to the corrugations. The hat-stiffened panels are optimized for axial loading and transverse loading. All panels are composed of elastic-perfectly plastic material with $\sigma_y=290$ MPa and $v=0.33$.

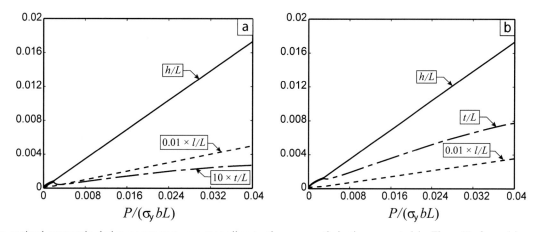

Fig. 12. The optimal geometric design parameters, corresponding to the mass optimization presented in Figure 11, for a (a) corrugated core sandwich column compressed perpendicular to the corrugations and (b) pyramidal core sandwich column, made of an elastic-perfectly plastic material.

and m). Examining these columns further shows flaws in the manufacturing process that result in slight variations in the face sheet geometry. The cause of these variations is primarily attributed to the welding process that was used to join individual extruded columns. This process resulted in specimens that were not perfectly flat prior to machining the faces to a desired thickness. While the exterior face surfaces were milled flat, the inner face surfaces retained the original offset resulting in local variations in face sheet thickness. The face wrinkling load, proportional to h^3 (Eq. 9), is sensitive to variations in face sheet thickness. In certain instances, the critical load for face wrinkling using the thinnest face sheet thickness measured can become less than the macro plastic buckling load calculated using the average face sheet thickness. Thus, localized face wrinkling becomes an active failure mechanism prior to macro buckling. The minimum measured face thickness was sufficient to activate plastic face wrinkling prior to the designed macro plastic buckling mode for several column geometries tested (g, i, j, l, m). In most instances, the calculated plastic face wrinkling load in the thinnest section was within 10% of the macro plastic buckling load using the average face thickness. Using the minimum face thickness, the predicted plastic face wrinkling load is in excellent agreement with the measured failure load and correctly captures the operative failure mechanism. The peak loads agree within 8% for all columns except point l, which is within 16%. This analysis can be applied to the columns designed in the plastic

face wrinkling region as well, bringing the agreement between the analytical and measured peak loads even closer—in many cases within 5%. For the columns that are observed to fail exclusively by macro plastic buckling, reasonable to very good agreement is seen between the analytical and measured peak loads (8%–20%), in spite of the potential manufacturing flaws present. Variations in geometry that result from manufacturing flaws can significantly alter both the predicted failure load and mechanism and must be mitigated to properly use the predictive tools (i.e., failure mode maps). However, when appropriate parameter values are used, the analytical predictions accurately capture the column response.

To show the detrimental effect that friction stir welding can have on the column response, sample k was tested without undergoing heat treatment post weld. These columns were designed with a geometry that placed them well within the macro plastic buckling region. However, the active failure mode observed was face wrinkling, which occurred at a friction stir weld joint. Upon examination of the tested columns, it was found that failure occurred at each of the friction stir weld joints. It should be noted that the minimum face sheet thickness was not sufficient to activate plastic face wrinkling prior to macro buckling as discussed above. These tests demonstrate how the friction stir welding process can have a significant affect on the column failure, dictating overall column response. This must be accounted for, through additional processing to prevent performance degradation.

(2) Optimal Designs

At very low load indices, corrugated core columns compressed perpendicular to the corrugations compete favorably with hat-stiffened panels and outperform extruded pyramidal core columns. However, the optimal panel performances converge for $\bar{P}^* > 0.003$, when yielding is active in all panels. In this region, there is no structural advantage between competing panels, illustrating the influence strain hardening characteristics can have on the optimal designs. For panels composed of this material, the benefit of sandwich constructions over hat-stiffened panels can be realized through their multifunctional capabilities. As elastic failure is unstable it is generally avoided as a possible failure mechanism in structural applications to prevent catastrophic failure. It can also be seen that for loading indices, $\bar{P}^* > 0.04$, neither the sandwich panels nor hat-stiffened panel offer a performance benefit over a monolithic plate.

For corrugated core columns compressed perpendicular to the corrugations the active failure mechanisms in the elastic regime ($\bar{P}^* < 0.002$) are elastic face wrinkling and macro elastic buckling. The same mechanisms dictate failure of an optimal pyramidal core column in the elastic region ($\bar{P}^* < 0.003$). Hat-stiffened panels in the elastic region ($\bar{P}^* < 2.5 \times 10^{-3}$) fail by simultaneous local elastic buckling and macro elastic buckling. For corrugated core columns compressed parallel to the corrugations, simultaneous local elastic face buckling, local elastic core buckling, and macro elastic buckling govern failure in the elastic region ($\bar{P}^* < 0.003$). Hat-stiffened panels loaded in the transverse direction behave as a monolithic plate and exhibit significantly inferior performance, making corrugated core columns an attractive alternative for combined loading scenarios. For all columns, failure in the plastic region occurs at the boundary between the failures described above and material yielding. The optimal geometry illustrates that the optimal face thickness, h, and ligament length, l, are similar, whereas the ligament thickness, t, for the corrugated core is much thinner than for a pyramidal core. The shear modulus for a corrugated core scales linearly with ligament slenderness ratio t/l, whereas for a pyramidal core, the shear modulus scales with $(t/l)^2$. Corrugated cores have continuous material through the width of the panel, whereas pyramidal cores have discrete ligaments that must be thicker to achieve similar core shear stiffness.

It is interesting to note the comparison between a corrugated panel loaded in the two transverse directions. In the elastic region, the corrugated core column loaded perpendicular to the corrugations is more weight efficient than when loaded parallel to the corrugations. This results from the sensitivity of the local failure mechanisms to the prescribed boundary conditions. Due to the panel geometry and fabrication method, a fixed boundary condition is assumed at the core/face sheet interface. This makes the local face wrinkling failure a strong failure mechanism when compressed perpendicular to the corrugations. If this condition is relaxed to a simply supported boundary at the interface, a corrugated panel loaded perpendicular to the corrugations becomes less weight efficient than a panel compressed parallel to the corrugations. Critically, when compressed parallel to the corrugations, there is an additional active failure mechanism, local core buckling, that is dependent on the core geometry. To suppress this mechanism, the thickness of the core ligaments must increase, adding mass to the structure. When compressed perpendicular to the corrugations, the core ligament thickness is set only by the shear stiffness of the core and can be much thinner.

Acknowledgments

The authors gratefully acknowledge the support of the Office of Naval Research through Grant No. N00014-06-1-0509 (Dr. David Shifler, Program Officer). This article is dedicated to the memory of Anthony G. Evans—his leadership, mentorship, and friendship are greatly missed.

References

[1]F. Côté, R. Biagi, H. Bart-Smith, and V. S. Deshpande, "Structural Response of Pyramidal Core Sandwich Columns," *Int. J. Solids Struct.*, **44** [10] 3533–56 (2007).

[2]D. T. Queheillalt, Y. Murty, and H. N. G. Wadley, "Mechanical Properties of an Extruded Pyramidal Lattice Truss Sandwich Structure," *Scr. Mater.*, **58** [1] 76–9 (2008).

[3]M. F. Ashby, A. Evans, N. A. Fleck, L. J. Gibson, J. W. Hutchinson, and H. N. G. Wadley, Metal Foams: A Design Guide. Butterworth Heinemann, Woburn, MA, 2000.

[4]V. S. Deshpande and N. A. Fleck, "Collapse of Truss Core Sandwich Beams in 3-Point Bending," *Int. J. Solids Struct.*, **38** [36–37] 6275–305 (2001).

[5]V. S. Deshpande, N. A. Fleck, and M. F. Ashby, "Effective Properties of the Octet-Truss Lattice Material," *J. Mech. Phys. Solids*, **49** [8] 1747–69 (2001).

[6]F. W. Zok, S. A. Waltner, Z. Wei, H. J. Rathbun, R. M. McMeeking, and A. G. Evans, "A Protocol for Characterizing the Structural Performance of Metallic Sandwich Panels: Application to Pyramidal Truss Cores," *Int. J. Solids Struct.*, **41** [22–23] 6249–71 (2004).

[7]A. G. Evans, J. W. Hutchinson, and M. F. Ashby, "Multifunctionality of Cellular Metal Systems," *Prog. Mater. Sci.*, **43** [3] 171–221 (1999).

[8]N. A. Fleck and V. S. Deshpande, "The Resistance of Clamped Sandwich Beams to Shock Loading," *J. Appl. Mech.*, **71**, 386 (2004).

[9]S. Gu, T. J. Lu, and A. G. Evans, "On the Design of two-Dimensional Cellular Metals for Combined Heat Dissipation and Structural Load Capacity," *Int. J. Heat Mass Transfer*, **44** [11] 2163–75 (2001).

[10]X. Qiu, V. S. Deshpande, and N. A. Fleck, "Impulsive Loading of Clamped Monolithic and Sandwich Beams Over a Central Patch," *J. Mech. Phys. Solids*, **53** [5] 1015–46 (2005).

[11]H. N. G. Wadley, "Multifunctional Periodic Cellular Metals," *Phil. Trans. Roy. Soc. A: Math. Phys. Eng. Sci.*, **364** [1838] 31–68 (2006).

[12]Z. Xue and J. W. Hutchinson, "A Comparative Study of Impulse-Resistant Metal Sandwich Plates," *Int. J. Impact Eng.*, **30** [10] 1283–305 (2004).

[13]C. Hamilton, S. Dymek, M. Blicharski, and W. Brzegowy, "Microstructural and Flow Characteristics of Friction Stir Welded Aluminium 6061-T6 Extrusions," *Sci. Technol. Weld. Join.*, **12** [8] 732– (2007).

[14]G. Liu, L. E. Murr, C. S. Niou, J. C. McClure, and F. R. Vega, "Microstructural Aspects of the Friction-Stir Welding of 6061-T6 Aluminum," *Scr. Mater.*, **37** [3] 355–61 (1997).

[15]R. S. Mishra and Z. Y. Ma, "Friction Stir Welding and Processing," *Mater. Sci. Eng. R Rep.*, **50** [1–2] 1–78 (2005).

[16]ASTM Standard E8, Standard Test Methods for Tension Testing of Metallic Materials. ASTM International, West Conshohocken, PA, 2001.

[17]ASTM Standard E18, Standard Test Methods for Rockwell Hardness of Metallic Materials. ASTM International, West Conshohocken, PA, 2008.

[18]H. G. Allen, Analysis and Design of Structural Sandwich Panels. Pergamon Press, Oxford, 1969.

[19]F. J. Plantema, Sandwich Construction. Wiley, New York, 1966.

[20]D. Zenkert, An Introduction to Sandwich Construction. Engineering Materials Advisory Services, London, 1995.

[21]Z. P. Bazant and L. Cedolin, Stability of Structures. Oxford University Press, New York, 1991.

[22]S. P. Timoshenko and J. M. Gere, Theory of Elastic Stability. McGraw-Hill, New York, 1961.

[23]F. Cote, V. S. Deshpande, N. A. Fleck, and A. G. Evans, "The Compressive and Shear Responses of Corrugated and Diamond Lattice Materials," *Int. J. Solids Struct.*, **43** [20] 6220–42 (2006).

[24]F. R. Shanley, "Inelastic Column Theory," *J. Aeronaut. Sci.*, **14** [5] 261–8 (1947).

[25]R. Biagi, "The Mechanical Response of Corrugated Core Sandwich Columns"; Thesis, University of Virginia, Charlottesville, VA, 2010.

[26]B. Budiansky, "On the Minimum Weights of Compression Structures," *Int. J. Solids Struct.*, **36** [24] 3677–708 (1999).

[27]F. Bloom and D. Coffin, Handbook of Thin Plate Buckling and Postbuckling. Chapman and Hall/CRC Press, Boca Raton, FL, 2001.

[28]Abaqus Analysis User's Manual. Hibbitt, Karlsson & Sorensen, Inc. 2002.

J. Am. Ceram. Soc., **94** [S1] S85–S95 (2011)
DOI: 10.1111/j.1551-2916.2011.04499.x
© 2011 The American Ceramic Society

journal

Lifetime Assessment for Thermal Barrier Coatings: Tests for Measuring Mixed Mode Delamination Toughness

Robert G. Hutchinson

Pratt & Whitney, MS 184-15, East Hartford, 06108 Connecticut

John W. Hutchinson[†]

School of Engineering and Applied Sciences, Harvard University, Cambridge, 02138 Massachusetts

Mechanisms leading to degradation of the adherence of thermal barrier coatings (TBC) used in aircraft and power generating turbines are numerous and complex. To date, robust methods for the lifetime assessment of coatings have not emerged based on predictions of the degradation processes due to their complexity. In the absence of mechanism-based predictive models, direct measurement of coating adherence as a function of thermal exposure must be a component of any practical approach toward lifetime assessment. This paper outlines an approach to lifetime assessment of TBC that has taken shape in the past few years. Most TBC delaminations occur under a mix of mode I and mode II cracking conditions, with mode II delamination being particularly relevant. Direct measurement of TBC delamination toughness has been challenging, but recent progress has made this feasible. This paper surveys a range of potentially promising tests for measuring the mode dependence of delamination toughness with particular emphasis on toughness under mode II conditions.

I. Introduction

A thermal barrier coating (TBC) is a miracle of materials engineering and science that, aided by internal cooling, reduces the temperature on the underlying metal alloy substrate while withstanding repeated thermal cycles with variations on the order of 1000°C.[1] Research on the durability of TBCs has been underway since they were employed to extend the lifetime of turbine blades in aircraft engines roughly two decades ago. Efforts to improve TBC durability are driven by longevity considerations and also by the quest to achieve higher efficiency via higher operating temperatures.[2]

TBC systems generally involve three components: a metal bond coat with one surface adhering to the substrate alloy and the other surface oxidizing under thermal exposure to create an impervious aluminum oxide layer called the thermally grown oxide (TGO), and a porous ceramic topcoat which serves as the thermal insulation. Details of the coating failures depend on the specific materials making up the coating system. There are several bond coats in widespread use and these are typically in the range from 50 to 100 μm thick. Most topcoats are yttria-stabilized zirconia. Depending on the application, the topcoat may be electron beam deposited with a columnar structure and typically on the order of 100 μm thick (for aircraft engine blades) or plasma sprayed with a splat structure and as

thick as a millimeter or more (for hot surfaces in aircraft engines other than blades or most hot surfaces in gas power turbines).

Multiple failure modes leading to TBC spallation have been observed through laboratory tests and examination of coatings that have experienced service conditions. These observations have motivated extensive efforts to quantitatively characterize the micromechanics of the failure processes.[3] Many failures originate and propagate along the interface between the TGO and the bond coat (e.g., for some NiCoCrAlY bond coats) or just above the interface between the TGO and the topcoat (e.g., for some Pt–aluminide bond coats and for many plasma spray systems). Under conditions where the TBC is subject to very high temperature gradients through its thickness, failures have been observed to originate and propagate within the topcoat and well away from its interface with the TGO.[4] The focus in this paper will be on the most widely observed delamination failures that originate and propagate at the bottom of the topcoat, either just above or below the TGO.

While the micromechanical studies have provided considerable understanding of how delamination toughness degrades as a function of thermal history and while they have pointed to property changes that can lead to improvements in TBC systems, they are not yet sufficiently mature to allow quantitatively reliable prediction of delamination toughness degradation. This state of affairs is no different from essentially all other areas where fracture mechanics is used to assess structural integrity: fracture toughness is a property that is measured, not predicted, because models of toughness are usually not sufficiently accurate for prediction. Thus, the working assumption in this paper is that an essential component of any lifetime assessment scheme is the experimental determination of the delamination toughness of the TBC as a function of the relevant thermal history.

A fracture mechanics approach to TBC lifetime assessment that has emerged in recent years is introduced in the next section. Several of the most common delaminations will be reviewed illustrating that delamination usually occurs under mixed mode conditions. Mode II, or near-mode II, delaminations appear to be especially common. A modified four-point bend test[5,6] has proven to be an effective means of measuring delamination toughness under conditions with a nearly equal mix of mode I and II components. After a brief review of several tests for measuring delamination toughness, including the modified four-point bend test, the body of the paper surveys a range of possible tests to measure toughness over the full range of mode mix relevant to coating delaminations. Each of the proposed tests uses the modification used in the four-point bend test wherein a stiffener is bonded to the coating to increase the elastic energy available for delamination. If successfully implemented, the suite of tests has the potential to generate data over the range of mode dependence relevant to TBC delamination.

R. McMeeking—contributing editor

Manuscript No. 28891. Received November 11, 2010; approved February 11, 2011.
This paper is dedicated to the memory of A. G. Evans. His contributions form the foundations for much of the ongoing research on thermal barrier coatings.
[†]Author to whom correspondence should be addressed. e-mail: hutchinson@husm.harvard.edu

II. Delamination Mechanics and Examples of TBC Delaminations

At temperatures representative of the highest temperatures experienced by coatings in service it is generally believed that stresses in the topcoat and the TGO are relaxed due to creep of the constituent materials. With this assumption, the largest in-plane stresses in the topcoat and the TGO are compressive and they occur during cool down due to thermal expansion mismatch between the ceramic layers and the metal substrate. Thus, while the degradation processes primarily occur at high temperatures, it is generally believed that the critical conditions for delamination occur during cool down. The coatings must be able to withstand the stresses at the lowest temperatures when the turbines are cool. For this reason, the relevant toughness is believed to be "room temperature" toughness, and tests to measure the delamination toughness of TBC systems have invariably been conducted at room temperature.

A variety of simplified models are considered in this paper, none of which account for either the bond coat or the TGO. The model results presented here are intended to illustrate basic ideas; they will need to be embellished in quantitative applications to account for additional layers such as the TGO. In the models considered here, the Young's modulus and Poisson's ratio of the coating are, E_C and v_C, and those of the substrate are E and v. For plane strain, the two Dundurs elastic mismatch parameters are

$$\alpha = \frac{\bar{E} - \bar{E}_C}{\bar{E} + \bar{E}_C} \text{ and } \beta = \frac{1}{2} \frac{\mu(1 - 2v_C) - \mu_C(1 - 2v)}{\mu(1 - v_C) + \mu_C(1 - v)} \quad (1)$$

with $\bar{E}_C = E_C/(1 - v_C^2)$, $\bar{E} = E/(1 - v^2)$, $\mu_C = E_C/[2(1+v_C)]$ and $\mu = E/[2(1+v)]$.

It is assumed that the interface plane is the weak link between the coating and the substrate and that the delamination crack stays within this plane. Kinking of an interface crack out of the interface region defeats the purpose of the test, and conditions that encourage the crack to stay within the interface will be discussed later in the paper. Denote the toughness of the controlling failure plane between the coating and the substrate (interface or interfacial layer) in its current state by $\Gamma(\psi)$ in units J/m^2 with the convention defined in Fig. 1 having the coating above the substrate. For most interfaces, the toughness depends on the mode mix as measured by ψ and as depicted in

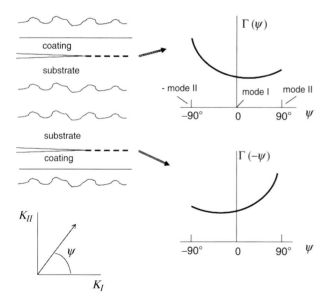

Fig. 1. Schematic of coating/substrate interface toughness as a function of mode mix. The convention in this paper defines the toughness function, $\Gamma(\psi)$, such that the coating lies above the substrate. When the coating lies below the substrate the toughness is $\Gamma(-\psi)$ when subject to the mode mix, ψ. The toughness function $\Gamma(\psi)$ can be symmetric in ψ but generally it should be assumed to be asymmetric.

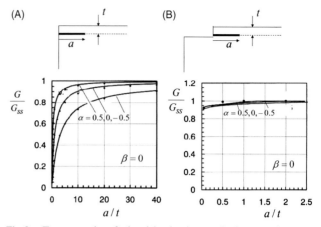

Fig. 2. Two examples of edge delamination cracks for a coating experiencing a tensile stress due to thermal expansion mismatch with the substrate. (A) An edge delamination emanating from an edge of the substrate. (B) An edge delamination emerging from a coating edge at an interior point of the substrate or possibly an open vertical sinter crack. For tensile stress, the crack is open with roughly equal components of mode I and mode II. For compressive stress, the crack is a mode II delamination, and the above results apply approximately if friction is neglected. The results reveal the inherently greater resistance to delamination initiation for coatings terminating at an edge compared with a coating terminating at an interior edge.

Fig. 1. If $\beta = 0$, $\psi = \tan^{-1}(K_{II}/K_I)$, where K_I and K_{II} are the conventional mode I and II stress intensity factors. With the convention adopted in this paper (c.f., Fig. 1), the interface toughness for situations in which the coating lies below the substrate is $\Gamma(-\psi)$. The toughness function, $\Gamma(\psi)$, is not necessarily symmetric in ψ and, specifically, the negative mode II toughness, $\Gamma(-90°)$, is not necessarily the same as the positive mode II toughness, $\Gamma(90°)$. If $\beta \neq 0$, the definition of ψ is slightly more complicated and is given in Appendix A. With G as the energy release rate (units of J/m^2) and ψ as the measure of the mode mix of the interface crack subject to the present loading, the condition for incipient advance of the delamination in the interface plane is

$$G = \Gamma(\psi) \quad (2)$$

The emphasis in this paper is on tests to measure $\Gamma(\psi)$ for a wide range of ψ relevant to coating delamination failures.

Edge delaminations are among the most common TBC delaminations observed on components where spallations occur. Two types of edge delaminations are illustrated in Fig. 2. In these examples, the coating is taken to be of uniform of thickness, t, and is bonded to an infinitely thick substrate. The film has a uniform thermal expansion mismatch with the substrate such that away from its edge the film is subject to a uniform equi-biaxial stress, σ. The steady-state, or long crack, energy release rate limit is independent of the delamination length:

$$G_{SS} = \frac{\sigma^2 t}{2\bar{E}_C} \quad (3)$$

If the thermal mismatch conditions produce tensile stress in the coating, the crack is completely open with mix of mode I and II given by $\psi \cong 55°$, depending somewhat on the elastic mismatch.[7] However, if the stress in the coating is compressive, as it would be for the TBC under cool down, the crack faces contact each other and the examples in Fig. 2 are *mode II (or near-mode II) delamination cracks* with $\psi \cong -90°$.[‡] Friction between the faces of the crack is neglected in these results.

[‡]The results in Fig. 2 have been taken from Yu *et al* [7] where a coating under tension due to a uniform thermal expansion mismatch is analyzed. The tensile loading produces an open crack. The results are only approximately valid for the compressive case due, in part, to neglect of contact that occurs between the crack faces. However, when friction is neglected, the energy release rate is only slightly affected by contact and the results in Fig. 2 are approximately applicable. An example where friction is taken into account in mode II delamination is considered later in this paper and a coating under compression is analyzed in Balint and Hutchinson.[8]

For an edge delamination emerging from an edge of the substrate in Fig. 2(A), the energy release rate, G, requires a delamination length, a, at least several times the coating thickness to approach the steady-state limit, even for a mismatch with $\alpha = 0.5$ ($\bar{E}_C/\bar{E} = 1/3$) which is representative of some TBC systems. By contrast, an edge delamination emanating from an interior edge of the coating as in Fig. 2(B) attains the steady-state limit when a is a small fraction of the coating thickness for all mismatches. An interior edge delamination arises, as illustrated, where the coating terminates abruptly away from the substrate edge or, for example, at an open vertical sinter crack in the coating. An interior coating edge is more susceptible to delamination initiation than a coating that extends all the way to the substrate edge. The latter has some built in protection against the initiation due to the extra compliance of the substrate edge that lowers the stress in the coating in that vicinity.

Edge delaminations can initiate from corners and from air holes in a substrate. As for delaminations emanating from a substrate edge, the local stress distribution in the coating will be affected by substrate compliance at such features. If the substrate thickness is comparable to that of the coating the compliance will be increased. A reduction in stored strain energy in the coating at potential initiation locations adds to the protection against delamination.

The mode mix of the delamination depends on the distribution of the compressive stress through the thickness of the coating. A uniform stress distribution, or a stress distribution which is more compressive at the surface than at the interface, produces mode II. If the stress is sufficiently more compressive just above the coating's interface than at its surface, then the crack may open.[9] For example, if the stress in the coating vanishes in a layer of thickness kt below the surface and is uniform compression in the remaining layer, then a mode I component will exist if $k < 0.449$; otherwise mode II prevails. This example assumes no elastic mismatch between the coating and substrate, but elastic mismatch is less important than the details of the stress distribution.

III. Tests for Measuring Delamination Toughness

Measuring the delamination toughness, $\Gamma(\psi)$, of TBC coatings has been challenging, especially so given the importance of a range of mixed mode toughness in applications. This section begins by citing several tests that have been used to measure delamination toughness and some representative results. The fracture mechanics approach to determining TBC durability is illustrated in conjunction with the tests.

Vastinonta and Beuth[10] used a conical brale C indenter to induce a circular delamination in a 100-μm-thick electron beam deposited TBC system. The indenter is pushed into the substrate through the topcoat and TGO. The indenter forces an outward radial plastic flow of the substrate under the coating that decays with distance from the central axis of symmetry. The substrate motion induces additional compression of the coating in the radial direction and increases the elastic energy density stored in the TBC in the vicinity of the indent. The indenter also initiates a delamination edge where it pushes through the coating. The delamination spreads axisymmetrically until the driving force falls below the toughness, i.e., until $G = \Gamma(\psi)$ can no longer be met. A difficult aspect of this test is the analysis of the delamination crack problem required to generate both G and ψ as a function of the radius delamination and the indenter depth. A sophisticated elastic-plastic finite element calculation is required. In addition, the induced radial compressive stress can be high enough to buckle the coating, further complicating the determination of G and ψ. The loading is a mix of mode I and mode II (with $\psi > 0$ if buckling does not occur) that varies with crack radius.

Jones et al.[11] sliced a planar section of a electron beam deposited TBC system on a turbine blade substrate. Then these authors used a focused ion beam to cut a broad notch under the coating system creating a trilayer bridge consisting of the bond coat, the TGO and the topcoat. The bridge was supported (effectively clamped) at its ends by the uncut substrate and coating system. A concentrated load normal to the surface was applied to the center of the bridge. In the first series of tests, the load was applied to the underside of the bridge pushing upward. This load produces a moment distribution in the bridge, which at sufficient force causes the topcoat to develop a vertical crack above the load that runs from the surface to the interface. This crack branches into the weak interface and spreads as a delamination crack. The geometry and preparation of the sample allow for clear visualization of the various stages, including the stable advance of the crack tip as the bridge is pushed upward. This is a highly sophisticated test that is not likely to be used routinely.

This in situ test allows for the interrogation of engine hardware but is a highly sophisticated test and is not likely to be used routinely. It also requires a detailed finite element analysis to obtain G and ψ. Plastic deformation of the metal bridge components can occur. The mode I component is somewhat larger than the mode II component when the load is applied under the bridge. To avoid cracking in brittle coatings and to obtain toughness data under conditions closer to mode II, Jones, Manning and Hemker (unpublished work) carried out a second set of tests in which they created a narrow open notch in the center of the top coat to the interface of interest and loaded the specimen by pushing down from above the bridge. This direction of loading adds to the compression in the coating from the thermal expansion mismatch and therefore more closely mimics the delamination conditions experienced in service. Kagawa and colleagues have developed two related tests, the barb test[12] and the push-out test,[13] to measure delamination toughness of TBCs. In each test, the coating is subject to additional compression by forcing it, but not the substrate, against a hard block. The force at which the coating delaminates is used to determine the critical energy release rate and the associated toughness. These tests add compression to any residual compression in the coating, but they impose a delamination displacement and shearing stress on the interface in the opposite direction from that experienced by a delamination in service driven by thermal expansion mismatch (i.e., $\psi > 0$ rather than $\psi < 0$). The energy release rate and mode mix of the barb test has been analyzed by finite element methods[14] with the finding that $\psi \cong 60°$. Like the aforementioned bridge test, the barb and push-out tests require highly refined specimen preparation and sophisticated testing which are likely to limit their use for routine toughness testing. Nevertheless, Kagawa and colleagues[13] have collected an extensive data set showing how the toughness

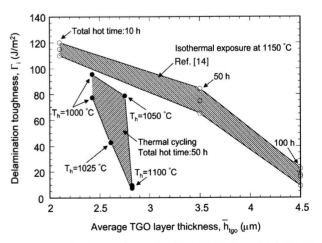

Fig. 3. Delamination toughness of an EB-PVD TBC on a NiCoCrAlY bond coat as a function of thermal exposure plotted as a function of the thickness of the TGO from Tanaka *et al.*[13] The upper band of data is for isothermal exposure and the lower band is for cyclic thermal exposure. The TGO thickness increases with thermal exposure time in accord with the markers on the upper band of data.

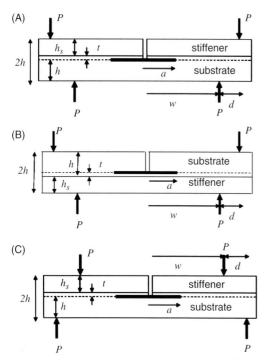

Fig. 4. Substrate/coating/stiffener trilayers. The coating thickness is t. The crack lies along the coating/substrate interface. (A) The four-point bend test with the loads configured to produce fully separated crack faces ($\psi \cong 40°$) giving $\Gamma(\psi)$. (B) The four-point bend test with the roles of the substrate and stiffener interchanged from that in (A) with $\psi \cong 40°$ giving $\Gamma(-\psi)$. (C) The inverted four-point test with the lines of action of the loading points interchanged or, equivalently, with specimen in (A) turned up-side down. The crack faces come into contact in the center of the specimen producing loading conditions with a larger component of mode II relative to mode I ($\psi \cong -60°$).

degrades with thermal exposure and with thermal cycling as illustrated in Fig. 3. The significant degradation of toughness with thermal exposure reflects the microcracking and other processes taking place at the delamination interface which lies either within the TGO or between the TGO and the bond coat, depending on the exposure time. Moreover, the toughness data for this system shows that thermal exposure under cycling is considerably more damaging than thermal exposure without cycling.

Generally, the popular four-point bend delamination test cannot be applied directly to measure the delamination toughness of coatings because the bending deformations required to create critical levels of stored energy in the coating become so large that extensive plastic deformation occurs in the substrate. To circumvent this difficulty, Hofinger et al.[5] proposed a modification of the four-point bend test wherein stiffeners are bonded to the surface of the coating with a gap cut at the center to allow delamination to occur (see Fig. 4(A) for an example). The loading bends the central section of the specimen upward so that when delamination occurs, the crack opens with a significant mode I component. The effectiveness of the test was demonstrated by measuring the delamination toughness of a plasma spray TBC coating.[5] The modified specimen enjoys the property of the conventional four-point bend test in that the delamination crack attains stable steady-state conditions when it is well within the central section between the inner loading points.[15] There are other advantages to the modified test. If the stiffener balances the substrate, the coating interface will lie near the neutral bending axis and, therefore, the bending load does not appreciably change the stress at the coating interface away from the crack tip. The energy for delamination is primarily provided by the elastic energy stored in the stiffener. In addition, if the coating is thin compared with the stiffener, most of the residual stress in the coating will not be released in the test because the coating remains bonded to the stiffener. Consequently, uncertainty in knowledge of the residual stress, which is

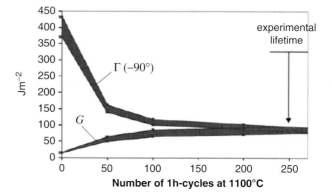

Fig. 5. A demonstration of the fracture mechanics approach for predicting TBC lifetime taken from Thery et al.[6] The system has a NiAlZr bond coat and a EB-PVD yttria-stabilized zirconia topcoat. The upper curve is the mode II toughness as a function of thermal cycles as inferred from four-point bend tests using a phenomenological conversion factor. The lower curve is the energy release rate predicted for a mode II edge delamination on the interface of the prototypical TBC system subject to the same thermal history.

common in many delamination tests, is not a serious disadvantage in the modified test because the residual stress makes very little contribution to either the energy release rate or the mode mix.

The modified four-point bend test was used by Thery et al.[6] at ONERA to conduct an extensive experimental study of the effect of thermal cycling on the delamination toughness of two TBC systems. The authors used this data in conjunction with estimates of the evolution of the energy release rate, G, for a prototypical demonstration of the efficacy of the fracture mechanics approach to delamination based on the fracture condition (2). The demonstration for one of the TBC systems considered by the ONERA group is reproduced in Fig. 5. One complication the authors faced was that the measured value of the toughness in their modified bend test corresponds to a mixed mode with $\psi \cong 40°$ while the toughness relevant to their prototypical test (which involved edge delamination) was close to pure mode II with $\psi \cong -90°$. The mode II toughness plotted in Fig. 5 was converted from the measured toughness using a phenomenological amplification factor, $\cong 3.7$, to account for the higher toughness in mode II, together with an implicit assumption that the toughness is not strongly dependent on the sign of ψ.[6] The degradation of toughness with thermal cycles seen in Fig. 5 is qualitatively similar to that displayed in Fig. 3. The increase in G with thermal cycles is due to the increase in stored residual elastic energy upon cool-down associated with the increases in the thickness of the TGO and modulus of the topcoat. Failure of the TBC in the prototypical demonstration was in reasonable agreement with attainment of condition (1) for both TBC systems in the study. The fact that the mode II toughness estimate used to obtain this agreement is so much higher than the toughness measured in their four-point bend test (with $\psi \cong 40°$) highlights the significance of mode dependence.

IV. Potential Tests for a Full Range of Mixed Mode Delamination Toughness

In this section, a selection of basic tests is reviewed in order to determine whether one or more may prove to be an effective means of generating delamination toughness data for TBCs and other coating systems. To focus attention on the most important details of the tests, consideration will be confined to specimens of type shown in Fig. 4 where the modulus and Poisson's ratio of the stiffeners will be taken to be identical to that of the substrate, E and v. The bond coat is considered to be part of the substrate and the TGO is not explicitly considered. The coating has thickness t and modulus and Poisson's ratio, E_C and v_C. The

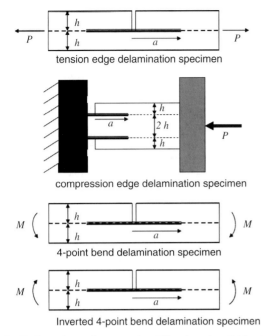

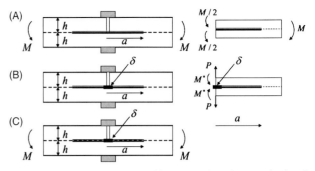

Fig. 8. (A) Mode II bend test (positive or negative M) constrained such that the upper layer does not rotate at the center but is free to slide laterally without friction. (B) Mode I wedge test constrained such that layers do not rotate at the center ($M^* = Pa/2$). (C) Mixed mode test under combined frictionless bending and wedging and constrained against rotation at the center. These specimens are shown without any coating.

Fig. 6. Basic specimens for measuring different modes of delamination toughness. These specimens are shown without any coating.

thickness of the stiffeners is h_S and, to focus just on the essential details, it will be assumed that this thickness has been chosen such that the bending stiffness of the coating/stiffener bilayer, B, is the same as that of the substrate layer, i.e., $B = \bar{E}h^3/12$. Thus, the neutral axis of the modified specimen under pure bending lies approximately along the interface between the coating and the substrate. As in the case of the modified four-point bend test, the purpose of the stiffener is to substantially boost the stored elastic energy available to drive the delamination. In fact, for specimens such as those in Fig. 6 subject to tension or compression, the stiffener is even more effective in storing energy than for the bend-type specimens.

It has been remarked that only a small fraction of the residual stress in the coating is released to drive delamination if the coating is thin and/or compliant compared with the stiffener. Under these circumstances the delaminated coating is constrained assuming it remains bonded to the stiffener. Suppose the coating has a uniform stress, σ, before bonding the stiffener. The fraction of the elastic energy in the coating that is released when the coating/stiffener bilayer is separated from the substrate is readily

calculated. The released fraction is the difference between the initial strain energy in the coating and the strain energy in the coating/stiffener bilayer after separation. The released fraction is plotted in Fig. 7 for combinations of $(h_S+t)/t$ and $\bar{E}_C/\bar{E}$. This plot can be used to assess if it is possible to ignore the role of the residual stress in the coating in the modified tests. If residual stress cannot be ignored, then Fig. 7 can be used to estimate the contribution from the residual stress to G assuming the residual stress is known. It is important to note that, while the residual stress contribution can be added to the contributions to G from the applied load, the phase angle, ψ, characterizing the mode mix must be computed by a linear superposition of the mode I and II stress intensity factors from the two contributions.

We begin by giving basic mechanics results for G and ψ for a number of potential specimens for the case where the coating is absent or, equivalently, with $E_C = E$. The loading cases are presented and labeled in Figs. 6, 8, and 9. The simple formulas presented in this section all apply under conditions when the crack length, a, exceeds several times the layer thickness, h, and has not yet begun to interact with the load points or with the ends of the specimen. Finite element results for some of the specimens will be used to demonstrate the validity of the basic results and to clarify issues related to crack face contact and friction. Following presentation of results for the homogeneous specimens, the mode mix ψ in the presence of the coating will be addressed.

(1) Homogeneous Specimens with no Coating Layer
For the specimens in Fig. 6:

Tension edge delamination specimen :

$$G = \frac{\sigma^2 h}{\bar{E}}, \quad \psi = 49.1° \quad (\sigma = P/2h) \tag{4}$$

Compression edge delamination specimen :

$$G = \frac{\sigma^2 h}{\bar{E}}, \quad \psi = -90° \quad (\sigma = -P/4h) \tag{5}$$

Four-point bend specimen :

$$G = \frac{7}{16}\frac{M^2}{B} = \frac{21}{4}\frac{M^2}{\bar{E}h^3}, \quad \psi = 40.9° \tag{6}$$

Inverted four-point bend specimen :

$$G = \frac{1}{4}\frac{M^2}{B} = 3\frac{M^2}{\bar{E}h^3}, \quad \psi = -60° \tag{7}$$

In each of these examples, the energy release rate and the mode mix are steady-state values, which are independent of crack length. For long cracks in long specimens in Fig. 6, (4) and (6) are results from exact two dimensional elasticity solutions[9] while (7) is the result of a beam analysis presented in Appendix

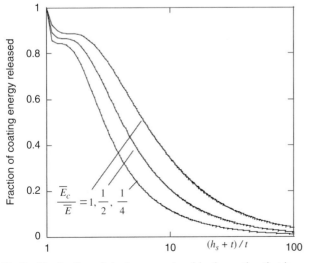

Fig. 7. The fraction of elastic energy stored in the coating that is released to delaminate a coating/stiffener bilayer from the substrate. The stress in the coating before delamination is assumed to be uniform.

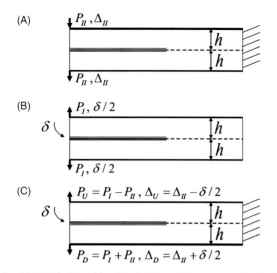

Fig. 9. (A) Mode II end-loaded double-cantilever beam. (B) Mode I wedge-loaded double-cantilever beam. (C) Mixed mode double-cantilever beam as superposition of (A) and (B).

A. Friction is neglected for the compression edge-delamination specimen in (5) but is considered in numerical simulations presented below. The energy release rate (5) is derived assuming that the two cracks in the specimen grow together maintaining symmetry. This will not necessarily occur and the possibility of growth of just one of the cracks should also be considered. Specimen dimensions needed to preclude elastic buckling and plastic deformation of the compression specimen are readily established and easily met. Elastic buckling is precluded if $L/h < (\bar{E}h/\Gamma)^{1/4}/(2\pi\sqrt{3})$, while avoidance of plastic deformation requires $h > \bar{E}\Gamma/\sigma_Y^2$, with Γ as the mode II toughness.

Contact between the crack faces occurs in the center of the *inverted four-point bend specimen* at the free ends of the upper layers; friction is neglected in (7). The accuracy of (7) for the inverted bend specimen is confirmed by finite element simulations presented in Fig. 10. The ANSYS code has been used in the calculations and the contact option has been invoked to account for crack face contact. In contact regions, Coulomb friction is invoked with a friction coefficient, μ_f. For the friction cases, the results apply for monotonically increasing M with fixed crack length. Details of the finite element modeling are

given in Appendix A. For the frictionless case, steady-state conditions (7) are attained when the crack length exceeds about $2h$, and for crack lengths below $2h$ the energy release rate is below the steady-state limit and the mode mix emerges from negative mode II ($\psi = -90°$). Friction lowers the energy release rate, but decreasingly so as the crack gets longer because the normal contact force diminishes. The mode mix is relatively unaffected by friction. It should be possible to reduce friction at the single point of contact by lubrication with a thin film of low friction material.

The *compression edge-delamination specimen* in Fig. 6 would appear to provide a relatively straightforward test for measuring mode II toughness relevant to edge delaminations, although there are complications that have to be addressed related to crack face contact, friction, and symmetric crack growth. Selected finite element simulations have been carried out to gain preliminary insights into some of these effects. The applied load, $P = -4\sigma h$, is increased monotonically such that for any fixed crack length, a, the stresses throughout the specimen increase linearly with σ, the mode mix is constant, and the energy release rate and the work dissipated in friction increase in proportion to σ^2.

For the frictionless limit, Fig. 11(A) displays the normalized pressure, p/σ, along the crack faces for three values of fixed a/h, while Fig. 11(B) presents the opening gap, δ, between the faces. Crack face contact occurs in the vicinity of the crack tip in all cases and thus the crack is mode II with $\psi = -90°$. The shortest crack with $a/h = 0.4$ is closed over its entire length and has high pressure between the faces at the end of the contact region ($x/a = 1$). The two longer cracks are effectively open at distances greater than about $2h$ from the tip. The peak pressure between the faces for the long cracks occurs at a distance of roughly $h/2$ from the tip. In the finite element model the crack in the right half of the specimen has length a originating from a notch at the center of the upper layer with half-width $h/10$.

The energy release rate and mode II stress intensity factor of the compression edge-delamination specimen are plotted as a function of the crack length in Fig. 12 for the frictionless case and for two values of the friction coefficient. Steady-state conditions are attained at crack lengths greater than about h for both the frictionless case (with $G = G_{SS}$ given by (5)) and the two cases with Coulomb friction. Friction clearly influences the energy release rate, but the effect is relatively modest for these examples. The reason that a steady state exists for the frictional cases is due to the fact that the contact region is confined to a

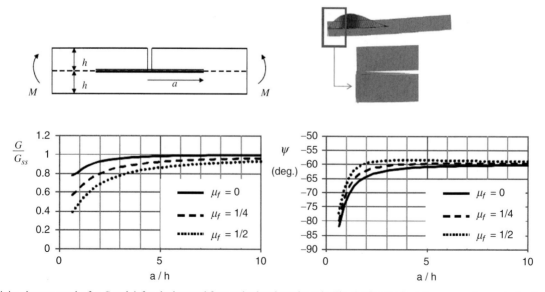

Fig. 10. Finite element results for G and ψ for the inverted four-point bend specimen in Fig. 6. The steady-state energy release rate for the frictionless limit ($\mu_f = 0$), G_{SS}, is given by (7) and derived in Appendix A based on beam analysis. This same analysis gives $\psi = -60°$. Results for the two cases with Coulomb friction have been computed under monotonically increasing M with a/h fixed. The analysis verifies that the crack is open except for a small contact region at the center of the specimen.

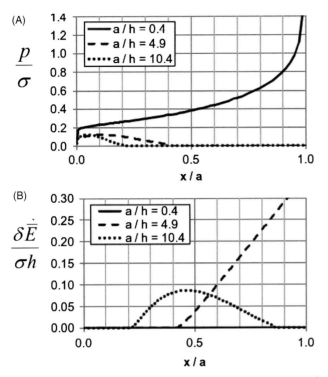

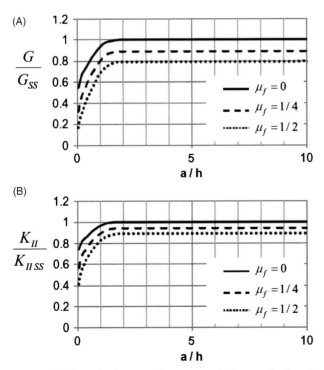

Fig. 11. (A) Normalized pressure between crack faces in the regions of contract, and (B) normalized gap between the faces in the regions where no contact occurs. These are for the compression edge-delamination specimen in Fig. 6 with no friction for three values of normalized crack length a/h. The coordinate x is measured from the crack tip along the crack faces. Friction makes only modest changes to the region of contact.

Fig. 12. (A) Normalized energy release rate and (B) normalized mode II stress intensity factor as a function of crack length for the compression edge-delamination specimen in Fig. 6 under monotonically increasing applied load for the frictionless case and two values of Coulomb friction. The steady state energy release rate, $G_{SS} = \sigma^2 h / \bar{E}$, and the steady state mode II stress intensity factor, $K_{IISS} = \sigma\sqrt{h}$, for the frictionless case have been used in the normalizations.

zone of width about $2h$ behind the tip. Thus, for a given friction coefficient, all long cracks have essentially the same zone of frictional sliding behind the tip dissipating the same amount of energy. The dissipation scales with σ^2 but is not strictly linear in μ_f. Under certain circumstances, it is possible that this frictional dissipation may be included as part of the fracture process energy forming the mode II toughness.[8] Further study of crack face contact and frictional sliding on the compression edge-delamination specimen is clearly required before it can be used to extract mode II delamination toughness. Nevertheless, the results presented in Figs. 11 and 12 are promising in the sense they suggest that it may be possible to obtain relatively simple characterizations of the specimen.

In principle, the *constrained bend-wedge mixed mode specimen* and loadings in Fig. 8(C) allow access to the entire range of mixed mode loading. It is imagined that a specially designed fixture, or guide, has been inserted at the center of the specimen such that the end of the coating–stiffener bilayer is constrained against rotation but is free to slide without friction relative to the substrate layer. This fixture is also assumed to be capable of forcing a separation, δ, of the crack faces at the center of the specimen.

Mode II constrained bend test (Fig. 8(A)):

$$K_I = 0, \quad K_{II} = \frac{3}{2} M h^{-3/2}, \quad G = \frac{3}{16}\frac{M^2}{B} \tag{8}$$

Mode I constrained wedge test (Fig. 8(B)):

$$K_I = \sqrt{3}Pah^{-3/2}, \quad K_{II} = 0, \quad G = \frac{1}{4}\frac{P^2a^2}{B} \tag{9}$$

These results have been determined using a beam theory analysis; (8) is exact plane strain elasticity for a long crack. The result in (9) is approximate but increasingly accurate for long cracks. For intermediate length cracks, its accuracy can be improved by the inclusion of an extra term depending on a/h.[16]

The *constrained bend-wedge mixed mode test* in Fig. 8(C) is obtained from the linear superposition of (8) and (9):

$$G = \frac{3}{16}\frac{M^2}{B} + \frac{1}{4}\frac{P^2a^2}{B}, \quad \tan\psi = \frac{\sqrt{3}M}{2Pa} \tag{10}$$

The crack faces will be open if $P > 0$ and the full range of ψ is accessed by reversing the sign of M. Crack growth is usually unstable if P is prescribed, but not if the opening displacement $\delta = Pa^3/6B$ is prescribed. With Θ as the rotation through which the moment M at the right end of the specimen works, beam theory gives: $\Theta = (w+a)M/8B$ where w is the distance from the center to the right end of the specimen. The results in (10) can be rewritten as

$$G = 12\frac{B\Theta^2}{(w+a)^2} + 9\frac{B\delta^2}{a^4}, \quad \tan\psi = \frac{2\Theta a^2}{\sqrt{3}\delta(w+a)} \tag{11}$$

The energy release rate decreases under prescribed rotation and opening displacement and crack growth will usually be stable. For example, if an opening, δ, is prescribed and held fixed while Θ is increased from zero, the mode mix will increase from pure mode I toward mode II. If the crack length can be measured, the results in (11) suggest a means of measuring the interface toughness over the entire range of ψ.

Various loadings of *a double-cantilever beam specimen* are depicted in Fig. 9. The *mode II double-cantilever beam specimen* in Fig. 9(A) has

$$K_I = 0, \quad K_{II} = 3P_{II}ah^{-3/2}, \quad G = \frac{3}{4}\frac{P_{II}^2a^2}{B} \tag{12}$$

The *mode I double-cantilever beam specimen* in Fig. 9(B) has

$$K_I = 2\sqrt{3}P_I a h^{-3/2}, \quad K_{II} = 0, \quad G = \frac{P_I^2 a^2}{B} \qquad (13)$$

Superposition of the above gives for the mixed mode double-cantilever beam specimen

$$G = \frac{P_I^2 a^2}{B} + \frac{3}{4}\frac{P_{II}^2 a^2}{B}, \quad \tan\psi = \frac{\sqrt{3}P_{II}}{2P_I} \qquad (14)$$

The accuracy of both contributions to G can be improved by including extra terms of order h/a, as has been done in Hutchinson and Suo.[9] Based on beam theory predictions, $\Delta_{II} = P_{II}(w^3 + 3a^3)/12B$ and $\delta = 2P_I a^3/3B$, such that one can re-write (14) as

$$G = 108\frac{B\Delta_{II}^2 a^2}{(w^3 + 3a^3)^2} + \frac{9}{4}\frac{B\delta^2}{a^4},$$
$$\tan\psi = \frac{4\sqrt{3}a^3\Delta_{II}}{(w^3 + 3a^3)\delta} \qquad (15)$$

The crack will be open if $\delta > 0$ and crack growth is expected to be stable under prescribed displacements.

In the literature of laminated composites, double-cantilever specimens are also called end-notched specimens and they have been widely used to the measure mixed mode toughness dependence with various loadings including those outlined above. An important recent development is a loading device for carrying out a mixed mode double-cantilever test by applying unequal moments to the two layers.[17] This system is capable of applying the full range of mode mix and it has the advantage that steady-state crack growth occurs under constant applied moments. This system holds great promise for TBC testing if it could be scaled down to an appropriate size.

The *Brazil nut specimen* has been used successfully to measure a substantial range of mixed mode toughness for specific interfaces[18] and for composite laminates.[19] It is possible this specimen could also be adapted to measure TBC coating delamination toughness by bonding one half of a circular metallic disk to the surface of the coating and the other half to the underside of the substrate. Unlike the test just described, the Brazil nut test does not exhibit a steady-state and crack growth can occur unstably. The range of energy release rates achievable in the Brazil nut test with a specimen having a radius on the order of 1 cm is sufficient to drive most coating delaminations. Friction is also a concern in this test.[20]

(2) Specimens with a Coating Layer
For all of the tests discussed above, the effect of the elastic mismatch of the coating layer on the mode mix of the interface crack can be estimated simply if the coating is thin compared with the thickness of the substrate and the stiffener. The delamination crack lies on the interface between the coating and the substrate. An important distinction is whether the coating lies above or below the substrate, as has been noted in Fig. 1. With the convention employed in this paper, the interface toughness for a coating lying above the interface is $\Gamma(\psi)$, while that for a coating lying below the interface is $\Gamma(-\psi)$. As in the previous section, the discussion which follows assumes that the bending stiffness of the coating/stiffener bilayer, B, is the same as that of the lower substrate layer, $\bar{E}h^3/12$. With this choice, all of the above expressions for the energy release rate are the same, to the accuracy to which they hold.[§] It will also be assumed that the

Table I. The Phase Angle Shift, $\omega(\alpha,\beta)$ in Degrees, for Combinations of the Elastic Mismatch Parameters[21]

β	α								
	−0.8	−0.6	−0.4	−0.2	0.0	0.2	0.4	0.6	0.8
−0.4	2.2	3.5							
−0.3	3.0	4.0	3.3	1.4					
−0.2	3.6	4.1	3.4	2.0	−0.3	−3.3			
−0.1	4.0	4.1	3.3	2.0	0.1	−2.3	−5.5	−10.8	
0.0	4.4	3.8	2.9	1.6	0.0	−2.1	−4.7	−8.4	−14.3
0.1			2.3	1.1	−0.5	−2.3	−4.5	−7.4	−11.6
0.2					−1.3	−3.0	−4.9	−7.3	−10.5
0.3							−5.8	−7.8	−10.4
0.4									−11.1

coating is sufficiently thin and/or compliant that the elastic energy released associated with the residual stress in the coating can be ignored, as discussed in conjunction with Fig. 7.

The mode mix for the crack on the interface between the coating and the substrate, as measured by ψ, can be estimated using a general asymptotic relationship for cases with $t/h \ll 1$.[21] Denote the mode mix for any of the specimens in the absence of a coating (i.e., with $t = 0$, or, equivalently, with $E_C = E$ and $\nu_C = \nu$) by ψ^0, and denote the mode mix for the crack on the interface between the coating and the substrate under the same loading by ψ. The relation between the two measures is

$$\psi = \psi^0 \pm \omega(\alpha, \beta) \qquad (16)$$

where the "+" applies if the coating lies below the substrate and the "−" applies if the coating lies above the substrate and where $\omega(\alpha,\beta)$ is given in Table I.[¶] For thin coatings ($t/h \ll 1$), this asymptotic result does not depend on t. The result (16) is based on an elasticity analysis of a crack on a bimaterial interface which, for $\beta \neq 0$, neglects the consequences of crack face interpenetration behind the crack tip on the assumption that the interpenetration is subsumed within the fracture process zone. Thus, (16) is limited to cases in which interpenetration does not invalidate application of the solution. If $\beta \neq 0$, the predicted zone of interpenetration increases as the loading becomes dominantly mode II, and thus (16) should be used with caution near mode II.

To illustrate the effect of elastic mismatch between the coating and the substrate for the specimens in Fig. 6, suppose $\bar{E}_C/\bar{E} = 1/4$ and $\nu_C = \nu = 0.3$: then, $\alpha = 3/5$, $\beta = 0.17$ and, from Table I, $\omega = -7.3°$. If the coating lies above the substrate, by (16), phase angle of the *tension edge-delamination specimen* increases from $\psi^0 = 49.1°$ to $\psi = 56.4°$; that of the *four-point bend specimen* increases from $\psi^0 = 40.9°$ to $\psi = 48.2°$; while mode II component of the *inverted four-point bend specimen* decreases by $\psi^0 = -60°$ to $\psi = -52.7°$. The mode II component of the *compression edge-delamination specimen* is also predicted to decrease from $\psi^0 = -90°$ to $\psi = -82.7°$, but this should only be considered as a trend rather than a quantitative estimate for reasons mentioned above. In general, the shift in mode mix due to the compliance of the coating layer is relatively small, as these results illustrate, except for coatings that are exceptionally stiff or compliant.

(3) Interchanging the Substrate and Stiffener
For any of the specimens considered, by turning the trilayer upside down and creating a notch down to the interface in the substrate and having a single continuous stiffener, one can access $\Gamma(\psi)$ in the opposite sign range of ψ. The interchange in Fig. 4 from (A) to (B) illustrates precisely what this entails. The

---

[§]Some of the energy release rates may vary slightly from those quoted if one undertook more accurate plane strain calculations. It is strongly recommended that finite element calculations of final specimen geometries and properties be performed. Such calculations also reveal when the crack begins to interact with the specimen ends. In this paper, to simplify the discussion, we have taken the bending stiffness of the bilayer to be the same as that of the lower layer, and we have not directly accounted for the TGO. These effects, and contributions from residual stress in the coating, can be included if one is prepared to compute the energy release rate and the mode mix for the specimen.

[¶]This table is reproduced from Suo and Hutchinson.[21] In Suo and Hutchinson,[21] the thin layer was taken to lie below substrate and thus "+" was used. Here, the convention of the earlier reference in defining the Dundurs parameter was followed wherein $\alpha > 0$ for systems with coatings more compliant than substrates. The reader should be alert to the fact that the usual convention has $\alpha > 0$ when the material above the interface crack is stiffer than the material below it.

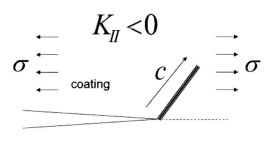

Fig. 13. Conventions related to kinking of an interface crack into the coating. For a coating lying above the substrate, as shown, a negative K_{II} encourages kinking into the coating. If the coating lies below the substrate a positive K_{II} encourages kinking into the coating. A tensile residual stress in the coating ($\sigma > 0$) increases the likelihood of kinking into the coating while a compressive residual stress ($\sigma < 0$) discourages kinking into the coating.

discussion, which follows continues to assume the interface of interest is that between the substrate and the coating. As noted in Fig. 1, inverting the substrate/coating bilayer changes the toughness from $\Gamma(\psi)$ to $\Gamma(-\psi)$ if the mode mix on the interface does not change. Let ψ^0 be the reference mode mix of the specimen under the particular loading in the absence of the coating as defined above. If the coating lies above the substrate (as in Fig. 4(A)), then $\psi = \psi^0 - \omega$ and the toughness is $\Gamma(\psi^0 - \omega)$. If the coating lies below the substrate (as in Fig. 4(B)), then $\psi = \psi^0 + \omega$ and the toughness is $\Gamma(-\psi^0 - \omega)$.

The possibility of interchanging the roles of the substrate and the stiffener in this manner has several advantages. The most obvious advantage is that it potentially provides a method to generate data for both positive and negative mode II contributions. Less obvious is the advantage that can accrue by changing the position of the coating relative to the substrate in helping to suppress kinking of the crack out of the interface. As discussed in the next subsection, the sign of the mode II stress intensity factor and the location of the coating relative to the interface can have a strong influence on the propensity of the crack to remain in the interface, especially if the interface toughness is comparable to that of the coating. Thus, a delamination crack that tends to stray from the interface for one substrate/coating/stiffener configuration might remain in the interface when the roles of the substrate and stiffener are interchanged in the manner suggested. In other words, if it is difficult to measure $\Gamma(\psi)$ for loadings with one sign of ψ, it may be possible to measure the toughness for the opposite sign of ψ. For toughness functions, which are symmetric in ψ, this would suffice. Generally, however, $\Gamma(\psi)$ need not be symmetric in ψ, as the set of data for an epoxy/glass interface illustrates.[22] For some interfaces $\Gamma(\psi)$ does appear to be symmetric.[19]

(4) Propensity for a Delamination Crack to Remain in or Near the Interface

Success in measuring delamination toughness requires that the path followed by the crack in a test is similar to the path followed by the crack in the application of interest. If the interface between the coating and the substrate has low toughness compared with the coating itself (and compared with the substrate, which for the discussion here will be assumed to be very tough), then crack propagation initiated in the interface is likely to remain in the interface for all loading conditions. Depending on the actual system, this interface might lie between the coating and the TGO or between the TGO and the bond coat. For systems without a TGO, this interface would lie between the coating and the bond coat. For some coating systems, delaminations occur within the coating but just above the coating/substrate interface (i.e., just above the TGO if one is present or above the bond coat if not). This may be a consequence of loading conditions which continually drive the crack toward the interface, as

discussed below, or it may be due to the existence of a very thin interfacial layer of less tough material. Residual stress in the coating can also play a role in determining whether the crack stays in or near the interface. A residual compressive stress in the coating acts to discourage cracks from turning into the coating, while tensile stress encourages deviations away from the interface or the low-toughness interfacial layer. A few brief observations related to these effects close out this section.

With the coating above the substrate and with the tip at the right end of the interface crack as in Fig. 13, a positive K_{II} ($\psi > 0$) would promote kinking of the crack downward into the substrate were it not too tough. Thus, if a crack is propagating in the coating just above the interface, a positive K_{II} will tend to cause it to hug the interface. Conversely, a negative K_{II} ($\psi < 0$) promotes upward kinking into the coating. If the coating lies below the interface the situation reverses: a positive K_{II} ($\psi > 0$) promotes downward kinking into the coatings while a negative K_{II} ($\psi < 0$) keeps the crack in or near the interface. Quantitative conditions based on the relative toughness of the interface to that of the coating are available for assessing the likelihood of kinking out of the interface when the sign of K_{II} promotes kinking into the substrate.[23]

The discussion thus far suggests that a test with the sign of K_{II} favoring kinking is less likely to succeed in delivering delamination data than one with the opposite sign of K_{II}. However, other factors must clearly be in play because some relatively brittle TBC coating systems appear to have a fracture path in the coating just above the interface with the TGO, or just above the bond coat if no TGO is present, due to compressive edge delaminations and buckle delaminations having negative mode II ($\psi = -90°$), or nearly so. These are conditions under which kinking into the coating should be most likely to occur. One possibility is that the thin layer of the coating material just above the interface has significantly lower toughness than the coating itself due to chemical or microstructural differences. Microcracking along, or just above, the interface under thermal exposure, which is thought to produce the degradation of toughness seen in Figs. 3 and 5, would be an example. Another possibility is that the in-plane compressive stress in the coating suppresses any tendency for cracks to wander away from the interface or the low-toughness layer. Quantitative mechanics results are also available to assess the role of residual stress.[23] The relevant dimensionless parameter is

$$\eta = \frac{\sigma \sqrt{c}}{\sqrt{\bar{E}^* G}} \qquad (17)$$

Here, σ is the residual stress, c is the putative kinked crack length (see Fig. 13), and $\bar{E}^* \cong 2(1/\bar{E} + 1/\bar{E}_C)^{-1}$. A compressive residual stress significantly reduces any tendency to kink into the coating if η is larger in magnitude than about 0.5, and, conversely, a residual tension of this level strongly promotes kinking. Assuming $c \cong t/10$ and representative values for the other quantities in (17), one concludes that compressive residual stress in the coating can indeed play a role in suppressing kinking even when the sign of K_{II} favors kinking into the coating.

V. Conclusions

Experimental measurement of mixed mode delamination toughness as a function of thermal history is an essential element of TBC durability assessment. Edge delaminations are among the most common types of TBC failures, and, therefore, test methods to measure mode II, and near-mode II, delamination toughness must be developed. To date, most toughness data acquired for TBC systems has fallen within the range of mode mix $0 \leq \psi < 60°$, using the convention adopted in this paper. Several mixed mode tests have been surveyed here which, in principle, could generate data over the entire range of mode mix, although serious obstacles to their implementation may exist. The compression edge-delamination specimen in Fig. 6 closely mimics edge delaminations experienced in service, and it may be the most promising test to generate mode II ($\psi = -90°$) toughness data. Friction must be considered in this test, but friction must also be accounted for in any attempt to predict the behavior of an in-service edge delamination. Indeed, friction will have to be considered in any test if contact of the crack faces occurs for mode mixes approaching mode II. Further work to account for the interplay between friction and toughness under near-mode II conditions is needed, especially when the elastic mismatch between the coating and the substrate is large. When the crack is open, the role of the elastic mismatch on the mode mix has been quantified. The shift in mode phase angle is modest as long as the mismatch is not large.

With the conventions adopted in this paper, edge delaminations driven by compressive stress in a coating lying above the substrate experience negative mode II conditions ($K_{II} < 0$, $\psi = -90°$). For this situation, one consequence of negative K_{II}, as opposed to positive K_{II}, is the greater tendency for an interface crack to kink out of the interface into the coating interior. This tendency is problematic in any delamination toughness test unless the interface is weak. A residual compressive in the coating helps to counteract the errant propensity. Edge delaminations occurring in service are often observed to be interfacial, or to lie with a layer just above the interface, suggesting that, for whatever the reason, kinking out of the interface plane is suppressed.

Appendix A

(A.1) Interfacial Crack Mechanics for General Elastic Mismatch

For a plane strain crack on a planar interface between two isotropic elastic solids, the stresses acting on the interface ($x_2 = 0$) a distance r ahead of the crack tip within the region dominated by the singular field are[24]

$$\sigma_{22} + i\sigma_{12} = \frac{(K_I + iK_{II})}{\sqrt{2\pi r}} r^{i\varepsilon} \qquad (A\text{-}1)$$

with $i = \sqrt{-1}$ and where

$$\varepsilon = \frac{1}{2\pi} \ln\left(\frac{1-\beta}{1+\beta}\right) \qquad (A\text{-}2)$$

If $\beta = 0$, $\varepsilon = 0$ and (A-1) reduces to the usual expression for a homogeneous solid and the mode mix definition in terms of the stresses just ahead of the crack tip is

$$\psi = \tan^{-1}(\sigma_{12}/\sigma_{22}) = \tan^{-1}(K_{II}/K_I)$$

If $\beta \neq 0$, $\varepsilon \neq 0$ and the crack tip field has an "oscillatory" nature. A number of complications must be considered, including the possibility of crack face interpenetration. In addition, the stress ratio, σ_{12}/σ_{22}, from (A-1) is not independent of r and a specific location on the interface must be identified to define the mode mix.[9,23] Identify a distance ℓ ahead of the crack tip within the zone governed by (A-1) characterizing the fracture process. If the fracture process depends on the relative amount of shear to normal traction on the interface, σ_{12}/σ_{22}, then the location $r = \ell$ is a sensible choice to evaluate the mode mix. With $r = \ell$ in (A-1),

$$\psi \equiv \tan^{-1}\left(\frac{\sigma_{12}}{\sigma_{22}}\right)_{r=\ell} = \tan^{-1}\left(\frac{\text{Im}((K_I + iK_{II})\ell^{i\varepsilon})}{\text{Re}((K_I + iK_{II})\ell^{i\varepsilon})}\right) \qquad (A\text{-}3)$$

By dimensional considerations, the plane strain solution to any interface crack problem necessarily has the form

$$(K_I + iK_{II}) = \text{Applied stress} \times F \times L^{1/2 - i\varepsilon}$$

where $F = |F|e^{i\phi}$ is a dimensionless complex function of the

dimensionless parameters in the problem and L is one of the lengths. By (A-3), it follows that

$$\psi \equiv \tan^{-1}\left(\frac{\sigma_{12}}{\sigma_{22}}\right)_{r=l} = \phi + \varepsilon \ln\left(\frac{\ell}{L}\right) \qquad (A\text{-}4)$$

It is obvious from (A-4) that ψ depends on the choice of ℓ if $\varepsilon \neq 0$. This equation also reveals how the mode mix changes when the choice of ℓ changes. With ψ_1 associated with ℓ_1 and ψ_2 associated with ℓ_2, (A-4) gives

$$\psi_2 = \psi_1 + \varepsilon \ln(\ell_2/\ell_1) \qquad (A\text{-}5)$$

To summarize, when $\varepsilon \neq 0$, the mode mix, ψ, depends on the choice of ℓ and, consequently, the interface toughness $\Gamma(\psi)$ also implicitly depends on the choice of ℓ. The transformation from one choice to another satisfies

$$\Gamma(\psi_2, \ell_2) = \Gamma(\psi_1 + \varepsilon \ln(\ell_2/\ell_1), \ell_1) \qquad (A\text{-}6)$$

Illustrations have been given in Hutchinson and Suo.[9]

Some authors have chosen ℓ so as to make the toughness function, $\Gamma(\psi, \ell)$, as symmetric as possible with respect to ψ when fitting data. This is not necessarily the most rational choice of ℓ. It is worth noting that even when $\beta = 0$ the function $\Gamma(\psi)$ need not be symmetric in ψ. Effects contributing to the fracture toughness such as crack tip plasticity and microcracking can produce significant asymmetry in $\Gamma(\psi)$.

(A.2) Beam Theory Solution for Inverted Four-Point Bend Specimen

With reference to the inverted four-point bend specimen in Fig. 6, let x be measured from the center of the beam and anticipate that an upward force/depth, P, is exerted by the lower beam on the upper beam at the point of contact just to the right of the center. Further, anticipate that there is no other contact between the beams in the interval $0 < x < a$. At $x = 0$, the lower beam has moment M while the upper beam sustains no moment. At $x = a$, the deflections and the slopes of the two beams must coincide. Under these assumptions, $P = 3M/(4a)$ and the difference between the deflections of the vertical deflections of the upper and lower beams is found to be

$$w_{\text{upper}} - w_{\text{lower}} = \frac{1}{4}\frac{Ma^2}{B}\frac{x}{a}\left(1 - \frac{x}{a}\right)^2$$

in accord with starting assumption. The energy release rate can be computed directly by the derivative of the total energy with respect to a giving (7). In addition, as $x \to a$, one finds $M_{\text{upper}} = 3M/4$ and $M_{\text{lower}} = M/4$. The mode mix, $\psi = -60°$, can be estimated using the exact results[9] for a infinite layer with a semi-infinite crack such that the equal thickness layers above and below the crack support moments $M_{\text{upper}} = 3M/4$ and $M_{\text{lower}} = M/4$, respectively, and the uncracked layer to the right supports M. It can also be noted that the energy release for this exact solution agrees with the direct calculation based on the beam solution in (7).

(A.3) Finite Element Modeling

ANSYS version 12.1 was used for the linear-elastic finite element modeling of the inverted four-point bend specimen and the compression edge-delamination specimen.[||] The general modeling approach was as follows: (i) the specimen geometry and loading was parameterized; and (ii) a customized ANSYS script was written to preprocess, solve, and postprocess the static solution for each specimen crack length independently, i.e., crack growth was not explicitly modeled.

Specimens dimensions were selected to ensure fairly slender layers, e.g., $L/h = 20$ and $a/L \leq 2$ where a is the crack half-length emerging from the "small" notch with half-width $b = h/10$ in the

||ANSYS v12.1 Mechanical & Mechanical APDL Documentation.

center of the top layer, and L is the total half-length (parallel to the crack) of the specimen. The inverted four-point bend specimen was subjected to a linear longitudinal stress distribution remote from the crack tip defined to give a "pure" bending moment M as shown in Fig. 6. The compression edge-delamination specimen was subjected to a prescribed, uniform longitudinal compressive displacement generated by the rigid platens depicted in Fig. 6. The load P was given by the sum of the corresponding axial nodal reactions.

Relatively coarse meshes comprised of plane-strain 2D, quadratic eight-node elements were used away from the crack tip, e.g., 32 elements through the total specimen thickness. Singular forms of these elements, wherein the mid-side nodes are placed at the quarter points to produce an asymptotic square-root stress/strain singularity, were used to mesh the first row of elements defining the crack tip such that the maximum element edge length did not exceed 2.5% of h. Relatively fine meshes were used to transition between the coarsely meshed regions and the crack tip, e.g., 200 quadrilaterals in the axial and transverse directions for a bounding box with a maximum (total) edge length of about one-quarter h; and a small number of quadratic six-node elements comprised the perimeter of this transitional region—well away from the crack tip as these are not permitted in the J-integral calculation mentioned below. At the crack faces, the initial coarse mesh was refined (e.g., initial edge length divided by four) using a combination of six-node and eight-node quadratic elements; then, three-node quadratic contact (a.k.a. slave) and target (a.k.a. master) elements were overlaid on the free faces of the existing elements comprising the crack faces.

Crack face contact was modeled using standard, unilateral contact along the entire crack length such that the crack faces could open, or separate, via a transverse (normal to the crack faces) displacement gap between the contact/target elements; or, the crack faces could close such that a nonzero contact pressure developed. Most of the default ANSYS standard-contact options were selected such that contact interference was minimized using an augmented Lagrangian approach with automatic solution control. Coulomb friction was specified with a coefficient of friction μ_f. Note that, because the crack faces were treated as initially co-linear (e.g., essentially zero initial gap to within numerical tolerances), the contact problem is one involving closely conforming surfaces such that, in general, many nonlinear Newton–Raphson iterations are required to solve for the crack-face contact pressure and contact gap distributions; the number of iterations needed increases with μ_f.

Once the static solution was obtained for each specimen crack length, the energy release rate was calculated using ANSYS' J-integral, the mode mix was determined using ANSYS' interaction integrals for the stress intensity factors. For comparison purposes, this result was checked via ANSYS' more approximate crack-tip displacement extrapolation for the stress intensity factors. In the range of crack lengths considered, the crack tip was closed and the mode II stress intensity was negative. Corresponding mode I stress intensities were positive for the inverted four-point bend specimen, whereas small-magnitude (i.e., up to about 10% of mode II) negative mode I stress intensities were calculated for the compression edge-delamination specimen. The latter result is probably due to the numerically approximate enforcement of the ideal point-wise penetration constraint. One would expect zero mode I if this constraint was perfectly satisfied.

For ease of postprocessing, only "average" contact pressure and contact gap data were tabulated, i.e., nodal pressures/gaps averaged at the centroid of each contact element. This is acceptable because these data are only used to gain qualitative insight into the effects of crack face contact. In addition, the contact element mesh is apparently not overly coarse given the relatively "smooth" appearance of much these data when plotted. However, for "small" a/h, e.g., $a/h = 0.4$ in Fig. 11(A), one expects the (elastic) contact pressure distribution to be

asymptotically singular as $x/a \to 1$ because of the "effective" reentrant corner formed at the (transverse) notch contact when closed; this detail is not captured here, nor do we anticipate a significant error in the crack-tip quantities calculated as a result of this approximation. The singular pressure behavior also shows up in the limit of $x/a \to 1$ for the inverted four-point bend specimen.

Acknowledgments

The authors are indebted to M. R. Begley, G. Ojard, G.V. Srinivasan, and F. Zok for helpful discussions.

References

[1]N. P. Padture, M. Gell, and E. H. Jordan, "Thermal-Barrier Coatings for Gas-Turbine Engine Applications," *Science*, **296**, 280–4 (2002).

[2]D. R. Clarke and C. G. Levi, "Materials Design for the Next Generation Thermal-Barrier Coatings," *Annu. Rev. Mater. Res.*, **33**, 383–417 (2003).

[3]A. G. Evans, D. R. Mumm, J. W. Hutchinson, G. H. Meier, and F. S. Pettit, "Mechanisms Controlling the Durability of Thermal-Barrier Coatings," *Prog. Mater. Sci.*, **46**, 505–53 (2001).

[4]S. Kraemer, S. Faulhaber, M. Chambers, D. R. Clarke, C. G. Levi, J. W. Hutchinson, and A. G. Evans, "Mechanisms of Cracking and Delaminations within Thick Thermal Barrier Systems in Aero-Engines Subject to Calcium–Magnesium–Alumino–Silicate (CMAS) Penetration," *Mater. Sci. Eng. A*, **490**, 26–35 (2008).

[5]I. Hofinger, M. Oechsner, H.-A. Bahr, and M. V. Swain, "Modified Four-Point Bend Specimen for Determining the Interface Fracture Energy for Thin, Brittle Layers," *Int. J. Fracture*, **92**, 213–20 (1998).

[6]P.-Y. Thery, M. Poulain, M. Dupeux, and M. Braccini, "Spallation of Two Thermal Barrier Coating Systems: Experimental Study of Adhesion and Energetic Approach to Lifetime During Cyclic Oxidation," *J. Mater. Sci.*, **44**, 1726–33 (2009).

[7]H. H. Yu, M. Y. He, and J. W. Hutchinson, "Edge Effects in Thin Film Delamination," *Acta Mater.*, **49**, 93–107 (2001).

[8]D. Balint and J. W. Hutchinson, "Mode II Edge Delamination of Compressed Thin Films," *J. Appl. Mech.*, **68**, 725–30 (2001).

[9]J. W. Hutchinson and Z. Suo, "Mixed Mode Cracking in Layered Materials," *Adv. Appl. Mech.*, **29**, 63–191 (1992).

[10]A. Vastinonta and J. L. Beuth, "Measurement of Interfacial Toughness in Thermal Barrier Coating Systems by Indentation," *Eng. Fract. Mech.*, **68**, 843–60 (2001).

[11]C. Eberl, D. S. Gianola, Xi Wang, M. Y. He, A. G. Evans, and K. J. Hemker, "In Situ Measurement of the Toughness of the Interface Between a Thermal Barrier Coating and a Ni Alloy," *J. Am. Ceram. Soc.* (this issue).

[12]S. Q. Guo, D. R. Mumm, A. M. Karlsson, and Y. Kagawa, "Measurement of Interfacial Shear Mechanical Properties in Thermal Barrier Coating Systems by a Barb Pullout Method," *Scr. Mater.*, **53**, 1043–8 (2005).

[13]M. Tanaka, Y. F. Liu, S. S. Kim, and Y. Kagawa, "Delamination Toughness of Electron Beam Physical Vapor Deposition (EB-PVD) Y2O3–ZrO2 Thermal Barrier Coatings by the Pushout Method: Effect of Thermal Cycling Temperature," *J. Mater. Res.*, **23**, 2382–92 (2008).

[14]Y. F. Liu, Y. Kagawa, and A. G. Evans, "Analysis of a "Barb Test" for Measuring the Mixed-Mode Delamination Toughness of Coatings," *Acta Mater.*, **56**, 43–9 (2008).

[15]P. G. Charalambides, J. Lund, A. G. Evans, and R. M. McMeeking, "A Test Specimen for Determining the Fracture Resistance of Bimaterial Interfaces," *J. Appl. Mech.*, **56**, 77–82 (1989).

[16]G. Bao, S. Ho, B. Fan, and Z. Suo, "The Role of Material Orthotropy in Fracture Specimens for Composites," *Int. J. Solids Struct.*, **29**, 1105–16 (1992).

[17]B. F. Sorensen, K. Jorgensen, T. Jacobsen, and R. C. Ostergaard, "DCB-Specimen with Uneven Bending Moments," *Int. J. Fracture*, **141**, 163–76 (2006).

[18]J. S. Wang and Z. Suo, "Experimental Determination of Interfacial Toughness Curves Using Brazil-Nut-Sandwiches," *Acta Metall.*, **38**, 1279–90 (1990).

[19]L. Bank-Sills, V. Boniface, and R. Eliasi, "Development of a Methodology for Determination of Interface Fracture Toughness of Laminate Composites—the 0°/90° Pair," *Int. J. Solids Struct.*, **42**, 663–89 (2005).

[20]A. Dorogoy and L. Banks-Sills, "Effective of Crack Face Contact and Friction on Brazilian Disk Specimens—A Finite Difference Solution," *Eng. Fract. Mech.*, **72**, 2758–73 (2005).

[21]Z. Suo and J. W. Hutchinson, "Sandwich Test Specimens for Measuring Interface Crack Toughness," *Mater. Sci. Eng.*, **A107**, 135–43 (1989).

[22]K. M. Liechti and Y.-S. Chai, "Asymmetric Shielding in Interfacial Fracture Under Inplane Shear," *J. Appl. Mech.*, **59**, 295–304 (1992).

[23]M. Y. He, A. Bartlett, A. G. Evans, and J. W. Hutchinson, "Kinking of a Crack Out of an Interface: Role of In-Plane Stress," *J. Am. Ceram. Soc.*, **74**, 767–71 (1991).

[24]J. R. Rice, "Elastic Fracture Concepts for Interfacial Cracks," *J. Appl. Mech.*, **55**, 98–103 (1988).   □

J. Am. Ceram. Soc., **94** [S1] S96–S103 (2011)
DOI: 10.1111/j.1551-2916.2011.04436.x
© 2011 The American Ceramic Society

journal

Delamination of Ceramic Coatings with Embedded Metal Layers

Matthew R. Begley[†]

Mechanical Engineering, University of California, Santa Barbara, CA 93106

Haydn N.G. Wadley

Materials Science and Engineering, University of Virginia, Charlottesville, VA 22904

This paper investigates the effects of thin plastically deformable layers embedded in an elastic coating upon debonding of a multilayer from its substrate. Such coatings are normally deposited at high temperature and cooled to ambient, resulting in significant stresses from thermal expansion mismatch. Other stresses can develop during subsequent thermal cycling if volumetric changes such as phase changes or oxidation occur in the system. We present an elastic–plastic model to calculate these stresses in the adhered state (after deposition and cooling but before debonding) and the released state (after delamination). These results are used to calculate the steady-state energy release rate (ERR) that drives debonding at the interface between the multilayer and the substrate. It is shown that plastic straining in the ductile layers can lead to significant reductions in crack driving force by dissipating energy both before and during coating release. Regime maps are developed to illustrate reductions in ERR in terms of the yield strength and volume fraction of the metal layers. As an example, the model is used to predict the impact of embedded platinum layers within an yttria-stabilized zirconica coating; the crack ERR for the composite coating is shown to be 30% lower than that for a uniform ceramic coating.

I. Introduction

Low-thermal-conductivity ceramic layers are now widely used as a component of thermal barrier coating (TBC) systems on superalloy airfoils used in gas turbines/engines. They reduce the metal's operating temperatures, slow the rate of metal oxidation and reduce the superalloy's susceptibility to hot corrosion.[1–4] Yttria-stabilized zirconia (YSZ) or YSZ/rare earth zirconate bilayer ceramics are usually used for this application because of their low thermal conductivity and high environmental durability. They are applied to superalloy components that have been coated previously with an aluminum-rich metallic bond coat, which is designed to oxidize slowly and form a protective thermally grown oxide (TGO) layer. Analogous environmental barrier coatings (EBCs) based upon barium strontium aluminum silicate or rare earth silicate top coats applied to mullite/silicon bilayers are also being developed to protect silicon-based ceramic matrix composites from oxidation and evaporative hydrolysis at elevated temperatures.[5] All of these coating systems are vulnerable to delamination failures due to the large stresses (and the associated stored elastic strain energy) during cooling that result from thermal expansion mismatch between the layers, as well as the volumetric strain associated with oxidation, phase change, or densification at high temperatures.[6]

Tony Evans and his many collaborators showed that delamination failure is driven by the difference between the stored energy of the adhered and debonded system; this energy is released during the propagation of (usually) an interface crack on the plane of minimum fracture toughness of the system. During the cooling of ceramics deposited at elevated temperature on metal substrates, the mismatch in coefficients of thermal expansion between coating layer components and the substrate can lead to large (usually) compressive coating system stresses and significant stored elastic energy.[6–8] The gradual growth of an oxide layer on the substrate surface during high temperature use of TBC systems also leads to significant stored elastic energy that can lead to time- and temperature-dependent spallation. Spallation can be reduced significantly by decreasing the thermal mismatch strain between the coating system components and by decreasing the elastic modulus of a TBC coating (e.g., by increasing its porosity), thereby reducing the stored elastic energy that drives debonding.[9] However, other coating modifications to suppress failure are relatively few, especially in comparison with the many approaches available to manipulate the metallic bond coat. Bond coat modifications enable control of oxidation, promote increases to the plastic dissipation in the bond coat near the delamination front,[10] or suppress the oxide rumpling that often initiates coating failure.[11,12] The dearth of ceramic coating modifications appears to be due to the difficulty of identifying suitable strategies for manipulating the coating's thermomechanical properties without sacrificing environmental resistance and thermal performance.

One potential strategy for suppressing coating failure utilizes composite coatings comprising multiple layers. The concept is to combine individual layers with different thermomechanical properties to produce a multilayer coating with more desirable properties than a comparable monolithic coating. The focus here is on multilayers comprising a planar stack of discrete thin films of different materials, one of which can plastically deform. Such coatings are now easily created with modern deposition techniques.[2,13] Significant analysis is required to understand the relationships between the layer properties, the stored elastic energy in the adhered, and debonded states and the interfacial fracture toughness. The analysis of systems that remain elastic are now abundant,[14] but situations in which one or more of the layers plastically deform are much lesser developed and require a different analysis than those set forth previously. While the focus here is on embedded thin layers of an elastic–plastic metal, the analysis is valid for arbitrary layer thicknesses and purely elastic response (e.g., a multilayer comprising two different ceramics[15]). It is worth noting that functionally graded materials (FGMs), which incorporate a continuous distribution of multiple phases, are an analogous concept.[16,17] However, because FGMs are typically more difficult to fabricate and require a different analysis approach, they are not addressed here.

Here, we explore the performance of a model multilayer with a symmetric stacking sequence as shown in Fig. 1. A micromechanical model is developed and used to evaluate potential improvements to delamination resistance and identify directions for future experimental assessment. The present study of ductile metal layers is motivated by a recent experimental study of multilayers comprising thin platinum layers embedded within a YSZ

R. McMeeking—contributing editor

Manuscript No. 28875. Received November 9, 2010; approved January 6, 2010.
[†]Author to whom correspondence should be addressed. e-mail: begley@engr.ucsb.edu

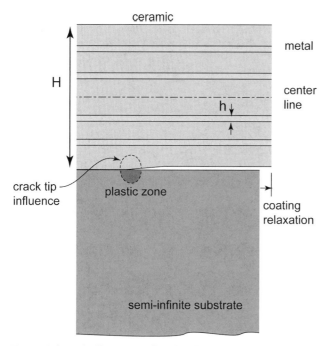

Fig. 1. Schematic illustration of a slice through a model multilayer/substrate system. Delamination occurs by extension of a crack at the interface between the substrate and the ceramic layer at the bottom of the stack. Debonding is driven by relaxation of the stored elastic strain energy in the multilayer. The present analysis is valid for any symmetric stacking sequence, with the results depending only on the volume fraction of the embedded layers $f = Nh/H$, where N is the number of embedded lines.

coating, whose intent was to reduce the radiative component of heat transport through the coating the surface of the metal substrate.[18] This study found that the thin platinum layers also appeared to delay spallation during thermal cycling. The micromechanical model presented here illustrates that metal layers can introduce three different effects that reduce the likelihood of spallation: (i) the metal layers can introduce constraint against ceramic expansion during debonding, (ii) yielding in the metal while adhered can reduce the stored elastic energy in the multilayer, and (iii) yielding in the metal during debonding can dissipate plastic work, which would otherwise be available to drive interface cracking.

The first effect is analogous to the role of the TBC coating when delamination occurs beneath a TBC/TGO bilayer; the TBC partially suppresses expansion of the highly compressed oxide during debonding, which raises the stored elastic strain energy in the debonded state. This lowers the driving force for interface delamination by reducing the strain energy released during debonding.[8] The second effect is analogous to reducing the modulus of a coating,[9] which lowers the stored elastic strain energy when the film is adhered; again, this reduces the strain energy released by debonding. The third effect is analogous to plastic yielding at the tip of a crack at a metal/ceramic interface, which dissipates energy that would otherwise be available to increase crack area.[19] While these analogous scenarios provide guidance to the relevant mechanics, previous models either assume elastic behavior for all the layers[14] or plastic yielding in a homogeneous metal film (or just the substrate).[19] The work presented here addresses situations where yielding occurs *within* a multilayered coating, and seeks to identify combinations of material properties and coating geometry that enhance the coating's resistance to delamination.

It is important to note that the analysis presented here does not take into account the role of a TGO layer beneath the multilayer coating; delamination is presumed to occur between the multilayer coating system and the underlying platform (which may consist of many layers, e.g., substrate, bond coat, and TGO). In this scenario, all layers beneath the coating system remain adhered and do not contribute to delamination driving forces because their stresses are not relieved by debonding. A complimentary analysis of the scenario where the delamination occurs *beneath* the TGO will be published elsewhere; qualitatively, the implications of the metal layers are the same, although the analysis is more complicated due to bending in the coating induced by the TGO. Finally, it is also worth noting that the microstructure of the multilayer is likely to be strongly dependent on the deposition process, and one can envision techniques that lead to a discontinuous metal layer (possibly surrounded by the ceramic phase); this is particularly true for small volume fractions of metal that could form islands on the tips of a columnar ceramic microstructure. In such cases, a completely different analysis is required, although it seems reasonable to expect that isolated metal islands (surrounded by either void space or ceramic phase) would be much less effective in altering the strain energy stored in the ceramic, due to more highly localized residual stresses and plastic straining in the metal phase.

II. Overview of the Model

Consider an isotropic elastic (ceramic) coating containing embedded elastic–plastic (metal) layers, which is bonded to a substrate (Fig. 1). The steady-state energy release rate (ERR) is computed for the extension of a semi-infinite crack at the interface between the substrate and the bottom of the multilayer.[14] It is assumed the embedded layers are distributed symmetrically in the multilayered coating such that the multilayer experiences pure extensional deformation (without bending), both while adhered to the substrate and after debonding.

The total multilayer thickness is defined as H and the layer thickness fraction of ductile layers is defined as $f = Nh/H$, where N is the number of ductile layers and h is the ductile layer thickness. The substrate is assumed to be semi-infinite in thickness; this implies (i) that there is no bending in the multilayer, and (ii) all layers in the adhered stack experience the strain imposed by the substrate. Because there is no bending, the results are independent of the layer thickness and positioning and depend only on the ductile layer fraction. It is also assumed that the embedded layers are elastic, perfectly plastic, and characterized by the uniaxial yield strength σ_Y, the elastic modulus E_m and the coefficient of thermal expansion α_m.

To further simplify the analysis, it assumed that the deformation in the bilayer after debonding is purely biaxial. Strictly speaking, debonding does not relieve constraint parallel to the crack front. However, imposing different in-plane deformations after debonding significantly complicates analyses involving metal yielding, because different plastic strains have to be tracked in the two in-plane directions. For elastic systems, the consequence of the assumptions regarding postrelease deformation is relatively modest, and it is reasonable to expect similar behavior for the present system. To simplify the presentation, the elastic moduli used here are the biaxial moduli, i.e., $E = E^*/(1-v)$ where E^* is the usual uniaxial loading modulus and v is the Poisson's ratio. This definition means that the yield strain ε_Y is the biaxial yield strain, i.e., $\varepsilon_Y = (1-v)\sigma_Y/E^*$, where σ_Y is the uniaxial yield stress.

Debonding is assumed to occur only between the bottom of the multilayer and the substrate, i.e., at an interface comprising a ceramic layer and the substrate. In the following, it is assumed that the critical ERR of the interface, G_c, is an intrinsic property of the interface, such that plastic deformation in the substrate is accounted for in G_c. Crack growth occurs when $G \equiv \Delta W_e - W_p \geq G_c$, where ΔW_e is the change in stored elastic strain energy (per unit area) in going from the adhered state to the debonded state, and W_p is the plastic work (per unit area) dissipated in the ductile layers (within the ceramic coating) in going from the adhered to the debonded state. We assume that the plastic work performed in the ductile layers during debonding is dominated by relaxation of the misfit strains, and not influenced by the presence of the crack tip. That is, the elevated stresses near the crack tip do not significantly influence plastic strains in the closest ductile layer, due to the constraint imposed by the ceramic layer immediately adjacent to the crack tip. This can be shown to be rigorously true if $E_c G_c / \sigma_Y^2 < d = H/N$, where d is the distance from the interface to the first ductile layer. Given the stiff elastic constraint of the bottom ceramic layer, and the fact that small metal layer fractions are considered here, it is reasonable to expect that any additional plastic work in the embedded layers due to the presence of the crack is much smaller than that in the substrate. Or, the dissipated work in the multilayer *due to crack tip fields* is small compared with that associated with the intrinsic toughness of the ceramic/substrate interface.

Finally, we do not explicitly discuss the possibility that delamination occurs between the ceramic layers and the embedded metal layers; however, the analysis presented here can be applied to this scenario, provided the multilayer above the interface crack is symmetric and the layer fractions and thickness values are adjusted accordingly. For example, the present analysis can be applied to calculate the delamination driving force for the metal/ceramic interface above the second metal line from the top in Fig. 1 (because the ceramic/metal/ceramic multilayer above this interface is symmetric).

The model is constructed in terms of layer strains generated in a stress-free state by extrinsic factors (such as temperature or volumetric swelling) defined as θ_i for each layer, with the subscripts denoting the layer: s-substrate, c-ceramic, and m-embedded (metal) layers. For strains due to thermal expansion, $\theta_i = \alpha_i \Delta T_i$, where α_i is the coefficient of thermal expansion of layer i, and $\Delta T_i = T - T_i$ is the change in temperature from a stress-free reference temperature T_i (e.g., the coating deposition temperature) to the current temperature T. More generally, the strains can be interpreted as those arising from other mechanisms, such as adsorption/desorption or oxidation. Multiple sources of stress-free straining can be included by superposition, as in $\theta_i = \alpha_i \Delta T_i + \theta_g$, where θ_g is the additional strain arising from a given extrinsic mechanism. Furthermore, intrinsic stresses that arise from deposition (e.g., due to epitaxial mismatch with the substrate) can be included by defining an appropriate value of θ such that the deposition stresses are reproduced when the substrate is not strained.

To simplify the subsequent discussion, the problem is cast in terms of the following specific scenario: the layers in the coating are deposited in a stress-free state at high temperature, and the multilayer is then cooled to ambient temperature. The state "following deposition" refers to the cooled system while the multilayer is still adhered to the substrate. The "released" state refers to the cooled system after the multilayer has debonded from the substrate. In this scenario, the discussion can be cast purely in terms of strains arising from thermal mismatch. As will be illustrated, the results depend only on the difference in thermal strains (misfit strains), defined here as $\Delta\theta_{ij} = \theta_i - \theta_j = \alpha_i \Delta T_i - \alpha_j \Delta T_j$ (no summation over repeated indices). In the following, the notation implies that the properties are independent of temperature; however, temperature dependence could be included simply by replacing the relevant property with a temperature-dependent function, e.g., replacing E_i with $E_i(T)$.

While the language of the paper is limited to the above scenario with only thermal strains, it should be emphasized that the model is applicable regardless of the meaning (or the signs) of the layer strains θ_i, such that the model can be used to describe debonding driven by elevated temperatures, intrinsic deposition stresses, swelling, etc.

III. Purely Elastic Systems

The analysis of purely elastic layers is relevant to the design of multilayers comprising layers of ceramics with different thermal expansions, and identifies the range of material properties and temperature changes that limit the response to purely elastic behavior. Examples include TBCs on superalloys, EBCs on sili-

con-based ceramics, or TiC/NbN coatings on cemented carbide cutting tools.[15] As discussed in the previous section, it is important to note that the moduli represent biaxial moduli, and the yield strain reflects the uniaxial yield strength divided by the biaxial modulus.

When adhered to the substrate, the cooled multilayer experiences a strain imposed by contraction of the semi-infinite substrate, θ_s. The stresses in the two components of the adhered coating are dictated by the misfit strains for each layer:

$$\sigma_c = E_c(\theta_s - \theta_c) = E_c \Delta\theta_{sc} \tag{1}$$

$$\sigma_m = E_m(\theta_s - \theta_m) = E_m \Delta\theta_{sm} \tag{2}$$

After release of the symmetric stacking sequence (and in the absence of external bending moments), a force balance in the horizontal direction dictates that

$$(1-f)E_c(\varepsilon_0 - \theta_c) + fE_m(\varepsilon_0 - \theta_m) = 0, \tag{3}$$

where ε_0 is the *total* strain of the multilayer after release. The total strain after release is

$$\varepsilon_0 = \frac{(1-f)E_c}{E_0}\theta_c + \frac{fE_m}{E_0}(\theta_m) \tag{4}$$

where $E_0 = (1-f)E_c + fE_m$ is the rule of mixtures modulus of the multilayer. The stresses in the layers after release are dictated by the ceramic and embedded metal layer misfit strains:

$$\sigma_c = -\frac{fE_cE_m}{E_0}\Delta\theta_{cm} \tag{5}$$

$$\sigma_m = \frac{(1-f)E_cE_m}{E_0}\Delta\theta_{cm} \tag{6}$$

In order for the system to remain elastic, the stresses in the embedded metal layers must be less than the yield stress both before and after release. In terms of misfit strains, these conditions are

$$|\Delta\theta_{sm}| < \varepsilon_Y \tag{7}$$

$$\left|\frac{(1-f)E_c}{E_0}\Delta\theta_{cm}\right| < \varepsilon_Y \tag{8}$$

If these conditions are met, the steady-state ERR *per unit thickness of the multilayer* for debonding at the bottom of the multilayer is then given by

$$G = \frac{1}{2}\left[(1-f)E_c(\Delta\theta_{sc})^2 + fE_m(\Delta\theta_{sm})^2\right] - \frac{f(1-f)E_mE_c}{2E_0}(\Delta\theta_{cm})^2 \tag{9}$$

The normalization "per unit thickness" implies that the actual numerical value of the ERR in units of J/m^2 is obtained by multiplying the above by the total multilayer thickness. This result can be used to evaluate the impact of the embedded layers by comparing it with the ERR for a uniform coating, which is simply $G = (1/2)E_c(\Delta\theta_{sc})^2$. The misfit strains and elastic moduli determine whether the embedded metal layers increase or decrease the ERR compared with that of a uniform coating. This is shown graphically in Fig. 2 for the case where $f = 0.2$ and $E_c = 4E_m$.

The ERR for an elastic multilayer will be lower than that of a uniform coating provided:

$$\frac{E_c}{fE_m}\left[-\sqrt{1-f+f\frac{E_m}{E_c}} - 1 + f\right]\Delta\theta_{sc} \leq \Delta\theta_{sm}$$

$$\leq \frac{E_c}{fE_m}\left[\sqrt{1-f+f\frac{E_m}{E_c}} - 1 + f\right]\Delta\theta_{sc} \tag{10}$$

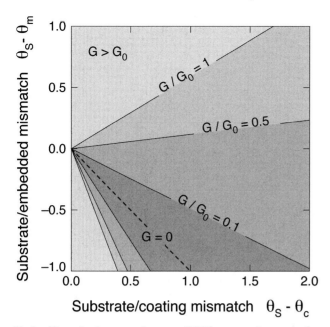

Fig. 2. Normalized energy release rate (ERR) contours for purely elastic systems as a function of thermal misfit strains for $f = 0.2$ and $E_c = 4E_m$. The regime map identifies multilayer property combinations that produce ERRs either larger or smaller than that associated with a uniform coating, $G_o = 0.5 E_c H (\Delta\theta_{sc})^2$.

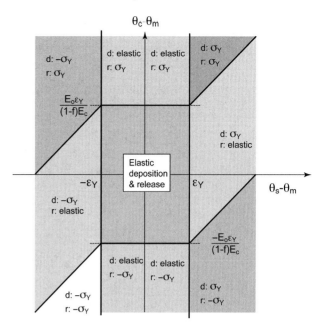

Fig. 3. Regime map indicating where yielding occurs following the deposition and cooling step (d) and the during debonding step (release, r), as a function of the mismatch strains between the substrate, metal, and ceramic. Plastic strains are generated for all combinations that fall outside of the green rectangle in the center.

The equalities in the above expression are represented by the two lines for $G/G_o = 1$ in Fig. 2. In this figure, the darker regions correspond to material property combinations and layer fractions for which the limits (10) are satisfied, such that $G/G_o < 1$. There is a specific set of properties and layer fraction for which the ERR of the multilayer $G = 0$. In these scenarios, the strain energy in the released state is equal to the strain energy in the adhered state. This occurs when the misfit between the substrate and embedded layers is of opposite sign to the misfit between the substrate and coating; upon release, the stresses in the layers change sign but are the same magnitude, implying equal amounts of stored elastic energy in the adhered and released states. From Eq. (9), $G = 0$ when

$$\Delta\theta_{sm} = -\frac{(1-f)E_c}{fE_m}\Delta\theta_{sc} \qquad (11)$$

Thus, if the CTE of the ceramic is larger than the substrate, the CTE of the embedded layers must be less than the substrate. This is difficult to achieve for a ceramic multilayer on a metal substrate; however, it is plausible for ceramic multilayers on ceramics such as EBCs on ceramic matrix composites. Moreover, this result may be useful in the design of composite multilayers with CTEs tailored by hierarchical geometry.[20]

IV. Theory for Elastic–Plastic-Embedded Layers

The introduction of elastic–plastic-embedded layers leads to four possible scenarios: (i) the metal remains elastic during both cool-down and debonding, (ii) the metal remains elastic during cool-down but yields during debonding, (iii) the metal yields during cool-down but experiences elastic recovery during debonding, and (iv) the metal experiences yielding during both cool-down and debonding. Figure 3 shows a regime map that indicates whether or not yielding occurs while adhered (d = "deposition and cooling step") and during debonding (r = "release step"); the boundaries between domains are determined from the analysis that follows. The four scenarios are covered below in separate sections for the adhered state (after deposition and cooling) and the released state (after debonding). The stresses, stored elastic energy, and plastic work dissipated for each step

are presented first. The results are then used to calculate the ERR in a separate section.

(1) Deposition Step: Adhered State

If the metal layers remain elastic during cooling from the deposition temperature, the results in Section II are valid. The yield conditions (7) and (8) define the rectangular box in the center of the regime map shown in Fig. 3. If any of these conditions is violated, the stress in the metal layer after deposition $\sigma_m = -\text{sign}[\theta_{sm}]\sigma_Y$; care must be taken to ensure the proper sign of the stress. Note that the sign[] function is defined as $\text{sign}[x] = -1$ for $x < 0$, $\text{sign}[x] = 1$ for $x > 0$, and $\text{sign}[0] = 0$. If yielding occurs, the following plastic strains are generated in the adhered state due to cooling following deposition:

$$\varepsilon_p^d = \Delta\theta_{sm} - \text{sign}[\Delta\theta_{sm}]\varepsilon_Y \qquad (12)$$

This simply says that the plastic strains generated by cooling from the deposition temperature are equal to the difference between the substrate/metal misfit strains and the yield strain. If yielding occurs during the deposition state, then the stored elastic energy in the *adhered state after cooling from deposition* is

$$\Phi_y^d = \frac{1}{2}(1-f)E_c(\Delta\theta_{sc})^2 + \frac{f\sigma_Y^2}{2E_m} \qquad (13)$$

where the latter term represents the maximum elastic energy that can be stored in an elastic/perfectly plastic material, scaled by the fraction of that material in the multilayer. Plastic work is dissipated during the deposition and cooling step if the metal yields. However, this work does not contribute to the ERR, which is defined as the change in energy/work between the adhered state (*after* the temperature change) and the released state. However, the plastic strains generated during cooling from the deposition temperature are needed to calculate the stresses in the layers after release.

(2) Released State

When the multilayer is released from the substrate, two possibilities exist. Either (i) the metal layers unload elastically or (ii) the metal layers yield; again, care must be taken to ensure the proper sign of the stresses after yield.

(A) Metal Layers Remain Elastic During Release: Once released, all layers experience the same *total* strain, and again a horizontal force balance can be used to solve for the total strain in the multilayer. Noting that the elastic strains (which generate stress) are the total strain minus the thermal and plastic strains (if relevant), equilibrium dictates that

$$(1-f)E_c(\varepsilon_o - \theta_c) + fE_m(\varepsilon_o - \theta_m - \varepsilon_p^d) = 0 \tag{14}$$

where ε_o is the *total* strain of the multilayer after release. This expression is valid even if the metal does not yield while adhered during the cool-down; in this case, $\varepsilon_p^d = 0$. The total elongation after release, assuming the metal layer does not yield during the release step is

$$\varepsilon_o = \frac{(1-f)E_c}{E_o}\theta_c + \frac{fE_m}{E_o}(\theta_m + \varepsilon_p^d) \tag{15}$$

where $E_o = (1-f)E_c + fE_m$ is the rule of mixtures modulus of the multilayer. If the metal does not yield during the release step, the stresses after release are given by

$$\sigma_c = -\frac{fE_cE_m}{E_o}(\Delta\theta_{cm} - \varepsilon_p^d) \tag{16}$$

$$\sigma_m = \frac{(1-f)E_cE_m}{E_o}(\Delta\theta_{cm} - \varepsilon_p^d) \tag{17}$$

The stress in the metal layer can then be used to determine the yield condition for the release step:

$$\left| \frac{(1-f)E_c}{E_o}(\Delta\theta_{cm} - \varepsilon_p^d) \right| \geq \varepsilon_Y \tag{18}$$

This result is represented by the slanted boundaries in the regime map shown in Fig. 3.

If the metal does not yield during release, the elastic strain energy in the released state is

$$\Phi_e^r = \frac{f(1-f)E_mE_c}{2E_o}(\Delta\theta_{cm} - \varepsilon_p^d)^2 \tag{19}$$

Note that the CTE of the substrate only affects the stored strain energy in the released structure by driving plastic strains during the deposition step. That is, the substrate's CTE is implicitly present in the plastic strain term in Eq. (19); the plastic strain scales with the substrate/metal misfit strain relative to the yield strain of the metal.

(B) Metal Layers Yield During Release: After release, horizontal force balance again dictates the total strain of the multilayer; the stress in the metal is equal to the compressive or tensile yield stress. Equilibrium dictates that

$$(1-f)E_c(\varepsilon_o - \theta_c) + \text{sign}[\Delta\theta_{cm} - \varepsilon_p^d]f\sigma_Y = 0 \tag{20}$$

where ε_o is the *total* strain of the multilayer after release. This expression is valid even if the metal does not yield during cooling from the deposition temperature (if it does not yield during cooling, $\varepsilon_p^d = 0$). The total strain after release, assuming the metal layer yields is

$$\varepsilon_o = \frac{-f\,\text{sign}[\Delta\theta_{cm} - \varepsilon_p^d]\sigma_Y}{(1-f)E_c} + \theta_c \tag{21}$$

The stresses in the layers after release, assuming the metal yields, is

$$\sigma_c = -\text{sign}[\Delta\theta_{cm} - \varepsilon_p^d]\frac{f\sigma_Y}{(1-f)} \tag{22}$$

$$\sigma_m = \text{sign}[\Delta\theta_{cm} - \varepsilon_p^d]\sigma_Y \tag{23}$$

These are the stresses in the layer provided the metal yields upon release, a condition that is met if Eq. (18) is satisfied. The stress in the metal is given by $\sigma_m = E_m(\varepsilon_o - \theta_m - \varepsilon_p) = \sigma_Y$, where ε_p is the total plastic strain after the release step. This implies that the plastic strain *increment* associated with the release step is given by

$$\Delta\varepsilon_p^r = \frac{-f\,\text{sign}[\Delta\theta_{cm} - \varepsilon_p^d]\sigma_Y}{(1-f)E_c} \\ + \Delta\theta_{cm} - \text{sign}[\Delta\theta_{cm} - \varepsilon_p^d]\varepsilon_Y - \varepsilon_p^d \tag{24}$$

The elastic strain energy in the multilayer assuming the metal yields upon release is given by

$$\Phi_y^r = \frac{f^2\sigma_Y^2}{2(1-f)E_c} + \frac{f\sigma_Y^2}{2E_m} \tag{25}$$

Note that *if the metal yields upon release*, the elastic strain energy in the released state does not depend on whether or not yielding occurs during following cooling from deposition. This arises because once the metal has yielded, its elastic strain energy is dictated by (limited to) that associated with the yield stress. Also, note that the strain energy in the ceramic layers is also dictated by the metal's yield stress, because the force in the ceramic must balance the force in the metal layers (which is governed by the their yield stress).

(3) Energy Release Rates

The ERR is given by the difference in elastic strain energy between the adhered state after deposition (Φ^d) and in the released state after debonding (Φ^r), minus the plastic work dissipated by plastic strains during the release step (W_p):

$$G = \Phi^d - \Phi^r - W_p \tag{26}$$

The dissipated plastic work (per unit thickness of multilayer) in going from deposition to release is given by $W_p = f\sigma_Y\Delta\varepsilon_p^r$. The plastic work performed during cooling from deposition step has no bearing on the energetics of release. The case of purely elastic systems (no yielding following cooling after deposition or during release) was considered in Section II.

If the metal yields following deposition, but remains elastic upon release, the ERR is given by

$$G = \Phi_y^d - \Phi_e^r = \frac{1}{2}\left[(1-f)E_c(\Delta\theta_{sc})^2 + f\frac{\sigma_Y^2}{E_m}\right] \\ - \frac{f(1-f)E_mE_c}{2E_o}(\Delta\theta_{cm} - \varepsilon_p^d)^2 \tag{27}$$

The conditions for this to be valid are

$$|\Delta\theta_{sm}| > \varepsilon_Y \tag{28}$$

$$\varepsilon_p^d = \Delta\theta_{sm} - \text{sign}[\Delta\theta_{sm}]\varepsilon_Y \tag{29}$$

$$\left| \frac{(1-f)E_c}{E_o}(\Delta\theta_{cm} - \varepsilon_p^d) \right| < \varepsilon_Y \tag{30}$$

If the metal remains elastic following the cooling from deposition, but yields upon release, the ERR is

$$G = \Phi_e^d - \Phi_y^r - W_p \\ = \frac{1}{2}[(1-f)E_c(\Delta\theta_{sc})^2 + fE_m(\Delta\theta_{sm})^2] \\ - \frac{f^2\sigma_Y^2}{2(1-f)E_c} - \frac{f\sigma_Y^2}{2E_m} - f\,\text{sign}[\varepsilon_p^r]\sigma_Y \cdot \Delta\varepsilon_p^r \tag{31}$$

In order for this result to be valid, the misfit strains must be such that

$$|\Delta\theta_{sm}| < \varepsilon_Y \tag{32}$$

$$\left|\frac{(1-f)E_c}{E_o}\Delta\theta_{cm}\right| > \varepsilon_Y \tag{33}$$

$$\Delta\varepsilon_p^r = \frac{-f\,\text{sign}[\Delta\theta_{cm}]\sigma_Y}{(1-f)E_c} + \Delta\theta_{cm} - \text{sign}[\Delta\theta_{cm}]\varepsilon_Y - \varepsilon_p^d \tag{34}$$

Finally, if the metal yields both following the deposition step and during the release step, the ERR is given by

$$\begin{aligned}G &= \Phi_y^d - \Phi_y^r - W_p \\ &= \frac{1}{2}(1-f)E_c(\Delta\theta_{sc})^2 - \text{sign}[\varepsilon_p^r]\cdot f\cdot\sigma_Y\cdot\Delta\varepsilon_p^r\end{aligned} \tag{35}$$

provided the following conditions are satisfied:

$$|\Delta\theta_{sm}| > \varepsilon_Y \tag{36}$$

$$\varepsilon_p^d = \Delta\theta_{sm} - \text{sign}[\Delta\theta_{sm}]\varepsilon_Y \tag{37}$$

$$\left|\frac{(1-f)E_c}{E_o}(\Delta\theta_{cm} - \varepsilon_p^d)\right| > \varepsilon_Y \tag{38}$$

$$\begin{aligned}\Delta\varepsilon_p^r &= \frac{-f\,\text{sign}[\Delta\theta_{cm} - \varepsilon_p^d]\sigma_Y}{(1-f)E_c} \\ &\quad + \Delta\theta_{cm} - \text{sign}[\Delta\theta_{cm} - \varepsilon_p^d]\varepsilon_Y - \varepsilon_p^d\end{aligned} \tag{39}$$

V. Results and Discussion

Because a comprehensive parameter study would be exhaustive, we illustrate the implications of the model by examining the example case of platinum layers embedded in an YSZ coating directly attached to a nickel superalloy (with no TGO layer). We examine the impact of the embedded metal layers by comparing the multilayer's crack driving force with that of a uniform coating with the same overall thickness; hence, the ERR results are normalized by $G_o = (1/2)E_c(\Delta\theta_{sc})^2$. The following properties are assumed (unless otherwise specified): $E_m = 168$ GPa, $E_c = 25$ GPa, $\alpha_m = 8.8$ ppm/°C, $\alpha_c = 12$ ppm/°C, and $\alpha_s = 13.5$ ppm/°C. This implies $\Delta\theta_{sc} = 1.5\cdot[\text{ppm}/°\text{C}]$, and $\Delta\theta_{sm} = 4.7\cdot[\text{ppm}/°\text{C}]$. The assumed temperature change used here is $\Delta T_i = -1000°\text{C}$.

Figure 4 shows the normalized ERR as a function of the metal yield stress (σ_Y) and the volume fraction of metal (f), using the properties defined above. For all combinations shown in Fig. 4, the metal layers yield in compression during cool-down from deposition, because the substrate has a larger CTE than the metal and shrinks to a greater degree during cooling. When the layer is released, the stresses are tensile in the metal and compressive in the ceramic, purely as a result of the compressive plastic strains generated in the metal during cool-down (note that for a purely elastic system, the stresses in the layer are of opposite sign). Above the red dashed line, the metal layers unload elastically during the debonding (release) step. Comparatively high yield stresses lead to comparatively high ERR because: (i) higher yield stresses mean higher levels of elastic strain energy in the metal before release, (ii) higher yield stresses mean smaller plastic strains due to cooling after deposition, which in turn lowers the multilayer stress in the released state, and (iii) higher yield stresses prevent plastic straining upon release, which would dissipate energy.

In contrast, for lower yield stresses (below the red dashed line), the ERR is comparatively lower because: (a) strain energy in the metal is limited in the adhered state due to yielding, (b) large plastic strains are generated during the cooling, which

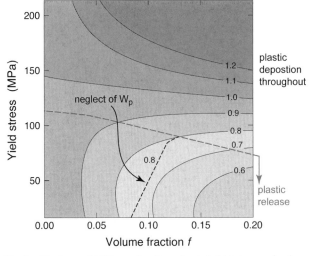

Fig. 4. Contours of G/G_o as a function of metal yield stress and volume fraction, $\Delta\theta_{sc} = -1.5\times10^{-3}$, and $\Delta\theta_{sm} = -4.7\times10^{-3}$. The elastic modulus ratio of the yttria-stabilized zirconia to the metal is ~ 0.15. For all property combinations shown, the metal layers yield during cooling from deposition. Beneath the red dashed line, the metal layers experience further plastic straining as the layer debonds. The dashed black line is the $G/G_o = 0.8$ contour obtained when plastic work during release is neglected, illustrating that plastic yield contributes about 30% of the total reduction (see text).

elevate stresses in the multilayer upon release, and (c) low yield strength implies the metal yields during debonding, which dissipates energy and lowers the ERR. The last effect is illustrated by the dashed black line, which corresponds to the ERR contour $G/G_o = 0.8$ that results if the plastic work term is neglected in the computation. For example, with $\sigma_Y \sim 50$ MPa and $f = 0.1$, the actual ERR reduction is $\sim 27\%$, as compared with $\sim 20\%$ if the plastic work during release is neglected. Hence, the dissipated plastic work comprises about 30% of the total reduction in ERR.

Because the ambient temperature yield strength of pure annealed platinum is low (~ 50 MPa),[21] the results in Fig. 4 strongly suggest that plastic yielding in the Pt layers of the Pt/YSZ system studied previously[18] will have a strongly beneficial effect in suppressing delamination failures. In order to be effective, the yield stress of the platinum layers must remain low, to ensure that plastic straining occurs and work is dissipated that would otherwise drive the interface crack. It is plausible that yield stresses will remain low during thermal cycling, because elevated service temperatures will anneal out the plastic strains from previous cooling cycles. ERR reductions of 30% are possible with layer fractions in the 10% range.

It should be emphasized that the yielding behavior (and associated tension or compression in the adhered and deboned states) is strongly dependent on the sign of the CTE differences and on the sign of the imposed temperature change. A broader view of the relationship between layer properties, regimes of various behaviors and their associated reductions in crack driving force is shown in Figs. 5 and 6. The role of substrate/metal mismatch ($\Delta\theta_{sm}$) and metal volume fraction (f) is shown in Fig. 5, for a fixed value of the substrate/coating mismatch ($\Delta\theta_{sc} = -1.5\times10^{-3}$) and a fixed yield stress ($\sigma_Y = 85$ MPa). Many of the possibilities illustrated in the regime map in Fig. 3 are evident in the part of the parameter space investigated in Fig. 5. The black dashed lines represent the boundaries between elastic and plastic behaviors associated with cooling from deposition. For property combinations above the top dashed line, the metal layer will yield in tension following cooling from deposition. For property combinations below the bottom black dashed line, the metal layer will yield in compression following cooling from deposition. In between the black dashed lines, the multilayer is purely elastic following cooling from deposition.

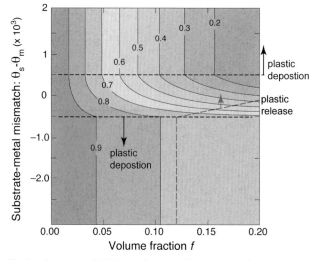

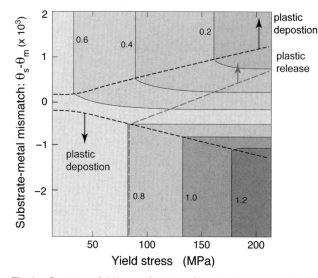

Fig. 5. Contours of G/G_o as a function of coating/metal mismatch and metal layer fraction, for $\Delta\theta_{sc} = -1.5 \times 10^{-3}$ and $\sigma_Y \sim 85$ MPa. The elastic modulus ratio of the coating to the metal is ~ 0.15. The vertical axis can be reinterpreted as $\Delta\theta_{cm} = \Delta\theta_{sm} - \Delta\theta_{sc} = \Delta\theta_{sm} + 1.5 \times 10^{-3}$.

Fig. 6. Contours of G/G_o as a function of coating/metal mismatch and yield stress, for $\Delta\theta_{sc} = -1.5 \times 10^{-3}$, and $f = 0.1$. The elastic modulus ratio of the coating to the metal is ~ 0.15. The vertical axis can be reinterpreted as $\Delta\theta_{cm} = \Delta\theta_{sm} - \Delta\theta_{sc} = \Delta\theta_{sm} + 1.5 \times 10^{-3}$.

The red dashed line indicates the boundary between elastic recovery during debonding (below the line) and plastic straining during debonding (above the line).

A striking feature in Fig. 5 is that the reduction in ERR is independent of the level of substrate/metal mismatch $\Delta\theta_{sm}$ when yielding occurs *both* during cooling from deposition and during debonding. Examining the expression for G in this regime, Eq. (35) indicates that the ERR is only a function of the substrate/metal mismatch through the the plastic strain increment term, $\Delta\varepsilon_p^r$. The combination of Eqs. (37) and (39) implies that $\Delta\varepsilon_p^r \propto \Delta\theta_{cm} - \Delta\theta_{sm}$ when yielding has occurred both during cool-down and debonding. Because $\Delta\theta_{ij} = \theta_i - \theta_j$, the plastic strain increment during release $\Delta\varepsilon_p^r$ scales as $\Delta\theta_{cs}$. This quantity is held fixed in Fig. 5, such that the ERR is independent of $\Delta\theta_{sm}$ whenever yielding has occurred during both steps. A key point is that the *total* plastic strain in the metal layers will scale with $\Delta\theta_{cm}$, but a portion of this misfit is offset by plastic strains generated during cool-down.

As shown at the top of Fig. 5, when $\Delta T_i < 0$ (i.e., cooling from deposition) and $\Delta\theta_{sc} < 0$, the reductions in ERR are the largest when $\alpha_m > \alpha_s > \alpha_c$. In this case, the stress in the metal while adhered is tensile, and equal to the yield stress above the dashed black line. After release, the stress in the metal is still tensile, which keeps the ceramic in compression, implying a smaller change in elastic strain energy in the ceramic during debonding. Moreover, plastic strains are generated during debonding, which further dissipates energy that would otherwise drive cracking. In the top of Fig. 5, the ERR reduction due to plastic work is much larger than that illustrated in Fig. 4, comprising nearly half of the total reduction. At the bottom of Fig. 5, the relative CTE's are $\alpha_s > \alpha_c > \alpha_m$ (for $\Delta T_i < 0$ and $\Delta\theta_{sc} < 0$). In this region, compressive plastic strains are generated during cooling from deposition, which partially accommodate the ceramic/metal mismatch and lead to smaller plastic strains during debonding and hence smaller reductions in ERR.

Similar behavior is evident in Fig. 6, which plots normalized ERR (G/G_o) contours as a function of substrate/metal mismatch ($\Delta\theta_{sm}$) and yield stress (σ_Y), for fixed layer fraction ($f = 0.1$). Once again, the dashed lines correspond to transitions in behavior, regarding yielding during cool-down and debonding. The maximum reductions in ERR correspond to property combinations for which the metal experiences tensile yielding both after cooling from deposition and during debonding. A critical feature of this plot is that there are property combinations for which the ERR is *increased* relative to the uniform coating, which occurs when the metal experiences compressive yielding during cool-down and tensile yielding during debonding. The

ERR can be large for high levels of yield stress because the elastic strain energy in the metal in the adhered state can be significant, and because plastic straining is prohibited during debonding.

It is worth emphasizing that the above discussion is for $\Delta\theta_{sc} < 0$ and $\Delta T_i < 0$; naturally, different conclusions are reached if the sign of the substrate/coating CTE mismatch changes or the system is heated above its assembly temperature.

VI. Summary

A model has been developed to investigate the effects upon the driving force for delamination of coating architecture and the thermomechanical properties for multilayers of two different materials. For elastic systems and layers that experience the same relative change in temperature, the ERR is minimized by the minimum value of an objective function (from Eq. (9)):

$$F = fE_m(\alpha_s - \alpha_m) + (1-f)E_c(\alpha_s - \alpha_m)$$

This expression may find utility for identifying favorable combinations of constituent properties and the volume fraction of the embedded layers that reduce the delamination driving force. We note that optimal solutions may not always include the use of an elastic–plastic layer; a ceramic multilayer with carefully chosen CTEs and layer fractions can have a delamination driving force of zero (at least theoretically). The fundamental underlying mechanism is that the strain energy in the adhered state should be close to that in the release state, because the strain energy density is transferred from one phase to the other during debonding.

For elastic–plastic-embedded layers with $\Delta\theta_{sc} < 0$ and $\Delta\theta_{sm} < 0$, the greatest benefit is reached when yielding occurs during release, as this dissipates energy that would otherwise have been released during crack advance. For layers with equivalent deposition temperatures (i.e., all layers experience the same temperature change from deposition), this occurs when

$$|\alpha_s - \alpha_m| \geq \varepsilon_Y$$

$$|\alpha_c - \alpha_s| \geq \frac{fE_m\varepsilon_Y}{(1-f)E_c}$$

The beneficial effect of plastic straining during release is most effective when these conditions are met, and the sign of $\alpha_s - \alpha_m$ is opposite to $\alpha_s - \alpha_c$. In this scenario, the compressive stress in the ceramic layer after deposition is maintained after debonding, thus decreasing the release of strain energy during debonding. Moreover, plastic strains generated during debonding are greater, leading to futher reductions. For the properties considered here, the reduction in the relative ERR (i.e., that compared with a uniform coating) is roughly twice the volume fraction of the metal; for example, for the Pt/YSZ system, 50% reductions of the elastic ERR are achieved with $f \sim 0.2$.

Finally, it is interesting to consider generalization of the central concept of an inelastic layer with alternative dissipative mechanisms. The underlying strategy would be to design systems for which the embedded layers experience additional damage during release; the work associated with this damage subtracts from the overall energy available to drive debonding. For example, it may be possible to design composite coatings comprising entirely brittle layers, in which the embedded layers experience microcracking (i.e., tunnel cracks in the embedded layers) or undergo martensitic transformations that are triggered only when the layers are released. Naturally, this would involve slightly different micromechanics than that outlined above, and the scaling factors associated with microcracking or phase change might not be as favorable as those associated with the strongly dissipative mechanism of metal plasticity, and the benefits might be limited to a single loading cycle. At this point, it is difficult to fully grasp the full range of possibilities afforded by metal toughening of ceramic coatings; for this, the remarkable insights and vision of of A. G. Evans are sorely missed.

Acknowledgments

H. N. G. W. is grateful for support of this research by the Office of Naval Research, under grant N00014-03-1-0297 (Dr. David Shiffler, Program Manager). M. R. B. is grateful for the support of the National Science Foundation under grant CMII0800790.

References

[1]R. A. Miller, "Current Status of Thermal Barrier Coatings—An Overview," *Surf. Coat. Technol.*, **30**, 1–11 (1987).

[2]C. Levi, "Emerging Materials and Processes for Thermal Barrier Systems," *Curr. Opin. Solid State Mater. Sci.*, **8**, 77–91 (2004).

[3]S. Stecura, "Advanced Thermal Barrier System Bond Coatings for Use on Nickel-, Cobalt- and Iron-Base Alloy Substrates," *Thin Solid Films*, **136**, 241–56 (1986).

[4]A. G. Evans, D. R. Mumm, J. W. Hutchinson, G. H. Meier, and F. S. Pettit, "Mechanisms Controlling the Durability of Thermal Barrier Coatings," *Prog. Mater. Sci.*, **46**, 505–53 (2001).

[5]K. N. Lee, D. S. Fox, and N. P. Bansal, "Rare Earth Silicate Environmental Barrier Coatings for SiC/SiC Composites and S₃N₄ Ceramics," *J. Eur. Ceram. Soc.*, **25**, 1705–15 (2005).

[6]I. T. Spitsberg, D. R. Mumm, and A. G. Evans, "On the Failure Mechanisms of Thermal Barrier Coatings with Diffusion Aluminide Bond Coatings," *Mater. Sci. Eng.*, **394**, 176–91 (2005).

[7]T. Xu, M. Y. He, and A. G. Evans, "A Numerical Assessment of the Propagation and Coalescence of Delamination Cracks in Thermal Barrier Systems," *Interface Sci.*, **11**, 349–58 (2003).

[8]M. R. Begley, D. R. Mumm, A. G. Evans, and J. W. Hutchinson, "Analysis of a Wedge Impression test for Measuring the Interface Toughness Between Films/Coatings and Ductile Substrates," *Acta Mater.*, **48**, 3211–20 (2000).

[9]H. Zhao, Z. Yu, and N. G. Wadley, "The Influence of Coating Compliance on the Delamination of Thermal Barrier Coatings," *Surf. Cout. Technol.*, **204**, 2432–41 (2010).

[10]A. G. Evans, M. Y. He, A. Suzuki, M. Gigliotti, B. Hazel, and T. M. Pollock, "A Mechanism Governing Oxidation-Assisted Low-Cycle Fatigue of Superalloys," *Acta Mater.*, **57**, 2969–83 (2009).

[11]V. K. Tolpygo and D. R. Clarke, "Surface Rumpling of a (Ni, Pt)Al Bond Coat Induced by Cyclic Oxidation," *Acta Mater.*, **13**, 3283–93 (2000).

[12]D. S. Balint and J. W. Hutchinson, "An Analytical Model of Rumpling in Thermal Barrier Coatings," *J. Mech. Phys. Solids*, **53**, 949–73 (2005).

[13]D. D. Hass, P. A. Parrish, and H. N. G. Wadley, "Electron Beam Directed Vapor Deposition of Thermal Barrier Coatings," *J. Vac. Sci. Technol.*, **16**, 3396–401 (1998).

[14]J. W. Hutchinson and Z. Suo, "Mixed Mode Cracking in Layered Materials," *Adv. Appl. Mech.*, **29**, 63–181 (1992).

[15]C. Subramanian and K. N. Stafford, "Review of Multicomponent and Multilayer Coatings for Tribological Applications," **165**, 85–95 (1993).

[16]W. Y. Lee, D. P. Stinton, C. C. Berndt, F. Erdogan, Y. D. Lee, and Z. Mutasim, "Concept of Functionally Graded Materials for Advanced Thermal Barrier Coating Applications," *J. Am. Ceram. Soc.*, **79**, 3003–12 (1996).

[17]S. Widjaja, A. M. Limarga, and T. H. Yip, "Modeling of Residual Stresses in a Plasma-Sprayed Zirconia/Alumina Functionally Graded-Thermal Barrier Coating," *Thin Solid Films*, **434**, 216–27 (2003).

[18]Z. Yu, H. Zhao, and H. N. G. Wadley, "The Vapor Deposition and Oxidation of Pt/YSZ Multilayers," *J. Am. Ceram. Soc.*, submitted (2010).

[19]Y. Wei and J. W. Hutchinson, "Nonlinear Delamination Mechanics for Thin Films," *J. Mech. Phys. Solids*, **45**, 1137–59 (1997).

[20]A. Kelly, R. J. Stearn, and L. N. McCartney, "Composite Materials of Controlled Thermal Expansion," *Compos. Sci. Technol.*, **66**, 154–9 (2006).

[21]Y. N. Ioginov, A. V. Yermakov, L. G. Grohovskaya, and G. I. Studenak, "Platinum Metals," *Review*, **4**, 178–84 (2007). □

J. Am. Ceram. Soc., **94** [S1] S104–S111 (2011)
DOI: 10.1111/j.1551-2916.2011.04494.x
© 2011 The American Ceramic Society

journal

Controlled Introduction of Anelasticity in Plasma-Sprayed Ceramics

Gopal Dwivedi,[†] Toshio Nakamura, and Sanjay Sampath[†]

Center for Thermal Spray Research, Stony Brook University, Stony Brook, New York 11794-2275

Recent studies on the mechanical compliance measurements of plasma-sprayed ceramic coatings have revealed anelastic response, i.e. the stress–strain relations are nonlinear and hysteretic, collectively. The anelasticity stems from the "brick" layered assemblage of spray-deposited droplets along with the presence of porosity and other geometric discontinuities. This anelastic response is reproducible and can provide a quantitative description on the mechanical properties of the sprayed ceramic coating. In this paper, we have examined strategies to manipulate and control these anelastic characteristics through plasma spray processing parameters as well as through extrinsic modifications of the defect interfaces. Plasma-sprayed ceramic coatings are fabricated under various conditions and their unique anelastic parameters are computed. These results reveal that the novel properties are tunable via process and material manipulation, opening up opportunities for microstructural design and optimization of mechanical compliance in industrially relevant coating systems.

I. Introduction

THERMAL spray coatings are produced by high-velocity impingement of thousands of micrometer-sized droplets, melted in a thermal flame (plasma or combustion) and accelerated toward a prepared substrate. The molten or semimolten particles that impinge on the substrate flatten and rapidly solidify to form thin "splats." Successive assembly of these splats results in the form of a coating as shown in Fig. 1. Typical splats are disk-shaped, 50–200 μm in diameter and 0.5–5 μm in thickness.[1,2] The chaotic assemblage of these spray droplets produces incomplete consolidation with broad-ranging porosity and a myriad array of interfaces. The gaps resulting from poor contact between splats are referred as interlamellar pores, while the incomplete filling among impacting particles results in voids or globular pores.[3] The ceramic splats also contain vertical mud cracks resulting from relief of residual stresses, associated with rapid quenching on a substrate.[4,5] In addition, there exist microscopic discontinuities among the splat boundaries with multitude of interfaces. These splat interfaces are nonplanar/rough, have multiscale (nm to μm) contact points with differential bonding characteristics providing significant implications on the macroscale thermo-mechanical deformation behavior. Figure 1 illustrates the various microstructural attributes resulting from a splat-based assemblage of the coating.

It has long been speculated that this assembly of pores, cracks and interfaces in sprayed coatings provide significant deviations in mechanical, thermal, and electrical (for conducting materials) properties from their bulk counter parts, i.e., effective properties are strongly dependent on morphology of the defect architecture which, in turn, are sensitive to processing conditions and feedstock materials.[6–9] This is particularly important for ceramic coatings where the volume fraction of these defects can be quite large (e.g., 5%–25%) and impart much needed mechanical compliance for thermo-structural coatings used in high temperature environments (e.g., thermal barrier coatings).[10–12]

There is considerable literature exploring quantitative attributes of the defected microstructure.[13–16] An indirect link among porosity, thermal property, and mechanical compliance has been established through correlations.[14,17–20]

Recent advances in the adaptation of *in situ* substrate curvature measurements[21–23] have revealed anelastic mechanical response of such splat-assembled ceramic coatings (i.e., the curvature–temperature relationship of a coated bilayer system is nonlinear) with heating and cooling curves exhibiting hysteresis. This implies nonlinear deformation mechanisms and dissipated energy during the loading–unloading cycle. The results exhibit nearly complete elastic recovery in these substrate–curvature experiments indicating anelastic response. Results are reproducible under repeated cycling, and in various experimental apparatuses (e.g., Tencor Flexus, Tencor Instruments, Mountain View, CA).

Extensive characterization of this anelastic behavior has been conducted over the last few years along with development of a quantitative framework to describe the nonlinear mechanical behavior.[23–26] It was concluded that the anelasticity is driven by geometric and microstructural factors: specifically, microcracks as well as weak splat interfaces acting collectively. The first phenomenon is assumed to be the outcome of asymmetrical responses from crack opening and closure under tension and compression loadings. The physical mechanism for the second behavior is perhaps more complex although it is likely due to the frictional sliding of *unbonded interfaces* or embedded crack surfaces. Recent studies have shown that this sliding mechanism is also responsible for damping behavior of plasma-sprayed ceramics.[27,28] Both mechanisms, crack opening/closure and frictional sliding, collectively operate within complex coating microstructures during loading as schematically illustrated in Fig. 2. A number of investigators have used crack/void models to quantitatively study the nonlinear and hysteresis effects. Kachanov *et al.*[29] considered the interacting crack effects through Mori–Tanaka scheme. Lawn and Marshall[30] modeled the crack face sliding with friction and derived the solutions via analysis of complementary energy density. These and other studies suggest the anelastic effects depend strongly upon geometry and chemistry in splat interfaces, and are thus highly sensitive to manufacturing process. Of further importance is that this behavior can be not only controlled but produced repeatedly and reliably. Such behavior has vast implications for robust and tough (energy-absorbing) thermo-mechanical design of ceramic coatings.

To elucidate the anelastic deformation mechanism in sprayed ceramic coatings, Liu *et al.*[23] explored analytical and numerical solutions based on both characteristic and real microstructures including the effects of initial crack opening under compressive mechanical loads. In the analytical scheme, single and multiple crack/interface system was investigated through a Coulomb friction model during crack closure to simulate the sliding across the splat interfaces. The model accurately captured the nonlinear and anelastic response and pointed to two key parameters to describe the hysteretic response.

C. Levi—contributing editor

Manuscript No. 28788. Received October 15, 2010; approved February 11, 2011.
This research is supported by the NSF GOALI program CMMI 1030492 and Industrial Consortium for Thermal Spray Technology.
[†]Author to whom correspondence should be addressed. e-mail: gopal.dvd@gmail.com, ssampath@ms.cc.sunysb.edu

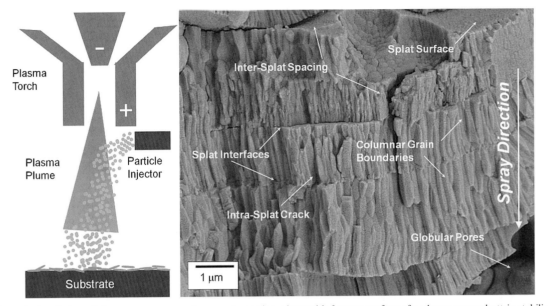

Fig. 1. Schematic description of plasma spray process to generate coating, along with fracture surface of a plasma-sprayed yttria-stabilized zirconia (YSZ) coating, depicting various features of the layered microstructure.

(1) Effects of the friction coefficient, μ. Without friction, no hysteresis is produced. The hysteresis is maximum with $\mu = 0.5$ and disappears at a large μ when sliding is inhibited.

(2) Effects of initial crack opening. The extent of hysteresis is influenced by the initial crack face separation (or splat interface gap). A smaller gap tends to produce larger hysteresis.

The above suggests that by varying the friction coefficient and the initial crack opening, it was feasible to synthesize or control the anelastic response. These results and recent progress led to several opportunities.

(1) A new temperature-controlled instrumented bilayer thermal cycling system allows for very accurate measurements of temperature–curvature behavior. Although the past studies offered a successful measure of the coating attributes and its repeatability, the precise hysteresis behavior was not obtainable. Our ability to quantify this property, which is a strong measure of internal friction, and manipulate this via judicious selection of processing and material selection, can offer new avenues to tailor mechanical compliance in thermo-structural coatings.

(2) The past work primarily focused on thermo-mechanical loading of the ceramic coating through substrate constraint during thermal cycling. It is of interest to examine the deformation behavior of the free-standing layered ceramic coating under mechanical loading (three-point-bend).

(3) Finally, it has come to light that the interfacial features (opening dimension, metrology, and chemistry) can be manipulated by both processing parameters as well as postspray chemical modification/infiltration. This imposes the question as to the level of microstructural tailoring that can be imparted to modify the anelastic response.

In this study, we seek both a successive investigation to the prior work through quantification of the thermo-mechanical hysteresis, and an exploration of new processing ideologies toward controlled introduction of anelasticity.

II. Experimental Procedure

(1) Specimen Preparation

All the coatings were sprayed using atmospheric plasma spray onto a 230 mm × 25.4 mm Al6061 plate with thicknesses ranging from 1.6 to 2.4 mm. More specific details of coating and substrate thicknesses are provided in their corresponding sections. For 8 wt% yittria-stabilized zirconia specimens (YSZ), plasma-densified hollow sphere (HOSP) powders are used in the fabrication. Other materials, cordierite, and alumina, are also used in the present analysis. In addition, in order to control the anelastic behavior, some of the processing conditions are varied to investigate their effects. These changes are noted within the specific sections.

(2) Bilayer Thermal Cycle Curvature Measurements

Owing to thermal mismatch between coating and substrate, the curvature of the coating–substrate system changes with temperature, representing the change in stress state within the coating. The bilayer thermo-mechanical test is based on the above principle. In the present analysis, the coated substrates were heated inside a furnace from room temperature to $\sim 230°C$ in approximate 30 min, and then the furnace was switched off. The specimens were left in the furnace for about 3 h to cool down to the room temperature. The curvature change was calculated from the beam displacements measured at three locations on the specimen using three separate laser displacement sensors with their measurement axes placed perpendicular to the specimen. The temperature variation was recorded using a self-adhesive thermocouple glued on the substrate. Detailed experimental procedure is described in earlier paper by Dwivedi *et al.*[31] Because the data processing scheme assumes no transient temperature conditions through thickness and across the specimen span, maintaining the isothermal state during the test is critical in accurately determining nonlinear behavior of the thermal spray coating. The flame torch and heat box methods used in the past did not assuredly provide isothermal conditions during both heating and cooling cycle.[21–23]

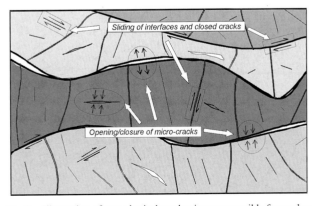

Fig. 2. Illustration of two physical mechanisms responsible for anelastic behavior. They are crack opening/closing and frictional sliding across interfaces/cracks occurring within the coating microstructure.

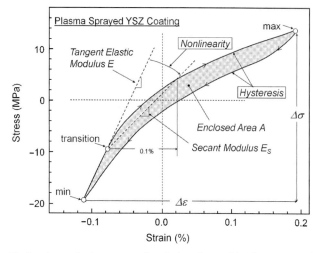

Fig. 3. An anelastic stress–strain relation observed in plasma-sprayed YSZ coating[23]. Two key features, nonlinearity and hysteresis during loading/unloading path and the parameters to quantify them are noted.

The measured curvature–temperature data is processed to identify a stress–strain relation of a given coating. The detailed procedures are described earlier[21,22] and only key descriptions are summarized below. First the phenomenological constitutive model based on modified Ramberg–Osgood nonlinear stress–strain relation is assumed. Then, the measured data is processed through Kalman filter algorithm to extract four unknown material parameters (defining the nonlinear stress–strain model). Owing to the residual curvature, the specimen is in a compressive state at room temperature (i.e., starting point of stress–strain relation is at negative stress). In the data processing, the temperature-dependent properties of aluminum substrate are used. For the YSZ coatings, their coefficient of thermal expansion (CTE) is set to be temperature dependent. A typical stress–strain loop for thermal-sprayed YSZ coating obtained via thermal cycle test[23] is shown in Fig. 3. In the figure, the minimum point generally corresponds to the room temperature values while the maximum point is attained at $T \sim 230°C$. Each loading and unloading path has linear and nonlinear parts with the transition point in the middle as indicated in the figure. The clockwise and closed stress–strain behaviors are consistent with the analytical and computational models shown earlier.[23,30]

From the stress–strain relation, three key parameters which quantify the anelastic behavior can be defined. The first parameter is the slope between the minimum point (at room temperature) and the transition point that is denoted as the tangent elastic modulus of coating E. The second is the extent of non-linearity which is quantified by the changes of stress–strain slope. Here "Nonlinear Degree (*ND*)" is defined from the ratio of E over a secant elastic modulus, E_s (i.e., $ND = E/E_s$). As shown in the figure, the secant modulus is chosen as the slope between the transitional point and the point at additional 0.1% strain. A larger *ND* indicates greater nonlinearity while $ND = 1$ represents a linear elastic response. In general, E and *ND* obtained from the loading/heating curve are reported. Although similar parameters can be determined from the unloading/cooling curve, they depend on the maximum stress level and vary under different load/thermal cycles. The third anelastic property is introduced as "Hysteresis Degree (*HD*)." This parameter is obtained from the ratio of enclosed area (*A*) of stress–strain curve over the rectangular area given by $\Delta\varepsilon \times \Delta\sigma$ (i.e., $HD = A/\Delta\varepsilon\Delta\sigma$) as shown in the figure. Although the enclosed area also depends on the maximum stress, the variability among different load cycles is reduced when it is normalized by $\Delta\varepsilon\Delta\sigma$. An exhaustive study has been performed in past to present a correlation between the two anelastic parameters, E and *ND*, with the coating processing conditions,[22,23] however for the newly introduced parameter, *HD*, yet to be explored more in order to establish a correlation with coating's processing conditions and

other properties. Nevertheless, these three parameters, E, *ND*, and *HD*, are valuable in quantifying the characteristics and differentiating anelastic responses of plasma-sprayed coatings.

(3) Three-Point-Bend Test

In order to confirm anelastic behavior of plasma-sprayed coatings, mechanical loading test using a three-point-bend fixture is also performed. Free-standing ceramic coupons were prepared by carefully removing the substrates from the coatings by dissolving the sprayed beam in Aqua Regia (a solution of 75 vol% HCl and 25 vol% HNO_3) for an extended period of 3–4 h. The procedure works well for the aluminum substrates and with ceramic coating without any deleterious effects on the deposited microstructure and properties. These free-standing samples range from 12 to 13 mm in width and are 40 mm in length. They were surface polished on both sides to ensure flatness. The samples were subjected to flexural bending on a TIRATEST-26500 tensile machine (TIRA GmbH, Schalkau, Germany). For all the experiments, support span was kept constant at 34 mm, and the crosshead movement rate was fixed at 0.1 mm/min. The specimens were loaded cyclically with predefined incrementally increasing load amplitudes until fracture. The displacement at the midspan was monitored using a laser displacement sensor with 1 μm resolution. Typically, the specimens are fractured in four to five cycles as the cyclic load amplitudes are increased.

III　Thermal Cycle Test Results

Our previous modeling studies[23] pointed to two attributes that control the hysteretic response of the spray-deposited ceramics: the opening dimension of cracks or interlamellar pores and the frictional coefficient associated with interfacial/crack surface sliding. In this study, three approaches were developed to control the anelastic characteristics of TS coatings. First, the processing/fabrications parameters can be modified to affect the overall coating porosity, type of defects, densities of interfaces/cracks, and their opening dimensions. There have been extensive studies in the literature including those conducted by the authors to control these attributes.[23] Second is the chemical modification after coatings are sprayed. The porous nature of the sprayed materials offers a secondary route to *infiltrate* the deposit to vary the frictional coefficient and openings of interfaces and cracks. In this approach, the coating was inserted into $2M$ NaCl solution. Alternatively, different feedstock powder of YSZ were considered and along with different ceramic compositions including cordierite ($Mg_2Al_4Si_5O_{18}$) and alumina (Al_2O_3). Cordierite was chosen so as to create a predominantly amorphous-sprayed coating compared with nanocrystalline structures in the case of alumina and zirconia.

(1) Effects of Processing Condition: Spray Distance

Earlier work by Liu *et al.*[22,23] had demonstrated that curvature–temperature measurements were dependent on the microstructure imposed by the processing parameters. It was shown both qualitatively and quantitatively that the elastic modulus and the nonlinearity were closely linked to the extent of porosity and type of the intersplat bonding within the samples. The quantification was primarily conducted through analysis of the cooling curve during thermal cycling,[21,22] however there was thermal gradient introduced during flame based heating of the sprayed beam. Our present approach to monitoring curvature changes during thermal cycling involves insertion of the specimen into a furnace while using noncontact displacement sensors positioned outside the furnace to monitor the curvature evolution. As such, it is feasible to obtain a high quality record of curvature evolution during both heating and cooling while maintaining isothermal conditions to analyze complete thermal cycling responses. From the curvature–temperature data and following the extraction of the stress–strain relations, the complete anelastic relations are

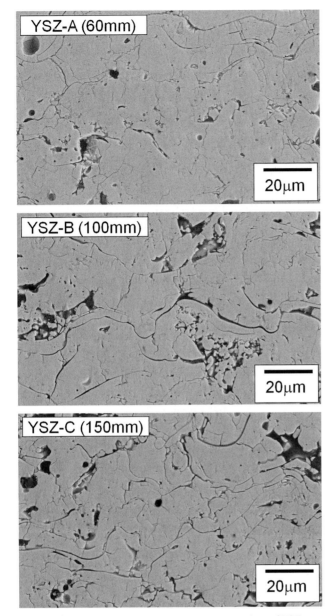

Fig. 4. SEM micrographs showing polished cross-sections of YSZ coatings sprayed at various standoff distances (60, 100, and 150 mm).

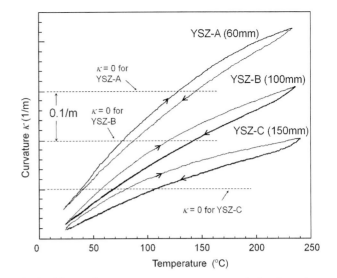

Fig. 5. Measured curvature–temperature responses during thermal cycling of the three YSZ coatings made with different spray distances. The results are staggered along the vertical axis for clarity.

followed by YSZ-B (0.82 W·(m·K)$^{-1}$) and YSZ-C (0.69 W·(m·K)$^{-1}$).

Each specimen is placed within the furnace and its curvature change under thermal cycling is measured as shown in Fig. 5. Note that all have negative residual curvatures at the room temperature, which are an indication of the compressive stress state within the YSZ coatings. As the temperature increases, the higher CTE of the aluminum substrate causes the stress state in the coatings to increase toward tension. After reaching the maximum temperature ($\sim 230°$C), the specimens were cooled inside the furnace until near room temperature. The effective curvature change is governed principally by the microstructure and resultant coating stiffness as the differences in thickness and thermal expansion among these coatings are minimal. The measured curvature–temperature clearly exhibits nonlinear behavior as well as hysteresis as the heating and cooling paths are different and nonlinear.

The curvature–temperatures records are processed through the inverse analysis technique described earlier and converted to stress–strain relations as shown in Fig. 6. Similar to the curvature–temperature data, all specimens exhibit distinct nonlinearity as well as hysteresis in their closed loop relations. However, unlike the curvature measurements, which are thickness dependent,

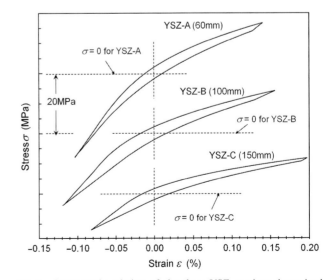

Fig. 6. Stress–strain relation of the three YSZ coatings determined from curvature–temperature data via inverse analysis. The results are staggered along the vertical axis for clarity and the zero stress location is indicated for each coating.

obtained following the procedure described in the experimental procedures section.

Three coatings with significantly different porosity as well as interface microstructures were produced for this investigation. Here the same feedstock material (YSZ HOSP) was used and the spray process conditions including the substrate thickness (2.24 mm) were kept constant except for the spray distance (i.e., nozzle–substrate separation). It was varied at 60, 100, and 150 mm, with corresponding coatings labeled as YSZ-A (0.39 mm), YSZ-B (0.36 mm), and YSZ-C (0.35 mm), respectively. As the spray distance is increased, several effects are manifested. The particle temperature and velocity are lowered resulting in coatings with greater pore content as shown in the cross-sectional micrographs in Fig. 4. The lower particle impact velocity also effectively reduces the interfacial interactions, and finally, and widened spray plume causes reduced particle flux with concomitant change in number of sprayed layers required to produce the same coating thickness. Figure 4 shows YSZ-A to be densest among the three coatings, while YSZ-C has the highest porosity. It can be also observed that the number of splat interfaces is the highest in YSZ-C and decrease with spray distance. The through-thickness thermal conductivity of the three coatings measured by laser flash technique shows YSZ-A to have the highest thermal conductivity (0.99 W·(m·K)$^{-1}$)

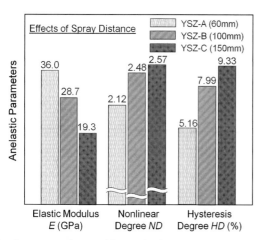

Fig. 7. Parameters characterizing anelastic stress–strain behaviors are shown for YSZ coatings with different spray distances. Corresponding values are noted on top of the bars. Note that the heights of *ND* are adjusted because $ND = 1$ corresponds to zero nonlinearity.

extracted stress–strain relations enable direct comparisons among various coatings. The room temperature residual stress ranges from $\sigma_{RT} = -28$ to -12 MPa with the largest value for YSZ-A. The maximum stress (at $T = \sim 230°C$) reaches $\sigma_{max} = 10–16$ MPa, with YSZ-A again being the highest. The strain variations during the thermal cycle were $\Delta\varepsilon = \sim 0.30\%$ in all cases.

In order to interpret the constitutive behaviors, the key anelastic parameters were computed for the three specimens as shown in Fig. 7. Based on numerous tests (including those not reported here), the error bounds of these parameters are estimated at 3%–5% of corresponding magnitudes (e.g., ± 0.5–1.5 GPa for E). In the figure, all three parameters show clear and consistent trends as the spray distance is changed. For the elastic modulus, YSZ-A is nearly twice as high as YSZ-C. The lower porosity and fewer interfaces in YSZ-A observed in Fig. 4 produce the stiffer response. Unlike the modulus, *ND* increases with spray distances. This parameter is primarily attributed to the density of thin microcracks and the results suggest the higher density of such cracks in YSZ-C. A larger variation is observed among the specimens in the calculation of *HD*, which is the normalized closed area of the stress–stress curves. The largest *HD* in YSZ-C specimen indicates greater presences of interfaces, intrasplat boundaries and cracks. Essentially a greater frictional sliding cause the larger difference between the loading and unloading stress–strain curves. Note that the values of *HD* can be also approximated from the curvature–temperature measurement. In fact, the values obtained from curves in Fig. 5 are within a few percent of those reported in Fig. 7.

(2) *Effects of Processing Condition: Particle Energy*

Controllability of anelastic behavior was also examined by varying particle state, temperature and velocity, during spray processing. This was accomplished by changing the plasma power and the total gas flow of the plasma plume. Three specimens YSZ-D, YSZ-E, and YSZ-F, were fabricated with three different plasma energies (input power) at 24.3, 33.9, and 45.5 kW, respectively. All other processing parameters, including the spray distance (100 mm) and the substrate thickness (3.2 mm) were kept the same. The coating thicknesses of the three coatings were 0.88, 0.98, and 0.95 mm, respectively. The measured curvature–temperature records are processed and their stress–strain relations were obtained. Owing to space constraints, only the computed anelastic parameters are shown in Fig. 8. For the elastic modulus, YSZ-F has the highest E as the high-energy particle impacts produce lower porosity and stiffer coating. The results of *ND* are more uniform among the specimens. However, as opposed to the trend shown in Fig. 7, it increases slightly as the modulus is increases (from YSZ-D to F). With larger *ND*, a higher density of thin microcracks is predicted in the high-energy specimen YSZ-F.

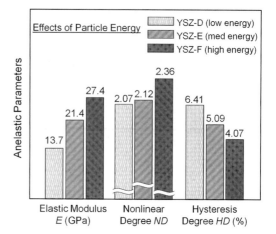

Fig. 8. Parameters characterizing anelastic stress–strain behaviors are shown for YSZ coatings deposited with different plasma energies.

The trend is clearer in *HD* results as the hysteresis is reduced in YSZ-F specimen. It is speculated that higher particle energy results in stronger interfacial bonding and inhibited frictional sliding during loading, and therefore, resulting in lower hysteresis.

(3) *Effects of Splat Surface Modifications: Addition of Secondary Phase*

Effects of friction coefficient on the mechanical response[23] suggest that coating's anelasticity, especially extent of hysteresis can be regulated by changing the surface's frictional properties. The anelastic behavior can be significantly modified by changing the frictional characteristics of the defect surfaces. Two approaches have been explored: (a) extrinsically, by introducing a secondary medium (salt) to coat the defect and interface surfaces, and (b) by changing the material itself so as to change the intersplat surface friction.

The porous nature of the sprayed materials allows for introduction of secondary phases via aqueous or chemical infiltration treatment. To accomplish this, a 0.27-mm-thick sprayed ceramic coating on an aluminum substrate of 1.45 mm thickness was dipped in a 2*M* NaCl solution for 36 h. This allowed the salt solution to effectively penetrate the interlamellar regions of the sprayed microstructure. The samples were then thermally cycled on the bilayer curvature measurement system. The response was as follows:

(1) During the first thermal cycle, the water evaporated from the specimen, leaving behind a residue of salt within the interfacial regions of the coating. Owing to water evaporation, this first thermal cycle was significantly different in terms of curvature–temperature relations compared with the as-sprayed materials, but was a transient phenomenon.

(2) The second, third, and additional cycles resulted in highly repeatable curvature–temperature relations and behaved in ways similar to the as-sprayed material. However notable differences were observed in both stiffness and the magnitude of the hysteresis loops compared with the untreated material.

The temperature–curvature results shown in Fig. 9 indicate stiffer response with the salt solution immersed coating. In addition, the hysteresis loop area was larger for the case of immersed coating. Figure 10 shows the quantified values for these changes. As noted in the figure, the elastic modulus, E, increased about 30% after immersion. Similarly, the hysteresis effect exhibited an increase of about 50%, indicating a greater frictional dissipation with salt phase deposited along the interfaces and cracks.

At this point, we do not have specific microstructural confirmation to quantity of the salt residue within the interfacial regions. Nevertheless, the curvature responses are highly repeatable and suggesting clear differences in the internal structures of the coating. Additional work is underway to explore different salt chemistries, treatment time, and solution characteristics. Furthermore,

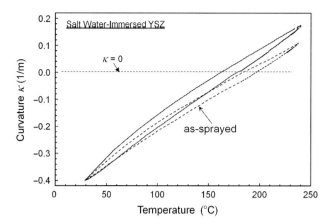

Fig. 9. Measured curvature–temperature response (second thermal cycle) of a YSZ coating after immersion in NaCl solution. As-sprayed results are also shown for comparison.

this infiltration method can be combined with microstructural changes as those described earlier. The implications of this observation are twofolds:

(1) It will allow preferential chemical modification of the interfacial structure to create hybrid composites that can be both structural and functional.

(2) From an application perspective, thermal barrier coatings in engines are exposed to salt laden moisture during service. It would be worthwhile to determine if the observed effects translate to performance degradation. Such studies are planned in the future.

(4) Effects of Splat Surface Modifications: Variation of Materials

The frictional component during sliding among the defects and interfaces will be affected by the surface texture of the interfaces. In the case of crystalline materials such as YSZ, splat surfaces comprise of nanoscale roughness associated with uneven grain terminations (as shown in Fig. 1). However, there are situations involving ceramic materials, which produce amorphous structures due to the rapid quenching occurring in plasma spray, where no such grain terminations exist. Cordierite is a good example of such a system. Micrographs shown in Fig. 11 indicate unique morphology of amorphous cordierite. At a lower magnification, the microstructure of cordierite shows assembly of splats with microcracks and interfaces. However, at higher magnification, the micrograph shows, smooth splat surfaces as well

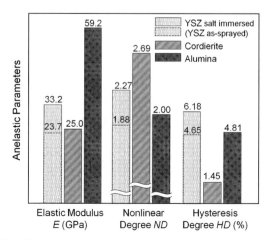

Fig. 10. Comparison of anelastic properties of various plasma-sprayed ceramic materials obtained from curvature–temperature measurements. For YSZ coatings, the as-sprayed results are shown with dashed outline.

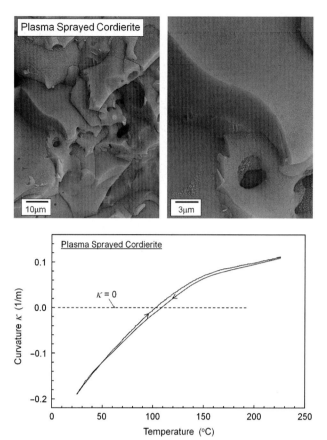

Fig. 11. SEM micrographs showing morphology of the cordierite coating at low and high magnifications. Note the smooth texture of the splat compared with the YSZ (Fig. 1). Measured curvature–temperature depicts large nonlinearity but limited hysteresis during thermal cycle.

as no intrasplat columnar grain boundaries and cracks generally observed in crystalline coatings such alumina or YSZ (Fig. 1).

The thermal cycle test produced an unusual curvature–temperature data for cordierite specimen as shown in Fig. 11. For the test, a 0.33 mm cordierite coating was deposited on a 2.25-mm-thick Al substrate Similar to curvature response of the YSZ specimens, it is also nonlinear with respect to the temperature change but the observed hysteresis is very limited. In fact, the cooling path almost follows the heating path after the temperature reversal. Our preliminary explanation for this response is that the low roughness of the glassy surfaces within the interlamellar interfaces in cordierite offers little or no friction during cooperative sliding, and thus, does not contribute significantly to hysteretic dissipation. However, the nonlinearity is still produced by existing many microcracks and unbonded splat interfaces.

As a verification study, we have also examined other crystalline materials such as alumina (coating and substrate thicknesses 0.18 and 2.26 mm, respectively), Fig. 12. Similar to plasma-sprayed YSZ, it has intrasplat boundaries but their density appears to be lower for this set of sample. Note they were processed under similar conditions, but because alumina has a lower melting point, denser coatings may have resulted.

The anelastic parameters of these materials are summarized in the bar graphs in Fig. 10. The results of cordierite are distinctly different from the others (also shown in Figs. 7 and 8). It has a moderate elastic modulus but the highest ND of all specimens reported here. Yet its HD is negligible compared with the others. The comparison of cordierite results and those of other specimens support the interfacial frictional sliding assumption as the primary contributor to the hysteresis. These results point interesting future possibility in controlling nonlinear and hysteretic response through combination of materials and process modifications.

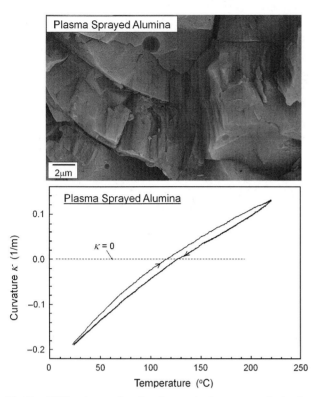

Fig. 12. SEM micrograph of a fractured plasma-sprayed alumina coatinga. Unlike YSZ, it has fewer intrasplat boundaries and cracks. Measured curvature–temperature during thermal cycle is shown below.

IV. Three-Point-Bend Test Results

Thermal cycling offers robust test data, is relatively easy to apply and enables exploration of wide ranging materials and process conditions. However, because in this test, coatings are constrained by the metallic substrate, the stress impositions on the coatings are always associated with thermal mismatches. In fact, the interpretations of measured curvature data require the knowledge of coupling conditions of coating and substrate. In order to isolate the response of plasma-sprayed materials, free-standing specimens have been prepared as described earlier and subject to mechanical loading using three-point-bending.

Two free standing coatings, YSZ and alumina (thickness 1.6 and 1.27 mm, respectively), were examined under three-point-bend test. These samples were prepared with same parameters as described earlier (YSZ-B and alumina). For the given condition, due to higher melting temperature the YSZ coating was more porous than the plasma-sprayed alumina. Measured midspan load–deflection records are shown in Fig. 13. Note that cyclic loading unloading was performed with successive load amplitude to increase until failure. Several observations can be made.

(1) The loading as well as unloading curves are not linear, and the nonlinearity increases with applied load until fracture. However, unlike the thermal cycle test, the through-thickness stress varies from tension to compression and the behavior reflects the net effect of both.

(2) The YSZ coating exhibits greater hysteresis than that of aluminum consistent with the thermal cycle results shown in Figs. 5 and 12. As verification, a sintered alumina plate of similar dimension was tested which displayed an ideally elastic response with no measurable nonlinearity or hysteresis.

(3) Unlike the curvature–temperature response, the unloading curves did not return back to its initial state, i.e. coatings had residual displacement after each cycle. This is due to the fact that the loading starts at zero stress and not at a residual compressive stress in the thermal cycle test. The residual stress in the specimen was removed as the substrate was separated.

(4) At failure, alumina specimen exhibits sudden drop in the load, which represents more brittle nature of denser alumina

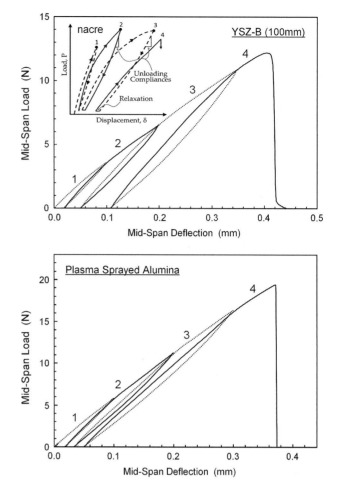

Fig. 13. Cycled load–deflection measurements under three-point-bend condition for shown for plasma-sprayed YSZ and alumina specimens. Amplitudes are increased after each cycle. Schematic of predicted behavior for nacre[32] is shown in the inset.

coating. Note that both *ND* and *HD* of YSZ-B are significantly larger than those of alumina (see Figs. 7 and 10), which imply greater density of interfaces and microcracks.

The anelastic responses are in many ways analogous to those reported and modeled for natural composites such as nacre. In fact our observations are very similar to those predicted by Evans *et al.*[32] for nacreous composites as shown inset of Fig. 13. They had speculated that in the case of nacre, frictional effects will likely result due to sliding across interfaces. However, there are differences. The sprayed coatings do not contain any organic layer at the interfaces and the internal friction and nonlinearity are purely geometric effects. Secondly, nacre is a much more ordered material compared with the chaotic assembly of the plasma-sprayed system. Nevertheless, the underlying similarities in the operating mechanisms are noteworthy.

V. Summary and Conclusions

In thermal-sprayed materials, their unique geometrical/microstructural features allow mechanical responses to be anelastic, namely, nonlinear and hysteretic. In ceramics, the former introduces a quasi-ductile response in otherwise brittle solids, and the latter is associated with energy-dissipative process. The mechanisms behind these characteristics are attributed to the large numbers of embedded interfaces and microcracks, resulting from the rapid quenching process. These geometrical attributes collectively produce the anelastic behavior through opening/closing and sticking/sliding of interfaces.

In this paper, we present strategies and results in controlling the anelastic response. First the anelasticity is quantitatively identified by three parameters, elastic modulus (E), ND, and HD which allow for comparison among the various sprayed samples. It was shown that for the plasma-sprayed yttria-stabilized zirconia, all of the anelastic parameters are sensitive to processing conditions, and the results indicate that it is feasible to tune the process for controlled introduction of anelasticity.

It is further demonstrated that the hysteretic response relates both to the dimensionality of the interfacial separations (intersplat porosity) as well as the roughness of the constituent splat surface. As such it is feasible to suppress the hysteretic behavior while maintaining the nonlinear response. Extrinsic factors such as chemical modification of interfaces will also affect the anelastic processes in these layered materials.

The observed anelastic behavior is similar to those predicted for natural brick and mortar structures by Evans and colleagues, although, compositionally, sprayed materials do not generally contain any polymeric interfacial binder. As such some of the anelastic attributes are intrinsic to layered architecture resulting from processing. The results point to strategies for imparting compliance and toughness to thermostructural coatings via microstructural engineering.

Acknowledgments

This paper is dedicated to Professor Tony Evans, who appreciated the significance of the initial observations some years ago and encouraged our group to further develop the idea. Tony's comment "people have been trying to make synthetic analogues of natural materials for some time, you guys just have it !" will continue to inspire our work in this arena. Authors are also thankful to Dr. K. Shinoda and Dr. Y. Tan for their assistance in processing some of the plasma-sprayed samples and to Prof. C. M. Weyant for his review of the manuscript and suggestions. We are also grateful to Dr. C. Johnson for identifying the issues of thermal gradient in our initial test, which led to our improvement of our experimental approach.

References

[1]P. Fauchais, "Formation of Plasma-Sprayed Coatings," *J. Therm. Spray Technol.*, **4** [1] 3–6 (1995).

[2]H. Herman, S. Sampath, and R. McCune, "Thermal Spray: Current Status and Future Trends," *MRS Bull.*, **25** [7] 17–25 (2000).

[3]A. Kulkarni, Z. Wang, T. Nakamura, S. Sampath, A. Goland, H. Herman, J. Allen, J. Ilavsky, G. Long, J. Frahm, and R. W. Steinbrech, "Comprehensive Microstructural Characterization and Predictive Property Modeling of Plasma-Sprayed Zirconia Coatings," *Acta Mater.*, **51** [9] 2457–75 (2003).

[4]S. Kuroda, T. Dendo, and S. Kitahara, "Quenching Stress in Plasma-Sprayed Coatings and its Correlation with the Deposit Microstructure," **4** [1] 75–84 (1995).

[5]J. Matejicek and S. Sampath, "Intrinsic Residual Stresses in Single Splats Produced by Thermal Spray Processes," *Acta Mater.*, **49** [11] 1993–9 (2001).

[6]R. Mcpherson, "The Relationship between the Mechanism of Formation, Microstructure and Properties of Plasma-Sprayed Coatings," *Thin Solid Films*, **83** [3] 297–310 (1981).

[7]R. Mcpherson, "The Structure and Properties of Plasma Sprayed Alumina Coatings," *Ceram.-Silik.*, **35** [3] 273–82 (1991).

[8]S. Sampath, "Thermal Sprayed Ceramic Coatings: Fundamental Issues and Application Considerations," *Int. J. Mater. Prod. Technol.*, **35** [3–4] 425–48 (2009).

[9]J. Ilavsky, A. J. Allen, G. G. Long, S. Krueger, C. C. Berndt, and H. Herman, "Influence of Spray Angle on the Pore and Crack Microstructure of Plasma-Sprayed Deposits," *J. Am. Ceram. Soc.*, **80** [3] 733–42 (1997).

[10]T. Nakamura, G. Qian, and C. C. Berndt, "Effects of Pores on Mechanical Properties of Plasma Sprayed Ceramic Coatings," *J. Am. Ceram. Soc.*, **83** [3] 578–84 (2000).

[11]Z. Wang, A. Kulkarni, S. Deshpande, T. Nakamura, and H. Herman, "Effects of Pores and Interfaces on Effective Properties of Plasma Sprayed Zirconia Coatings," *Acta Mater.*, **51** [18] 5319–34 (2003).

[12]A. Kulkarni, A. Vaidya, A. Goland, S. Sampath, and H. Herman, "Processing Effects on Porosity-Property Correlations in Plasma Sprayed Yttria-Stabilized Zirconia Coatings," *Mater. Sci. Eng.A—Struct. Mater. Prop. Microstruct. Process.*, **359** [1–2] 100–11 (2003).

[13]Y. Tan, J. P. Longtin, and S. Sampath, "Modeling Thermal Conductivity of Thermal Spray Coatings: Comparing Predictions to Experiments," *J. Therm. Spray Technol.*, **15** [4] 545–52 (2006).

[14]W. Chi, S. Sampath, and H. Wang, "Ambient and High-Temperature Thermal Conductivity of Thermal Sprayed Coatings," *J. Therm. Spray Technol.*, **15** [4] 773–8 (2006).

[15]A. Cipitria, I. O. Golosnoy, and T. W. Clyne, "Sintering Kinetics of Plasma-Sprayed Zirconia TBCs," *J. Therm. Spray Technol.*, **16** [5–6] 809–15 (2007).

[16]W. G. Chi, S. Sampath, and H. Wang, "Microstructure-Thermal Conductivity Relationships for Plasma-Sprayed Yttria-Stabilized Zirconia Coatings," *J. Am. Ceram. Soc.*, **91** [8] 2636–45 (2008).

[17]Y. Tan, J. P. Longtin, S. Sampath, and H. Wang, "Effect of the Starting Microstructure on the Thermal Properties of As-Sprayed and Thermally Exposed Plasma-Sprayed YSZ Coatings," *J. Am. Ceram. Soc.*, **92** [3] 710–6 (2009).

[18]J. Matejicek and S. Sampath, "In situ Measurement of Residual Stresses and Elastic Moduli in Thermal Sprayed Coatings—Part 1: Apparatus and Analysis," *Acta Mater.*, **51** [3] 863–72 (2003).

[19]J. Matejicek, S. Sampath, D. Gilmore, and R. Neiser, "In Situ Measurement of Residual Stresses and Elastic Moduli in Thermal Sprayed Coatings—Part 2: Processing Effects on Properties of Mo Coatings," *Acta Mater.*, **51** [3] 873–85 (2003).

[20]S. Sampath, V. Srinivasan, A. Valarezo, A. Vaidya, and T. Streibl, "Sensing, Control, and In Situ Measurement of Coating Properties: An Integrated Approach Toward Establishing Process–Property Correlations," *J. Therm. Spray Technol.*, **18** [2] 243–55 (2009).

[21]T. Nakamura and Y. J. Liu, "Determination of Nonlinear Properties of Thermal Sprayed Ceramic Coatings Via Inverse Analysis," *Int. J. Solids Struct.*, **44** [6] 1990–2009 (2007).

[22]Y. Liu, T. Nakamura, V. Srinivasan, A. Vaidya, A. Gouldstone, and S. Sampath, "Non-Linear Elastic Properties of Plasma-Sprayed Zirconia Coatings and Associated Relationships with Processing Conditions," *Acta Mater.*, **55** [14] 4667–78 (2007).

[23]Y. J. Liu, T. Nakamura, G. Dwivedi, A. Valarezo, and S. Sampath, "Anelastic Behavior of Plasma-Sprayed Zirconia Coatings," *J. Am. Ceram. Soc.*, **91** [12] 4036–43 (2008).

[24]E. F. Rejda, D. F. Socie, and T. Itoh, "Deformation Behavior of Plasma-Sprayed Thick Thermal Barrier Coatings," *Surf. Coat. Technol.*, **113** [3] 218–26 (1999).

[25]S. R. Choi, D. M. Zhu, and R. A. Miller, "Mechanical Properties/Database of Plasma-Sprayed ZrO$_2$—8 wt% Y$_2$O$_3$ Thermal Barrier Coatings," *Int. J. Appl. Ceram. Technol.*, **1** [4] 330–42 (2004).

[26]F. Kroupa, "Nonlinear Behavior in Compression and Tension of Thermally Sprayed Ceramic Coatings," *J. Therm. Spray Technol.*, **16** [1] 84–95 (2007).

[27]N. Tassini, S. Patsias, and K. Lambrinou, "Ceramic Coatings: A Phenomenological Modeling for Damping Behavior Related to Microstructural Features," *Mater. Sci. Eng. A—Struct. Mater. Prop. Microstruct. Process.*, **442** [1–2] 509–13 (2006).

[28]P. J. Torvik, "A Slip Damping Model for Plasma Sprayed Ceramics," *J. Appl. Mech.-Trans. ASME*, **76** [6] 061018, 8pp (2009).

[29]M. Kachanov, I. Tsukrov, and B. Shafiro, "Effective Moduli of Solids with Cavities of Various Shapes," *Appl. Mech. Rev.*, **47** [1] 151–74 (1994).

[30]B. R. Lawn and D. B. Marshall, "Nonlinear Stress–Strain Curves for Solids Containing Closed Cracks with Friction," *J. Mech. Phys. Solids*, **46** [1] 85–113 (1998).

[31]G. Dwivedi, T. Wentz, S. Sampath, and T. Nakamura, "Assessing Process and Coating Reliability Through Monitoring of Process and Design Relevant Coating Properties," *J. Therm. Spray Technol.*, **19** [4] 695–712 (2010).

[32]A. G. Evans, Z. Suo, R. Z. Wang, I. A. Aksay, M. Y. He, and J. W. Hutchinson, "Model for the Robust Mechanical Behavior of Nacre," *J. Mater. Res.*, **16** [9] 2475–84 (2001). □

J. Am. Ceram. Soc., **94** [S1] S112–S119 (2011)
DOI: 10.1111/j.1551-2916.2011.04496.x
© 2011 The American Ceramic Society

journal

Damage Evolution in Thermal Barrier Coatings with Thermal Cycling

Bauke Heeg

Lumium, Leeuwarden, The Netherlands

Vladimir K. Tolpygo

Honeywell Aerospace, Phoenix, Arizona 85034

David R. Clarke[†]

School of Engineering and Applied Sciences, Harvard University, Cambridge, Massachusetts 02138

Thermal barrier coatings typically fail on cooling after prolonged thermal cycling or isothermal exposure. The mechanics of spalling requires that first a critical sized portion of the coating separates from the underlying material, then buckles and finally spalls away. The critical size for buckling depends on the thickness of the coating but is several millimeters for typical zirconia coatings 150 µm thick. As-deposited coatings do not have interface separations but they form on thermal cycling as described in this work based on observations of coating cross-sections combined with the stress redistribution in the thermally grown oxide imaged using a piezospectroscopic luminescence method. Analysis of the images reveals that small, isolated regions of damage initially form and then grow, linking up and coalescing to form percolating structures across the coating until the buckling condition is attained, the buckle extends and failure occurs by spallation. The piezospectroscopic imaging of the stresses in the thermally grown oxide formed by oxidation beneath thermal barrier coatings provides a form of "stress tomography" enabling the subcritical separations to be monitored.

I. Introduction

ONE of the major themes of Tony Evans' research in the last decade, and one which we were privileged to collaborate with him, was the prime reliance of thermal barrier coating systems. Thermal barrier coatings have been in widespread use in commercial and military gas turbine engines for several decades providing thermal protection to superalloy components in the hottest sections of engines and in the last decade or so have been deemed "prime reliant." In some advanced engines, the coatings are in contact with gases that exceed the melting temperature of the superalloy blades and vanes.[1] In many respects, thermal barrier coatings must withstand the most demanding conditions that any ceramic component is subject to in today's technology but being ceramic materials, the coatings are prone to fail. The central scientific and engineering issue is to understand the failure mechanisms under different engine operating conditions and use that information to help identify conditions under which the coatings can be considered prime-reliant. This was the focus of an ONR Multi-University Research Initiative that Tony Evans led. The research described in this manuscript is a part of that quest.

Failure of TBCs is not well defined but is usually taken to have occurred when a portion of the coating has buckled and spalled away to expose a portion of the underlying component. Usually, this originates from edges but can also occur, especially on 1 in. diameter test samples, away from an edge. Irrespective of the origin, local buckling occurs before spallation as illustrated in the optical micrographs in Fig. 1. From images such as these and cross-sections of partially spalled coatings, failure occurs by an accumulation of damage until at some critical condition, yet poorly defined, the coating buckles, the buckle propagates and the coating then spalls away, leaving the underlying superalloy unprotected. The damage usually forms in the vicinity of the interface region between the thermal barrier coatings and the superalloy. Several mechanisms have been identified that can cause the onset of local damage but the mechanics of buckling[2–4] requires that a tensile stress perpendicular to the interface is necessary to form local separations or delaminations. In this contribution, we describe and quantify observations of the subcritical damage evolution obtained using microstructural observations made of polished cross-sections and by a nondestructive imaging method based on photo-stimulated luminescence piezospectroscopy (PSLS). The latter reveals the spatial distribution of the mean stress in the thermally grown oxide that forms between the thermal barrier oxide and the bond-coat. This, in turn, is sensitive to the extent of local damage in and around the TGO, including the TBC itself.

II. Materials

The 7 wt% yttria-stabilized zirconia (YSZ) coatings investigated in this work were deposited by electron beam evaporation onto flat platinum-modified nickel aluminide coated René N5 superalloy coupons by Howmet Research Center. The coatings were all similar to other state-of-the-art coatings made the same way but they were distinguished by all being from the same large batch of coatings deposited on 25 mm diameter, 3 mm thick superalloy disks all wafered from the same single crystal casting.

All the coatings were subjected to thermal cycling between room temperature and 1150°C with 1 h holds at the high temperature as has been described previously.[5] A sufficiently large number of coatings were tested that the average life of the coating batch under these thermal cycling conditions was established. This was determined to be 185 cycles. Some of the coatings were cycled to intermediate fractions of life rather than to spallation failure. Most of these were sectioned to characterize the damage at different fractions of life while four were subject to further analysis by PSLS[6] described in this work. The thermal diffusivities of these same coatings were also evaluated using a novel test based on the time delay in luminescence.[7]

A. Heuer—contributing editor

Manuscript No. 28809. Received October 21, 2010; approved February 11, 2011.
Part of the experimental work was conducted while at MetroLaser Inc., Irvine, CA.
[†]Author to whom correspondence should be addressed. e-mail: clarke@seas.harvard.edu

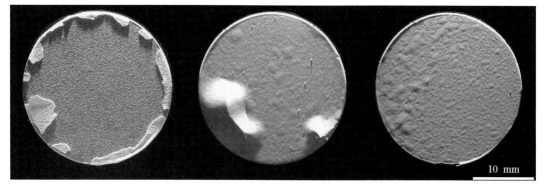

Fig. 1. Optical micrographs of EB-PVD deposited TBCs illustrating different stages in failure. Before macroscopic buckling (right), the formation of a macroscopic buckle (center) and after the TBC has everywhere spalled away except around the edges (left). The coated superalloy buttons are 25.4 mm (1 in.) in diameter.

III. Luminescence Imaging Modality

Microstructural examination of both cross-sections and fracture surfaces of thermal barrier coatings reveal that several types of local damage can be produced by thermal cycling. Figure 2, which is a composite of cross-sections recorded after deposition and after 25, 100, and 180 cycles, illustrates several of these damage types. In some cases, local separation or delamination occurs between the YSZ coating and the thermally grown oxide, such as at locations A. In other cases, local separation can occur between the thermally grown oxide and the metal although the majority of these separations may be caused by the cutting the cross-sections and subsequent polishing.

In still other cases, cracking takes place in the thermally grown oxide, such as at location B in Fig. 2(c). This occurs mainly where the TGO is locally deformed in bending. In the majority of cases, as well as in under isothermal testing, separation occurs within the YSZ coating in the vicinity of its interface with the thermally grown oxide. An example is seen at location C in Fig. 2(b). All of these separations are well below the outer surface of the coating and so none of these can be discerned through the coating itself under normal imaging conditions since the coatings are optically turbid because of scattering from internal pores. (It is possible to image some of these damaged regions with rather poor spatial resolution by mid-infra-red imaging at 3–6 μm where the scattering is weakest, as has been shown at NASA.[8] For thinner TBC, it is also possible to see separations in light microscopy.[9]) The depth beneath the coating surface over which the damage forms and localizes is also relatively narrow, substantially smaller than the coating thickness, and so cannot currently be resolved by X-ray tomography. While useful for visualization of the damage, direct imaging is nevertheless expected to be of limited value since it only provides areal information and no information about the local

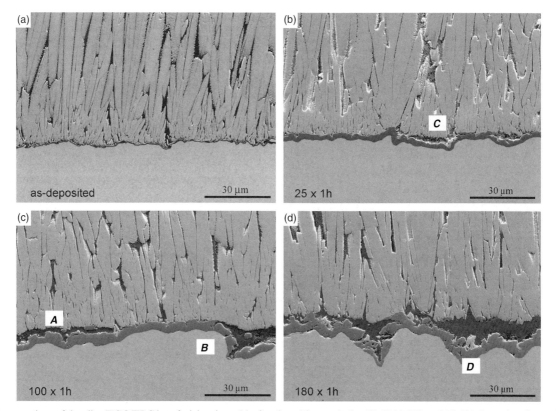

Fig. 2. Cross-sections of the alloy/TGO/TBC interfacial regions. (a) after deposition, and after (b) 25 (c) 100, and (d) 180 thermal cycles, each 1 h long, illustrating the thickening of the thermally grown oxide (TGO), the roughening of the interfaces, interface separations and the cracking of the TGO associated with the roughening. The TGO, which appears dark gray, is located between the columnar thermal barrier top-coat, above, and the aluminide bond-coated superalloy, below the TGO.

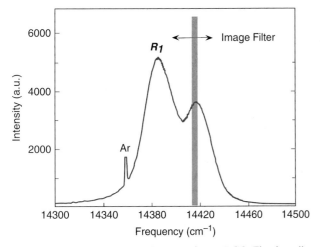

Fig. 3. Characteristic R-line luminescence from a TGO. The sharp line labeled Ar is from an argon discharge lamp used to calibrate frequency.

stresses or elastic strain energy that motivate failure. To provide this information, we have developed an imaging system for measuring the spatial distribution of the local strains in the TGO, by PSLS through the TBC.[6] In essence, the method uses the TGO as a strain sensor of the damage in its vicinity to monitor the stress re-distribution on thermal cycling.

(1) Piezospectroscopy

The physical basis of the piezospectroscopy technique has been described in detail previously so is only briefly summarized here.[10–14] The thermally grown oxide is aluminum oxide and contains trace concentrations of Cr^{3+} incorporated during its growth at high temperature. The Cr^{3+} ions luminesce when excited optically, producing a very sharp emission doublet at 694 nm, the R_1 and R_2 lines as illustrated in Fig. 3. The radiative and nonradiative transitions following the excitation of electrons from the ground state are shown in Fig. 4 with the R-lines resulting from emission from the split E2 state. Because the energy of the $2E \rightarrow 4A_2$ transition responsible for the R-line emission is sensitive to the overlap of the $3d^3$ orbitals, and consequently the distance between the Cr^{3+} ions and the surrounding O^{2-} ions, the wavelength of the R lines is sensitive to strain.[15] Measurement of the wavelength can thus be used to nondestructively monitor strain.[11] Although the stress dependence of the shift of the R_1 and R_2 lines is slightly different and is also dependent on both the crystallographic orientation of the alumina[11] as well as on polarization,[12,16] the frequency shift reduces to being proportional to the trace of the stress tensor, and hence the mean stress, σ_m

$$\bar{\Delta \nu} = \frac{\Pi_{ii}}{3} \langle (\sigma_{11} + \sigma_{22} + \sigma_{33}) \rangle = \Pi_{ii} \sigma_m \qquad (1)$$

for a polycrystalline alumina having a grain size that is much smaller than the probe size and where the grains are randomly oriented.[11,17] The trace of the piezospectroscopic coefficient tensor, Π_{ii}, has a value of 7.61 $(\text{cm} \cdot \text{GPa})^{-1}$ for the R_2 line so by inverting Eq. (1), the mean stress can be obtained from the measured change in frequency from its' stress-free value. The intensity of the R-line luminescence is not dependent on stress and so variations in intensity due to, for instance, variations in TBC thickness, local optical scattering lengths or incident illumination, do not affect the value of the measured stress.

(2) Imaging System

Until recently, the stress distribution in the TGO could only be monitored point by point moving a focused laser beam across a coating and subsequently forming maps pixel by pixel, a laborious process ill-suited to evaluation of coatings on thermal cycling. In the imaging system shown schematically in Fig. 5, an area of interest on a coating is illuminated by a uniform laser beam and the luminescence is collected by a CCD camera, after filtering through a wavelength-tunable filter. A series of images, each at a slightly different narrow wavelength band, is recorded, thus mapping the R_1 and R_2 spectra for each pixel on the CCD. From this a map of the strain distribution is assembled by analyzing the spectra by software. Details of the algorithm for curve fitting and extracting the strain are described elsewhere.[18,19] The tunable filter is a motorized Fabry–Perot filter that only passes a very narrow band of the luminescence as illustrated schematically in Fig. 3. The observations described in this contribution were formed using a 1 W frequency doubled Nd:YAG laser (532 nm) and a low noise CCD camera (Santa Barbara Imaging Group, Goleta, CA). The mean stress is taken to be over the volume of the resolution element, i.e., the volume of the TGO corresponding to the measurement over an area projected onto a pixel of the CCD.

IV. Observations and Analysis

Images of the spatial distribution of the mean stress in four different coatings at the different fractions of thermal cycles are reproduced in Fig. 6. The values of the stresses are color coded (note that the stress-scale is adjusted in each case to optimize the color contrast). Each data point in the stress image corresponds to approximately 16 μm spatial resolution, such that the total imaged area is about 4.4 by 5.6 mm. Most measurements allow determination of the stress, except in a few locations where either luminescence is too weak or too strong. These areas are colored white. Apart from the evident variations in color there are no obvious spatial correlations in the stress variations that might indicate the size or extent of damage or its evolution. Analysis of the frequency distribution of the stresses extracted

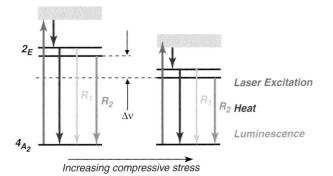

Fig. 4. Schematic of the energy levels of Cr^{3+} in alumina, the decay paths and the energy shift caused by a compressive stress.

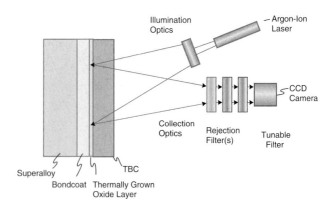

Fig. 5. Schematic of multispectral imaging system in which a TBC-coated sample is uniformly illuminated with an argon ion laser beam and the luminescence imaged onto a CCD through a variable wavelength tunable Fabry–Perot filter.

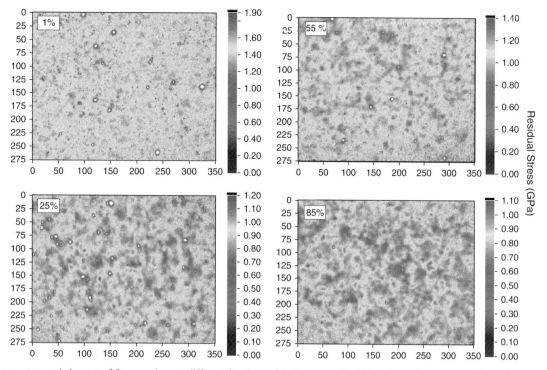

Fig. 6. Piezospectroscopic images of four coatings at different fractions of their average life. The color coding corresponds to the mean stress levels shown on the vertical color stripes on the right hand side of each panel. The numbers on the left hand ordinate and horizontal are pixels, each 16 μm in size, so the images correspond to areas of 4.4 mm × 5.6 mm.

from the images shown in Fig. 6 is more revealing. This data, Fig. 7, indicates that the average value of the mean stress decreases with thermal cycling, consistent with the release of elastic strain energy density. Furthermore, the stress variation over the imaged area more closely fits a Gaussian rather than a Weibull distribution, although this data is for the overall stresses rather

than those in the immediate vicinity of the damaged regions. A Gaussian distribution indicates that stress values across the image can be interpreted as independent or at least of short lateral order, i.e., with little correlation between different subregions. This lack of correlation is also evident from an overall featureless two-dimensional Fourier transform (not shown).

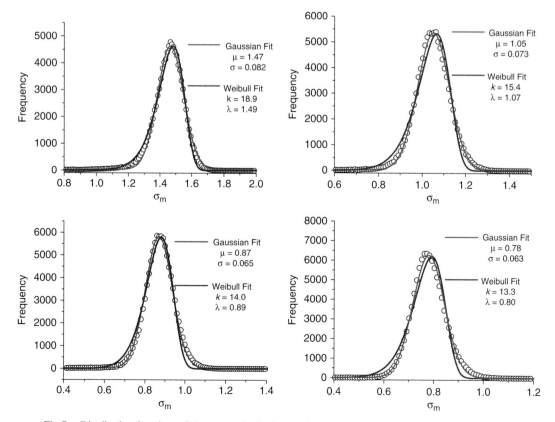

Fig. 7. Distribution functions of the stresses in the images shown in Fig. 6 as a function of the mean stress.

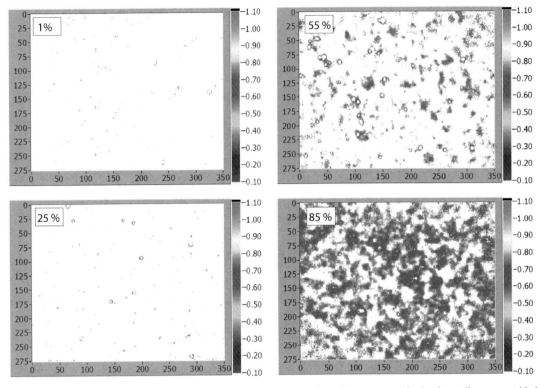

Fig. 8. Filtered piezospectroscopic images of Fig. 6 showing in green those regions where the mean stress in the thermally grown oxide is between 0.6 and 0.8 GPa. The white regions correspond to regions where the local mean stress is above 0.8 GPa. The numbers on the left hand ordinate and horizontal are pixels, each 16 μm in size.

Spatial correlations in the stress variation at particular stress intervals are more revealing. For instance, Fig. 8 are images formed by mapping out those regions of the four samples where the stresses lie between 0.6 and 0.8 GPa, an arbitrary stress interval. As can be expected from the stress distribution functions shown in Fig. 7, only small regions of the coatings at early stages of life are below this threshold. By 55% of life, 15.5% of the area lies within this stress interval and by 85% of life, about 60% of the area. While these numbers provide an additional quantification not obvious in the stress images, the image at 85% life suggests a new insight, namely that percolation of the damage appears to occur toward the end of life. Another way of representing the same data is by plotting the fractional area within this stress interval as a function of the life as in Fig. 9. This

suggests that the fractional area grows as a power law function of the number of thermal cycles.

To quantify the size and number of local separations, cross-sections of the coatings were analyzed. In each case, the number and length of the individual separations along the coating interface were measured and recorded. From this quantification procedure, carried out on coatings exposed for a number of different thermal cycles, a number of variables were analyzed: proportion of the interface that was separated, interface roughness, the average length of the separations, and the size of the largest separation. In addition, the luminescence shift was recorded through the TBC before cross-sectioning. The results of these analyzes are summarized in Fig. 10. Figure 10(a) shows that the number of separations per unit length of interface at first increases, reaches a maximum and then decreases with the number of cycles until, at failure, it approaches one. Before cycling begins, a small number of separations are already present, seemingly randomly distributed along the interface. The initial increase indicates that new separations form, and after reaching a maximum, the number decreases as individual separations coalesce at a greater rate than fresh separations nucleate. No unique function describes the data but its functional form is similar to that of nucleation and growth in two-dimensions. Complementary is the data in Fig. 10(b) of the proportion of the interface that is separated as a function of the number of cycles. This is a monotonic function that asymptotes to unity as failure is approached. The piezospectroscopic shift decreases monotonically with thermal cycle life as shown in Fig. 10(c). Data from both the measurements as well as the PS imaging analysis results are shown. The last set of data obtained from analysis of the cross-section images is the lateral size of the separations shown in Fig. 10(d). Interestingly, both the average separation size and the size of the largest separation increase as a power-law function of the number of thermal cycles. This data suggests that separations do grow by coalescence. Furthermore, although there is greater statistical variation and fewer measurements could be made, the size of the largest separation increases at a faster rate than does the average size. The

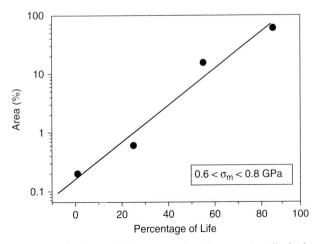

Fig. 9. Fractional area of the coatings where the mean stress lies in the interval 0.6–0.8 GPa as a function of the fractional life. Data obtained from the piezospectroscopic images in Fig. 6.

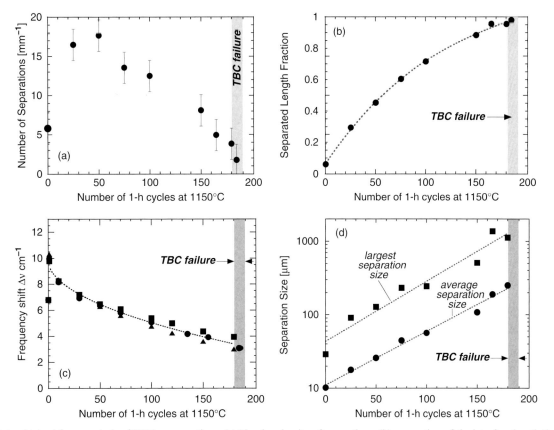

Fig. 10. Data obtained from analysis of SEM cross-sections. (a) Number density of separations, (b) proportion of the interface length that has separated, (c) piezospectroscopic shift, and (d) lateral size of interface separations, all as a function of the number of thermal cycles.

implication is that the largest separation grows by linking up larger separations.

V. Discussion

The cross-sectional microstructural analysis and the piezospectroscopy images provide complementary information concerning the damage evolution with thermal cycling. Comparison of the results also provides confidence in the interpretation of the piezospectroscopy images. Together, these results clearly, and unambiguously, show that well before macroscopic buckling and spallation occurs, the EB-PVD coatings on Pt-modified aluminized superalloy coupons begin to fail by a sequence of events in which local separations nucleate in the vicinity of the TGO, grow and link up. In this and several previous studies by different groups,[20–22] it has been found that the mean stress in the TGO is highest immediately after deposition of the TBC, when the alloy surface is flattest, and then monotonically decreases with thermal cycling as illustrated in Fig. 10(d). The large, mean compressive stress is attributed to the thermal expansion mismatch between the initially flat TGO and the underlying alloy as well as a compressive growth stress in the TGO. With thermal cycling the compressive stress decreases. This is a result of damage to the TGO, such as shown in Fig. 2, and change in shape as the TGO deforms to conform to the shape of the evolving bondcoat surface and constrained by the TBC on top of it. The change in shape causes the initial approximately biaxial stress state in the TGO to become increasingly tri-axial with out-of-plane tensile and compressive stresses. (The stress state is only approximately biaxial because the alloy is not entirely flat but has some initial roughness.) Insufficient data yet exists to determine the most appropriate physically based mathematical description of the decrease in mean stress but it can be fit to a form

$$\sigma_{\mathrm{m}} = a - b\,t^n \tag{2}$$

where a and b are constants and the exponent n is approximately 0.5. As the strain energy density in the TGO is proportional to the square of the mean stress, this would imply that the overall driving force in the TGO for further damage is also decreasing with thermal cycling. However, as will be described later in this section, the crack driving force for TBC failure is likely to be more complicated since there is strain energy in the TBC as well as in the TGO.

Analysis of the cross-section images indicates that local separations can exist even before thermal cycling begins. Whether these are the result of local contamination, the thermal stresses associated with cooling from deposition or shadowing during deposition is not known but it is evident that during the early stages of thermal cycling, other separations form along the interface. At the same time, as evident in both the PSLS images and the cross-sectional images, these local separations grow and begin to link up and coalesce to form clusters of local damage. Throughout the life, the average size of the separated regions, as well as their largest size, C_{L}, increase with the number of thermal cycles as a power law function

$$C_{\mathrm{L}} = C_0\,N^n \tag{3}$$

where C_0 is an initial size. Furthermore, although the average value of the mean stress decreases with the number of thermal cycles as mentioned above, the probability distribution in stress across the coatings appears to remain Gaussian. However, further analysis of the "low stress" portion of the stress distributions is needed to ascertain more about how the actual statistics of the coalescence process.

The current study does not provide any new insights into the origin of the local interface separations but it is assumed that separations form in response to the development of out-of-plane tensile stresses. This is consistent with an inability of the TGO and TBC to conform, by creep or plasticity, to the displacements of the bond-coat surface produced by a combination of rumpling and swelling that occur with thermal cycling. Rumpling causes local bending of the TGO, creating out-of-plane strains and stresses, which not only decrease the mean stress in the TGO but also can produce TGO cracks by local tension or tearing. An example of this form of cracking is shown at B in Fig. 2. Inhomogeneous swelling, in particular, of the bond-coat from preferential nickel diffusion into the bond-coat from the underlying superalloy will cause nonuniform spatial displacements of the surface, promoting rumpling. This is believed to be responsible for separations of the type shown at D in Fig. 2. If the TBC is thick enough and remains in contact with the TGO, it suppresses rumpling. However, preexisting separations at the TBC/TGO interface or local regions where there are gaps in the TBC columns, either as a result of shadowing or very small grains, would result in less constraint and allow local rumpling and hence the development of out-of-plane tensile stresses. In some cases, as at location D in Fig. 2(d), the local TBC constraint is lost as a divot of TBC, a small wedge shaped grain formed at the intersection of two larger misoriented columns, appears to have been pulled away by the differential bond-coat motion.

Our finding that TBCs fail by a progressive nucleation, growth and linkage process until linked damaged regions extend across an area corresponding to the macroscopic buckling condition, provide some insight into why the thermal cycle life is sensitive to so many thermal cycling conditions.[5,23] Long hot times combined with slow heating and cooling rates and relatively low maximum temperature leave the bond-coat surface relatively flat and unchanged roughness and so the stresses normal to the interfaces are small and few local separations can occur. The lives are consequently relatively long. Under these conditions, it is speculated that processes that lower the interface fracture toughness cause failure such as might occur by long-range diffusion of species, for instance, S, from the underlying superalloy. Under these conditions, failure tends to occur once a critical thickness of the TGO is formed by oxidation and separation occurs preferentially by delamination along the bond-coat/TGO interface. The life can then be reasonably well predicted by oxidation life codes, such as COATLIFE,[24,25] embodying a critical thickness based on attaining a critical strain energy in the TGO and TBC. At the other extreme is rapid, short cycling at high temperatures, above about 1100°C. The lives under these conditions do not fit the oxidation life codes as has been remarked by Chan *et al.*,[24] being both much shorter and having a much larger statistically variability. These conditions, more typically encountered in military engines, are known to produce pronounced rumpling of bare PtNiAl aluminide coatings and, as illustrated in this work, several different types of interface damage but few bond-coat/TGO delaminations. Experiments and models indicate that at temperatures of 1100°C and above, the surface displacements are dependent on not only temperature but heating and cooling rates in the thermal cycles, too.[5,23,26] Consequently, there remains an outstanding problem of how to relate thermal cycle life to both maximum temperature and thermal cycle time, reconciling Eqs. (2) and (3) introduced above.

The complexity of the damage observed makes it difficult to describe failure in terms of detailed crack driving forces and fracture resistances. Not only does the propagation of a single crack seem an inappropriate description for the failure until the latest stages of the failure but also the separations are not strictly coplanar. Furthermore, although the luminescence shift is a measure of the mean stress it does not provide any information about the thickness of the TGO and hence is not a direct measure of the elastic strain energy, U_{SE}. Neither does it explicitly include the strain energy in the TBC. Nevertheless, as fracture of the TGO does not decrease its elastic modulus, the stored elastic strain energy in the TGO is proportional to the square of the mean stress. The TGO thickens by oxidation and so its thickness increases approximately parabolically with time. Consequently, the total elastic strain energy will be a slowly varying function of

time. As the fracture resistance is unlikely to change very rapidly with thermal cycling, the critical condition for fracture at which the total strain energy area density equals the fracture resistance is likely to be poorly defined and variable. Consequently, one might expect that the failure life exhibits a large statistical variation about a mean value determined by the thermal cycling conditions. Furthermore, as described by Hutchinson in a companion article, failure (as well as buckling) occurs under mixed mode conditions, so the statistical variability is expected to be even greater than just discussed. Further complicating any detailed energy balances is the fact that the TBC sinters with time increasing its elastic modulus and with its strain energy. This emphasizes the importance of knowing where the separations lie; if they lie largely above the TGO, it's the strain energy in the TBC that provides the driving force for failure, whereas if the separations are principally at the alloy/TGO interface, then the strain energy in both the TGO and TBC must be included.

Finally, the imaging system used in this work remains a prototype. Further improvements in the optics as well as image processing hardware and software, currently being developed, can be expected to yield faster imaging of the stresses and make it possible to follow the damage evolution in even greater detail. For instance, the instrumentation should enable images of the residual stresses to be generated after each thermal cycle of the same test object, so that the growth and coalescence of individual separation events can be followed with cycling. The ability to generate large databases of residual stress data also allows better estimates of variance in residual stress data between samples for a given fractional life, in comparison with the current results based on a few samples. With better knowledge of the conditions for delamination coalescence, the image system promises to provide nondestructive monitoring of the health of individual blades and vanes. This would also facilitate comparison of our laboratory scaling findings to those on engine hardware as well as other types of coatings, including overlay coatings.

Acknowledgments

The authors are grateful to Ken Murphy of Alcoa Howmet Research Center for providing the coatings studied in this work. The authors are also grateful for support from the Office of Naval Research (D. R. C., V. K. T.) and the Air Force Research Laboratory (B. H., while at MetroLaser).

References

[1]T. Editors, "Britain's Lonely High-Flier"; The Economist, 2009
[2]A. G. Evans and J. W. Hutchinson, "On the Mechanics of Delamination and Spalling in Compressed Films," *Int. J. Solids Struct.*, **2**, 455–66 (1984).
[3]M. D. Thouless, H. M. Jensen, and E. G. Liniger, "Delamination from Edge Flaws," *Proc. Roy. Soc.*, **A447**, 271–9 (1994).
[4]J. W. Hutchinson and Z. Suo, "Mixed Mode Cracking in Layered Materials," *Adv. Appl. Mech.*, **29**, 63–191 (1992).
[5]V. K. Tolpygo and D. R. Clarke, "Rumpling of CVD (Ni,Pt)Al Diffusion Coatings Under Intermediate Temperature Cycling," *Surf. Coat. Technol.*, **203**, 3278–85 (2009).
[6]R. J. Christensen, D. M. Lipkin, D. R. Clarke and K. Murphy, "Nondestructive Evaluation of the Oxidation Stresses Through Thermal Barrier Coatings Using Cr^{3+} Piezospectroscopy," *Appl. Phys. Lett.*, **69** [24] 3754–6 (1996).
[7]B. Heeg and D. R. Clarke, "Optical Measurement of the Thermal Diffusivity of Intact Thermal Barrier Coatings," *J. Appl. Phys.*, **104**, 113119, 7pp (2008).
[8]J. I. Eldridge, C. M. Spuckler, and R. E. Martin, "Monitoring Delamination Progression in Thermal Barrier Coatings by Mid-Infrared Reflectance Imaging," *Int. J. Appl. Ceram. Technol.*, **3** [2] 94–101 (2006).
[9]V. K. Tolpygo and D. R. Clarke, "Morphological Evolution of Thermal Barrier Coatings Induced by Cyclic Oxidation," *Surf. Coat. Technol.*, **163–164**, 81–6 (2003).
[10]Q. Ma and D. R. Clarke, "Stress Measurement in Single-Crystal and Polycrystalline Ceramics Using their Optical Fluorescence," *J. Am. Ceram. Soc.*, **76** [6] 1433–40 (1993).
[11]Q. Ma and D. R. Clarke, "Piezospectroscopic Determination of Residual-Stresses in Polycrystalline Alumina," *J. Am. Ceram.c Soc.*, **77** [2] 298–302 (1994).
[12]S. H. Margueron and D. R. Clarke, "The Use of Polarization in the Piezospectroscopic Determination of the Residual Stresses in Polycrystalline Alumina Films," *Acta Mater.*, **54** [20] 5551–7 (2006).
[13]B. Heeg and D. R. Clarke, "Non-Destructive Thermal Barrier Coating (TBC) Damage Assessment Using Laser-Induced Luminescence and Infrared Radiometry," *Surf. Coat. Technol.*, **200** [5–6] 1298–302 (2005).

[14]J. A. Nychka and D. R. Clarke, "Damage Quantification in TBCs by Photo-Stimulated Luminescence Spectroscopy," *Surf. Coat. Technol.*, **146**, 110–6 (2001).

[15]B. Henderson and G. F. Imbusch, *Optical Spectroscopy of Inorganic Solids.* Claredon Press, Oxford, 1989.

[16]D. J. Gardiner, M. Bowden, S. H. Margueron and D. R. Clarke, "Use of Polarization in Imaging the Residual Stresses in Polycrystalline Alumina Films," *Acta Mater.*, **55**, 3431–56 (2007).

[17]D. R. Clarke and D. J. Gardiner, "Recent Advances in Piezospectroscopy," *Int. J. Mater. Res.*, **98**, 1–8 (2007).

[18]B. Heeg and J.B Abbiss, "Piezospectroscopic Imaging with a Tunable Fabry-Perot Filter and Tikhonov Reconstruction," *Opt. Lett.*, **32** [7] 859–61 (2007).

[19]J. B. Abbiss and B. Heeg, "Imaging Piezospectroscopy," *Rev. Sci. Instrum.*, **79**, 123105, 9pp (2008).

[20]M. Wen, E. H. Jordan, and M. Gell, "Evolution of Photostimulated Luminescence of EB-PVD (Ni,Pt)Al Thermal Barrier Coatings," *Mater. Sci. Eng.*, **A398**, 99–107 (2005).

[21]V. K. Tolpygo, D. R. Clarke, and K. S. Murphy, "The Effect of Grit Blasting on the Oxidation Behavior of a Platinum-Modified Nickel Aluminide Coating," *Metall. Mater. Trans. A-Phys. Metall. Mater. Sci.*, **32** [6] 1467–78 (2001).

[22]G. Lee, A. Atkinson, and A. Selcuk, "Development of Residual Stress and Damage in Thermal Barrier Coatings," *Surf. Coat. Technol.*, **201** [7] 3931–6 (2006).

[23]V. K. Tolpygo and D. R. Clarke, "Temperature and Cycle Time Dependence of Rumpling in Platinum-Modified Diffusion Aluminide Coatings," *Scri. Mater.*, **57**, 563–6 (2007).

[24]K. Chan, S. Cheruvu, and R. Viswanathan, "Development of a Thermal Barrier Coating Life Model"; pp. 591–5 in *ASME Turbo Expo 2003.* ASME, Atlanta, 2003.

[25]T.A Cruse, S. E. Stewart, and M. Ortiz, "Thermal Barrier Coating Life Prediction Model Development," *J. Eng. Gas Turbines Power*, **110**, 610–6 (1988).

[26]D. S. Balint and J. W. Hutchinson, "An Analytical Model of Rumpling in Thermal Barrier Coatings," *J. Mech. Phys. Solids*, **53** [4] 949–73 (2005).   □

J. Am. Ceram. Soc., **94** [S1] S120–S127 (2011)
DOI: 10.1111/j.1551-2916.2011.04588.x
© 2011 The American Ceramic Society

journal

In Situ Measurement of the Toughness of the Interface Between a Thermal Barrier Coating and a Ni Alloy

Christoph Eberl,[†,‡,§] Xi Wang,[‡,¶] Daniel S. Gianola,[‡,∥] Thao D. Nguyen,[‡] Ming Y. He,[††] Anthony G. Evans,[††]
and Kevin J. Hemker[‡]

[‡]Department of Mechanical Engineering, Johns Hopkins University, Baltimore, Maryland 21218-2682

[§]Institute for Applied Materials, Karlsruhe Institute of Technology, Karlsruhe 76131, Germany

[¶]Institute of Mechanics, Chinese Academy of Science, Beijing 100190, China

[∥]Department of Materials Science & Engineering, University of Pennsylvania, Philadelphia 19104-6272, Pennsylvania

[††]Department of Materials, UCSB, Santa Barbara 93106-5050, California

A flexural test for the *in situ* measurement of the delamination toughness of the interface between a thermally grown oxide and a bond coat in the presence of a thermal barrier coating (TBC) has been implemented. To accomplish the testing, a section of the substrate was removed by microelectro-discharge machining and a precrack introduced through the TBC by center point loading. This was followed by application of an asymmetric single-point load to extend a delamination along the interface. A displacement and strain mapping method was used to locate the delamination and to ascertain its extension. To relate the energy release rate and mode mixity to the crack extension and the loads, a finite-element method was implemented. The ensuing fracture resistance was found to vary along the interface with values in the range of 25–95 J/m².

I. Introduction

THE integrity of metal/oxide interfaces governs the viability of many multilayer systems and governs the lifetime of such applications. The mechanics dictating the energy release rates and mode mixities have been comprehensively established for many salient problems.[1] However, implementation has often been limited by the difficulty in measurement of the interface fracture toughness at the relevant mode mixity. There are literally hundreds of different methods for characterizing adhesion, and a recent review of interfacial toughness measurements for thin films on substrates has identified six different categories of tests for measuring critical values of stress intensity or energy release rates.[2] These categores include: superlayer, indentation, combined superlayer indentation, scratch, bulge and blister, and sandwich specimen tests. Each test comes with inherent advantages and disadvantages and no one methodology has come to be known as the gold standard for measuring thin film or coating adhesion. For the thermal barrier coatings (TBCs) considered in the current study, techniques involving notch flexure sandwich specimens are most salient.

Variants on a notch flexure test have been widely exploited, particularly for interfaces in multilayer semiconductor devices.[3,4] This test typically measures the mixed mode toughness, at a phase angle of approximately $\Psi \approx 42°–50°$. The method is

applicable to interfaces within planar devices, primarily on thin substrates. It cannot be applied to either thick or curved substrates. This paper presents a new flexure test method that obviates this limitation and allows *in situ* measurement on films and coatings attached to substrates; the mechanics-based design for which is described in a companion article.[5] In the current study, this method has been used to measure the toughness between an oxide coating and a relatively small diameter cylindrical substrate, the configuration for which is depicted in Fig. 1.

The test design is guided by beam theory solutions for a planar four-point bending configuration.[5] A three-point bending configuration with asymmetric loading was used in this study but is difficult to solve analytically. For this reason, the simpler case, namely the four-point bending configurations, will be used to explain the experiment. In the absence of residual stress, the energy release rate, G_P, for this configuration can be expressed in the normalized form:

$$G_P E_2 h_2^3 / (PL)^2 \equiv \prod \left(\frac{a}{L}, \frac{h_1}{h_2}, \frac{E_1}{E_2}, b/L \right)$$

where P is the load/width, a is the delamination length, L is the span, b is the separation of the inner load points in the four-point bending setup, and h_1, h_2 are the thickness of the coating and substrate, respectively, and E_1, E_2 the corresponding values of the Youngzs modulus. The functional form for $\prod$ has been elaborated elsewhere.[5] One desirable feature of this test is the fact that the energy release rate diminishes with increase in crack length, so that the delamination progresses *only upon increasing the load* (stable behavior). In the presence of residual stress, σ_R, another nondimensional group influences G_P, given by: $\Re \equiv \sigma_R h_1^2 / PL$. The influence of residual stress is often overlooked but can be substantial. One demonstrated advantage of the proposed test methodology is that fabrication of the specimen relaxes the residual stresses in the bilayer and greatly simplifies analysis of the experimental data.

In the aero-turbine industry, cylindrical test configurations are routinely used to examine the thermo-mechanical integrity of coated components upon thermal cycling. Such assessment methods are referred to as burner-rig tests. Developing an interface toughness test that can be performed on such configurations as a function of exposure to thermal cycling is of great interest. Extension of this test protocol to commercial engine hardware such as blades and vanes is straightforward. In this study, specific measurements have been performed on a columnar TBC generated by electron beam physical vapor deposition (EB-PVD) on a single crystal, Ni-based superalloy substrate with a NiCoCrAlY bond coat.[6] During deposition of the

T. Pollock—contributing editor

Manuscript No. 28992. Received December 2, 2010; approved March 26, 2011.
This work was supported by AFOSR under the MEANS-2 Program (Grant No. FA9550-05-1-0173) and by ONR (Grant No. N00014-08-1-0454).
[†]Author to whom correspondence should be addressed. e-mail: chris.eberl@kit.edu

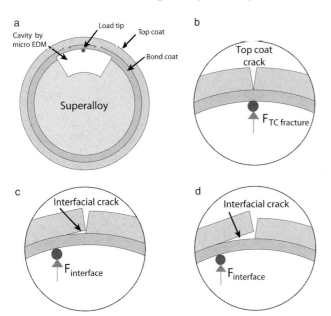

Fig. 1. A schematic of the specimen design and test procedure (a) for the example of a coating on the periphery of a circular substrate. The precracking method is illustrated on the top right (b) and the subsequent asymmetric loading for ascertaining the delamination toughness is on the lower left (c) and right (d).

coating, a thin thermally grown oxide (TGO) layer develops, consisting largely of α-Al_2O_3.

To create the test configuration (Fig. 1), a section of the substrate is removed by microelectro-discharge machining (μ-EDM) to leave an intact bilayer beam along an arc of the surface. The beam segment consists of the coating on the outside and the bond coat on the inside. The specimen is placed within a microtensile test system with the loads applied as indicated on Fig. 1.

The article is organized as follows. The protocol used for the measurements is described. The interface crack extension measurements are presented. Finite-element (FE) results for the energy release rates and mode mixity are summarized and used to ascertain the interface toughness. The results are interpreted in terms of the mechanisms of interface crack extension.

II. Experimental Protocol

The specimens, provided by Pratt & Whitney, included a TBC on the surface that consisted of a 110 μm 7%-yttrium-stabilized zirconia (7-YSZ) coating, deposited by a EB-PVD process onto a low-pressure plasma-sprayed NiCoCrAlY bond coat. The substrate was a Ni-base superalloy substrate (PWA 1484). Cross-sections were cut and polished from the provided cylindrical burner rig bars and subsequently μ-EDM was used to carve out the substrate and parts of the bond coat underneath the TBC coating (Fig. 1). The final test geometry (Fig. 2(a)) consisted of a doubly end-supported beam of roughly 1 mm length. A bond coat layer of 35–65 μm (depending on the specimen) was left to support the 110-μm-thick TBC, resulting in a bilayer beam. The thickness of the cross-section was roughly 550 μm. Accordingly, the specimens (Fig. 2) have the relative span to thickness ratio, $L/h = 6.7$ (Beam 2, Table I) and 7.4 (beam 1, Table I), coating to substrate thickness ratio, $h_1/h_2 \approx 3$, and modulus ratio, $E_2/E_1 \approx 8$.

The experiment is carried out on a custom-built microtensile setup consisting of a horizontal load train supported by an air bearing.[7] This configuration provides the alignment of the sample while minimizing friction on the 5 lb load cell. Steel microtips are connected to a piezoelectric stepper motor fixed to a five-axis stage to enable the precise alignment of the loading axis. Images are captured during testing by a high-resolution camera attached to an optical microscope. To reduce the influence of

vibrations the setup is mounted to an air table. More detailed information on the experimental configuration in provided in Eberl *et al.*[7]

The test is conducted in two steps. Initially, single-center-point loading is used to induce a vertical crack that extends through the ceramic top coat to the interface (Fig. 1(b)). A single-point load provides a focused stress field with maximum tensile stress at the surface of the TBC. Therefore the TBC is forced to crack at the selected region.[8] The loading position was chosen to be the center of the beam for all precracking experiments.

Thereafter, an asymmetric single-point load is used to propagate the crack stably along the interface (Figs. 1(c) and (d)). The loading tip is positioned beyond the leading edge of the interface crack away from the center TBC crack during all experiments. This configuration forces the interface crack to extend in the direction of the loading site. The loads at which this happens provide a measure of the toughness as the crack only extends if the load exceeds a critical value.

The key experimental feature needed to analyze the measurements with high fidelity is the ability to use a digital image correlation and tracking (DICT) technique to observe *in situ* the local displacements. For this purpose, the sample deformation is monitored by a high-resolution CMOS camera (PL-782A, Pixelink, Ottawa, ON, Canada) mounted to an optical microscope (Nikon, Tokyo, Japan) with a field of view ranging from 3.2 mm × 2.3 mm to 400 μm × 290 μm. The displacement field is calculated by a digital image correlation technique from a series of images, such as those shown in Figs. 2(b) and (c). The achievable displacement resolution is limited by the vibrations, the quality of the optical system, the camera resolution and the signal to noise ratio of each pixel in the image. The experimental configuration used here results in a displacement noise floor with a root mean squared value of roughly 30 nm.

A spatially dense displacement field is calculated from the acquired image sequence in a post process. This is achieved using a virtual mesh of markers with a pitch size of 1–2 μm, corresponding to roughly 10–20 pixels at the highest magnification, which are defined and tracked by the DICT functions. An example of a mesh used for these measurements is shown in Fig. 2(b). The size of the tracked markers is between 20 × 20 and 40 × 40 pixels^2, which is equal to 2.5 μm × 2.5 μm and 5 μm × 5 μm at the highest optical resolution used in these experiments. The local displacement gradient is used to calculate the axial strain along the length of the beam, $\varepsilon = \partial u/\partial x$ and the differential displacement along the interface is used to acquire the opening displacement along the crack. The strain and displacement resolutions are increased by averaging the displacement gradient over a finite number of tracked markers and by taking multiple images per load step.

Crack propagation is tracked by analyzing the displacement field using custom Matlab® functions, and the relevant quantities needed for data analysis are defined in Fig. 2(c). One advantage of the DICT method is the output of full-field displacements, providing local information about deformation, as shown in Fig. 3. These computed displacement fields show the precise location of cracks and their evolution. After defining the interface in the acquired images, the crack opening displacement along the interface is measured and correlated to the applied load.

III. Experimental Results

As described in previous sections, a precrack is induced close to the geometric center by single-point symmetric loading. This crack follows the columnar microstructure and proceeds from the surface down to the interface. Two different methodologies have been used to extend the interface crack: designated asymmetric single-point loading and symmetric two-point loading. The latter is motivated by the analysis detailed in He *et al.*[5] and the tests conducted in this manner are the most straightforward to interpret. However, asymmetric loading was found to provide much broader experimental flexibility, albeit that the

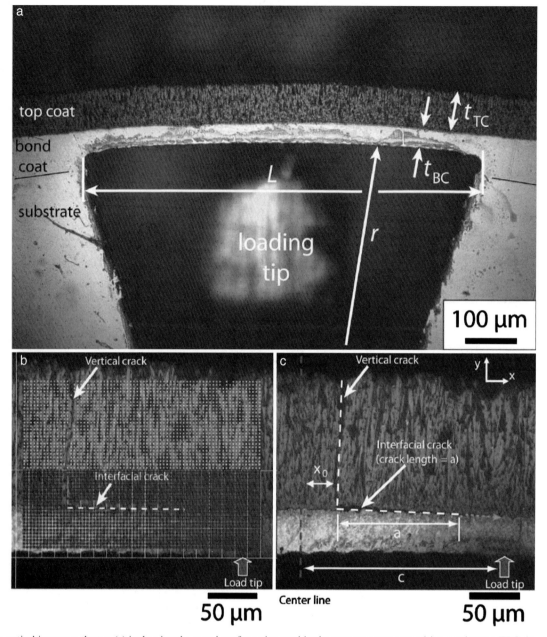

Fig. 2. The optical image on the top (a) is showing the actual configuration used in the measurements comprising a columnar TBC deposited onto a Ni-based alloy and the cut out section below the bond coat. The grid used for the digital image correlation is shown in the lower left (b) where a vertical and interfacial crack has been formed. The geometry is shown in the lower right (c).

interpretation then requires FE solution. In the end, asymmetric loading emerged as the method of choice for the current study.

In this method, a dominant interface crack was generated from the precrack by relocating the load point off-center. This mode of loading caused an interface crack to form and propagate toward the load point in a stable manner. The ensuing interface crack path is shown in SEM images of the cracks in Fig. 4(a). It penetrates the TGO and then follows the interface with the bond coat. The tip region comprises a series of *en echelon* microcracks traversing the TGO (Figs. 4(b) and (c)), indicative of a mode II mechanism, as elaborated later. To prevent the load/unload/load sequence from influencing the interface at the crack tip, all subsequent tests were conducted in consecutive load steps without unloading in between.

The experimental crack propagation events for two different beams are summarized in Table I. The results from these series of critical events are the basis for the FE-based fracture toughness calculation described in the subsequent section. For beam 1, the aforementioned precrack through the top coat formed at a distance of $x_0 = 29$ μm away from the geometric center of the

beam. The crack proceeded from the TBC surface at the critical load and arrested at the interface between top and bond coat. In the next step the loading tip was offset to $c = 36$ μm from the center of the beam and then loaded again. The loading was conducted in steps to $F = 4.9$ N and after unloading the analysis showed that an interfacial crack had nucleated at a load of $F_c = 3.1$ N. During this event (not listed in Table I), the interface crack had propagated to a length of $a = 29$ μm and extended beyond the position of the loading tip. This was the only time this occurred in all experiments. After nucleating the interface crack, the load tip was again offset to a new location ahead of the interface crack tip at a position $c = 78$ μm. After loading to a maximum force of $F = 3.7$ N, the analysis revealed that the crack had extended after applying a critical force of $F_c = 3.3$ N to a length of $a = 50$ μm (not listed in Table I). Continuing the experiment, the tip was then offset to $c = 171$ μm relative to the beam center. Increasing the load in a stepwise fashion led to several consecutive crack propagation events for this configuration. As documented in Table I, the crack propagated by amounts of $a = 16, 18, 16, 14,$ and 3 μm when the critical loads

Table I. Event table

Crack event #	Schematic	x_0 (μm)	a (μm)	c (μm)	Critical load, P_c (N)	Propagation, Δa (μm)	Interfacial toughness, Γ (J/m²)	Phase angle, ψ (°)
Beam 1: $L = 1.1$ mm; W $= 550$ μm; $h_{bc} = 36$ μm; $h_{tbc} = 112$ μm; $E_{bc} = 155$ GPa, $E\|_{tbc} = 20$ GPa; $E^{\perp}_{tbc} = 180$ GPa								
1		29	60	171	2.3	16	27.1	17.9
2		29	76	171	3.0	18	42.5	18.0
3		29	94	171	3.5	16	53.6	18.1
4		29	110	171	4.1	14	67.8	18.9
5		29	124	171	5.1	3	95.2	20.5
Beam 2: $L = 1.1$ mm; W $= 550$ μm; $h_{bc} = 51$ μm; $h_{tbc} = 112$ μm; $E_{bc} = 155$ GPa, $E\|_{tbc} = 20$ GPa; $E^{\perp}_{tbc} = 180$ GPa								
1		0	14	50	3.2	6	51.1	18.9
2		0	20	50	3.5	7	63.8	19.1

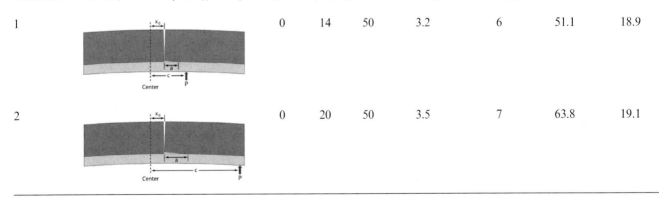

of $F_c = 2.3$, 3, 3.5, 4.1, and 5.1 N were reached (crack events 1–5, beam 1).

The second sample (beam 2) was treated in a similar fashion and the crack propagation and critical loads and corresponding fracture energy and phase angle values as calculated by FEA can be found in Table I. First, the beam was center loaded till a crack was induced in the top coat at the very center of the beam with $x_0 = 0$ μm. In a second step, the load was applied at an off center position $c = 50$ μm and the load was increased. As a result, the crack channeled to the interface and extended and to a length of $a = 14$ and 20 μm at corresponding critical loads of $F_c = 3.2$ and 3.5 N (crack events, beam 2).

IV. Energy Release Rates and Interface Toughness

The mechanical response of the bilayer beam has been analyzed by the FE method using the commercial code ABAQUS Standard. Only the bilayer beam and part of the support material was simulated in the analysis and the boundary conditions were assumed as rollers. No energy exchanges are allowed between the modeled specimen and the rest of the ring. The finite-element model and the mesh are depicted in Fig. 5. The eight-node biquadratic plane strain elements with reduced integration were used. At the crack tip focused elements were used and the midsize node parameter is chosen to be 0.25. In these simulations the

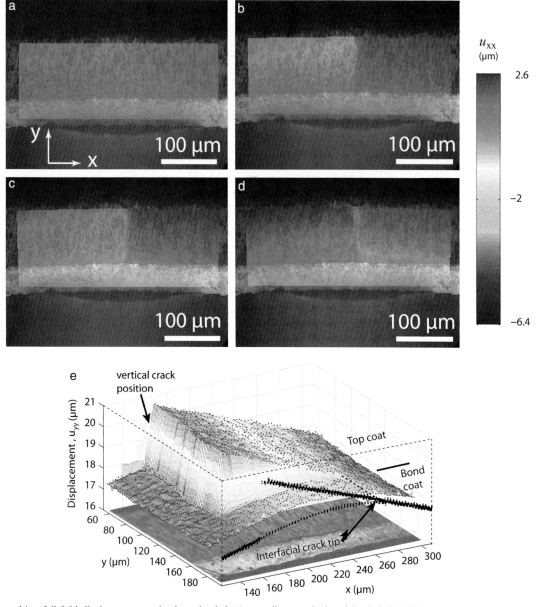

Fig. 3. The resulting full-field displacement u_{xx} in the unloaded (a) as well as cracked and loaded (b)–(d) is shown in the upper images. The 3D deformation plot (e) shows how the crack tip can be measured from the full-field displacement data.

TBC is assumed aniostropic in material property: the in-plane modulus, $E_{\parallel}$, is considered to be 30 GPa according to an elasticity study of the same material,[8] and the out-of-plane modulus, $E_{\perp}$, is assumed to be 150 GPa (compared with $E_{\perp} \approx 200$ GPa for dense YSZ). Poisson's ratio of TBC is considered to be 0.2. The bond coat is considered to be elastic/plastic with power law hardening having the stress/strain curve obtained in Kim *et al.*[9] (Young's modulus $E = 155$ GPa, Poisson's ratio $\nu = 0.25$, yield strength $\sigma_Y = 750$ MPa, and strain hardening exponent $n = 0.2$).

(1) Basic Features

ABAQUS calculations based on the appropriate specimen dimensions and material properties allows the energy release rate and mode mixity to be determined as a function of the crack length. The energy release rate, i.e. the *J*-integral, is calculated in ABAQUS for the interface crack as the loading increases. The results are presented in the normalized coordinates suggested by beam theory,[5] $G_p E_{bc} h_{bc}^3 / (P_c)^2$ as a function of a/L (Fig. 6). Note that, at fixed load, the energy release rate decreases as the crack extends, *reaffirming the stable nature of this test configuration.* The stress fields and mode mixity of an interfacial crack has

been shown to have an oscillatory character in a bimaterial specimen. These oscillations were observed in the FE simulations and the associated mode mixity varied with distance from the crack tip.[1] At a distance of 1 µm, which corresponds with the plastic zone size in the FE simulation, the mode mixity was calculated to be 20°. The mode mixity increased to approximately 40° at a distance of 8–10 µm. This larger value approaches the mode mixity that is traditionally reported for four-point flexure measurements, but is still slightly smaller[1,2] for the other methods that have been used to measure TBC interfacial toughness.[9–12]

(2) Experimental Findings

Specific values of fracture resistance, Γ_R, have been calculated for each of the critical crack growth events reported in the previous section. The calculated *J*-integral at the critical load is considered as the interface toughness. These fracture resistance values are included in Table I, and inspection of the data indicates that the measured toughness values fall into a rather broad range (25–95 J/m²), mimicking perhaps the tortuosity of the TGO/bond coat interface. It is worth noting that the measured

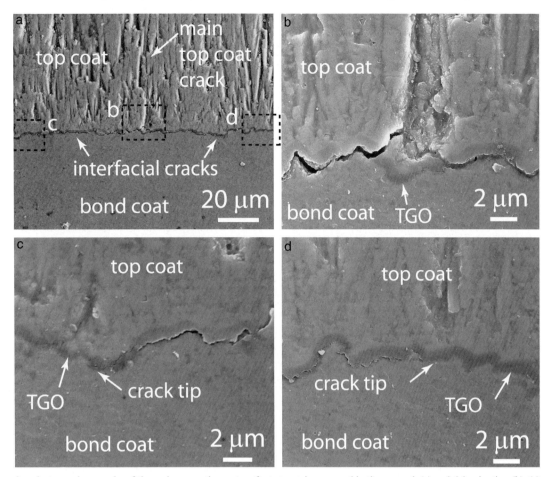

Fig. 4. Scanning electron micrographs of the region near the center of a test specimen reveal both, precrack (a) and delamination (b), (c) and (d) at the interface between the TGO and the bond coat.

fracture resistance appears to rise with crack length in both data sets. This rise cannot be attributed to changes in mode mixity and may be indicative of a rising resistance with crack extension. This could be explained if the crack may arrest at selectively tougher features on the tortuous interface. Alternatively, the role of permanent plastic deformation in the bond coat has not been completely ruled out. It is also possible that the TBC was

damaged when the sample was initially loaded. No evidence of damage was noted, but it is possible that undetected damage resulted in the lower value of interfacial toughness for the first event in Table I. Given this uncertainty, we prefer not to ascribe any significance to the variation in toughness with mode mixity and to represent our measurement with the mean of all measurements. The mean values of the toughness for each individual

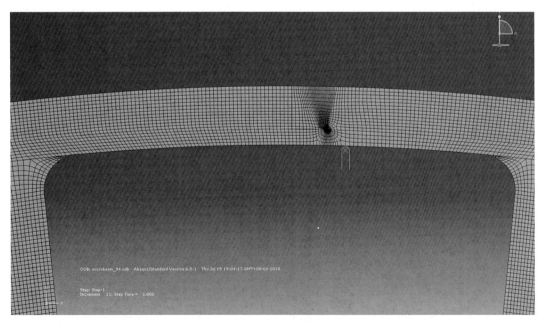

Fig. 5. The finite-element model and the mesh, with the refined mesh close at the crack tip.

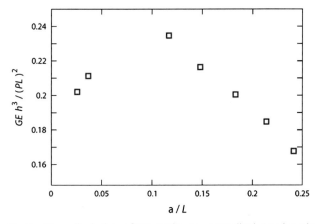

Fig. 6. Normalized plots of FE results, e.g. normalized J and mode mixity as a function of a/L. (a) The nondimensional energy release rate as a function of relative interface crack length for the asymmetric single-point test, see event table for geometric details. The phase angle shows a constant value until the crack tip is closing in on the off center load tip position (b).

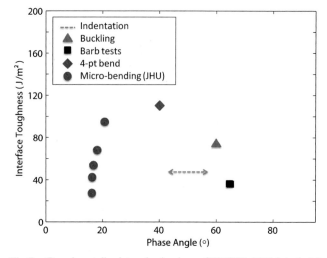

Fig. 7. Experimentally determined values of EBPVD TBC interfacial toughness plotted as a function of mode mixity. The average value of toughness measured in the current study is in reasonable agreement with data published previously. The scatter in this data, which were obtained by a variety of techniques with differing assumptions of modulus and residual stress, masks any possible dependence of toughness on mode mixity.

beam were measured to be nearly identical and the overall mean value for all toughness measurements was determined to be 57.3 ± 21.5 J/m^2.

(3)　Influence of Residual Stress

One important but often overlooked caveat in the measurement of interfacial toughness is the fact that potential residual stresses in the TBC will influence the energy release rate, as well as the mode mixity. These residual stresses could result from fabrication processes in which the sample is subjected to different temperatures. Because the materials involved have different thermal expansion coefficients, significant thermal stresses arise in the individual layers with different signs. Furthermore, in service TBC layers are also exposed to stresses from growth, sintering and changes in chemical composition leading to a phase evolution. Theses stresses are additive to those from thermo-mechanical cycling and can result, together with creep processes, in a difficult to predict stress distribution. One advantage of the test methodology developed in this study is the fact that the residual stresses in the bilayer beam are relaxed during specimen preparation, during μ-EDM removal of the underlying substrate. This salient feature was illustrated in Eberl *et al.*,[8] where one edge of a double-end-clamped bilayer beam was cut through and the relative displacement between the shoulder and the free-end of the beam was undetectable. This finding holds practical significance in that it obviates the need to correct the measured toughness data to account for the effects of residual stress in the test specimens.

V.　Discussion

A new test methodology that can be used to measure *in situ* the delamination toughness of coatings or films attached to both test specimens and components has been developed. The use of μ-EDM to remove a section of the substrate and create a coating/substrate bilayer beam provides a robust specimen that can be easily and stably loaded. Digital image correlation provides for direct high-fidelity measurement of crack length, which can be used to measure the critical load for crack extension. FE analysis can and has been used to estimate the associated energy release rates and mode mixity from these measurements.

The efficacy of this new methodology is demonstrated through *in situ* measurement of the toughness of the interface between an as-deposited TBC and its underlying NiCoCrAlY bond coat on a standard burner rig bar. The toughness values measured in this study (57.3 ± 21.5 J/m^2) are compared with literature values of toughness that were obtained using other

methodologies (Fig. 7). Vasinonta and Beuth[13] used the size of the delamination that occurs upon indentation to estimate the interfacial toughness of as-deposited EB-PVD TBC samples to be 49 J/m^2, assuming a residual stress of 50 MPa and anisotropic TBC modulus of 44 GPa. Faulhaber *et al.*[11] analyzed the shape of the coating in buckled regions adjacent to spalled areas of burner rig bars identical to those examined in the current study. Establishing the mechanics of ridge-cracked buckle delaminations for multilayers on curved substrates allowed them to predict delamination toughness; assuming a TBC modulus of 50 GPa they predicted the interfacial toughness to be 75 J/m^2 for a mode mixity of 60°. Guo and colleagues implemented a barb pullout test for EB-PVD TBC and initially reported an interfacial toughness of 45 J/m^2 and a phase angle of 90°, but Liu and colleagues reevaluated these tests by taking into account residual stresses and reported an interfacial toughness of 36 J/m^2 and a phase angle of 65° for Young's modulus of 44 GPa. Thery *et al.*[14] used a double cantilever experiment to measure fracture toughness[3] of EB-PVD and reported a value of 110 J/m^2 but did not mention the modulus or residual stresses. The phase angle for such a test should be roughly 40°.[15] Similar experiments have been carried out by Bahr *et al.*[16] who reported a value of >81 J/m^2 at which point their stiffening layer detached.

One additional benefit of the new test methodology is that it provides a measure of the interfacial toughness for values of mode mixity that are different than measured previously. The values of mode mixity predicted by the FE analysis of the microbend experiments fall in the range of 20°–40°, which is slightly closer to mode I than has be reported for other test methods. Hutchinson and Hutchinson[17] have shown that the delamination of a TBC occurs as a result of mode II loading. In this regard the ability to measure the mode II interfacial toughness would be most straightforward; nevertheless, the ability to measure interfacial toughness over a broad range of mode mixity is also desirable.

Two additional benefits of the new specimen geometry and test methodology include: (i) the relaxation of the residual stresses in the test specimen and (ii) the natural extension of this technique to the characterization of engine hardware. With regard to the former, the realization that the residual stresses in the bilayer beam are relaxed when the underlying superalloy section is removed by μ-EDM simplifies the analysis of the test. Although often overlooked residual stress can have a very large influence on measured values of interfacial toughness, and in is helpful to have the importance of residual stresses in reduced during specimen preparation. The results obtained for the study outlined in this article were for as-deposited TBC coatings, but the use of this technique to measure the influence of thermal cycling, external deposits (e.g., CMAS), TBC sintering and phase evolution are envisioned. Of particular interest is the ability to use this in situ technique to characterize engine hardware. Slicing, μ-EDMing, and characterizing commercial components provide an opportunity to measure location specific properties as well as the influence of the turbine environment.

Acknowledgments

The authors would like to thank J. Hutchinson for creative and enlightening discussions that allowed this technique to come to fruition. We also thank M. Maloney and D. Litton of Pratt and Whitney (USA) for providing the specimens, S. Faulhaber for preparing the disk specimens, J. Mraz (Smaltec) for fabricating the microbeam specimens, and W. Sharpe, Jr. and J. Sharon for technical support. C. E. would like to acknowledge financial support from the German Science Foundation (SFB499/3-2007 N01).

References

[1]J. Hutchinson and Z. Suo, "Mixed Mode Cracking in Layered Materials," *Adv. Appl. Mech.*, **29**, S63–191 (1992).

[2]A. A. Volinsky, N. R. Moody, and W. W. Gerberich, "Interfacial Toughness Measurements for Thin Films on Substrates," *Acta Mater.*, **50**, S441–66 (2002).

[3]P. G. Charalambides, J. Lund, A. G. Evans, and R. M. McMeeking, "A Test Specimen for Determining the Fracture Resistance of Bimaterial Interfaces," *J. Appl. Mech.*, **56**, 77–83 (1989).

[4]P. G. Charalambides, H. C. Cao, J. Lund, and A. G. Evans, "Development of a test method for measuring the mixed mode fracture resistance of bimaterial interfaces," *Mech. Mater.*, **8**, S269–83 (1990).

[5]M. Y. He, J. W. Hutchinson, and A. G. Evans, "A Stretch/Bend Method for In Situ Measurement of the Delamination Toughness of Coatings and Films Attached to Substrates," *J. Appl. Mech.*, **78** [1] S011009, 5pp 2010.

[6]K. Hemker, B. Mendis, and C. Eberl, "Characterizing the Microstructure and Mechanical Behavior of a Two-Phase NiCoCrAlY bond coat for thermal barrier systems," *Mater. Sci. Eng. A*, **483–484**, S727–30 (2008).

[7]C. Eberl, and K. Hemker, "Mechanical Characterization of Coatings Using Microbeam Bending and Digital Image Correlation Techniques," *Exp. Mech.*, **50**, S85–97 (2010).

[8]C. Eberl, D. S. Gianola, X. Wang, M. Y. He, A. G. Evans, and K. J. Hemker, "A Method for In Situ Measurement of the Elastic Behavior of a Columnar Thermal Barrier Coating," *Acta Mater.*, **59** [9] 3612–20 (2011).

[9]S. Kim, Y. Liu, and Y. Kagawa, "Evaluation of Interfacial Mechanical Properties under Shear Loading in EB-PVD TBCs by the Pushout Method," *Acta Mater.*, **55**, S3771–81 (2007).

[10]Y. Liu, Y. Kagawa, and A. Evans, "Analysis of a "Barb Test" for Measuring the Mixed-Mode Delamination Toughness of Coatings," *Acta Mater.*, **56**, S43–9 (2008).

[11]S. Faulhaber, C. Mercer, M. Moon, J. Hutchinson, and A. Evans, "Buckling Delamination in Compressed Multilayers on Curved Substrates with Accompanying Ridge Cracks," *J. Mech. Phys. Solids*, **54**, S1004–28 (2006).

[12]S. Guo, Y. Tanaka, and und Y. Kagawa, "Effect of Interface Roughness and Coating Thickness on Interfacial Shear Mechanical Properties of EB-PVD Yttria-Partially Stabilized Zirconia Thermal Barrier Coating Systems," *J. Eur. Ceram. Soc.*, **27**, S3425–31 (2007).

[13]J. L. Vasinonta and A. Beuth, "Measurement of Interfacial Toughness in Thermal Barrier Coating Systems by Indentation," *Eng. Fracture Mech.*, **68**, S843–60 (2001).

[14]P. Thery, M. Poulain, M. Dupeux, and M. Braccini, "Adhesion Energy of a YPSZ EB-PVD Layer in Two Thermal Barrier Coating Systems," *Surf. Coat. Technol.*, **202**, S648–52 (2007).

[15]R. H. Dauskardt, M. Lane, Q. Ma, and N. Krishna, "Adhesion and Debonding of Multi-Layer Thin Film Structures," *Eng. Fracture Mech.*, **61**, S141–62 (1998).

[16]H. Bahr, H. Balke, T. Fett, I. Hofinger, G. Kirchhoff, D. Munz, A. Neubrand, A. S. Semenov, H.-J. Weiss, and Y. Y. Yang, "Cracks in Functionally Graded Materials," *Mater. Sci. Eng. A*, **362**, S2–16 (2003).

[17]R. G. Hutchinson and J. W. Hutchinson, "Lifetime Assessment for Thermal Barrier Coatings: Tests for Measuring Mixed Mode Delamination Toughness," *J. Am. Ceram. Soc.*, 2011, doi:10.1111/j.1551-2916.2011.04499.x.   □

J. Am. Ceram. Soc., **94** [S1] S128–S135 (2011)
DOI: 10.1111/j.1551-2916.2011.04539.x
© 2011 The American Ceramic Society

journal

Thermomechanical Fatigue Damage Evolution in a Superalloy/Thermal Barrier System Containing a Circular Through Hole

Makoto Tanaka,[‡,§] Christopher Mercer,[¶,‖] Yutaka Kagawa,[†,‡,¶] and Anthony G. Evans[‖]

[‡]Research Center for Advanced Science and Technology, The University of Tokyo, 4-6-1 Komaba, Meguro-ku, Tokyo, Japan

[§]Materials Research and Development Laboratory, Japan Fine Ceramics Center, 2-4-1 Mutsuno, Atsuta-ku, Nagoya 456-8587, Japan

[¶]National Institute for Materials Science, 1-2-1 Sengen, Tsukuba, Ibaraki, 305-0047, Japan

[‖]Materials Department, University of California Santa Barbara, Santa Barbara, CA 93106

The results of an investigation of thermomechanical fatigue (TMF) on a superalloy specimen, with an applied thermal barrier coating (TBC) and a circular through hole, are presented. Tensile loads were applied in phase with increasing temperature. Damage evolution in the form of cracks develops in the TBC adjacent to the hole. These cracks run perpendicular to the loading axis. Stress mapping of the thermally grown oxide (TGO) using luminescence spectroscopy determined an increase in compressive residual stress with increasing TMF cycling. Scanning electron microscopy examination, following cross sectioning, determined the TBC cracks to be vertical separations of the columnar yttria-stabilized zirconia (YSZ) top coat. Microscopic damage mechanisms in the form of plasticity (bending of YSZ columns) and TGO cracking were observed. Imperfections in the bond coat are associated with these vertical separations. Energy-dispersive element mapping of these imperfections indicated a composition of alumina and mixed Cr, Co, and Ni oxides.

I. Introduction

AIRFOILS in aeroturbines experience a wide range of thermomechanical fatigue (TMF) loadings that can adversely influence the durability of the system.[1–3] The loadings can be in phase or out of phase in both tension and compression. The range of responses is diverse. The intent of the present article is to examine one of these TMF modes to ascertain whether it presents a concern for the performance of the airfoil. The scenario envisages an airfoil subject to unidirectional inertial loads imposed in phase with the temperature. Namely, it addresses regions of the airfoil where the temperature increases in phase with the inertial loads so that the maximum load is imposed simultaneously with the maximum temperature. To be representative of an actual airfoil, the system to be investigated incorporates a bond coat and a thermal barrier coating (TBC). Prior in-phase TMF experiments have been conducted on hollow cylindrical (axisymmetric) and plate type (uniaxial) TBC specimens.[1–4]

In the design of the test, the goal has been to replicate as closely as possible the thermomechanical circumstances expected in the engine. Specifically, the test scheme imposes a thermal gradient through the TBC, while cycling the temperature in phase with a tensile load. The temperatures are chosen to be large enough to induce creep strains in the superalloy and the bond coat, so that elongation occurs as the system cycles. Indeed, one objective of the study is to investigate the combined influence of creep and thermal strains on the response of a system with a brittle external coating.

During that portion of the thermal cycle when the interface between the TBC and bond coat is at its hottest, a thermally grown oxide (TGO) forms, comprised primarily of fine grained, polycrystalline α-Al$_2$O$_3$.[5,6] This material experiences creep at elevated temperatures. Absent an applied stress, the creep balances growth-induced strains, resulting in a small compressive (growth) stress as it thickens.[7,8] On cooling, substantial thermal expansion misfit between the α-Al$_2$O$_3$ and the substrate causes it to develop large in-plane compressive stress (of order 4 GPa) at ambient.[9,10] This stress can be measured by using a laser to excite fluorescence.[10–12] Moreover, because the TBC is transparent at certain laser wavelengths, the stress can be determined even when a TBC is present.[13–15] Probing this stress at various stages of TMF cycling provides an independent assessment of the state of the system.

In this investigation, TMF tests have been used to evaluate stress evolution in a TBC system. In service, TBCs are subjected to various loading modes, stress concentrations, temperature/load conditions etc. To simulate complicated degradation behavior in actual gas turbine blades, one idea is to use a holed specimen, because such a specimen allows well-controlled stress/strain distributions. When a through-holed specimen is used, the stress around the hole varies between $3\sigma_a$ and $-\sigma_a$, where σ_a is the remote applied stress. Although they are local stresses, rather than uniform stresses, the test is useful for the observation of the effects of superimposed applied cyclic stress and temperature on the degradation behavior of a TBC system. The test is also interesting because actual gas turbine blades have many through holes for air cooling, and there is concern that coating failure may initiate at holes. However, no work has been published on the effect of stress concentration on the degradation of TBC systems.

The article is organized as follows. Initially, and the material system and specimen preparation procedures are presented, and the thermomechanical test system and measurement methods are described. Then the experimental measurements and observations are presented and discussed. Finally, the experimental results are summarized and some concluding remarks are given.

II. Experimental Procedure

(1) TBC Specimen

Before application of the bond coat and TBC layers, the substrate superalloy Inconel 738LC (Mitsubishi Materials Corp., Tokyo, Japan) (16.0-Cr, 8.5-Co, 3.4-Al, 3.4-Ti, 1.7-Mo, 2.6-W, 1.75-Ta, 0.9-Nb, 0.11-C, balance-Ni, and comprised of a ~3 mm thick

C. Levi—contributing editor

Manuscript No. 28700. Received September 30, 2010; approved March 4, 2011.
[†]Author to whom correspondence should be addressed. e-mail: kagawa@rcast.u-tokyo.ac.jp

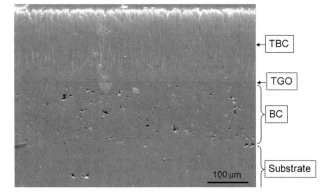

Fig. 1. A typical polished transverse section of the EB–PVD TBC system. EB–PVD, electron-beam physical vapor deposition; TBC, thermal barrier coating.

plate), was machined into flat tensile test specimens. Both of the flat surfaces were then polished to allow good bonding between the bond coat and the substrate. The entire coated material system will hence be referred to as the "TBC system". The bond coat layer was 150 μm thick and was applied by low-pressure plasma spray. The chemical composition of the CoNiCrAlY bond coat (wt%) was: 32.0-Ni, 21.0-Cr, 8.0-Al, 0.5-Y, with the remainder Co. Following application of the bond coat, a 4 mol% Y_2O_3–ZrO_2 TBC layer, with a thickness of ∼200 μm, was applied via electron-beam physical vapor deposition (EB–PVD). The yttria-stabilized zirconia (YSZ) material was supplied by Japan Fine Ceramics Center (JFCC, Nagoya, Japan). Figure 1 shows a typical example of a polished transverse section of the as-deposited EB–PVD TBC system: the three different layers, i.e., the TBC, TGO, and bond coat, are clearly distinguished in the cross section. The TBC layer consists of columnar grains with a rough feather-like surface appearance: this surface appearance is typical of an EB–PVD coating, as reported elsewhere.[16]

A center hole, 1 mm in diameter, was introduced by an ultrasonic machining process. To avoid damage to the TBC layer, the entire specimen was embedded in an epoxy resin before machining. Once the hole had been machined, the embedded specimen was treated with CH_2Cl_2 solution (at 20°C) for 24 h to

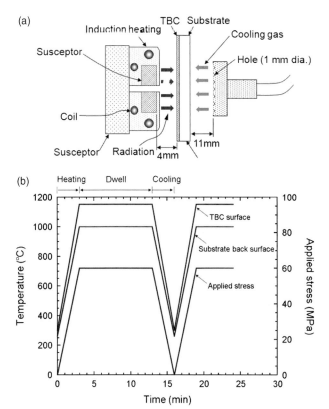

Fig. 3. (a) Heating/cooling system used in the thermomechanical fatigue (TMF) test, and (b) temperature-time relations during the in-phase TMF test.

remove the epoxy resin. Subsequently, the specimen was heat treated for 30 min at 400°C to completely remove all traces of the epoxy. No additional coating was applied inside the hole, i.e., the as-machined surface was employed. The final shape and dimensions of the TMF specimen, as well as the surface appearance of the specimen (TBC side), are shown in Fig. 2. In this investigation, the applied stress at maximum temperature was 60 MPa, which is smaller than the yield stress of the substrate alloy (345 MPa at 980°C [17]). The estimated in-plane stress distribution in the specimen under the applied remote stress, σ_a, is presented in Appendix A (Fig. A1). As previously mentioned, the maximum stress concentration in the holed specimen varies between $3\sigma_a$ and $-\sigma_a$. That is, a stress distribution is present, making the observation of damage evolution under various loading conditions possible.

(2) TMF Test, Stress Measurement, and Observation

Cyclic, in-phase thermomechanical fatigue (TMF) tests were performed in ambient air using a specially designed TMF test system, the details of which are reported elsewhere.[18,19] The surface temperature of the TBC layer and substrate were measured using two R-type thermocouples. The heating/cooling arrangement is shown in Fig. 3(a), and the temperature-time relationship during the test is presented in Fig. 3(b). The temperature of the substrate was controlled by compressed cooling air blown from a T-shaped, flat-type nozzle, containing pinholes. Using this procedure, the surface of the substrate was kept at a constant 1000°C, while the maximum temperature at the TBC surface was held at ≈1150°C. Temperature variation using this procedure was ±5°C. Preliminary temperature measurement confirmed that a uniform surface temperature region exists in the ≈10 mm × 10 mm central region of the specimen. This indicates that a temperature gradient of $\Delta T \approx 150°C$ between the TBC surface and the substrate, was achieved using this heating/cooling system. The total time period of a TMF cycle was 940 s, with heating and cooling rates of 5°C/s and a maximum temperature dwell time of 600 s.

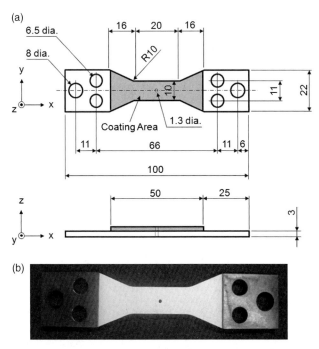

Fig. 2. (a) Shape and dimensions of the thermomechanical fatigue test specimen, and (b) surface appearance of the specimen (thermal barrier coating surface).

During the maximum temperature hold, a uni-axial tensile load of 1.8 kN was applied to the specimen, which corresponded to a maximum applied stress of $\sigma_a = 60$ MPa. Applied stress was calculated using: $\sigma_a = P_a/((h_{tbc}+h_{bc}+h_s)w)$, where P_a is the applied load, h is the thickness, w is the width of the gage section of the specimen, and the subscripts "tbc", "bc" and "s" refer to the TBC, bond coat and substrate, respectively. The specimen was attached to the test machine via a water-cooled mechanical grip. Axial strain in the specimen was measured by a contact type displacement gage (nominal strain resolution: 0.008%). After selected numbers of applied thermomechanical cycles ($N = 120, 240, 360,$ and 480 cycles), the test was interrupted and the specimen removed from the test machine. The TBC surface of the tested specimen was then examined using scanning electron microscopy (SEM). After 480 cycles, the test was terminated. Stress in the thermally grown oxide (TGO) layer was measured using Cr^{3+} luminescence spectroscopy. The measurement was performed through the TBC layer, at a fully controlled temperature of 20°C, in ambient air using a microconfocal laser luminescence spectroscopy system (NRS-1000: Special Version, JASCO Corp., Tokyo, Japan). The same experimental equipment has already been used successfully for prior studies of TGO stress in the same TBC system.[20] The frequency shift of the R_2 peak is known to have a linear relationship with stress. The measured R_2 peak shift, $\Delta\nu$, and average stress in the TGO layer, $\bar{\sigma}_{tgo}$, has the following relationship[11]:

$$\bar{\sigma}_{tgo} = \frac{1}{\Pi}\Delta\nu = \frac{1}{3}(\sigma_{tgo}^x + \sigma_{tgo}^y + \sigma_{tgo}^z) \quad (1)$$

where $\bar{\sigma}_{tgo}$ is the average stress in the TGO layer, Π is the trace of the piezospectroscopic coefficient tensor, and has a value of 7.62 cm/GPa,[21] $\Delta\nu$ is the frequency shift (cm^{-1}) of the Ruby line at 14430 cm^{-1} (R_2) Assuming that the entire TBC specimen is under plane stress conditions, i.e., $\sigma_{tgo}^z = 0$, the in-plane TGO stress is given by

$$\bar{\sigma}_{tgo} = \frac{1}{3}(\sigma_{tgo}^x + \sigma_{tgo}^y) = \frac{2}{3}\sigma_{tgo}^x \left(\equiv \frac{2}{3}\bar{\sigma}_{tgo}^{x,y}\right) \quad (2)$$

This relation is valid when luminescence spectroscopy measurements are performed through the TBC layer, and the laser incident direction is perpendicular to the TGO. Though the laser beam diameter is ≈ 15 μm, the spatial resolution is ~ 150 μm because of scattering through the TBC layer.[22]

Following luminescence spectroscopy, the specimen was sectioned both parallel to and perpendicular to the loading direction. The cross-sections were initially ground using SiC paper before polishing with diamond suspension. The specimens were then coated with gold and examined using SEM. Compositional information of the TGO and bond coat was obtained using backscattered electron imaging and energy-dispersive spectroscopy (EDS) element mapping.

III. Experimental Results and Discussion

(1) Deformation Behavior of the TBC System

Maximum and minimum tensile strains in the specimen (along the stress axis) evolving with the number of applied TMF cycles, are plotted in Fig. 4(a). The strain is defined as elongation measured by the gage, Δl_g, divided by the initial gage length, l_g, i.e, $\varepsilon_a = \Delta l_g/l_g$. The effect of local strains caused by the existence of the hole is neglected. The maximum strain (ε_{max}) is defined as the value measured at the end point of the maximum temperature hold ($T_{tbc} \approx 1150$°C) during each cycle, and the minimum strain (ε_{min}) is defined as the value at the minimum temperature ($T_{tbc} \approx 300$°C). Steady state creep behavior occurs up to about $N \approx 400$ with a creep rate $d\varepsilon/dN \approx 1.78 \times 10^{-3}N^{-1}$. The ratio of load applied to the substrate versus that applied to the entire specimen (P_s/P_a) is given by

$$P_s/P_a = E_s h_s/(E_{tbc}h_{tbc} + E_s h_s) \quad (3)$$

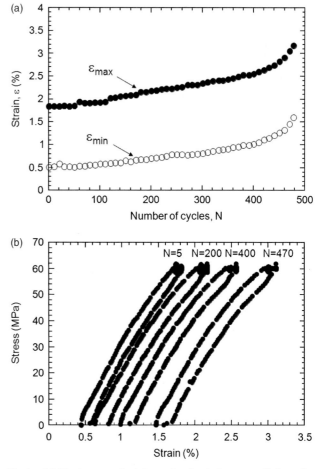

Fig. 4. (a) Maximum and minimum tensile strains vs. applied number of cycles, and (b) ratcheting behavior during the thermomechanical fatigue test.

where E is Young's modulus and h is the thickness, subscripts "s" and "tbc" refer to the substrate and TBC layer, respectively. Ratcheting behavior develops during the test (Fig. 4(b)) and the behavior becomes more apparent with increasing number of TMF cycles. The material properties derived from this estimation are presented in Table I The estimated P_s/P_a is ≈ 0.99. Therefore, the observed steady state creep behavior of the TBC system seems reasonable, because the substrate is much thicker than the TBC layer, and almost all of the applied load is supported by the superalloy substrate. An accelerated creep stage follows, with a creep rate $d\varepsilon/dN \approx 7.83 \times 10^{-3}/N$, in the range of $400 \leq N \leq 480$. The creep rate in the steady state stage coincides with that of similar specimens without holes.[19] However, the specimen in this TMF test did not fracture even though the total number of TMF cycles exceeded the number of cycles to failure observed in specimens without holes. The ultimate cycle in this TMF test seems to correspond to the onset of the final creep stage, when compared to creep curves of specimens without holes.

Table I. Mechanical and Physical Properties of Materials Used in the Investigation

Materials	Layer Thickness (μm)	Young's Modulus (GPa)	Coefficient of Thermal Expansion (10^{-6} K^{-1})
TBC	200	20	11
TGO	3	280[25]	8
Substrate	3000	200	15

TBC, thermal barrier coating; TGO, thermally grown oxide.

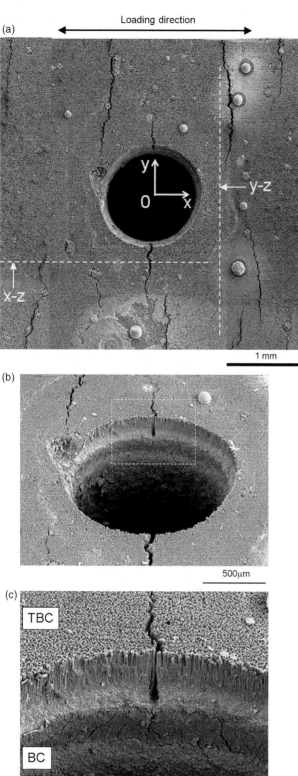

(a) Loading direction

Fig. 5. (a) Appearance of the specimen surface near a hole after N = 480 cycles, and cross section markers (dashed lines) for cross sections shown in Fig. 6. (b) Micrograph showing inside of hole after N = 480 cycles. (c) Higher-magnification image indicated by dashed box in Fig. 5(b).

(2) Damage Evolution Observed at the TBC Surface

The appearance of the specimen surface near the hole after 480 cycles is shown in Fig. 5. The applied loading direction and assigned coordinates are also shown. Transverse cracks, which are shorter than 1 mm in length, and initiate from a maximum

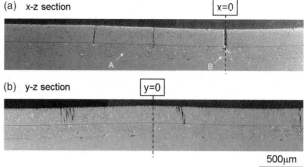

Fig. 6. Scanning electron microscopy micrographs of polished cross sections: (a) parallel to loading direction (*x–z* section in Fig. 5(a)), (b) perpendicular to loading direction (*y–z* section in Fig. 5(a)).

tensile stress site ($3\sigma_a$), occur symmetrically with respect to the center of the loading axis. Formations of other transverse cracks, which are not directly connected to the free surface of the hole, are also observed inside the hole (Figs. 5(b) and (c)). Conversely, no visible damage formation in compressive stress sites ($-\sigma_a$) is observed. A notable characteristic of these cracks is that the crack mouths are open even when the applied load is completely removed. Note also that the spheroidal material (diameter 100∼200 μm) observed on the TBC surface in Fig. 5 are actually the tops of conical defects formed in the YSZ during the EB–PVD process and are not related to the cracking behavior of the TBC layer. Cracks in the TBC layer do not extend the entire width of the specimen. The spacing of the transverse cracks is about 1∼1.3 mm, except in the case of cracks formed near the edge of the hole. This suggests that tensile stress concentration during maximum temperature hold, $3\sigma_a$, is responsible for the formation of the transverse cracks, and compressive stress is effective at arresting the cracks.

(3) Observation of Microscopic Damage Evolution

Polished cross sections, in both *x–z* and *y–z* orientations, are shown in Figs. 6(a) and (b), respectively. Parallel to the loading

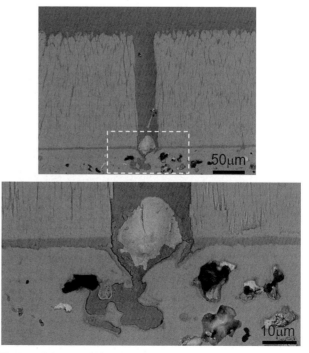

Fig. 7. Higher-magnification images showing the transverse crack nucleated near the edge of the hole.

(a)　　　　　　　　　Area A

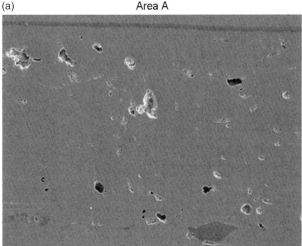

(b)　　　　　　　　　Area B

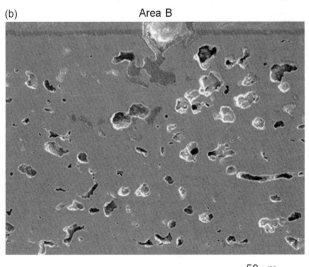

50 μm

Fig. 8. Creep voids in the bond coat layer at areas A and B in Fig. 6.

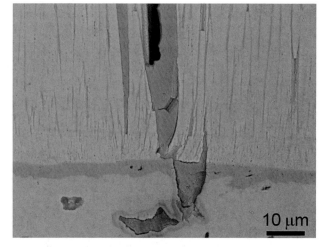

Fig. 9. Scanning electron microscopy micrograph of vertical separation in the thermal barrier coating layer and imperfection in the thermally grown oxide layer.

direction (Fig. 6(a), *x*–*z* section), cracks or vertical separations in the TBC layer are observed at points that coincide with a stress of $3\sigma_a$. These TBC cracks generally stop at the TGO interface. A detailed image of the widest separation is shown in Fig. 7. The crack tip is blunted in the bond coat layer and an irregularly shaped imperfection extends into the bond coat. Within the bond coat below the crack, the number of voids has increased, compared with that observed in the as-deposited state, or in an undamaged region (Fig. 8). This is expected, because the voids are formed via bond coat creep in the regions adjacent to the TBC crack, where localized plastic deformation will occur. Within the voids, new oxide formation is clearly observed in the polished section (Fig. 7(b)), indicating that the voids are connected to the central breakthrough feature. Perpendicular to the loading direction (Fig. 6(b), *y*–*z* section), the specimen shows a slight convex appearance with the apex at the maximum compressive stress, $-\sigma_a$. The vertical separations in the TBC are also apparent in this section, but have a different appearance. This is due to the fact that the TBC cracks are not straight, but somewhat wavy.

Another vertical separation in the TBC layer is shown in Fig. 9. An imperfection in the TGO layer is clearly visible at the root of the crack. Microcracking, both parallel and perpendicular to the TGO interface is associated with the imperfection. Bending of the TBC columns to the right of the separation is also visible. This implies that the YSZ columns are undergoing plastic deformation during the maximum temperature phase of the TMF cycle. No such bending of the TBC columns is asso-

ciated with the TBC crack in Fig. 7. The reason for this not entirely clear, but it should be noted that the crack shown in Fig. 7 lies at the bottom of the concave undulation of the TBC, where bending stresses may be at a minimum. In any case, it would appear that two potentially competing microdamage mechanisms are operative within the TBC system; YSZ plasticity and TGO cracking. Further work needs to be carried out to assess the relative effects of these mechanisms on the overall integrity and service life of the TBC system.

Figure 10 shows an EDS element map taken in the region of the TGO imperfection shown in Fig. 9. The map indicates that walls of the imperfection and nearby void are covered by a thin Al_2O_3 layer. The presence of Cr, Co, and Ni may indicate that mixed Cr, Co, and Ni oxides are also associated with the entrained void, although this is not completely clear.

Another interesting observation is the difference in the TGO morphology between the two sections. Parallel to the loading direction, the TGO layer is flat (Fig. 11(a)). However, perpendicular to loading direction, the TGO shows clear undulation behavior (Fig. 11(b)). The average thickness of the TGO layer is ~ 4 μm. The undulation amplitude is approximately $2\delta \approx 22.7$ μm in the *y*–*z* section, while that in the *x*–*z* section is $2\delta \approx 0$. The large undulation amplitude in the *y*–*z* section would be associated with the compressive stress, σ_y and is attributed to cyclic plasticity caused by bond coat creep in compression. Meanwhile, the flat TGO in the *x*–*z* section is associated with elongation of the bond coat. Similar anisotropic TGO undulation behavior has also been observed in previous studies,[23,24] including under compressive thermomechanical testing on a (Ni,Pt)-Al coated superalloy.[23] The present result indicates that the anisotropic TGO undulation behavior occurs despite the presence of the YSZ layer, which suggests that such behavior probably occurs in actual TBC systems used in gas turbine engines.

(4) Change of TGO Stress Distribution with Number of TMF Cycles

Visible damage evolution and associated in-plane TGO stress are shown in Fig. 12. The surface appearance shows a sequence of transverse cracking events. After 360 cycles, transverse cracks appear at the edge of the hole. Further initiation and growth of transverse cracks is also observed with increasing number of applied cycles. Generally, the TGO compressive stress increases with number of cycles, *N*. However, detailed observation suggests the TGO stress change depends on location, i.e., the stress change depends on the remote applied stress. Far from the hole, the TGO compressive stress continuously increases with increasing number of cycles. The TGO stress at sites of maximum compressive stress, $-\sigma_a$, is slightly lower than that far from the

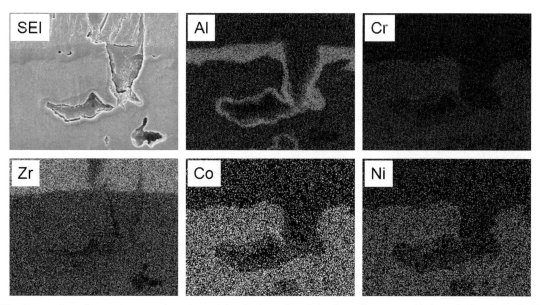

Fig. 10. Energy-dispersive spectroscopy element map taken in the region of the thermally grown oxide imperfection shown in Fig. 9.

hole (σ_a). The stress at $3\sigma_a$ shows nearly the same variation as that at σ_a. However, after 360 cycles, the TGO stress in the maximum tensile stress region clearly decreases, and this tendency continues with increasing number of TMF cycles.

Plots of average in-plane compressive stress in the TGO layer far from the hole vs. applied number of TMF cycles, N, are presented in Fig. 13, with the total time at maximum temperature also shown for comparison purposes. The result shows gradually increasing compressive stress with increasing number of cycles. Typically, TGO stress is estimated assuming that all the stress originates from thermal expansion mismatch. The estimated thermal stress is given by:

$$\sigma_{tgo}^T = \frac{E_{tgo}^* \Delta T [E_{tbc}^* h_{tbc}(\alpha_{tbc}^* - \alpha_{tgo}^*) + E_s^* h_s(\alpha_s^* - \alpha_{tgo}^*)]}{E_{tbc}^* h_{tbc} + E_{tgo}^* h_{tgo} + E_s^* h_s} \quad (4)$$

where for plane strain conditions, $E^* = E/(1 - v^2)$, $v^* = v/(1 - v)$, and $\alpha^* = (1 + v)\alpha$, E is Young's modulus, v is Poisson's ratio, α is the thermal expansion coefficient, h is thickness, and the subscripts represent materials. At the beginning of the TMF test, the measured compressive stress is 3 GPa and the estimated value using Eq. (4) is 2.8 GPa, so they are in good

agreement. During TMF cycling, sintering of the TBC layer and growth of TGO layer occur, and these factors affect the TGO stress. Considering the contribution of individual materials properties, the increase of Young's modulus of the TBC layer and increase in the thickness of the TGO layer, lead to the stress decrease. In contrast, an increase of the Young's modulus of the TGO causes an increased in-plane compressive stress. The increase of Young's modulus of the TGO with increasing thermal exposure is reported in,[25] and is attributed to the closure of microcracks. Assuming a 10% increase of Young's modulus, the stress increase is also $\approx 10\%$. The estimation only incorporates linear elastic considerations; creep deformation behavior, temperature gradient, ratcheting effect, growth strain of the TGO, etc. are not considered. The increase of the Young's modulus of the TGO is only one possibility; the reason for the increase of compressive stress with increasing number of TMF cycles is still an open problem.

IV. Summary and Concluding Remarks

In-phase TMF tests on a center-holed plate type specimen with an applied TBC system have been conducted. Damage evolution in the form of cracks is observed in the TBC adjacent to the hole. These cracks run perpendicular to the direction of the applied load. The cracking initiates at the edge of the hole exhibiting the highest tensile stress. This suggests that the cracking behavior is strongly correlated with tensile stress at maximum temperature. Changes in the TGO stress state with applied number of TMF cycles have been measured by luminescence spectroscopy. The results indicate that the compressive residual stress in the TGO increases with increasing number of cycles. The Increase in Young's modulus of the TGO layer, possibly as a result of the closure of microcracks, is believed to be the explanation for the increasing TGO stress. Anisotropic TGO growth behavior is also observed. Parallel to the loading direction, the TGO is flat; however, perpendicular to loading direction, the TGO exhibits marked undulations. Employing holed specimens in TMF tests is an effective way to observe the effect of applied stress on damage evolution in a TBC system, because the specimen has a stress distribution of $-\sigma_a$–$3\sigma_a$, for an applied tensile stress, σ_a.

Detailed SEM analysis of the cross-sectioned specimen has determined that the cracks in the TBC are vertical separations in the YSZ columnar structure, that extended to the TGO interface. Imperfections in the TGO and increased incidence of bond coat voids occur at the roots of these separations. These voids

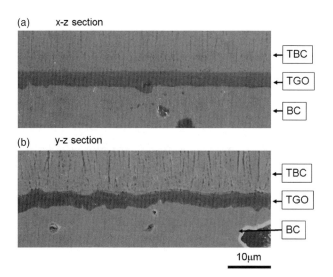

Fig. 11. Differences in the thermally grown oxide (TGO) morphology in the two orthogonal directions, (a) *x–z* and (b) *y–z*.

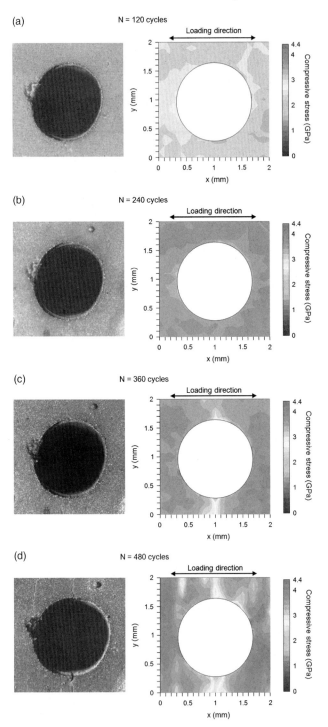

Fig. 12. In-plane biaxial thermally grown oxide stress maps after selected numbers of cycles, (a) $N = 120$, (b) $N = 240$, (c) $N = 360$, and (d) $N = 480$ cycles.

oxidize with further TMF cycling. Microdamage mechanisms occurring at the tips of the TBC separations include TBC plasticity (bending of YSZ columns), and TGO cracking, both parallel and perpendicular to the interface.

Appendix A

When a plate containing a circular through hole is subjected to a uniform, remote tensile stress, a stress distribution develops around the hole. Coordinate definitions and stress components of the plate are shown in Fig. A1. Under linear elastic deformation, the stresses at the point at r distance from the center of

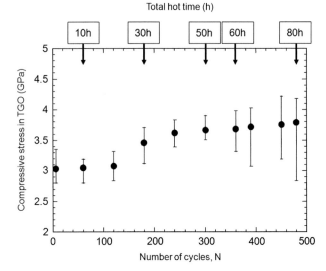

Fig. 13. Variation of the thermally grown oxide (TGO) stress far from the hole with number of thermomechanical fatigue cycles.

the hole using r–θ coordinates, are given by:

$$\sigma_r = \frac{\sigma_a}{2}\left(1 - \frac{a^2}{r^2}\right) + \frac{\sigma_a}{2}\left(1 + \frac{3a^4}{r^4} - \frac{4a^2}{r^2}\right)\cos 2\theta \quad \text{(A-1)}$$

$$\sigma_\theta = \frac{\sigma_a}{2}\left(1 + \frac{a^2}{r^2}\right) - \frac{\sigma_a}{2}\left(1 + \frac{3a^4}{r^4}\right)\cos 2\theta \quad \text{(A-2)}$$

$$\tau_{r\theta} = -\frac{\sigma_a}{2}\left((1 - \frac{3a^4}{r^4} + \frac{2a^2}{r^2}\right)\sin 2\theta \quad \text{(A-3)}$$

where a is the radius of hole, and σ_a is the remote applied tensile stress. These equations can then be converted to the stress components in the x and y directions as follows:

$$\sigma_x = \sigma_r \cos^2\theta + \sigma_\theta \sin^2\theta - 2\tau_{r\theta}\sin\theta\cos\theta \quad \text{(A-4)}$$

$$\sigma_y = \sigma_r \sin^2\theta + \sigma_\theta \cos^2\theta + 2\tau_{r\theta}\sin\theta\cos\theta \quad \text{(A-5)}$$

$$\tau_{xy} = (\sigma_r - \sigma_\theta)\sin\theta\cos\theta + \tau_{r\theta}(\cos^2\theta - \sin^2\theta) \quad \text{(A-6)}$$

The stress distributions obtained from equations (A-4)–(A-6) are presented in Fig. A2, σ_x^{x-z} and σ_y^{x-z}, are the stresses along the x–z line, and σ_x^{y-z} and σ_y^{y-z} are the stresses along the y–z line. When the applied tensile stress is σ_a, the maximum tensile stress is $3\sigma_a$ and the maximum compressive stress is $-\sigma_a$.

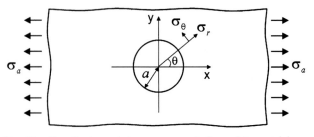

Fig. A1. Coordinates and stress components for a plate containing a circular through hole.

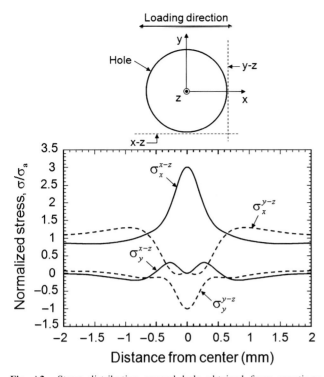

Fig. A2. Stress distribution around hole obtained from equations (A-4)–(A-6).

References

[1] P. K. Wright, "Influence of Cyclic Strain on Life of a PVD TBC," *Mater. Sci. Eng.* **A245**, 191–200 (1998).

[2] E. Tzimas, H. Müllejans, S. D. Peteves, J. Bressers, and W. Stamm, "Failure of Thermal Barrier Coating Systems under Cyclic Thermomechanical Loading," *Acta Mater.*, **48**, 4699–707 (2000).

[3] B. Baufeld, E. Tzimas, P. Hähner, H. Müllejans, S. D. Peteves, and P. Moretto, "Phase-Angle Effects on Damage Mechanisms of Thermal Barrier Coatings under Thermomochanical Fatigue," *Scr. Mater.*, **45**, 859–65 (2001).

[4] M. Bartsch, B. Baufeld, S. Dalkilic, L. Chernova, and M. Heinzelmann, "Fatigue Cracks in a Thermal Barrier Coating System on a Superalloy in Multiaxial Thermomechanical Testing," *Int. J. Fatigue*, **30**, 211–8 (2008).

[5] A. G. Evans, D. R. Mumm, J. W. Hutchinson, G. H. Meier, and F. S. Pettit, "Mechanisms Controlling the Durability of Thermal Barrier Coatings," *Prog. Mater. Sci.*, **46**, 505–53 (2001).

[6] A. G. Evans, D. R. Clarke, and C. G. Levi, "The Influence of Oxides on the Performance of Advanced Gas Turbines," *J. Eur. Ceram. Soc.*, **28**, 1405–19 (2008).

[7] V. K. Tolpygo, J. R. Dryden, and D. R. Clarke, "Determination of the Growth Stress and Strain in α-Al₂O₃ Scales During the Oxidation of Fe–22Cr–4.8Al–0.3Y Alloy," *Acta Mater.*, **46**, 927–37 (1998).

[8] V. K. Tolpygo and D. R. Clarke, "Wrinkling of a-Alumina Films Grown by Thermal Oxidation—I. Quantitative Studies on Single Crystals of Fe–Cr–Al Alloy," *Acta Mater.*, **46**, 5153–66 (1998).

[9] D. R. Clarke and F. Adar, "Measurement of the Crystallographically Transformed Zone Produced by Fracture in Ceramics Containing Tetragonal Zirconia," *J. Am. Ceram. Soc.*, **65**, 284–8 (1982).

[10] D. M. Lipkin and D. R. Clarke, "Measurement of the Stress in Oxide Scales Formed by Oxidation of Alumina-Forming Alloys," *Oxid. Met.*, **45**, 267–80 (1996).

[11] Q. Ma and D. R. Clarke, "Stress Measurement in Single-Crystal and Polycrystalline Ceramics using their Optical Fluorescence," *J. Am. Ceram. Soc.*, **76**, 1433–40 (1993).

[12] V. Sergo, D. R. Clarke, and W. Pompe, "Deformation Bands in Ceria-Stabilized Tetragonal Zirconia/Alumina: I. Measurement of the Internal Stresses," *J. Am. Ceram. Soc.*, **78**, 633–40 (1995).

[13] R. J. Christensen, D. M. Lipkin, and D. R. Clarke, "Nondestructive Evaluation of the Oxidation Stresses through Thermal Barrier Coatings using Cr³⁺ Piezospectroscopy," *Appl. Phys. Lett.*, **69**, 3754–6 (1996).

[14] D. R. Clarke, R. J. Christensen, and V. Tolpygo, "The Evolution of Oxidation Stresses in Zirconia Thermal Barrier Coated Superalloy Leading to Spalling Failure," *Surf. Coat. Technol.*, **94–95**, 89–93 (1997).

[15] K. W. Schlichting, K. Vaidyanathan, Y. H. Sohn, E. H. Jordan, M. Gell, and N. P. Padture, "Application of Cr³⁺ Photoluminescence Piezo-Spectroscopy to Plasma-Sprayed Thermal Barrier Coatings for Residual Stress Measurement," *Mater. Sci. Eng.*, **A291**, 68–77 (2000).

[16] M. Tanaka, R. Kitazawa, T. Tomimatsu, Y. F. Liu, and Y. Kagawa, "Residual Stress Measurement of an EB–PVD Y₂O₃–ZrO₂ Thermal Barrier Coating by Micro-Raman Spectroscopy," *Surf. Coat. Technol.*, **204**, 657–60 (2009).

[17] A. Thakur, "Microstructural Responses of a Nickel-Base Cast IN-738 Superalloy to a Variety of Pre-weld Heat-Treatments, Master Thesis, The University of Manitoba, 1997.

[18] M. Tanaka and Y. Kagawa, "Development of ThermoMechanical Fatigue Test Machine for Plate Type Thermal Barrier Coating Specimen," *J. Gas Turbine Soc. Jpn.*, **37**, 93–6 (2009).

[19] R. Kitazawa, M. Tanaka, Y. Kagawa, and Y. F. Liu, "Damage Evolution of TBC System under in-Phase Thermo-Mechanical Tests," *Mater. Sci. Eng*, **B173**, 130–4 (2010).

[20] M. Tanaka, Y. F. Liu, and Y. Kagawa, "Identification of Delamination through TGO Stresses Due to Indentation Testing of an EB–PVD TBC," *J. Mater. Res*, **24**, 3533–42 (2009).

[21] J. He and D. R. Clarke, "Determination of the Piezospectroscopic Coefficients for Chromium-Doped Sapphire," *J. Am. Ceram. Soc*, **78**, 1347–53 (1995).

[22] M. Tanaka, Ph.D. Thesis, The University of Tokyo (2009).

[23] S. Dryepondt and D. R. Clarke, "Effect of Superimposed Uniaxial Stress on Rumpling of Platinum-Modified Nickel Aluminide Coatings," *Acta Mater*, **57**, 2321–7 (2009).

[24] D. S. Balint, S-S. Kim, Y-F. Liu, R. Kitazawa, Y. Kagawa, and A. G. Evans, "Anisotropic TGO Rumpling in EB-PVD Thermal Barrier Coatings under in-Phase Thermo-Mechanical Loading," *Acta mater*, **59**, 2544–2555 (2011).

[25] S. Guo and Y. Kagawa, "Young's Moduli of Zirconia Top-Coat and Thermally Grown Oxide in a Plasma-Sprayed Thermal Barrier Coating System," *Scripta Mater*, **50**, 1401–6 (2004). □

J. Am. Ceram. Soc., **94** [S1] S136–S145 (2011)

DOI: 10.1111/j.1551-2916.2011.04578.x

© 2011 The American Ceramic Society

Journal

Oxide-Assisted Degradation of Ni-Base Single Crystals During Cyclic Loading: the Role of Coatings

Tresa M. Pollock,[†,‡,§] Britta Laux,[‡] Clinique L. Brundidge,[§] Akane Suzuki,[¶] and Ming Y. He[‡]

[‡]Materials Department, University of California, Santa Barbara, CA 93106

[§]Department of Materials Science and Engineering, University of Michigan, Ann Arbor, MI 48105

[¶]GE Global Research, Niskayuna, NY 12309

The role of oxidation-induced layers in the failure process of aluminide-coated nickel base single crystals subject to high-temperature fatigue cycling has been investigated experimentally and via finite element analysis. Isothermal strain-controlled compressive fatigue experiments ($R = -\infty$) with 120 s holds in compression were conducted at 982° and 1093°C. Surface-initiated cracks containing a layer of alumina progressively grew through the coating layers into the superalloy substrate, ultimately causing failure. Growth stresses in the oxide provided a driving force for extension of the oxide into the softer coating and substrate layers. Finite element modeling shows the rate of growth of the oxide-filled cracks is sensitive to the strength of the constituent layers and the magnitude of the oxide growth strains. Implications for design of failure-resistant coating–substrate systems are discussed.

I. Introduction

Turbine blades are among the most critical components of advanced aircraft engines and power generation turbines and the properties of their constituent materials pose significant constraints to engine performance and efficiency. In the extreme combustion environments of turbine engines the nickel-base single crystal turbine airfoils require coatings for enhancement of oxidation, corrosion, and for extension of maximum temperature capabilities.[1–3] Coatings, including aluminide and/or yttria-stabilized zirconia (YSZ)-based thermal barrier layers often constitute a significant fraction of the airfoil wall thickness[1,4] and may therefore strongly influence the overall thermomechanical response of the turbine blade. However, to date, coatings have not been designed to enhance the mechanical performance of the system, since the coating-substrate degradation mechanisms under relevant mechanical cycling conditions are not well understood and coating properties are often unknown.

Damage accumulation in turbine airfoils during service is driven by the superposition of centrifugal and vibratory stresses with thermal stresses that arise due to internal air-cooling.[5] In the absence of a thermal barrier, the impingement of hot gases on the surface of the airfoil and the constraint caused by the cooler inner surfaces typically results in surface compressive stresses.[5–7] Under these conditions, the failure of nickel-base single crystals subjected to thermomechanical fatigue cycling typically occurs by surface crack initiation and subsequent propagation inward through the coating into the substrate.[6] This damage growth process is influenced by the simultaneous oxi-

dation of the intermetallic coating and/or the superalloy substrate.[6–17] When a thermal barrier coating (TBC) is present, these processes often occur following local spallation of the YSZ, resulting in "conservative" maximum design temperatures being imposed on the TBC. Material degradation is further accelerated by sustained compressive holds present in the fatigue cycle.[6,14] Herein fatigue cycling with holds will be referred to as sustained peak low cycle fatigue (SPLCF). During the compressive hold, creep deformation occurs, resulting in tension upon returning to zero strain at the end of the cycle, Fig. 1. Cracks, if present, then open at the end of the cycle, permitting oxidation along the crack faces and at the crack tip. As degradation due to SPLCF cycling often limits the performance of aircraft and power generation turbines,[6] a better understanding of the mechanics and materials issues resulting in SPLCF failure is needed to motivate development of more optimal substrate–coating combinations.

Evans and colleagues[7,18] recently proposed a model for oxide-assisted crack growth during SPLCF cycling. The model is unique in that it embodies the same mechanics and phenomena previously used to successfully predict rumpling of the thermally grown oxide (TGO) on bond coats[19–24] in the absence of fatigue cycling. In the model, alumina forms on the superalloy sample surface as well as on the faces of incipient cracks. The early stages of this process on the surface of a Ni-base single crystal sample are shown in Fig. 2, where an incipient crack filled with oxide is penetrating into the superalloy substrate.

During continued oxidation in these high-temperature systems, while most of the new α-Al_2O_3 forms at the surface by inward diffusion of oxygen, an outward counter-flux of Al causes some new oxide to form along the transverse grain boundaries.[19,20] The alumina formed on the grain boundaries is accommodated by lateral straining of the neighboring grains, at a strain-rate $\dot{\varepsilon}_{growth}$.[22–24] The oxide responds by creeping and a steady state a compressive growth stress, σ_{growth} is quickly established, wherein creep relaxation balances the stresses caused by the growing oxide.[25,26] The magnitude of this stress has been measured *in situ* for the TGO on several different bond coats and is of order $\sigma_{growth} \approx -300$ MPa,[25,26] consistent with deformation mechanisms for α-Al_2O_3 and with stress relaxation rates measured in a typical TGO.[27] When the surface is nonplanar, Fig. 2, the stress σ_{growth} exerts a downward pressure on the softer substrate, causing it to creep,[7,16,22,27] resulting in the development of crack-like features. Continued growth of the oxide as the crack opens in tension upon returning to zero strain, Fig. 1, and deformation of the oxide and substrate during the compressive fatigue loading portion of the cycle result in a continued penetration of the oxide-filled crack deep into the substrate to the point where the crack becomes physically long and begins to grow in a manner consistent with that observed in conventional cyclic crack growth experiments.[7,12] In this paper, the oxide-assisted damage growth model is further developed for analysis of the key features of the superalloy–coating system

C. Levi—contributing editor

Manuscript No. 29073. Received December 21, 2010; approved March 19, 2011.
This work was supported by the ONR Grant N00014-08-1-0331.
[†]Author to whom correspondence should be addressed. e-mail: pollock@engineering.ucsb.edu

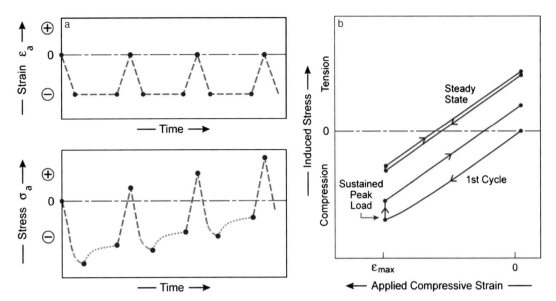

Fig. 1. Schematics of the applied straining conditions associated with sustained peak load cyclic fatigue. (a) The compressive strains imposed and trends in the stresses induced as cycling proceeds. (b) The stress/strain loops and their evolution with cycling.

that influence the rate of damage development. The SPLCF degradation process is considered for the intermetallic coating, the interdiffusion zone and the superalloy substrate (without the presence of a thermal barrier). Corresponding SPLCF experiments are also presented for uncoated and coated variants of René N5 superalloy single crystal.

II. Experimental Approach

To study crack advance mechanisms, experimental thermomechanical cycling experiments have been conducted on single crystal substrates of René N5 with a nominal composition of 7.5Co–7.0Cr–1.5Mo–5.0W–6.5Ta–6.2Al–3.0Re–0.15Hf–Bal Ni (wt%). Tests were conducted on bare, uncoated crystals, and on samples coated with either a standard Pt aluminide (PtAl) or a vapor phase (Pt-free) nickel aluminide (VPA) coating.[28] Coatings were applied directly to [001] single crystal fatigue samples with 5 mm gage diameter and 19 mm gage length. The initial thickness of the β-NiAl layer in the coating was approximately 50 μm. Since the coating structure develops by diffusion, an interdiffusion layer of comparable thickness to the β-layer is also formed as a part of the process. A summary of sample testing conditions is given in Table I.

High-temperature SPLCF cycling experiments were conducted isothermally in air at an *R*-ratio of $R = -\infty$ ($A = -1.0$) with

120-s compressive holds combined with a 3-s loading–unloading cycle, Fig. 1. Tests were conducted at 982° and 1093°C with total strain ranges varying from 0.20% to 0.96%. Cycling was conducted in a standard servohydraulic system in strain-controlled mode and hysteresis loops were analyzed. Following cycling, cracking on the surfaces of samples was characterized by optical and scanning electron microscopy. Samples were then sectioned longitudinally for analysis of crack depths and the development of the oxide on the surface and within the cracks.

III. Experimental Results

At 1093°C the majority of the experiments were conducted at a constant total strain range of $\Delta\varepsilon_t = 0.35\%$. Experimentally measured hysteresis loops for an uncoated sample with a total number of cycles to failure of $N_f = 6755$ cycles are shown in Fig. 3. In the first cycle, the sample is compressively loaded and held at point "A" for 120 s. During the 120-s hold, creep deformation causes a reduction in the stress from about −240 to −160 MPa. Upon returning to zero strain, the sample experiences a tensile

Table I. Experimental SPLCF Testing Conditions

Sample #	Coating	Temperature (°C)	Total strain range (%)	# Cycles[†]
1	PtAl	982	0.94	801
2	PtAl	982	0.88	1260
3	PtAl	982	0.77	1544
4	PtAl	982	0.72	2234
5	PtAl	982	0.65	3409
6	PtAl	982	0.60	4051
7	PtAl	982	0.56	6483
8	PtAl	982	0.56	6534
9	VPA	1093	0.35	2000[†]
10	VPA	1093	0.35	4000[†]
11	VPA	1093	0.35	6000[†]
12	VPA	1093	0.35	8000[†]
13	VPA	1093	0.35	10 200
14	Bare	1093	0.35	1000[†]
15	Bare	1093	0.35	2000[†]
16	Bare	1093	0.35	4000[†]
17	Bare	1093	0.35	6000[†]
18	Bare	1093	0.35	8000[†]
19	Bare	1093	0.35	6755

[†]Interrupted tests.

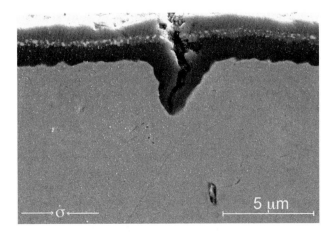

Fig. 2. Cracks in the surface oxide extending into the Ni-base superalloy during fatigue cycling with compressive holds at 1093°C.

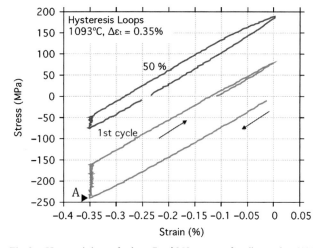

Fig. 3. Hysteresis loops for bare René N5 at start of cycling and at 50% of total SPLCF life.

stress approximately equal to the magnitude of the stress relaxation. The tensile stresses at the end of the cycle continue to increase until approximately 20% of the total life, where they stabilize. At 50% of life for $\Delta\varepsilon_t = 0.35\%$ at 1093°C, a maximum tensile stress of approximately 190 MPa is reached upon returning to zero strain, Fig. 3. Hysteresis loops are similar for the vapor phase aluminide-coated sample, however, the total cycles to failure increased to $N_f = 10\,200$. Thus the rate of damage accumulation is apparently higher in the uncoated sample.

The dependence of SPLCF life on total strain range for Pt-aluminide-coated samples at 982°C is shown in Fig. 4. The SPLCF life increases by a factor of approximately $10\times$ as the strain range is reduced from 0.96% to 0.56%. Under the conditions studied, the compressive hold can degrade the fatigue life (with no compressive hold) by a factor of up to $100\times$, depending on the temperature and strain range.

SPLCF cycling of samples to failure as well as experiments interrupted at various fractions of the average failure life have been conducted at 1093° and 982°C. At 1093°C a series of bare and vapor phase aluminide-coated samples were interrupted at various fractions of total fatigue life. Vapor phase aluminide-coated samples were removed from SPLCF tests after 2000, 4000, 6000, and 8000 cycles (approximately 20%, 40%, 60% and 80% of life, respectively) at 1093°C and $\Delta\varepsilon_t = 0.35\%$ and sectioned longitudinally. While the details of these sectioning studies have been reported elsewhere,[6] it was observed that cracks do not penetrate beyond the interdiffusion zone until

approximately 80% of the life is consumed by cycling. Similarly, testing of a Pt aluminide-coated sample at 982°C with $\Delta\varepsilon_t = 0.56\%$ demonstrates that at failure many cracks have not progressed beyond the interdiffusion zone, Fig. 5. Among the 75 cracks examined along the length of the gage section after failure, only three cracks extended to depths of 150–250 μm, not including the one longer crack that ultimately caused failure of the sample, Fig. 5.

The experiments at two temperatures (982° and 1093°C) with two different coatings (Pt aluminide and vapor phase aluminide) collectively reveal four successive stages of the failure process (Figs. 6 and 7): crack extension from the surface through the (I) bond coat (BC), (II) interdiffusion zone (IDZ), (III) superalloy substrate, and (IV) long crack growth. In the early stages of cycling, many cracks form in the bond coat along the length of the gage section. These cracks gradually progress through the bond coat and interdiffusion zone and into the superalloy substrate. The cracks are filled with an alumina layer a few micrometers thick, Fig. 8, that is relatively uniform in structure and thickness along the surface of the crack. After growth into the superalloy to a depth of a few hundred micrometers, Stage IV is approached and the cracks grow at an inclined angle with respect to the axis of the applied compressive stress, Fig. 7. In many cases coarse striations are visible on the fracture surface in Stage IV, indicating that rapid mixed-mode crack growth occurs in this final stage.

To experimentally assess the role of coatings, interrupted SPLCF cycling tests were also conducted on bare René N5 samples. SPLCF experiments were interrupted at 1000, 2000, 4000, 6000, and 8000 cycles at 1093°C with $\Delta\varepsilon_t = 0.35\%$. The average and maximum crack depths are given in Table II. Similar to the coated samples, the cracks do not extend to depths beyond 100 μm until very late in life.

IV. Finite Element Modeling of the SPLCF Process

A series of ABAQUS-based finite element models were developed to elucidate the properties of the substrate, interdiffusion zone, β-coating and oxide layer that most strongly influence the failure process. The details of a finite element mesh that contains an oxide layer of thickness h on the superalloy surface and within a shallow crack of depth a are shown in Fig. 9. Initial analyses of the crack growth process for an oxide penetrating into the superalloy in this configuration have been presented elsewhere.[7] These analyses are extended here to consider coatings and the conditions under which damage development is strongly affected by their presence. To study the role of the β-coating and interdiffusion zone, additional meshes were developed with oxide-containing cracks within the successive layers, as shown schematically in Fig. 10.

High-temperature properties of the substrate have been measured[29] and these are considered in the model. Two versions of the model have previously been developed[7] for the oxide on the superalloy: one where creep properties are explicitly considered and a second where an equivalent plasticity analysis is implemented for numerical efficiency. Steady-state creep can be adequately represented by power law behavior of the form: $\dot{\varepsilon} = \dot{\varepsilon}_o \left(\frac{\sigma}{\sigma_o}\right)^n$, where σ is the local Mises stress, σ_o is a reference stress, $\dot{\varepsilon}_o$ a reference strain rate, and n the creep exponent. In the present calculations $\sigma_o = 100$ MPa and $n = 10$. The high-temperature properties of the Pt aluminide bond coat measured in samples removed directly from coatings have been incorporated into the model.[30] While the properties of the interdiffusion zone have not been directly measured, the properties of alloys with compositions similar to those observed in the interdiffusion zone indicate that the creep rates are likely to be intermediate to those of the weak β-coating and the precipitation-strengthened superalloy.[31] To capture the lateral straining in the TGO, in the model, $\dot{\varepsilon}_{\text{growth}}$ is imposed during those stages of the strain cycle when the crack is open. It is imposed at a uniform rate governed by the TGO thickness, h. This strain-rate causes creep of the

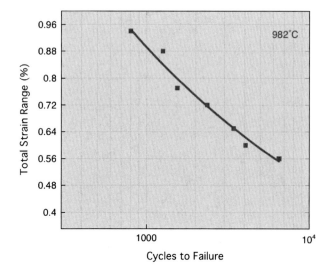

Fig. 4. SPLCF failure data for Pt-aluminide coated René N5 at 982°C cycled with 120-s compressive holds.

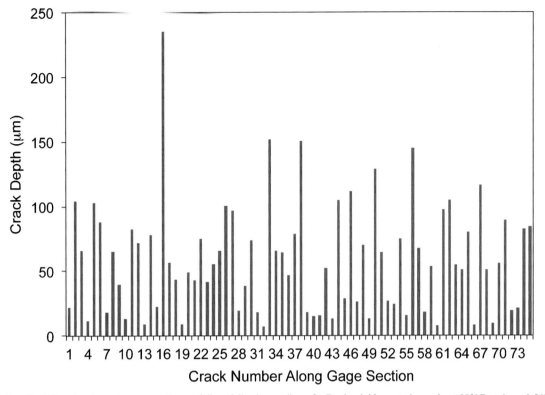

Fig. 5. Crack lengths along the gage section at failure following cycling of a Pt aluminide-coated sample at 982°C at $\Delta\varepsilon_t = 0.56\%$.

TGO,[27] replicated by ensuring that the Mises stress does not exceed the growth stress, $\sigma_{eq} \leq \sigma_{growth}$. The ensuing response of the TGO is governed by the creep deformation of the surrounding material (bond coat, IDZ, or substrate). Namely, creep deformation occurring around the tip accommodates elongation of the TGO along the "y" direction (Fig. 9), leading to a crack extension per cycle. The five parameters affecting the fatigue crack growth rate are thus: the TGO growth stress, σ_{growth}, the lateral strain-rate experienced by the TGO as it grows, $\dot{\varepsilon}_{growth}$, the oxide thickness, h, the creep characteristics of the surrounding material, $\dot{\varepsilon}_{sub} = \dot{\varepsilon}_o(\frac{\sigma}{\sigma_o})^n$, and the crack length, a. The oxide thickness was assumed to be $h = 3$ μm in all analyses presented here.

The principal features of the damage growth process can also be elucidated by using a plasticity model for the substrate and bond coat (with power law hardening) and by invoking the reference stress method to convert to power law creep.[7,18] Results presented here are primarily from plasticity analyses. The sequential stages of damage development are considered for bare

and coated materials and the conditions where the degradation is dominated by the superalloy, bond coat, and IDZ properties, respectively, are highlighted.

(1) Superalloy Dominated Growth

Initial calculations considered damage growth for a TGO formed on bare superalloy, to establish a baseline, since the properties of the substrate are well established. Figure 11 considers the early stages of the TGO penetration ($a_o = 20$ μm) into the superalloy substrate as a function of the strength of the substrate. For a moderate applied compressive strain of 0.4% and a superalloy strength of 100 MPa (comparable to René N5 at 1093°C) the oxide elongates at a rate of approximately 6.8 nm/cycle, Fig. 11(b). Across the range of conditions examined, the TGO penetration rate scaled linearly with the oxide thickness, so penetration rates are normalized by the oxide thickness, Fig. 11(a). As the strength of the superalloy approaches approximately 2/3 that of the TGO, the resistance to deformation along

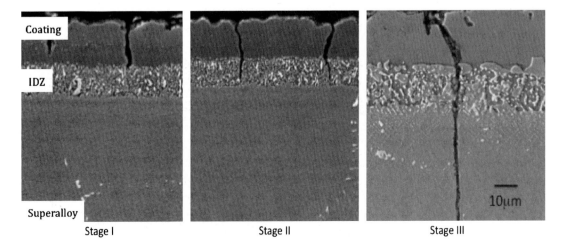

Fig. 6. The first three stages of crack progression with cracking in the Pt aluminide bond coat (Stage I), interdiffusion zone (Stage II), and superalloy substrate (Stage III). René N5 tested at 982°C.

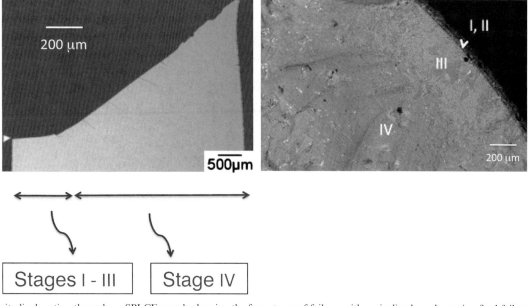

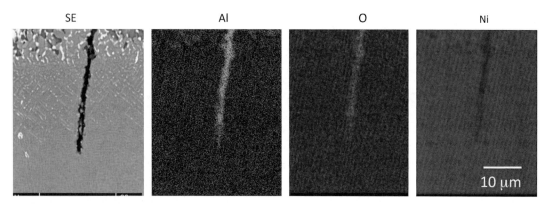

Fig. 7. Longitudinal section through an SPLCF sample showing the four stages of failure, with an inclined crack causing final failure in Stage IV.

Fig. 8. SEM secondary electron (SE) image and corresponding X-ray maps showing alumina in the crack tip after SPLCF testing of René N5 with Pt aluminide coating at 982°C.

the surface of the oxide-lined crack face and at the crack tip rises to a sufficiently high level to almost completely suppress crack growth. For an applied compressive strain of 0.35% and a substrate strength of 100 MPa, the TGO penetration rate is 6 nm/cycle, Fig. 11(a) and rises to 12 nm/cycle for 50 MPa strength. Experimentally, for an applied compressive strain of 0.35%, uncoated René N5 samples were interrupted after 2000 and 4000 cycles, Table I. The maximum crack depth was measured as 36 μm at 2000 cycles and 19 μm at 4000 cycles (two distinct samples), corresponding to a crack growth rates in the range of 5 nm/cycle–18 nm/cycle, in good agreement with the model. A limited number of high-temperature, *in-situ* experimental investigations on growth strains in the TGO formed on aluminide coatings have been conducted in the temperature range of

1000°–1100°C,[26,32] however, no measurements of growth strains are available for the TGO on superalloys. Bond coats experience compressive growth strains varying from −0.01% to −0.10%, so growth strains over this range were used in the modeling for both the bond coat and the superalloy. Figure 12 shows the influence of oxide growth strains on the rate of penetration of the TGO into the substrate. With no growth strains in the oxide, there is still a tendency for the oxide to push into the substrate due to transverse compression of the oxide during the hold period. Growth strains as low as 0.03%/cycle accelerate the rate of crack growth by more than a factor of 5× as the cracks grow beyond 30–40 μm in length. The influence of the growth strains on the damage development rate, however, are less pronounced at $a = 20$ μm and below. Comparing again with experimental results, for the 20 μm crack length in Fig. 12, the predicted growth rate for growth strains of 0.03%/cycle is 12 nm/cycle; again this is within the range of rates observed experimentally. These results suggest that the model is capturing some of the essential details of the crack growth process in the early stages.

It is worth considering more deeply penetrating oxides, since cracks that pass through the bond coat and interdiffusion zone will subsequently be embedded in the superalloy substrate. Figure 13 shows the TGO penetration rate normalized by the oxide thickness into the superalloy for an initial crack depth of 200 μm, which is well beyond the combined 100 μm thickness of the bond coat and IDZ. The predicted crack growth rate in Fig. 13 for $\Delta\varepsilon_t = 0.35\%$ is approximately 25 nm/cycle, assuming

Table II. Crack Length Data for Bare René N5 Sample Cycled at 1093°C with $\Delta\varepsilon_t = 0.35\%$

Cycles	Average depth (μm)	Maximum depth (μm)
1000	4.1	21.2
2000	3.8	36.0
4000	4.1	19.0
6000	7.2	49.7
6755	5.9	36.7
8000	7.3	209.7

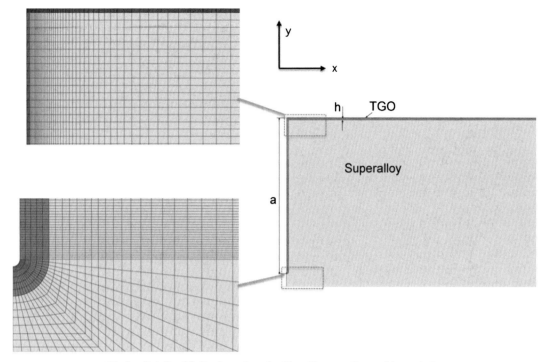

Fig. 9. Details of finite element mesh with oxide on surface and in crack tip.

a growth strain of 0.1%/cycle and bond coat and IDZ strengths of 50 MPa. In the interrupted tests on uncoated samples, a maximum crack depth of 200 μm was reached at approximately 80% of life, Table I. At this point the crack growth rate was ≈150 nm/cycle. Earlier experimental measurements at 80% of life with a vapor phase aluminide-coated sample indicate a crack growth rate of about 40 nm/cycle at 80% life,[6] so again the model predicts growth rates within a factor of 5. It should be noted that the crack growth rate increases rapidly in the later stages of life, rising from 40 nm/cycle at 80% life to 200 nm/cycle at 95% of life, so comparison of experiments with the model is more challenging in the later stages of damage growth. Ultimately, as the crack penetrates deeply into the superalloy substrate, the rate of growth depends on the difference in the substrate and TGO strengths, Fig. 13(b).

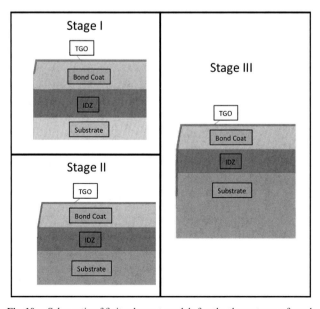

Fig. 10. Schematic of finite element models for the three stages of crack growth.

The formation of alumina along the crack faces by depletion of Al from the superalloy substrate can lead to a layer of soft γ-Ni adjacent to the oxide-filled crack. For this reason simulations were also conducted with soft layers surrounding the crack tip with thicknesses in the range of 1–3 μm with growth strains ranging 0.02%–0.20%. Decreasing the strength of the soft layer from 100 MPa (the same as the superalloy strength) to 10 MPa accelerated the crack growth rate by an additional factor of 2–3 ×.

(2) Role of the Bond Coat and IDZ

For the purposes of modeling the influence of coating and IDZ properties, the β-NiAl and interdiffusion zone are each considered to be 50 μm in thickness with a 3-μm-thick TGO. The oxide penetration rate in each of the three stages has been addressed separately by generating a series of meshes for varying crack depths. The properties of the β-NiAl coating, IDZ, and superalloy substrate are varied along with oxide growth strains over bounds that might reasonably be expected in present and future systems, with bond coat and IDZ strengths varied parametrically from 10 to 100 MPa.

Figure 14 shows the influence of the bond coat strength on the TGO crack tip lengthening rate for Stage I growth with a crack depth of 20 μm and a TGO thickness of 3 μm. At this early stage of the process where the cracks are shallow and the bond coat is softer than the substrate, the lengthening of the oxide causes not only penetration into the depth of the β-NiAl, but also a slight upward displacement of the bond coat surface of the order of 0.1 μm in the "y" direction, (direction indicated in Fig. 9). Determination of the overall crack growth rate in this early stage is complicated by the uplift as well as the included angle of the crack. Nevertheless, stronger bond coats do inhibit the rate of TGO lengthening as they increase in strength from 20 to 100 MPa. At this early stage in life, interdiffusion zone strengths have very limited influence on the overall response, Fig. 14, since the crack has not yet reached this zone. The overall crack growth rate within this stage for a bond coat strength close to β-NiAl is calculated to be about 4.1 nm/cycle, which is in reasonable agreement with the experimentally observed growth rate of 9 nm/cycle.[6] It is important to note that about half of the overall life is spent in Stage I, so strengthening the bond coat is likely to have an overall beneficial effect on SPLCF life.

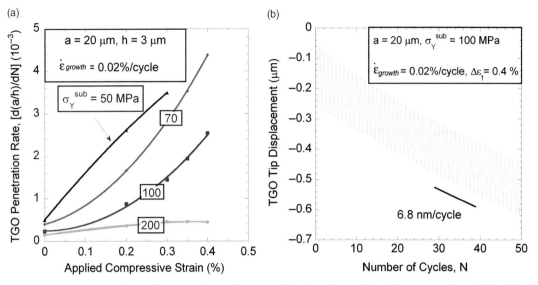

Fig. 11. (a) Influence of applied compressive strains on crack growth rates as a function of superalloy substrate strength and (b) cyclic displacement of the thermally grown oxide (TGO).

The crack tip extension rate in Stage II (crack depth of 70 μm, in IDZ) is shown in Fig. 15. A weak interdiffusion zone will clearly accelerate the rate of damage growth, particularly if the bond coat is weak (in this case with a strength less than approximately 60–70 MPa). Interestingly, even though the crack tip has progressed beyond the bond coat layer, the bond coat strength still influences the lengthening, since roughly half of the length of the oxide on the crack face is located along the bond coat layer. As both the bond coat and IDZ strengths approach that of the superalloy substrate, the TGO penetration rate is minimized. However, it is important to recall that bond coat and IDZ growth strains are likely different from those of the superalloy and higher growth strains or thicker oxides (more rapid oxidation kinetics) in the substrate could drive faster crack growth. Note in Fig. 16, that the TGO growth rate in Stage II is substantially enhanced by the fatigue cycling, particularly if the bond coat strength falls below 60 MPa. Also, as oxidation kinetics increase (the thickness of the TGO increases), the rate of damage growth increases. Finally, as Stage III is entered, the TGO extension rate is strongly influenced by the applied compressive strain in the fatigue cycle, Fig. 13, with crack extension rates increasing by more than a factor of 10 as the applied compressive strain increases from 0.2% to 0.8%.

As the crack progresses just beyond the IDZ, there is still an influence of the bond coat and interdiffusion zone properties, again because a substantial portion of the crack face is still located within these layers. There is a complex change in rates as the cracks extend from soft layers to harder layers and the reverse, Fig. 17. Apparently a soft interdiffusion layer is particularly detrimental to crack growth rates. Nevertheless, damage growth rates all converge to a common value as the crack progresses into the superalloy to a depth of 250–300 μm, which is 2.5–3× deeper than the combined thickness of the bond coat and interdiffusion zone, Fig. 17. As the crack extends well past the bond coat and TGO, the rate of damage growth ultimately depends on the difference in strength of the TGO compared with the substrate and the bond coat and IDZ properties become less important, Fig 13(b). However, it should be recalled that only a small fraction of life (<20%) is spent in these later "long crack" growth stages.

V. Discussion

Failure occurs during SPLCF cycling with compressive holds due to progressive growth of surface-initiated cracks containing a layer of alumina through the coating layers into the superalloy substrate. Finite element modeling demonstrates a fundamental role for growth stresses in the oxide and their relaxation and accommodation by creep in the bond coat, IDZ, and superalloy as a mechanism for propagating fatigue damage during SPLCF cycling. Growth strains during oxidation have been shown to develop due to counter-balanced outward fluxes of Al and inward fluxes of oxygen along grain boundaries during oxidation, permitting elongation of the oxide normal to the boundary in addition to oxide film thickening.[19–21] This induces compressive stresses in the oxide that have been measured for oxides on several intermetallic coatings.[25,26] Relaxation of these stresses during elevated temperature cycling[22,27] is a key element of the "rumpling" process that occurs during cyclic oxidation, resulting in oxide spallation.[21,22,24] During SPLCF cycling the role of the growth stresses is considerably more complex, since the oxide not only covers the surface of the sample, but penetrates deeply into the substrate with the assistance of the stresses induced by compressive fatigue loading and creep deformation that occurs during compressive holds. The creep deformation results in a net tension as the sample is returned to zero strain, permitting crack opening and further oxidation, Figs. 2 and 3. A complicating feature of the multilayered system is the fact that the growth strains or the kinetics of oxidation are likely to

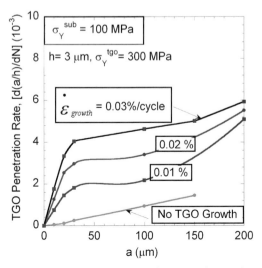

Fig. 12. Influence of oxide growth strains on crack growth rates at various crack depth.

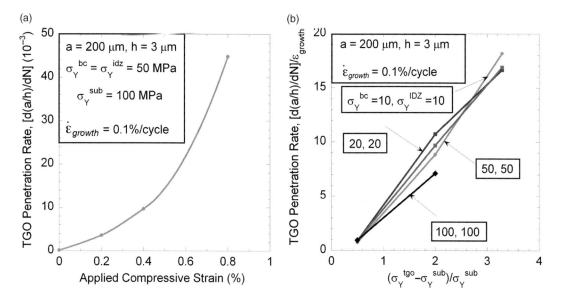

Fig. 13. (a) Rate of thermally grown oxide (TGO) penetration into the superalloy for an initial crack length of 200 μm and (b) the role of substrate yield strength (b).

fluctuate in magnitude as the crack passes from the bond coat to interdiffusion zone into the substrate; indeed the model underestimates the rate of crack growth observed experimentally as the cracks penetrate into the superalloy to depths of the order of 200 μm, Fig. 13. Experimental studies of the development of growth stresses in oxides that form directly on the single crystal superalloys would clearly be useful.

Since the alumina has a high strength relative to the metallic coating layers and substrate in the temperature range investigated here, changes in the strength of the coating layers and substrate strongly influence crack lengthening rates. As the strength of the bond coat and IDZ approach the superalloy substrate strength, the rate of crack advance is substantially reduced, Fig. 15. Considering that approximately 80% of total SPLCF life is consumed in propagating cracks through the bond coat and IDZ, strengthening these layers should substantially improve SPLCF life. State of the art bond coats such as the Pt aluminide bond coat investigated here have strengths that are only 1/4–1/3 of the superalloy substrate in the vicinity of 1000°C[29,30] and the modeling suggests that these bond coats are soft enough to significantly assist damage growth in the coating in Stages I and II. The properties of the interdiffusion zone are relatively unexplored and likely to continue to evolve during high-temperature cycling

as the composition of this zone changes. More detailed measurements of IDZ properties and interdiffusion models that couple with property models would clearly be useful. An alternative approach would be to minimize the thickness or completely eliminate the IDZ zone by tailoring coating composition. Ultimately, a bond coating that would maximize SPLCF resistance would be comprised of a fine-scale two-phase system (for strengthening) that is in near-thermodynamic equilibrium with the superalloy and forms a slowly growing oxide with low growth stresses.

Modeling of the constituent layers demonstrates that damage growth rates all converge to a common value as the crack progresses to a depth that is 2.5–3× greater than the combined thickness of the bond coat and interdiffusion zone, Fig. 17, due to the fact that bond coat and IDZ influence the stress state in the crack face oxide for some distance after the crack tip moves beyond these layers. For the present system, the depth at which the bond coat and IDZ no longer influence the crack front processes, Fig. 17, is about 0.3 mm, which in many airfoil designs could be equivalent to the entire wall thickness.[4] *This suggests that the properties of the oxide, coating, and IDZ may have an unexpectedly strong effect on airfoil life.*

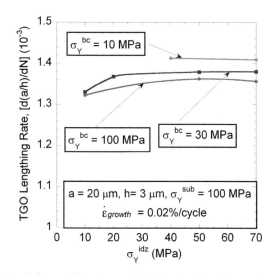

Fig. 14. Influence of bond coat and IDZ on Stage I crack extension rate.

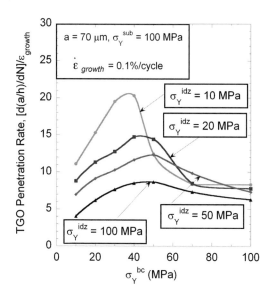

Fig. 15. Influence of bond coat and IDZ on Stage II crack extension rate.

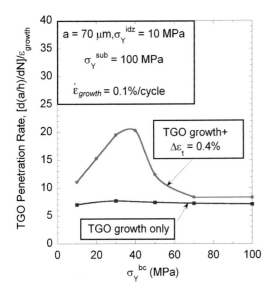

Fig. 16. Influence of fatigue cycling on crack growth through the interdiffusion zone in Stage II.

As the crack is well into Stage III growth, the crack growth rate, da/dN, is dependent on the difference in the TGO-substrate strength such that, Fig. 13(b):

$$\frac{da}{dN} = \lambda h \varepsilon_{growth} \left(\frac{\sigma_y^{TGO} - \sigma_y^{Sub}}{\sigma_y^{Sub}} \right)$$

where σ_y^{TGO} and σ_y^{Sub} are the yield strengths of the TGO and substrate at the test temperature, h is the thickness of the oxide, ε_{growth} is the oxide growth strain per cycle in the superalloy and a constant,[18] $\lambda \approx 4.5$ for the René N5 substrate. The crack is expected to grow at a steady rate until the threshold for macroscopic fatigue crack propagation is exceeded in the tension portion of the cycle, thereby entering Stage IV. For René N5 at $\Delta \varepsilon_t = 0.35\%$, the maximum tensile stress is 190 MPa, Fig. 3. Considering a typical long crack threshold of $\Delta K_{th} = 7$ MPa·m$^{1/2}$, this transition would occur for a crack depth of 430 μm, in rough agreement with the crack angle transition observed in Fig. 7. Beyond this, rapid crack growth is expected, independent of coating properties. However, since this represents only a small fraction of overall SPLCF life, strategies for

improving SPLCF resistance should focus on slowing the rate of damage growth during Stages I–III.

VI. Conclusions

1. Failure due to fatigue cycling with compressive holds in single crystal samples with or without aluminide coatings occurs due to oxide-assisted crack growth from sample surfaces at 982° and 1093°C.

2. Oxide-filled cracks do not penetrate the bond coat and interdiffusion zone until approximately 80% of the cyclic life.

3. A model that accounts for relaxation of oxide growth stresses as the primary driving force for extension of the oxide through the coating and into the substrate provides reasonable predictions of the rate of crack growth through the bond coat, interdiffusion zone, and superalloy single crystal substrate.

4. The model predicts that cyclic life can be enhanced by strengthening of the bond coat and interdiffusion zone or by reduction of oxide growth stresses.

Acknowledgments

This paper is dedicated to the memory of Tony Evans, whose original insights to this problem motivated this research. The authors also acknowledge the support of the General Electric Company and useful discussions with D. Konitzer, J. Rigney, B. Hazel, and M. F. X. Gigliotti.

References

[1]A. G. Evans, D. R. Mumm, J. W. Hutchinson, G. H. Meier, and F. S. Pettit, "Mechanisms Controlling the Durability of Thermal Barrier Coatings," *Prog. Mater. Sci.*, **46**, 505–53 (2001).

[2]T. M. Pollock and S. Tin, "Nickel-Based Superalloys for Advanced Turbine Engines: Chemistry, Microstructure and Properties," *AIAA J.: Propulsion Power*, **22** [2] 361–74 (2006).

[3]A. G. Evans, D. R. Clarke, and C. G. Levi, "The Influence of Oxides on the Performance of Advanced Gas Turbines," *J. Eur. Ceram. Soc.*, **28**, 1405–919 (2008).

[4]Y. L. Bihan, P.-Y. Joubert, and D. Placko, "Wall Thickness Evaluation of Single-Crystal Hollow Blades by Eddy Current Sensor," *NDT&E Int.*, **34**, 363–8 (2001).

[5]R. C. Reed, *The Superalloys: Fundamentals and Applications*. Cambridge University Press, Cambridge, 2006.

[6]A. Suzuki, M. F. X. Gigliotti, B. T. Hazel, D. G. Konitzer, and T. M. Pollock, "Crack Progression During Sustained Peak Low Cycle Fatigue in René N5," *Metall. Mater. Trans.*, **41A**, 948–56 (2010).

[7]A. G. Evans, M. Y. He, A. Suzuki, M. Gigliotti, B. Hazel, and T. M. Pollock, "The Mechanism Governing Sustained Peak Low Cycle Fatigue of Coated Superalloys," *Acta Mater.*, **57**, 2969–83 (2009).

[8]T. E. Strangman, "Thermal-Mechanical Fatigue Life Model for Coated Superalloy Turbine Components"; pp. 795–804 in *Superalloys 1992*, Edited by S. D. Antolovich., *et al* TMS, Warrendale, PA, 1992.

[9]J. W. Holmes and F. A. McClintock, "The Chemical and Mechanical Processes of Thermal Fatigue Degradation of an Aluminide Coating," *Metall. Trans.*, **21A**, 1209–22 (1990).

[10]P. Moretto and J. Bressers, "Thermomechanical Fatigue Degradation of a Nickel-Aluminide Coating on a Single Crystal Superalloy," *J. Mater. Sci.*, **31**, 4817–29 (1996).

[11]T. C. Totemeier and J. E. King, "Isothermal Fatigue of an Aluminide-Coated Single-Crystal Superalloy: Part II. Effects of Brittle Precracking," *Metall. Mater. Trans.*, **27A**, 353–61 (1996).

[12]J. S. Crompton and J. W. Martin, "Crack Growth in a Single Crystal Superalloy at Elevated Temperature," *Metall. Trans.*, **15A**, 1711–9 (1984).

[13]H. Zhou, H. Harada, Y. Ro, and I. Okada, "Investigations on the Thermo-Mechanical Fatigue of Two Ni-Based Single Crystal Superalloys," *Mater. Sci. Eng.*, **A**, 161–7 (2005).

[14]N. Isobe and S. Sakurai, "Compressive Strain Hold Effect on High Temperature Low-Cycle Fatigue Crack Growth in Superalloys," *Mater. Sci. Res. Int.*, **9**, 29–33 (2003).

[15]R. Nutzl, E. Affeldt, and M. Goken, "Damage Evolution During Thermomechanical Fatigue of a Coated Monocrystalline Nickel-Base Superalloy," *Intl. J. Fatigue*, **30**, 314–7 (2008).

[16]Y. H. Zhang, D. M. Knowles, and P. J. Withers, "Micromechanics of Failure of Aluminide Coated Single Crystal Ni Superalloy Under Thermomechanical Fatigue," *Scr. Mater.*, **37**, 815–20 (1997).

[17]E. Fleury and L. Remy, "Behavior of Nickel-Base Superalloy Single Crystals Under Thermomechanical Fatigue," *Metall. Mater. Trans.*, **25A**, 99–109 (1994).

[18]M. Y. He and A. G. Evans, "A Model for Oxidation-Assisted Low Cycle Fatigue of Superalloys," *Acta Mater.*, **58**, 583–91 (2010).

[19]J. A. Nychka and D. R. Clarke, "Quantification of Aluminum Outward Diffusion During Oxidation of FeCrAlY Alloys," *Oxidation Metals*, **63** [5–6] 325–52 (2005).

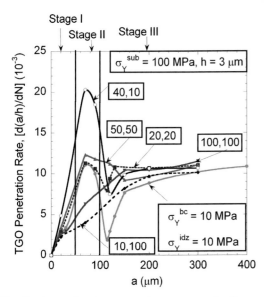

Fig. 17. Crack growth rates converge to a common rate after penetrating 250–300 μm into the substrate.

[20]V. K. Tolpygo and D. R. Clarke, "Microstructural Evidence for Counter-Diffusion of Aluminum and Oxygen during the Growth of Alumina Scales," *Mater. High Temp.*, **20**, 261–71 (2003).

[21]V. K. Tolpygo and D. R. Clarke, "Oxidation-Induced Failure of BPVD Thermal Barrier Coatings," *Surf. Coat. Tech.*, **146–147**, 124–31 (2001).

[22]D. S. Balint and J. W. Hutchinson, "An Analytical Model of Rumpling in Thermal Barrier Coatings," *J. Mech. Phys. Solids*, **53**, 949–73 (2005).

[23]A. W. Davis and A. G. Evans, "Some Effects of Imperfection Geometry on the Cyclic Distortion of Thermally Grown Oxides," *Oxidation Metals*, 1573–4889 (2006).

[24]A. M. Karlsson, J. W. Hutchinson, and A. G. Evans, "A Fundamental Model of Cyclic Instabilities in Thermal Barrier Systems," *J. Mech. Phys. Solids*, **50**, 1565–89 (2002).

[25]A. H. Heuer, A. Reddy, D. B. Hovis, B. Veal, A. Paulikas, A. Vlad, and M. Rühle, "The Effect of Surface Orientation on Oxidation-Induced Growth Strains in a Model NiCrAlY Bond Coat Alloy," *Scr. Mater.*, **54**, 1907–12 (2006).

[26]D. Hovis, L. Hu, A. Reddy, A. H. Heuer, A. P. Paulikas, and B. W. Veal, "In-Situ Studies of TGO Growth Stresses and the Martensitic Transformation in the B2 Phase in Commercial Pt-Modified NiAl and NiCoCrAlY Bond Coat Alloys," *Intl. J. Mater. Res.*, **98**, 1209–13 (2007).

[27]B. W. Veal, A. P. Paulikas, B. Gleeson, and P. Y. Hou, "Creep in α-Al$_2$O$_3$ Thermally Grown on β-NiAl and NiAlPt Alloys," *Surf. Coat. Tech.*, **202**, 608–12 (2007).

[28]B. Tryon, K. S. Murphy, C. G. Levi, J. Yang, and T. M. Pollock, "Hybrid Intermetallic Ru/Pt-modified Bond Coatings for Thermal Barrier Systems," *Surf. Coat. Tech.*, **202**, 349–61 (2007).

[29]T. M. Pollock and R. D. Field, "Dislocations and High Temperature Plastic Deformation of Superalloy Single Crystals"; pp. 547–618 in *Dislocations in Solids*, Vol 11, Edited by F. R. N. Nabarro, and M. S. Duesbery. Elsevier, Amsterdam, 2002.

[30]D. Pan, M. W. Chen, P. K. Wright, and K. J. Hemker, "Evolution of a Diffusion Aluminide Bond Coat for Thermal Barrier Coatings During Thermal Cycling," *Acta Mater.*, **51**, 2205–17 (2003).

[31]J. S. Van Sluytman, A. Suzuki, R. Helmik, A. Bolcavage, and T. M. Pollock, "Gamma Prime Morphology and Creep Properties of Nickel-Base Superalloys with Platinum Group Metal Additions"; pp. 499–507 in *Superalloys 2008*, Edited by R. C. Reed, K. A. Green, P. Caron, T. P. Gabb, M. G. Fahrmann, E. S. Huron, and S. R. Woodard TMS, Warrendale, PA, 2008.

[32]P. Y. Hou, A. P. Paulikas, and B. W. Veal, "Growth Strains in Thermally Grown Al$_2$O$_3$ Scales Studied Using Synchrotron Radiation," *JOM*, **61**, 51–5 (2009). ☐

J. Am. Ceram. Soc., **94** [S1] S146–S153 (2011)

DOI: 10.1111/j.1551-2916.2011.04573.x

© 2011 The American Ceramic Society

journal

Alumina Scale Formation: A New Perspective

Arthur H. Heuer[†] and David B. Hovis

Department of Materials Science and Engineering, Case Western Reserve University, Cleveland, Ohio 44106

James L. Smialek

NASA Glenn Research Center, Cleveland, Ohio 44135

Brian Gleeson

Department of Mechanical Engineering and Materials Science, University of Pittsburgh, Pittsburgh, Pennsylvania 15260

Due to a publishing error, this paper, "Alumina Scale Formation: A New Perspective," by Arthur H. Heuer, David B. Hovis, James L. Smialek and Brian Gleeson (DOI: 10.1111/j.1551-2916.2011.04573.x) was inadvertently omitted from the original printed and online versions of *J. Am. Ceram. Soc.* **94** [S1], and an unrelated paper, "Plastic Deformation of ⟨001⟩ Single-Crystal SrTiO₃ by Compression at Room Temperature," by Kai-Hsun Yang, New-Jin Ho, and Hong-Yang Lu (DOI: 10.1111/j.1551-2916.2011.04473.x) was included in its place. This error was corrected online on June 30, 2011.

Oxidation of Al₂O₃ scale-forming alloys is of immense technological significance and has been a subject of much scientific inquiry for decades. The oxidation reaction is remarkably complex, involving issues of alloy composition, kinetics, thermodynamics, microstructure, mechanics and mechanical properties, crystallography, etc. A brief overview of the formation of passivating, thermally grown oxide Al₂O₃ scales will be given in the light of recent findings on the defect structure and associated transport behavior of α-Al₂O₃. It is inferred that the electronic structure of Al₂O₃ is of direct relevance to understand the Al₂O₃ scale growth. We also discuss the effect of the so-called "reactive" elements (REs)—Y, Zr, and Hf—on reducing the rate of Al₂O₃ scale thickening by reducing the outward flux of aluminum. An important aspect of the "new perspective" is the suggestion that the REs change the electronic structure of Al₂O₃—the relevant near-band-edge defect (grain boundary) states that are crucial to vacancy creation both at the scale/gas and scale/metal interfaces.

I. Introduction

OXIDATION of Al-containing Fe-base (FeCrAl, FeCrAlY, and FeAl) and Ni-base (NiCrAl, NiCrAlY, and NiAl) alloys often results in the formation of a thin, protective (passivating), α-Al₂O₃ scale. Such a thermally grown oxide (TGO) scale can provide excellent oxidation resistance for high-temperature structural alloys in oxidizing environments, and understanding the factors that impart oxidation resistance has been an active area of research for decades. Recent reviews on Al₂O₃ scale formation include those by Stott,[1,2] Chevalier,[3] Jedlinski and Borchardt,[4] Prescott and Graham,[5] Pint and Alexander,[6] and Hou.[7]

High-temperature (i.e., *T* > 700°C) formation of a protective α-Al₂O₃ scale is actually one of the most complex phase transformations in materials science. The kinetics and thermodynamics of the oxidation reaction are obviously of interest, but many complications exist. This is partly because Al melts at 660°C,

and hence Al₂O₃ scale formation at high temperatures must be studied using Al-containing alloys. Within a multicomponent alloy, the selective oxidation of Al will change the composition of the alloy subsurface, which can induce phase transformations that, in turn, may cause volume changes. While α-Al₂O₃ (corundum) is the only stable polymorph of aluminum oxide at ambient pressure, a host of so-called transition aluminas exist (Table I), which often form preferentially during the early stages of oxidation, before transforming to the stable α polymorph. These transition aluminas all have cubic or pseudo-cubic defect spinel structures. Important aspects of the phase transformations from various transition aluminas to α-alumina have been studied[8–10]; however, detailed mechanistic understandings of the transformations are lacking. Furthermore, the transformation to α-Al₂O₃ involves unit cell volume changes of ∼8% (the α phase is more dense) and gives rise to substantial, if sometimes transitory, tensile stresses within the TGO.[8,9]

The mechanical properties of the TGO, both at service temperatures (up to 1200°C) and during the inevitable thermal cycling to which these high-temperature structural components are subject, are a matter of great concern to engineers and designers (of gas turbines for example). Scale spallation during thermal cycling can have catastrophic consequences, to the extent that scale adhesion and the fracture mechanics of the scale/substrate interface may be crucial to system performance and life.

Finally, metal chemistry dramatically affects scale performance. For instance, sulfur impurities in the alloys in even ppm quantities can promote scale spallation. This occurs because sulfur segregates to oxide/metal interfaces and causes a concomitant weakening and loss of adhesion. On the other hand, a group of so-called "reactive" elements (REs)—Y, Zr, and Hf are the most common—can substantially improve oxidation resistance, even if present in small quantities.[6,7,11] They are known to counteract the negative effect of trace impurities (e.g., yttrium getters sulfur in even low-sulfur alloys), but their beneficial role in enhancing the oxidation resistance of the base alloy must involve other factors as well.

The oxidation kinetics can be equally complex. Basic treatments of diffusion-controlled scaling find that the thickening kinetics should follow the parabolic rate law. Wagner[12] many years ago elucidated the transport processes involved by considering the combination of ionic and electronic diffusion through a growing oxide scale. By assuming that a neutral

M. Rühle—contributing editor

Manuscript No. 28854. Received November 2, 2010; approved March 19, 2011.
[†]Author to whom correspondence should be addressed. e-mail: ahh@case.edu

Table I. Polymorphs of Al$_2$O$_3$

Polymorph	Crystal system	Unit cell dimensions (nm)
α (corundum)	Trigonal	$a = 0.4758$, $c = 1.2991$
δ	Tetragonal	$a = 0.7943$, $c = 2.350$
ε	Hexagonal	$a = 0.7849$, $c = 1.6183$
γ	Cubic	$a = 0.7949$
κ′	Hexagonal	$a = 0.5544$, $c = 0.9024$
θ	Monoclinic	$a = 0.5620$, $b = 0.2906$, $c = 1.1790$, $\beta = 103°20′$
X	Hexagonal	$a = 0.577$, $c = 0.864$

oxide evolves during growth and that local equilibrium exists at any point within the oxide, Wagner showed that the oxidation rate is dependent on the environmental oxygen partial pressure, P_{O_2}, if negatively charged defects (i.e., metal vacancies or oxygen interstitials) predominate in the oxide, whereas the rate is independent on P_{O_2} if positively charged defects (i.e., oxygen vacancies or metal interstitials) predominate. The defect charges in this context are formal charges in accordance with the Kröger–Vink notation[13] and are therefore effective charges, relative to an ion on a normal lattice site. As such, it is assumed that the ionic defect charge does not change during migration, which may be an oversimplification.

Wagner further assumed the oxide scale to be compact and free of short-circuit migration paths. Thus, any effect of grain boundaries and other extended defects were ignored. However, extrapolation of lattice diffusion data[14] of oxygen and Al from elevated temperatures to the temperature range of interest for high-temperature structural alloys in service—750°–1200°C—make clear that the oxidation kinetics are almost certainly controlled by grain-boundary diffusion; for oxygen, D_{gb}^{Oxy} is significantly enhanced compared with lattice diffusion. (Tracer data for D_{gb}^{Al} do not exist.) Furthermore, while the oxidation kinetics should be controlled by the faster of oxygen or Al grain-boundary diffusion, there are numerous examples where countercurrent inward diffusion of oxygen and outward diffusion of Al contribute more or less equally to scale growth.[15] In such cases, oxidation-induced compressive stresses arise from formation of "new" alumina within the scale, as well as at the oxide/gas and oxide/metal interfaces.[16]

The REs have been shown to segregate to oxide grain boundaries and inhibit outward Al diffusion, causing an overall reduction in the oxidation kinetics.[6,7] However, the actual mechanism by which a RE inhibits Al diffusion is not well understood. The explanations usually offered in the literature are qualitative[17] and include "site blocking," in which Y (or other RE) cations on Al sites in grain boundaries block critical diffusive paths, or a "swamp out" mechanism, in which preferential Y segregation to boundaries prevent segregation of other aliovalent solutes, which would otherwise affect the local point defect populations in the boundaries and hence enhance grain-boundary diffusion. This latter mechanism implies that grain boundaries are similar to bulk material in that they can have well-defined equilibrium point defect populations; this is probably a doubtful assumption for general (high angle) unstructured boundaries.

Furthermore, the microstructures of the TGOs matter and it is a complicating factor in understanding transport processes during oxidation. The microstructure of a relatively thick scale on FeCrAlY formed during a 2000 h/1200°C exposure is shown in Fig. 1.[18] The microstructure is largely columnar throughout most of its thickness, but equiaxed near the scale/gas interface, which is typical of most intact oxide scales.[6] Inasmuch as oxidation occurs primarily by grain-boundary diffusion, this effect has to be considered when analyzing diffusion processes during oxidation. Equally important, a significant gradient in the oxygen potential (i.e., the oxygen chemical activity) exists across the thickness of a growing Al$_2$O$_3$ scale.

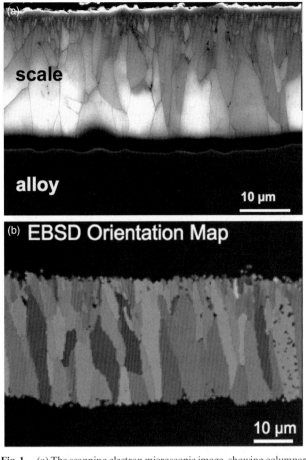

Fig. 1. (a) The scanning electron microscopic image, showing columnar growth in a thermally grown oxide (TGO) formed on an FeCrAlY after 2000 h at 1200°C, (b) an EBSD map of the TGO showing strong texturing of the TGO.[18]

II. Bulk and Grain-Boundary Diffusion in Al$_2$O$_3$

It is useful at this point to describe our present understanding on what is known about bulk and grain-boundary diffusion in Al$_2$O$_3$. The literature on this subject was reviewed recently by Heuer[14]; the salient conclusions of his review are shown in Figs. 2 and 3. Firstly, there have been at least eight[17,19–25] reliable studies of oxygen lattice diffusion, $D_{lattice}^{Oxy}$, in Al$_2$O$_3$, mostly using ^{18}O as a tracer, extending over four decades and involving crystals of different provenance and using several different techniques for determination of $D_{lattice}^{Oxy}$. Considering the low values of $D_{lattice}^{Oxy}$ (the oxygen penetration depth, $x \approx \sqrt{D_t}$, where t is time, is <1 μm in all these studies), the agreement is quite good among these diverse investigations.

This good agreement is actually surprising. There is now near-universal agreement that oxygen lattice diffusion in α-Al$_2$O$_3$ occurs by vacancy migration. Table II shows the results of theoretical calculations of the energy of intrinsic point defects, taken from three recent studies.[26–28] The empirical shell model results of Lagerlof and Grimes[26] suggested that anion Frenkel and Schottky disorder should dominate point defect equilibria in a hypothetically pure single crystal of Al$_2$O$_3$. The density functional theory (DFT) *ab initio* first principles quantum mechanical calculations of Matsunaga *et al.*[27] and Hine *et al.*[28] point instead to Schottky and cation Frenkel disorder dominating point defect disorder. We note that the Matsunaga and colleagues calculations were at 0 K, whereas Hine and colleagues considered temperatures up to the melting point of Al$_2$O$_3$, and variable P_{O_2} in their work. Thus, the more extensive DFT calculations[28] will be relied upon in further discussion.

As will be discussed in more detail in Section IV, the electronic structure of α-Al$_2$O$_3$ needs to be considered when

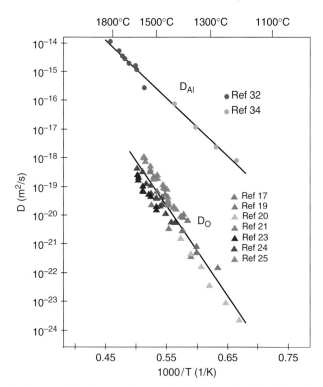

Fig. 2. An Arrhenius plot showing the best available bulk diffusivity data for oxygen and aluminum diffusion in Al_2O_3.

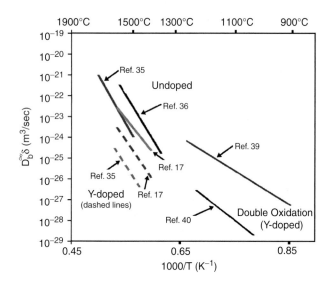

Fig. 3. An Arrhenius plot showing the best available grain-boundary diffusivity data for oxygen in Al_2O_3.

Table II. Calculated Intrinsic Defect Energies (Intrinsic Point Defects in Sapphire (Energy per Defect in kJ/mol))

	Lagerlof and Grimes[26]	Matsunaga et al.[27]	Hine et al.[28]
Schottky disorder $\varnothing \Leftrightarrow 3V_O^{\bullet\bullet} + 2V_{Al}'''$	499	387[†]	312[†]
Anion Frenkel disorder $O_O^\times \Leftrightarrow V_O^{\bullet\bullet} + O_i''$	470[†]	629	638
Cation Frenkel disorder $Al_{Al}^\times \Leftrightarrow V_{Al}''' + Al_i^{\bullet\bullet\bullet}$	636	478	394

[†]Dominant defect reaction.

tion, $[V_O^{\bullet\bullet}]$, as expected with uncontrollable (accidental) changes in impurity concentrations, do not occur; the system must be "buffered" with regard to oxygen vacancy concentration. A discussion of this interesting topic is beyond the scope of this paper.

The data on Al diffusion, $D_{lattice}^{Al}$, shown in Fig. 2, present other difficulties, namely a sparcity of independent measurements, which results from the lack of a convenient and available tracer for Al. There have been only three investigations[32–34] separated by nearly 50 years, only two of which[32,34] were used to produce the Arhennius plot of Fig. 2 (the data in the third study[33] appears to be problematical[14]). These data confirm the long-standing belief in the ceramic science community that $D_{lattice}^{Oxy} \ll D_{lattice}^{Al}$. It should be noted, however, that the lower temperature data were for Ti-doped crystals, so that, given the relatively high defect energies for the Schottky and cation Frenkel reactions in Table II, Ti^{4+} doping would be expected to increase the concentration of Al vacancies, V_{Al}''', and thus enhance $D_{lattice}^{Al}$. (It is likely that the higher temperature data are for specimens unintentionally doped with Si^{4+}.)

The most reliable tracer data[17,35,36] on oxygen grain-boundary diffusion, D_{gb}^{Oxy}, in Al_2O_3, are shown in Fig. 3 (δ is the width of the region where enhanced diffusion occurs, which cannot be discerned solely from tracer data). Considering the undoped data first, and assuming a grain-boundary width δ of 1 nm, there is marked enhancement of oxygen grain-boundary diffusion; $D_{gb}^{Oxy}/D_{lattice}^{Oxy}$ is about 10^6 at $1500°C$. Secondly, and quite surprisingly, the activation energy for grain-boundary diffusion is greater than that for lattice diffusion; the three studies shown reported Q values of 627, 825, and 884 kJ/mol, compared with Q for oxygen lattice diffusion of 585 kJ/mol.[14] There is a strong effect of Y doping, with D_{gb}^{Oxy} decreasing by a factor of 10–100 (compared with Y-free samples).[17] However, one of the data sets[17,37] refers to highly structured bicrystal, and data soon to be published by Nakagawa and Ikuhara shows that a suite of similar manufactured bicrystals showed $D_{gb}^{Oxy}\delta$ values varying by $\sim 10^3$; furthermore, Q values for $D_{gb}^{Oxy}\delta$ varied from 607 to 921 kJ/mol. This variation in $D_{gb}^{Oxy}\delta$ by a factor of 10^3 confirms earlier qualitative SIMS data,[38] which showed a similar variation in oxygen grain boundary diffusion in general high angle grain boundaries in polycrystalline alumina. Clearly, not all grain boundaries are created equal!

Finally, Fig. 3 includes $D_{gb}^{Oxy}\delta$ data derived from so-called double oxidation experiments during the course of Al_2O_3 scale formation.[39,40] Using alloy MA 956, an oxide dispersion-strengthened FeCrAl alloy containing fine particles of Y_2O_3, both studies involved oxidation of this alloy first in $^{16}O_2$ and then in $^{18}O_2$ and determined oxygen depth profiles using secondary ion mass spectroscopy. Not only are there large differences between the two data sets, the data seem unrelated to the higher temperature tracer-derived $D_{gb}^{Oxy}\delta$ data.

III. Recent Oxygen Permeability Experiments and Resulting Implications for Al_2O_3 Scaling

The inspiration for the present paper was motivated by two recent publications[41,42] reporting permeation of oxygen in P_{O_2} gradients in sapphire and in dense alumina polycrystals at

determining the energy of point defects assumed to be important in bulk diffusion (i.e., oxygen and aluminum vacancies, $V_O^{\bullet\bullet}$ and V_{Al}'''). Using the calculations of Hine et al.[28,29] it can be shown that the formation energy of oxygen vacancies at 1750 K and a P_{O_2} of 2×10^4 Pa is 350 kJ/mol and that of Al vacancies 336 kJ/mol.[28] Thus, none of the crystals used for the $D_{lattice}^{Oxy}$ data of Fig. 2 should show intrinsic behavior, as their native point defect populations would be too low to allow any appreciable intrinsic diffusion. Such an interpretation is in accordance with the extensive experimental work conducted by Kröger.[30] The corollary is that small quantities of background aliovalent impurities should cause large differences in the concentration of oxygen vacancies and hence poor reproducibility of $D_{lattice}^{Oxy}$. The "tight" scatter of the data in Fig. 2, however, attests to the "corundum conundrum," first noted by Heuer and Lagerlöf[31] in 1998. The issue is that significant variations in oxygen vacancy concentra-

elevated temperatures (1650°–2000°C). Permeation experiments were conducted in P_{O_2} gradients at both high and low oxygen pressures—with P_{O_2} varying between 10^{-8} and 1 Pa in one set of experiments and between 1 and 10^5 Pa in another. It is noteworthy that no oxygen permeation could be observed through 0.25-mm-thick sapphire plates, suggesting that the oxygen lattice diffusivity in Al_2O_3, even close to the melting point, is very low, which is consistent with the $D_{lattice}^{Oxy}$ data shown in Fig. 2.

In the polycrystal experiments at low oxygen pressures, thermal grooving was observed at both surfaces of the Al_2O_3 polycrystals, i.e. at the faces exposed to P_{O_2} at both 10^{-8} and at 1 Pa. Similar thermal grooving was observed in the high P_{O_2} experiments at the low P_{O_2} (1 Pa) surface of the polycrystals. Significantly, however, ridges rather than grooves were observed at the high P_{O_2} (10^5 Pa) surface of the polycrystal in the high P_{O_2} experiments (Fig. 4).

These data were interpreted as indicating oxygen permeation by oxygen grain-boundary diffusion at low oxygen pressures, but a more complex process must have been involved in the oxygen permeation at high oxygen pressures. Kitaoka *et al.*[41] postulated that dissociative adsorption of O_2 molecules occurs in the low P_{O_2} experiment such that oxygen *vacancies* (and *electrons*) are generated at the low P_{O_2} (10^{-8} Pa) interface and migrate to the 1 Pa interface; there is a counter flow of oxygen *ions* to the low P_{O_2} surface, where the oxygen ions recombine to form O_2 molecules. These reactions clearly allow an effective permeation of oxygen molecules from the 1 to the 10^{-8} Pa interface.

In the high P_{O_2} experiments, it was suggested that O_2 molecules absorb on the high P_{O_2} surface and dissociate such that the oxygen *atoms* become incorporated as lattice oxygen ($O_O^\times$), thereby creating Al *vacancies* (and electron *holes*) at this surface; the Al vacancies then diffuse to the low P_{O_2} (1 Pa) surface. This means that Al ions migrated from the low P_{O_2} to the high P_{O_2} surface. At this latter surface, the Al ions react with atmospheric

oxygen to create "new" Al_2O_3. At the low P_{O_2} surface, the Al vacancies (and holes) react with lattice oxygen to form O_2 molecules, which are then discharged into the local ambient; in effect, there is a displacement of oxide from the low P_{O_2} to the high P_{O_2} surface via grain-boundary diffusion, and hence the formation of grain-boundary ridges.

This interpretation of the permeability experiments implies that at the low P_{O_2} interface, oxygen dissociation must occur following the reaction,

$$O_O^\times \rightarrow \frac{1}{2}O_2 + V_O^{\bullet\bullet} + 2e' \qquad (1)$$

while at the high P_{O_2} interface, O_2 molecules are absorbed subject to the reaction

$$\frac{1}{2}O_2 \rightarrow O_O^\times + \frac{2}{3}V_{Al}''' + 2h^\bullet \qquad (2)$$

This clearly requires that charge transport involves *n*-type behavior at low P_{O_2} (Eq. (1)) and *p*-type behavior at high P_{O_2} (Eq. (2)), indicating that the nonstoichiometry and, hence, the electronic structure of Al_2O_3 are crucial to grain-boundary permeation.

Assuming these reactions are correct as written, and using the law of mass action together with the assumption of electroneutrality, it can be shown that the permeation rate should vary as $P_{O_2}^{-1/6}$ at low P_{O_2} and as $P_{O_2}^{+3/16}$ at high P_{O_2}, as was in fact observed.[41] The permeation rates allow grain-boundary diffusion coefficients, $D_{gb}\delta$, to be determined for both oxygen and Al. These data are shown by the filled and open circle data points in Fig. 5.

The second paper from this group[42] involved identical experiments on Al_2O_3 doped with 0.05 and 0.2 m/o Lu_2O_3. There was essentially no effect on $D_{gb}^{Al}\delta$ but a threefold decrease in $D_{gb}^{Oxy}\delta$ for the 0.2 m/o Lu_2O_3 sample (square data points in Fig. 5). Assuming Y behaves similarly to Lu in polycrystalline Al_2O_3, these data also provide insight into the dramatic effect of Y and other rare earths in decreasing the creep rate of polycrystalline Al_2O_3[43,44]; this must occur through reducing the grain-boundary diffusivity of oxygen.

It is important to remember that oxidation differs from creep, sintering, etc., in that oxidation is controlled by the *faster* of the two species (oxygen or Al) diffusing on its fastest path, whereas creep and sintering are controlled by the *slower* of the two species diffusing on its fastest path. Thus, attempting to gain insight from conventional experiments on polycrystalline Al_2O_3 (say creep for example) to oxidation may be problematical. Nevertheless, the permeability experiments just discussed strongly suggest that the electronic structure of Al_2O_3 must be considered

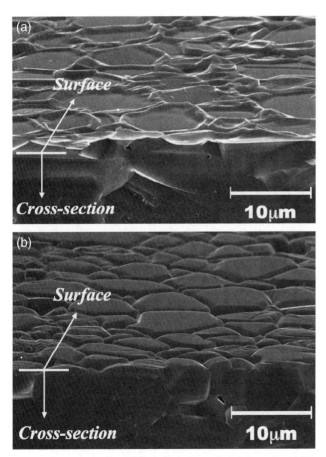

Fig. 4. Scanning electron microscopic images of a polycrystalline Al_2O_3 sample after exposure to an oxygen potential gradient during an oxygen permeability experiment. (a) The side exposed to $P_{O_2} = 10^5$ Pa. (b) The side exposed to $P_{O_2} = 1$ Pa.[41]

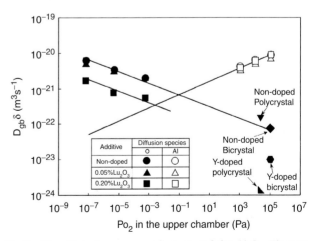

Fig. 5. The effect of pressure gradient on $D_b\delta$ for Al_2O_3. The pressure in the lower chamber is always $P_{O_2} = 1$ Pa. When the pressure in the upper chamber is high, aluminum diffusion dominates. When the pressure in the upper chamber is low, oxygen diffusion dominates. Doping with lutetium affects only the diffusion of oxygen.[42] See text for further discussion.

when discussing transport through an oxide scale in any oxidation experiment, as there must be a P_{O_2} gradient from $\sim 10^{-31}$ Pa (the dissociation pressure of Al_2O_3 at the metal/oxide interface at $\sim 1000°C$) to 2×10^4 Pa at the oxide/air interface. Whether the $p \to n$ transition occurs near the metal/oxide interface—say in the first few atomic layers of oxide formed—or somewhere further toward the oxide/gas interface depends on how the Al_2O_3 stoichiometry and, hence, the interdiffusion behavior varies across the scale. Based on conductivity measurements on an Al_2O_3 scale formed on NiAl at 1100°C for 96 h, Nicolas-Chaubet et al.[45] concluded that the major portion of the scale thickness is governed by a shallow P_{O_2} gradient, between $\sim 10^{-3}$ and 10^{-5} Pa, with abrupt P_{O_2} changes occurring at the alloy/scale and scale/gas interfaces. Thus, based on the data shown in Fig. 5, the results of Nicolas-Chaubet and colleagues suggest that the major portion of a growing Al_2O_3 scale is *n*-type, with $V_O^{\bullet\bullet}$ being the principal defect, and the $p \to n$ transition occurs quite close to the scale/gas interface. Considering Wagner's theory of oxidation,[12] the predominance of *n*-type behavior explains Ramanayanan et al.'s[46] finding that the parabolic rate constant for Al_2O_3 scale growth on Y_2O_3-dispersed FeCrAl(Ti) at 1100° and 1200°C is constant over the P_{O_2} range $\sim 10^5$ to $\sim 10^{-8}$ Pa. These authors had adopted the conventional wisdom of that time by assuming that the Al_2O_3 scale is invariably stoichiometric, and subject to the Schottky defect equilibrium (as opposed to the dominance of anion Frankel disorder, which was assumed by others[12]).

It is also possible to reinterpret some recent experiments conducted by Clarke and coworkers[47,48] assessing the extent of outward Al diffusion versus inward oxygen grain-boundary diffusion during oxidation. The experiment in question is shown in Fig. 6.[48] A relatively thick oxide scale was formed on an FeCrAlY alloy at 1100°C for 120 h. A taper section was then formed (Fig. 6a) and the sample reoxidized at the same temperature for 4 h in either air or a P_{O_2} of 10^{-12} Pa. Grain-boundary ridges developed during this reoxidation in air but not in the low P_{O_2} ambient. The authors attributed the suppression of ridge

formation to the lack of sufficient oxygen in the chamber. Based on the current discussion, however, it is instead suggested that the lack of ridges in the low-P_{O_2} environment could have resulted from the suppression of *p*-type behavior at the low-P_{O_2} environment in which Al^{3+} diffusion would predominate.

That the $p \to n$ transition lies somewhere within the TGO provides a ready explanation for the oft-reported oxidation-induced compressive stresses. A typical *in situ* data set is shown in Fig. 7 for oxidation of single crystal NiAl[49] and Ni–25Al–20Pt,[50] with and without additions of Zr and Hf, respectively. The initial tensile stresses are caused by the volume change associated with the first-formed θ-Al_2O_3 transforming to the stable α polymorph. The steady-state compressive stresses in the RE-free alloys are significant (100–200 MPa) and are attributed to a measurable level of "new" Al_2O_3 forming within the TGO, due to inward grain-boundary diffusion of oxygen and outward diffusion of Al. More specifically, the thickening of the scale under fixed boundary conditions necessitates that, at a given location within the scale, vacancy supersaturation occurs to the extent that reaction (3) below is favored to the right

$$2V_{Al}''' + 3V_O^{\bullet\bullet} + 2Al_{Al}^\times + 3O_O^\times \to Al_2O_3 \qquad (3)$$

The existence of a $p \to n$ transition and the defects associated with reactions (1) and (2) for each portion of the scale ensure that there are relatively high levels of both cation and anion

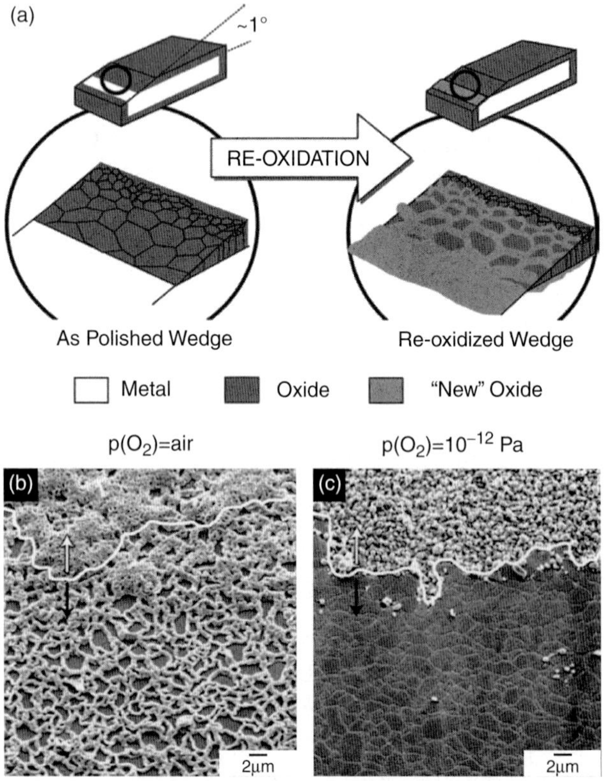

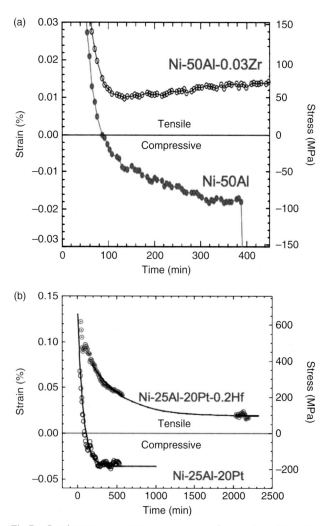

Fig. 6. (a) Schematic diagram showing a taper section of a preoxidized sample and then reoxidizing to observe the formation of grain-boundary ridges. (b) A sample reoxidized at $P_{O_2} = 2 \times 10^4$ Pa, showing ridge formation. (c) A sample reoxidized at $P_{O_2} = 10^{-12}$ Pa, showing no ridge formation.[47,48]

Fig. 7. Synchrotron *in situ* stress measurements for growing oxide scales. (a) Ni–50Al with and without 0.03Zr.[49] (b) Ni–25Al–20Pt with and without 0.2 Hf.[50] The initial tensile stress is due to the $\theta \to \alpha$ transformation in the Al_2O_3. In both cases, the addition of the reactive element (Zr or Hf) makes the intrinsic growth stress go from compressive to tensile.

defects, such that a measurable level of "new" Al_2O_3 formation within the scale is possible.

The exact location of the $p \rightarrow n$ transition within the scale does not affect whether or not reaction (3) proceeds to the right. Moreover, this mechanism for generating a compressive growth stress within the scale does not involve a plating reaction between aluminum and oxygen interstitials, as was postulated by Veal and Paulikas.[49] As will be shown in the next section, the energetics of defect formation do not favor the presence of any appreciable amount of interstitials in Al_2O_3 and, hence brings into question the mechanism proposed by Veal and Paulikas. The origin of the tensile stresses in the scales on alloys containing REs is still not clear.

Figure 8 shows two depth profiles obtained by secondary neutral mass spectroscopy obtained by Quadakkers et al.[15,18] after double oxidation experiments. The first involved oxidation of an FeCrAl alloy for 2.5 h in $^{16}O_2$ at 1000°C, followed by 5 h in $^{18}O_2$ at the same temperature. The amount of inward diffusion of oxygen and outward diffusion of Al realized in this experiment was about equal (as indicated by the shoulders on the ^{18}O profile). By contrast, the depth profile of an FeCrAlY alloy after 2 h oxidation in $^{16}O_2$ at 1200°C, followed by 6 h in $^{18}O_2$ at the same temperature indicated no outward diffusion of aluminum. Again assuming Y and Lu behave similarly as dopants in Al_2O_3, the oxidation resistance of the Y-containing alloy was much improved—Y clearly decreased $D_{gb}^{Oxy}\delta$ in accordance with the oxygen permeability data shown in Fig. 5 (although the decrease in oxidation rate was greater than the change in $D_{gb}^{Oxy}\delta$). Similar to Lu, Y is inferred to decrease the

mobility of oxygen along the Al_2O_3 scale grain boundaries (e.g., the Y–O bond is considerably stronger than the Al–O bond, as indicated by the higher melting point of Y_2O_3, 2425°C, compared with Al_2O_3, 2040°C). However, Fig. 5 also shows that adding Lu_2O_3 has little influence on the diffusion of aluminum, which is at variance with the finding that Y inhibits $D_{gb}^{Al}\delta$ during Al_2O_3 scale growth. An alternative explanation not heretofore considered, by the high-temperature oxidation community, is that REs affect the electronic structure of Al_2O_3, in particular grain-boundary acceptor or donor states. Specifically, the new perspective considers whether there are relevant near-band-edge defect (grain-boundary) states available to provide electrons and holes to allow reactions (1) and (2) to proceed. We consider this point further in the following section.

IV. Electronic Structure of Al_2O_3

The Hine et al.[28] DFT calculations referred to earlier provide valuable insight on aspects of the point-defect energetics of α-Al_2O_3, which will now be discussed. Before doing so, it must be realized that one cannot ask what is the energy of a charged defect in Al_2O_3 such as the oxygen or Al vacancy, $V_O^{\bullet\bullet}$ or V_{Al}''', without specifying the magnitude of the Fermi energy (the chemical potential of the electrons). This obtains because in performing a self-consistent calculation of the defect energetics, all charged species, including the electrons, must be considered.

Figure 9(a) shows how these energies vary with the Fermi energy. In the initial analysis, defects in all charge states (even some that may to a ceramist seem unusual) must be considered—for example, singly charged, doubly charged, and triply charged Al interstitials ($Al_i^{\bullet}$, $Al_i^{\bullet\bullet}$, and $Al_i^{\bullet\bullet\bullet}$, respectively). The familiar triply charged species, $Al_i^{\bullet\bullet\bullet}$, is the stable Al interstitial

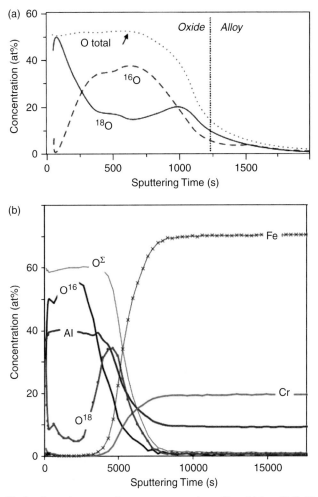

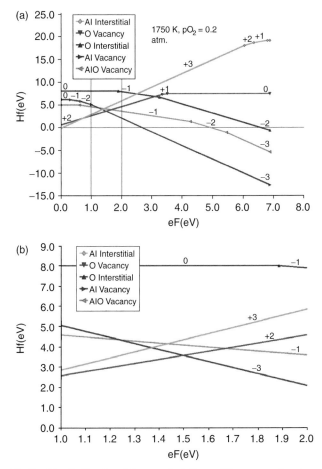

Fig. 8. Secondary neutral mass spectroscopic profiles of (a) an FeCrAl alloy and (b) an FeCrAlY alloy after double oxidation experiments. The ^{18}O profile shows new oxide formation throughout the scale in the FeCrAl alloy, but predominately inward growth in the FeCrAlY alloy. The addition of yttrium appears to shut off the outward diffusion of aluminum. (Quadakkers et al.[15] and D. Naumenko et al. (unpublished data)). The sputtering rate is ∼0.3 nm/s.

Fig. 9. Density functional theory calculations of the energy of various intrinsic defects in α-Al_2O_3 as a function of Fermi level. (a) Over the full range of the calculation, (b) the accessible range of Fermi levels for Al_2O_3.

point defect, but only for Fermi energies <6 eV. Similar considerations pertain for the other point defects shown in this figure.

The "accessible" range of Fermi energies are between 1 and 2 eV and the data for this range of the Fermi level are shown in Fig. 9(b). There are several features of note. Firstly, oxygen interstitials are uncharged (!) but have a very high energy, 8 eV. They also have an unusual "dumb-bell" shape, being covalently bonded to a lattice oxygen.[51] For a hypothetically pure stoichiometric crystal, the Fermi level will be near the intersection of the $V_O^{\bullet\bullet}$ and V_{Al}''' lines, close to 1.5 eV, and determined by the condition that the Schottky and cation Frankel equilibria have to be simultaneously satisfied.

Secondly, the Al–oxygen-bound divacancy, $V_{AlO}^{\bullet}$, has a reasonably low energy, less than that of the aluminum interstitial, $Al_i^{\bullet\bullet\bullet}$. This divacancy defect has not been discussed in much of the oxidation literature or in the literature on point defects in Al_2O_3. However, Doremus recently suggested[52,53] that AlO molecular ions could be important in the formation of Al_2O_3 scales. However, regardless of their relatively low energy, preliminary calculations suggest that their migration energy is sufficiently large that they do not contribute significantly to mass transport processes in bulk Al_2O_3. We believe it is also unlikely that such defects are involved in grain-boundary transport during TGO formation.

Regarding the results of Y and other REs apparently reducing $D_{gb}^{Al}\delta$ during Al_2O_3 scale growth (which was not observed for Al in the Lu_2O_3-doped oxygen permeability data (see Fig. 5)), we suggest that these elements act to inhibit Al ionization by modifying the donor/acceptor grain-boundary states relative to those present in undoped Al_2O_3. Such an effect would serve to reduce the outward migration of Al to the oxide/gas interface to the extent that, depending on the RE content, the Al_2O_3 scale could be predominantly inward growing (see Fig. 8). This is because a reduction in the extent of Al ionization would necessarily cause a reduction of V_{Al}''' relative to undoped Al_2O_3 (a nuance revealed only by Al_2O_3 scale growth and not by Al_2O_3 permeability measurements like those in Fig. 5).

V. Summary and Conclusions

High-temperature oxygen permeability experiments show that polycrystalline Al_2O_3 can exhibit either *n*-type or *p*-type behavior, depending on the local oxygen potential. Because of the large P_{O_2} gradient across a growing Al_2O_3 scale, a $p \rightarrow n$ transition must occur somewhere within that scale. The electronic structure of Al_2O_3 is thus of direct relevance to the oxidation resistance of Al_2O_3 scale-forming alloys.

Inasmuch as grain-boundary transport of both Al and oxygen appears to be the rule in Al_2O_3 scale-forming Fe-base and Ni-base alloys, the donor or acceptor states of importance must be those associated with the grain boundaries and not those of the lattice.

REs such as Y, Zr, and Hf have profound effects on scale formation when added to such structural high-temperature alloys as FeCrAl and NiCrAl. In particular, these additions tend to suppress the outward diffusion of Al^{3+}, reduce the kinetics of scale formation, and (although not discussed here) can retard the kinetics of the θ-$Al_2O_3 \rightarrow \alpha$-$Al_2O_3$ phase transformations. Compressive stresses in these alloys, which are thought to arise from "new" Al_2O_3 forming within the scale due to both inward grain-boundary diffusion of oxygen and outward grain-boundary diffusion of Al, becomes tensile in the presence of these REs.

In all these cases, the REs may affect the oxidation behavior not by geometric/structural effects—"site blocking" at grain boundaries due to grain-boundary segregation—but by modifying grain-boundary donor and acceptor states to the extent that Al ionization at the substrate/scale interface is reduced, which in turn reduces the extent of Al vacancy injection into the scale. This has the effect of causing the doped Al_2O_3 scale

to grow in a predominantly inward direction, due to migrating oxygen ions reacting with Al at the Al_2O_3/alloy interface. Thus, REs reduce the rate of Al_2O_3 scale growth by reducing both $D_{gb}^{Oxy}\delta$ and the extent of Al ionization. The latter drives the V_{Al}''' injection process needed for oxidation to take place at the Al_2O_3/gas boundary (i.e., reaction (2)).

Using this new insight, it has been possible to reinterpret a number of phenomena of importance to the oxidation resistance of high-temperature Al_2O_3 scale-forming alloys, including the beneficial effects of REs, oxidation-induced stresses, and the P_{O_2} dependence of oxidation kinetics.

It is fortunate that modern DFT techniques are quite capable of determining how REs affect the electronic structures of α-Al_2O_3, recognizing, however, that the electronic structures of grain boundaries present special problems. Work has just begun by A. H. Heuer *et al.* (unpublished data) exploring the effect of REs on the electronic structure of the defect states at highly structured α-Al_2O_3 grain boundaries using DFT.

Acknowledgments

This paper is dedicated to the memory of Tony Evans, whose loss to the materials community is immense. In particular, Tony's role in focusing the community's attention on Prime Reliant Coatings was key to AHH's efforts on studying TGO formation. We thank N. Hine, M. Finnis, and M. Foulkes (Imperial College, London) and T. Nakagama and Y. Ikuhara (University of Tokyo) for permission to quote unpublished data. This research was supported by ONR grants N00014-02-0479 (A. H. Heuer) and N000014-0911127 (B. Gleeson), Dr. D. Shifler, Program Manager.

References

[1]F. H. Stott, "The Oxidation of Alumina-Forming Alloys," *High Temp. Corros. Prot. Mater.*, **4 (Part 1)**, 251–2 (1997).

[2]F. H. Stott, "The Oxidation of Alumina-Forming Alloys," *High Temp. Corros. Prot. Mater.*, **4 (Part 2)**, 19–32 (1997).

[3]S. Chevalier, "Formation and Growth of Protective Alumina Scales"; in *Developments in High Temperature Corrosion and Protection of Materials*, Edited by W. Gao. CRC Press, Boca Raton, FL, 290–329 (2008).

[4]J. Jedlinski and G. Borchardt, "On the Oxidation Mechanism of Alumina Formers," *Oxid. Met.*, **36** [3–4] 317–37 (1991).

[5]R. Prescott and M. J. Graham, "The Formation of Aluminum Oxide Scales on High-Temperature Alloys," *Oxid. Met.*, **38** [3–4] 233–54 (1992).

[6]B. A. Pint and K. B. Alexander, "Grain Boundary Segregation of Cation Dopants in Alpha-Al_2O_3 Scales," *J. Electrochem. Soc.*, **145** [6] 1819–29 (1998).

[7]P. Y. Hou, "Impurity Effects on Alumina Scale Growth," *J. Am. Ceram. Soc.*, **86** [4] 660–8 (2003).

[8]J. Doychak, J. L. Smialek, and T. E. Mitchell, "Transient Oxidation of Single-Crystal Beta-NiAl," *Metall. Trans. A*, **20** [3] 499–518 (1989).

[9]P. Y. Hou, A. P. Paulikas, and B. W. Veal, "Growth Strains in Thermally Grown Al_2O_3 Scales Studied Using Synchrotron Radiation," *JOM*, **61** [7] 51–5 (2009).

[10]G. C. Rybicki and J. L. Smialek, "Effect of the Theta$\rightarrow$Alpha-Al_2O_3 Transformation on the Oxidation Behavior of Beta-NiAl+Zr," *Oxid. Met.*, **31** [3–4] 275–304 (1989).

[11]J. C. Yang, E. Schumann, I. Levin, and M. Ruhle, "Transient Oxidation of NiAl," *Acta Mater.*, **46** [6] 2195–201 (1998).

[12]C. Wagner, "Contributions to the Theory of the Tarnishing Process," *Z. Phys. Chem.*, **B21**, 25–41 (1933).

[13]F. A. Kröger and H. J. Vink, "Relations between Concentrations of Imperfections in Crystalline Solids"; pp. 307–435 in *Solid State Physics: Advances and Applications*, Vol. 3, Edited by F. Seitz, and D. Turnbull. Academic Press, New York, 1956.

[14]A. H. Heuer, "Oxygen and Aluminum Diffusion in Alpha-Al_2O_3: How Much Do We Really Understand?" *J. Eur. Ceram. Soc.*, **28** [7] 1495–507 (2008).

[15]W. J. Quadakkers, A. Elschner, W. Speier and H. Nickel, "Composition and growth mechanisms of alumina scales on FeCrAl-based alloys determined by SNMS," *Appl. Surf. Sci.*, **52** [4] 271–87 (1991).

[16]D. R. Clarke, "The Lateral Growth Strain Accompanying the Formation of a Thermally Grown Oxide," *Acta Mater.*, **51** [5] 1393–407 (2003).

[17]T. Nakagawa, I. Sakaguchi, N. Shibata, K. Matsunaga, T. Mizoguchi, T. Yamamoto, H. Haneda, and Y. Ikuhara, "Yttrium Doping Effect On Oxygen Grain Boundary Diffusion in Alpha-Al_2O_3," *Acta Mater.*, **55** [19] 6627–33 (2007).

[18]D. Naumenko, B. Gleeson, E. Wessel, L. Singheiser, and W. J. Quadakkers, "Correlation between the Microstructure, Growth Mechanism, and Growth Kinetics of Alumina Scales on a FeCrAlY Alloy," *Metall. Mater. Trans. A*, **38A** [12] 2974–83 (2007).

[19]J. D. Cawley, J. W. Halloran, and A. R. Cooper, "Oxygen Tracer Diffusion in Single-Crystal Alumina," *J. Am. Ceram. Soc.*, **74** [9] 2086–92 (1991).

[20]K. P. D. Lagerlof, T. E. Mitchell, and A. H. Heuer, "Lattice Diffusion Kinetics in Undoped and Impurity-Doped Sapphire (Alpha-Al_2O_3)—A Dislocation Loop Annealing Study," *J. Am. Ceram. Soc.*, **72** [11] 2159–71 (1989).

[21]T. Nakagawa, A. Nakamura, I. Sakaguchi, N. Shibata, K. P. D. Lagerlof, T. Yamamoto, H. Haneda, and Y. Ikuhara, "Oxygen Pipe Diffusion in Sapphire Basal Dislocation," *J. Ceram. Soc. Jpn.*, **114** [1335] 1013–7 (2006).

[22]Y. Oishi, K. Ando, and Y. Kubota, "Self-Diffusion of Oxygen in Single-Crystal Alumina," *J. Chem. Phys.*, **73** [3] 1410–2 (1980).

[23]Y. Oishi, K. Ando, N. Suga, and W. D. Kingery, "Effect of Surface Condition on Oxygen Self-Diffusion Coefficients for Single-Crystal Al₂O₃," *J. Am. Ceram. Soc.*, **66** [8] C130–1 (1983).

[24]D. Prot and C. Monty, "Self-Diffusion in Alpha-Al₂O₃. 2. Oxygen Diffusion in 'Undoped' Single Crystals," *Philos. Mag. A*, **73** [4] 899–917 (1996).

[25]K. P. R. Reddy and A. R. Cooper, "Oxygen Diffusion in Sapphire," *J. Am. Ceram. Soc.*, **65** [12] 634–8 (1982).

[26]K. P. D. Lagerlof and R. W. Grimes, "The Defect Chemistry of Sapphire (Alpha-Al₂O₃)," *Acta Mater.*, **46** [16] 5689–700 (1998).

[27]K. Matsunaga, T. Tanaka, T. Yamamoto, and Y. Ikuhara, "First-Principles Calculations of Intrinsic Defects in Al₂O₃," *Phys. Rev. B*, **68** [8] 085110, 9pp (2003).

[28]N. D. M. Hine, K. Frensch, W. M. C. Foulkes, and M. W. Finnis, "Supercell Size Scaling of Density Functional Theory Formation Energies of Charged Defects," *Phys. Rev. B*, **79** [2] 024112 (2009).

[29]N. D. M. Hine, P. D. Haynes, A. A. Mostofi, and M. C. Payne, "Linear-Scaling Density-Functional Simulations of Charged Point Defects in Al₂O₃ Using Hierarchical Sparse Matrix Algebra," *J. Chem. Phys.*, **133** [11] 114111, 12pp (2010).

[30]F. A. Kröger, "Defect Related Properties of Doped Alumina," *Solid State Ionics*, **12**, 189–99 (1984).

[31]A. H. Heuer and K. P. D. Lagerlof, "Oxygen Self-Diffusion in Corundum (Alpha-Al₂O₃): A Conundrum," *Philos. Mag. Lett.*, **79** [8] 619–27 (1999).

[32]A. E. Paladino and W. D. Kingery, "Aluminum Ion Diffusion in Aluminum Oxide," *J. Chem. Phys.*, **37**, 957–62 (1962).

[33]M. LeGall, B. Lesage, and J. Bernardini, "Self-Diffusion in Alpha-Al₂O₃. 1. Aluminum Diffusion in Single-Crystals," *Philos. Mag. A*, **70** [5] 761–73 (1994).

[34]A. Fielitz, G. Borchardt, S. Ganschow, R. Bertram, and A. Markwitz, "Al-26 Tracer Diffusion in Titanium Doped Single Crystalline Alpha-(Al₂O₃)," *Solid State Ionics*, **179** [11–12] 373–9 (2008).

[35]D. Prot, M. LeGall, B. Lesage, A. M. Huntz, and C. Monty, "Self-Diffusion in Alpha-Al₂O₃. 4. Oxygen Grain-Boundary Self-Diffusion in Undoped and Yttria-Doped Alumina Polycrystals," *Philos. Mag. A*, **73** [4] 935–49 (1996).

[36]K. P. R. Reddy, "Oxygen diffusion in close packed oxides"; Ph.D. thesis, Materials Science And Engineering, Case Western Reserve University, 1979.

[37]T. Nakagawa, "Atomic Diffusion Along Lattice Defects in Alumina Ceramics"; Ph.D. thesis, Department of Materials Science, The University of Tokyo, 2009.

[38]I. Sakaguchi, V. Srikanth, T. Ikegami, and H. Haneda, "Grain-Boundary Diffusion of Oxygen in Alumina Ceramics," *Journal of the American Ceramic Society*, **78** [9] 2557–9 (1995).

[39]D. Clemens, K. Bongartz, W. J. Quadakkers, H. Nickel, H. Holzbrecher, and J. S. Becker, "Determination of Lattice and Grain-Boundary Diffusion-Coefficients in Protective Alumina Scales on High-Temperature Alloys Using SEM, TEM and SIMS," *Fresenius J. Anal. Chem.*, **353** [3–4] 267–70 (1995).

[40]K. Messaoudi, A. M. Huntz, and B. Lesage, "Diffusion and Growth Mechanism of Al₂O₃ Scales on Ferritic Fe–Cr–Al Alloys," *Mater. Sci. Eng. A*, **247** [1–2] 248–62 (1998).

[41]S. Kitaoka, T. Matsudaira, and M. Wada, "Mass-Transfer Mechanism of Alumina Ceramics Under Oxygen Potential Gradients at High Temperatures," *Mater. Trans.*, **50** [5] 1023–31 (2009).

[42]T. Matsudaira, M. Wada, T. Saitoh, and S. Kitaoka, "The Effect of Lutetium Dopant on Oxygen Permeability of Alumina Polycrystals Under Oxygen Potential Gradients at Ultra-High Temperatures," *Acta Mater.*, **58** [5] 1544–53 (2010).

[43]H. Yoshida, Y. Ikuhara, and T. Sakuma, "High-Temperature Creep Resistance in Rare-Earth-Doped, Fine-Grained Al₂O₃," *J. Mater. Res.*, **13** [9] 2597–601 (1998).

[44]J. Cho, C. M. Wang, H. M. Chan, J. M. Rickman, and M. P. Harmer, "Role of Segregating Dopants on the Improved Creep Resistance of Aluminum Oxide," *Acta Mater.*, **47** [15–16] 4197–207 (1999).

[45]D. Nicolas-Chaubet, A. M. Huntz, and F. Millot, "Electrochemical Method for the Investigation of Transport Properties of Alumina Scales Formed by Oxidation," *J. Mater. Sci.*, **26**, 6119–26 (1991).

[46]T. A. Ramanarayanan, M. Raghavan, and R. Petkovic-Luton, "The Characteristics of Alumina Scales Formed on Fe-Based Yttria-Dispersed Alloys," *Mater. High Temp.*, **20** [3] 261–71 (1984).

[47]V. K. Tolpygo and D. R. Clarke, "Microstructural Evidence for Counter-Diffusion of Aluminum and Oxygen During the Growth of Alumina Scales," *Mater. High Temp.*, **20** [3] 261–71 (2003).

[48]J. A. Nychka and D. R. Clarke, "Quantification of Aluminum Outward Diffusion During Oxidation of FeCrAl Alloys," *Oxid. Met.*, **63** [5–6] 325–52 (2005).

[49]B. W. Veal and A. P. Paulikas, "Growth Strains and Creep in Thermally Grown Alumina: Oxide Growth Mechanisms," *J. Appl. Phys.*, **104** [9] 093525, 15pp (2008).

[50]B. W. Veal, A. P. Paulikas, B. Gleeson, and P. Y. Hou, "Creep in α-Al₂O₃ Thermally Grown on β-NiAl and NiAlPt Alloys," *Surf. Coat. Technol.*, **202** [4–7] 608–12 (2007).

[51]A. A. Sokol, A. Walsh, and C. R. A. Catlow, "Oxygen Interstitial Structures in Close-Packed Metal Oxides," *Chem. Phys. Lett.*, **492** [1–3] 44–8 (2010).

[52]R. H. Doremus, "Oxidation of Alloys Containing Aluminum and Diffusion in Al₂O₃," *J. Appl. Phys.*, **95** [6] 3217–22 (2004).

[53]R. H. Doremus, "Reply to Comment by Pint and Deacon on Oxidation of Alloys Containing Aluminum and Diffusion in Al₂O₃," *J. Appl. Phys.*, **97** [11] 116112, 1pp (2005). □

J. Am. Ceram. Soc., **94** [S1] S154–S159 (2011)

DOI: 10.1111/j.1551-2916.2011.04405.x

© 2011 The American Ceramic Society

journal

Interfacial Stoichiometry and Adhesion at Metal/α-Al₂O₃ Interfaces

Hong-Tao Li,[‡,§,¶] Lian-Feng Chen,[‡,§] Xun Yuan,[‡,§] Wen-Qing Zhang,[†,‡] John R. Smith,[†,∥] and Anthony G. Evans[✠]

[‡]State Key Laboratory of High Performance Ceramics and Superfine Microstructures, Shanghai Institute of Ceramics, Chinese Academy of Sciences, Shanghai 200050, China

[§]Graduate School of the Chinese Academy of Sciences, Beijing 200049, China

[¶]Entry-Exit Inspection & Quarantine Bureau, Shanghai 200135, China

[∥]Department of Materials Science and Engineering, University of Michigan, Ann Arbor, Michigan 48109

First-principles studies of metal/α-Al₂O₃ interfaces have revealed strong interfacial stoichiometry effects on adhesion. The metals included Al, Ni, Cu, Au, Ag, Rh, Ir, Pd, Pt, Nb, and β-NiAl. Metallic and ionic-covalent adhesive bonding effects were found in varying amounts depending on whether the interfacial stoichiometry is stoichiometric, oxygen-rich, or aluminum-rich in a qualitative way. A semiempirical but physically sensible understanding ensues for the effects of interfacial stoichiometry and reveals the underlying strong correlation of the interfacial adhesion with the physical properties of the bulk materials that join and form the interface. The metallic component of the bonding was found to be related to the ratio of $(B/V)^{1/2}$, where B and V are the bulk modulus and molar volume of the metal, respectively. In like manner, the ionic-covalent component of the bonding could be related to the enthalpy of oxide formation of the bulk metal. A unified model is proposed to describe the adhesion of metal/alumina interfaces with interfacial stoichiometry effects, and the model is also expected to be valid for other metal–oxide interfaces.

I. Introduction

ADHESION at metal/alumina interfaces is important in applications such as thermal barrier coatings for high-temperature gas-turbine engines,[1,2] heterogeneous catalysis,[3–5] microchip packaging,[6] and corrosion protection.[7–10] This has resulted in experimental and theoretical research in the fundamentals of adhesion at metal/alumina interfaces.[11–37] It has been found[14–16,23,38–40] that interfacial stoichiometries as well as interfacial adhesion can vary with the temperature and the environmental oxygen partial pressure. That is, the stable metal/alumina interface can vary from stoichiometric (two Al atoms for every three O atoms), to aluminum-rich or oxygen-rich.[32–36]

Recently we (H. Li *et al.*, unpublished data) have analyzed interfacial stoichiometry effects for a series of metal/α-Al₂O₃ interfaces via the results of first-principles computations. The metals considered included Al, Ni, Cu, Au, Ag, Rh, Ir, Pd, Pt, Nb, and β-NiAl. For each of the metal/α-Al₂O₃ interfaces, the first-principles computations were carried out not only for the

M. Rühle—contributing editor

Manuscript No. 28583. Received September 9 2010; approved December 30 2010.
This work is supported by the NSFC (Grant Nos. 50825205, 50821004, and 50820145203), the National 973 Project under Grant No. 2007CB607500, and the CAS Project under Grant No. KGCX2-YW-206.
[†]Authors to whom correspondence should be addressed. e-mails: wqzhang@mail.sic.ac.cn and johnjrspop@aol.com
[✠]Materials Department, University of California, Santa Barbara, California 93106 (deceased).

stoichiometric interface but also for an aluminum-rich and an oxygen-rich interface.

The adhesive bonding for each stoichiometry was found (H. Li *et al.*, unpublished data) to have features common to all eleven metal/α-Al₂O₃ interfaces. Aluminum-rich interfaces exhibit metallic bonding. On the other hand, bonding in oxygen-rich interfaces is primarily ionic, with an admixture of covalent bonding. Stoichiometric interfaces have mixed ionic-covalent and metallic adhesive bonding. It was found that the work of separation, W_{sep}, is sensitive to the interfacial stoichiometry. The work of separation, W_{sep}, of an interface is defined as the energy needed to separate the interface into two free surfaces,[12] and can be expressed by Dupré equation:

$$W_{\text{sep}} = \sigma_A + \sigma_B - \gamma_{A/B} \tag{1}$$

Here, $\gamma_{A/B}$ is the interfacial energy of the interface A/B. σ_A and σ_B are the surface energies of slabs A and B after the cleavage of the interface A/B, respectively. Oxygen-rich interfaces were found to have W_{sep} values approaching an order of magnitude larger than those of stoichiometric interfaces in some cases, with W_{sep} values of aluminum-rich interfaces lying in between. This was understood in terms of the type of bonding for each stoichiometry, as will be discussed below. It is qualitatively understood via Eq. (1) as follows. In general, α-Al₂O₃ surface energies for oxygen-rich and aluminum-rich surfaces are larger than that for the stoichiometric α-Al₂O₃ surface. Also, generally the aluminum-rich and oxygen-rich interfacial energies are smaller than the stoichiometric interfacial energy.

Most of the earlier work including our work mainly devoted to determining the effects of interfacial stoichiometry on the nature of the interfacial bonds for different metal/α-Al₂O₃ interfaces and connecting that knowledge to the works of separation in a qualitative picture. Interfacial bonds are characterized there via first-principles results for interfacial atomic structures, electron density distributions, and local densities of electronic states, as well as with works of separation. In contrast, no correlation between interfacial adhesion and the physical properties of the bulk materials forming interface has been investigated in a quantitative and systematic way, even with a few reported relationship linking the measured adhesion data to bulk properties of host materials for the one specific interface that were most likely observed experimentally. As a matter of fact, we now understood that metal/oxide interface show stoichiometry-dependent stable structures and that the adhesion are also critically related to interfacial stoichiometry. Therefore, classifying those interfaces into different categories and analyzing their adhesion properties are truly required but can only be done with the help of a systematic theoretical approach. This is the main purpose of the current work. Generally, the physically sensible analytical relationships ensue between works of separation for each interfacial stoichiometry and measured bulk properties of metals and metal oxides, and provide not only enhanced understanding of interfacial stoichiometric effects but also the potential for semi-empirical predictions of stoichiometry effects on works of separation for interfaces for which first-principles results are not available.

Here it will be seen for the first time that the dependence of the metal/alumina works of separation on typical stoichiometries can be expressed in terms of a semiempirical but physically sensible analytical relationships in a quantitative way. Specifically, the works of separation are expressed in terms of the bulk properties of the metal and metal oxide. These relationships will be shown to accurately represent metal/alumina bonding for all metals considered, which in fact reveals clearly the underlying strong correlation of the interfacial adhesion with the physical properties of the bulk materials that join and form the interface.

II. First-Principles Works of Separation

W_{sep} is determined via first-principles computations[35,36] according to:

$$W_{sep} = (E_{metal} + E_{Al_2O_3} - E_{metal/Al_2O_3})/2S \tag{2}$$

where E_{metal} and $E_{Al_2O_3}$ are the total energies of the isolated metal and α-Al₂O₃ slabs, respectively. The term E_{metal/Al_2O_3} represents the total energy of the system when the metal and the α-Al₂O₃ surfaces have been allowed to come together and adhere at equilibrium separation. For the first-principles computations, the Perdew–Burke–Ernzerhof generalized gradient approximation[41] (GGA) was used within the Vienna *ab initio* simulation package[42,43] (VASP). The projector augmented plane wave method[44,45] with a plane wave energy cut off of 500 eV and an energy convergence criterion of 10^{-4} eV is used. Readers interested in additional details about the density-functional computational details used here can read about them in Zhang and colleagues.[33–36,46] All atoms of the system have been allowed to relax, although the stoichiometry of the metal/α-Al₂O₃ interface is retained in the α-Al₂O₃ surface. This applies to fracture experiments, where bond separation rates are sufficiently high that the surfaces cannot relax to the ground-state stoichiometry before bonds rupture.[35] For the free α-Al₂O₃ (0001) surface, the ground-state stoichiometry has been shown[47,48] to be stoichiometric. For steady-state experiments like sessile drop experiments,[14–19] the works of adhesion, W_{ad}, can be obtained from Eq. (2) by using a stoichiometric surface for computations of $E_{Al_2O_3}$.

Here, the stoichiometric metal/alumina interface refers to an α-Al₂O₃ surface terminated by a single Al atomic layer and then joined to a metal (Al-terminated), denoted as $M/(Al_2O_3)_{Al1}$. The aluminum-rich case corresponds to an α-Al₂O₃ surface terminated by two Al atomic layers (Al₂-terminated), denoted as $M/(Al_2O_3)_{Al2}$. Finally the oxygen-rich case corresponds to the α-Al₂O₃ surface terminated by an O atomic layer (O-terminated), denoted as $M/(Al_2O_3)_O$. The Al₂-, O-, and Al-terminated interfaces are representative types that span the variety of expected chemistries.[14–16,38–40]

First-principles results for W_{sep} are found in Table I. Note the smallest W_{sep} values are found for the stoichiometric interface for all the metals in general. Also in all cases the oxygen-rich interfaces have the largest values of W_{sep}. This difference due to stoichiometry is large, in some cases an order of magnitude. These can perhaps be understood in terms of the dangling bonds available from the $(Al_2O_3)_O$ surface and not found at the $(Al_2O_3)_{Al1}$ surface. The aluminum-rich interfaces have W_{sep} intermediate values between those for the oxygen-rich and stoichiometric interfaces, respectively.

Łodziana and Norskov[28] performed density functional computations of the adhesion of a monolayer (ML) of Pd or Cu atoms on α-Al₂O₃ surfaces including aluminum-rich, oxygen-

Table I. Calculated Works of Separation, W_{sep}, for Metal (Alloy)/α-Al₂O₃ Interfaces

Interface	Works of separation, W_{sep} (J/m²) Calculated results		
	$M/(Al_2O_3)_{Al2}$	$M/(Al_2O_3)_O$	$M/(Al_2O_3)_{Al1}$
Rh/α-Al₂O₃			
Type-I	4.32[†]	5.91[†]	1.05[†]
Ir/α-Al₂O₃			
Type-I	4.47[†]	6.51[†]	0.89[†]
Pd/α-Al₂O₃			
Type-I	4.38[†]	4.35[†]	0.76[†]
ML	5.14[‡]	4.21[‡]	0.99[‡]
Pt/α-Al₂O₃			
Type-I	4.65[†]	4.76[†]	0.74[†]
Ni/α-Al₂O₃			
Type-I	3.78[§]	6.84[§]	1.30[§]
Type-II	4.05[§]	6.55[§]	1.13[§]
Type-III	3.64[§]	6.75[§]	1.09[§]
Cu/α-Al₂O₃			
Type-I	2.66[§]	5.94[§], 5.62[¶]	0.58[§]
Type-II	2.71[§]	5.42[§]	0.74[§]
Type-III	2.69[§]	6.07[§]	0.86[§]
ML	3.26[‡]	6.18[‡]	0.91[‡]
Au/α-Al₂O₃			
Type-I	2.31[‖]	2.78[‖]	0.29[‖]
Type-II	2.42[‖]	3.08[‖]	0.21[‖]
Ag/α-Al₂O₃			
Type-I	1.83[‖]	3.93[‖]	0.33[‖]
Type-II	1.85[‖]	4.10[‖]	0.32[‖]
Al/α-Al₂O₃			
Type-I	1.43[††]	10.10[††], 8.67[¶], 9.73[‡‡]	1.08[††], 1.36[¶], 1.06[‡‡]
Nb/α-Al₂O₃			
Type-III	2.80[§§]	9.80[§§], 10.60[¶¶]	2.70[§§], 2.60[¶¶]
β-NiAl/(Al₂O₃)_{Al2}			
Al₂*term*–1Al	2.85[‖‖‖]		
β-NiAl–Ni^{rich}/ (Al₂O₃)_{Al2}			
O*term*–8Al	3.43[‖‖‖]		

The type-I, type-II, and type-III are for the metal/α-Al₂O₃ interfaces using different interfacial matchings and strains as discussed in H. Li *et al.* (unpublished data). The data denoted by "ML" are from the first-principles results for the adsorption of a monolayer of metal atoms on alumina.[28] The values from GGA calculations[29,36] are adopted for Al, Au, and Ag/α-Al₂O₃ interfaces. [†]This work. [‡]Łodziana and Norskov.[28] [§]Zhang *et al.*[35] [¶]Batyrev and Kleinman.[30] [‖]Feng *et al.*[36] [††]Zhang and Smith.[33] [‡‡]Siegel *et al.*[29] [§§]Batyrev *et al.*[32] [¶¶]Zhang and Smith.[34] [‖‖‖]Li *et al.*[46] ML, monolayer; GGA, generalized gradient approximation.

rich, and stoichiometric configurations. Most of the calculated W_{sep} data for 1 ML of metal atoms on alumina are larger than those for thick metal slabs on alumina, but one can still see a similar trend of W_{sep} with interface chemistry, i.e., the Al₂- to O-, to Al-terminated cases, as shown in Table I for the ML on alumina results denoted as ML.

III. Analytical Relationships for Stoichiometry-Dependent Works of Separation

As adhesion of the metal/α-Al₂O₃ interfaces has been shown to vary significantly with interfacial stoichiometry, it is of interest to seek an enhanced understanding and a further connection with experiment of this effect. It's difficult to carry out direct experimental probes of these stoichiometry-dependent interfacial bonds. One can attempt to develop semiempirical relationships between experimental quantities and stoichiometry-dependent works of separation, however. Armed with the knowledge of how the nature of the bonding varies with stoichiometry, one can proceed. That is, bonding is predominantly

Table II. The Electron Density Parameter at the Boundary of the Wigner–Seitz Atomic cEll, n_{ws} [Density Units (d.u.)], Estimated for the Pure Element Phase in the Metallic State[49,50]

Metal	n_{ws}	Metal	n_{ws}	Metal	n_{ws}	Alloys	n_{ws}
Rh	5.45	Cu	3.18	Nb	4.41	β-NiAl	4.02
Ir	6.13	Au	3.87	Co	5.36	β-NiAl–Ni^{rich}	5.36
Pd	4.66	Ag	2.52	Fe	5.55		
Pt	5.64	Al	2.69	Mn	4.17		
Ni	5.36						

For the β-Ni$_{1-x}$Al$_x$/α-Al$_2$O$_3$ interfaces,[46] n_{ws} is approximately taken to be the average value of the parameters for pure Ni and Al to evaluate the W_{sep} of the Al$_2$term–1Al interface, and the n_{ws} of Ni is used to evaluate W_{sep} of the Oterm–8Al interface due to the presence of Ni-rich layer in the β-NiAl–Ni^{rich} slab (see Table I).

metallic for the aluminum-rich interfaces, ionic-covalent for the oxygen-rich interfaces, and mixed ionic–covalent and metallic for the stoichiometric interfaces.

(1) Aluminum-Rich M/α-Al₂O₃ Interfaces

Due to the predominately metallic bonding of aluminum-rich interfaces, one might look to relate W_{sep} to the electron density of the interface. Boer and colleagues[49,50] once used an electron density parameter, n_{ws}, to evaluate the thermodynamic quantities of binary alloys. They defined n_{ws} as the electron density at the boundary of the Wigner-Seitz atomic cell estimated for the pure element phase in the metallic state. Those authors[49,50] estimated n_{ws} as the ratio of experimental bulk modulus B and molar volume V of pure metals, i.e.,

$$n_{ws} = \left(\frac{B}{V}\right)^{1/2} \tag{3}$$

in which the unit of n_{ws} is defined as density units (d.u.) in Boer et al.[49] and Niessen and Miedema[50] as shown in Table II.

Getting back to the predominantly metallic bonding of the Al$_2$-terminated M/α-Al$_2$O$_3$ interface, one might look for the adhesion to scale with the electron density parameter of the corresponding metal, i.e.,

$$W_{sep-Al2} = A n_{ws} \tag{4}$$

in which A is a scaling parameter. Indeed, Fig. 1 is a plot of the calculated works of separation, $W_{sep-Al2}$, as a function of n_{ws} for Al$_2$-terminated interfaces. The solid line is obtained by fitting to the adhesion results for the type-I M/(Al$_2$O$_3$)$_{Al2}$ interfaces of Table

I, and the corresponding parameter A is estimated to be 0.76 [(J/m^2)/(d.u.)]. It is very interesting to notice that there is an approximate proportionality between n_{ws} and $W_{sep-Al2}$ for the Al$_2$-terminated interfaces, even for the relatively broad range of interfaces of Fig. 1. Physically, this is due to the metallic character of the aluminum-rich interface bonding. Figure 1 shows that the higher n_{ws} is, the stronger the $W_{sep-Al2}$ of an aluminum-rich interface.

Considering that the chemical bonding of the stable β-Ni$_{1-x}$Al$_x$/α-Al$_2$O$_3$ interfaces is also primarily metallic in both the Al$_2$term–1Al and Oterm–8Al interfaces,[46] the corresponding works of separation are also given in Table I and plotted in Fig. 1. In a simple approximation, the n_{ws} for the β-NiAl alloy is taken to be the average value of the parameters for pure Ni and Al to evaluate the W_{sep} of the Al$_2$term–1Al interface via Eq. (4). The n_{ws} of Ni is used to estimate the adhesion of the Oterm–8Al interface due to the presence of Ni-rich layer in the β-NiAl–Ni^{rich} slab (Table II).[46] Fig. 1 shows that the works of separation for both the β-NiAl/(Al$_2$O$_3$)$_{Al2}$ and β-NiAl–Ni^{rich}/(Al$_2$O$_3$)$_{Al2}$ interfaces also follow Eq. (4) very well, with the same scaling parameter A. In fact, an approximate proportionality between surface energies of metals and electron density parameters was also observed by Boer and colleagues.[49–51] This is consistent with the picture that both the surface energies of metals and adhesive energies for interfaces between metals and aluminum-rich α-Al$_2$O$_3$ (0001) surface involve predominantly metallic bonding. Therefore, it is reasonable to expect that Eqs. (3) and (4) are appropriate for estimating adhesion of other aluminum-rich M/α-Al$_2$O$_3$ interfaces. At this time, the physical meaning of the scaling parameter A is not understood. However, as the correlation exhibited in Fig. 1 covers many metal/α-Al$_2$O$_3$ systems, it is speculated to be closely linked with the metallic bonding of the Al$_2$-terminated interfaces as well as the character of electronic structure of the aluminum-rich α-Al$_2$O$_3$ (0001) surface.

(2) Oxygen-Rich M/α-Al₂O₃ Interfaces

For the oxygen-rich, clean α-Al$_2$O$_3$ (0001) surface, through Mulliken population analyses,[47,52] it is found that one of the surface O atoms per surface unit cell carries an electronic charge of only −0.58, compared with −1.0 for an O atom in the bulk oxide. This leads to unsaturated dangling bonds on the O-terminated α-Al$_2$O$_3$ surface. As discussed earlier, strong ionic-covalent interfacial bonds form for the oxygen-rich M/α-Al$_2$O$_3$ interface, and this is the primary reason why W_{sep} of such M/(Al$_2$O$_3$)$_O$ interfaces are always the largest in comparison with those of the aluminum-rich and stoichiometric interfaces.

This feature of the metal–O interaction at the oxygen-rich interface is similar to that found in the corresponding metal oxides. This is suggestive that the work of separation of the O-terminated interface, W_{sep-O}, might be expected to correlate with the experimental enthalpy of formation of the bulk metal oxide per oxygen atom, $|EOF|$ (Table III), in the following way:

$$W_{sep-O} = C + D|EOF| \tag{5}$$

here, C and D are scaling parameters. Indeed, the W_{sep-O} correlate with the $|EOF|$ of metal oxides well as shown in Fig. 2. By fitting all the adhesion data for the type-I M/(Al$_2$O$_3$)$_O$ interfaces in Table I, the C and D are estimated to be 3.87 J/m^2 and 1.01

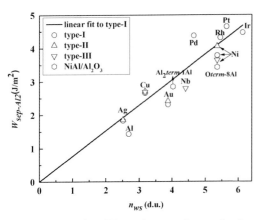

Fig. 1. Work of separation ($W_{sep-Al2}$) versus electron density parameter (n_{ws}) for the aluminum-rich metal (alloy)/alumina interfaces. The $W_{sep-Al2}$ and n_{ws} are given in Tables I and II, respectively. The solid line corresponds to the linear fit to all the data for type-I M/(Al$_2$O$_3$)$_{Al2}$ interfaces. The W_{sep} corresponding to the cleavage β-NiAl/(Al$_2$O$_3$)$_{Al2}$ for the Al$_2$term–1Al interface, and β-NiAl-Ni^{rich}/(Al$_2$O$_3$)$_{Al2}$ for the Oterm–8Al are also plotted in the figure.[46]

Table III. Experimental Enthalpies of Formation [EOF in eV/(O Atom)] of Metal Oxides

Metal oxide	EOF [eV/(O Atom)]	Metal oxide	EOF [eV/(O Atom)]	Metal oxide	EOF [eV/(O Atom)]
Rh$_2$O$_3$	−1.23	CuO	−1.68	NbO	−4.35
IrO$_2$	−1.29	Au$_2$O$_3$	−0.01	CoO	−2.46
PdO	−1.20, −0.89	Ag$_2$O	−0.32	Fe$_2$O$_3$	−2.85
PtO$_2$	−0.87	Al$_2$O$_3$	−5.79	MnO	−3.99
NiO	−2.49				

The data are derived from Kubaschewski et al.[53] and Lide.[54]

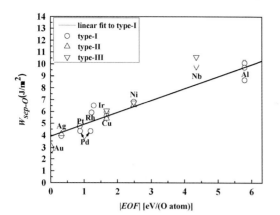

Fig. 2. Work of separation (W_{sep-O}) versus enthalpy of formation ($|EOF|$) for the oxygen-rich metal/alumina interfaces. The $|EOF|$ are listed in Table II. All W_{sep-O} are taken from Table I, and the solid line is a linear fit to all the adhesion results for type-I $M/(Al_2O_3)_O$ interfaces.

$[(J/m^2)/(eV/O \text{ atom})]$, respectively. Figure 2 shows that the W_{sep-O} of an interface increases with the increasing of $|EOF|$ of the corresponding metal oxide. This is consistent with the physical picture that the metal–oxygen interaction is important for the adhesion of oxygen-rich interfaces. Note, however, that W_{sep-O} via Eq. (5) is $\geq 3.87 \text{ J/m}^2$, even for small $|EOF|$. This is likely due to Al–O–metal effects beyond simple metal–O effects.

(3) Stoichiometric M/α-Al₂O₃ Interfaces

The chemistry of the stoichiometric metal/alumina interfaces falls in between the aluminum-rich and oxygen-rich cases. It contains one Al layer between the adhered metal atoms and the oxygen layer. The single layer of Al atoms does not make the Al-terminated metal/alumina interface metallic like the aluminum-rich interface. The $M/(Al_2O_3)_{All}$ interfaces in fact show a mix of metallic and ionic-covalent bonding. Based on the above discussion of the adhesion of the aluminum-rich and oxygen-rich interfaces, especially the two models described by Eqs. (4) and (5), one might reasonably expect that a combination of n_{ws} and $|EOF|$ could be used to describe the works of separation,

$W_{sep-All}$, for the $M/(Al_2O_3)_{All}$ interfaces. Therefore $W_{sep-All}$ is formulated by the following equation:

$$W_{sep-All} = En_{ws} + F|EOF| \qquad (6)$$

By fitting all the stoichiometric adhesion results from first-principles calculations in Table IV, the scaling parameters E and F are estimated to be 0.13 [(J/m²)/(d.u.)] and 0.15 [(J/m²)/(eV/O atom)], respectively. One can see the expression of Eq. (6) for the stoichiometric interface is not a simple linear combination of those expressions of Eqs. (4) and (5) for aluminum-rich and oxygen-rich interfaces because of the relatively large constant (3.87 J/m²) of Eq. (5) not found in Eq. (6). Again it is suggested that Al–metal–O interactions are not captured by n_{ws} and $|EOF|$.

In order to check the validity of the model, results from Eq. (6), from first-principles calculations, and from experimental measurements are given in Table IV. It should be pointed out that the metals in Table IV are not only of fcc structure, but also hcp (Co), bcc (Fe and Nb), and complex (Mn).[57] The results are also compared in Figs. 3 and 4 for clarity. The above results show clearly that there is a correlation described by Eq. (6), and this model reproduces well the variation of works of separation, no matter whether comparing to the first-principles calculations for the Al-terminated metal/alumina interfaces or to experimental measurements.

Comparing the scaling parameters obtained for the differently terminated interfaces, the parameter multiplying the n_{ws} term in Eq. (6) is smaller than the equivalent term in Eq. (4), i.e., $E < A$. This is consistent with the physical picture that the metallic bonding in the aluminum-rich interface is stronger than that of the stoichiometric interface. Likewise, the coefficient for the $|EOF|$ term in Eq. (6) is also smaller than the corresponding parameter in Eq. (5), i.e., $F < D$, which is consistent with the picture that the ionic-covalent bonding is stronger in the oxygen-rich interface.

Models[19,55,58–61] have been proposed to estimate the adhesion of Al-terminated metal/alumina interfaces, although none treated the effects of interfacial stoichiometry. Chatain et al.[19] proposed a phenomenological expression for estimates of W_{sep} of the Al-terminated $M/(Al_2O_3)_{All}$ interfaces, as follows:

$$W_{sep} = -\frac{c}{N^{1/3}V_{Me}^{2/3}}[\Delta \bar{H}_{O(Me)}^{\infty} + \frac{2}{3}\Delta \bar{H}_{Al(Me)}^{\infty}] \qquad (7)$$

Table IV. Calculated Works of Separation for Stoichiometric $M/(Al_2O_3)_{All}$ Interfaces, $W_{sep-All}$

Interfaces	By Eq. (6)	By Eq. (7)	By Li's model	Experimental data		By first-principles
Rh/(Al₂O₃)₍All₎	0.87	0.95	1.21		Type-I	1.05[†]
Ir/(Al₂O₃)₍All₎	0.97	0.82	1.36		Type-I	0.89[†]
Pd/(Al₂O₃)₍All₎	0.72,0.77	0.81	1.03	0.74[‡]	Type-I	0.76[†]
Pt/(Al₂O₃)₍All₎	0.85	0.53	1.25	0.90[§]	Type-I	0.74[†]
Ni/(Al₂O₃)₍All₎	1.04	1.16	1.19	1.19[‡]	Type-I	1.30[¶]
					Type-II	1.13[¶]
					Type-III	1.09[¶]
Cu/(Al₂O₃)₍All₎	0.65	0.58	0.70	0.49[‡]	Type-I	0.58[¶]
					Type-II	0.74[¶]
					Type-III	0.86[¶]
Au/(Al₂O₃)₍All₎	0.48		0.86	0.27[‡]	Type-I	0.29[††]
				0.56[‖]	Type-II	0.21[††]
Ag/(Al₂O₃)₍All₎	0.36	0.20	0.56	0.32[‡]	Type-I	0.33[††]
					Type-II	0.32[††]
Al/(Al₂O₃)₍All₎	1.20		0.60	0.95[‡], 1.13[‡‡]	Type-I	1.08[§§], 1.36[¶¶], 1.06[‖‖]
Nb/(Al₂O₃)₍All₎	1.20		0.98		Type-III	2.70[†††], 2.60[‡‡‡]
Co/(Al₂O₃)₍All₎	1.04	0.98	1.19	1.14[‡]		
Fe/(Al₂O₃)₍All₎	1.12	0.84	1.23	1.21[‡]		
Mn/(Al₂O₃)₍All₎	1.12	1.05	0.93	0.86–1.29[‡]		

First-principles calculations and model estimations using Eqs. (6) and (7), and Li's model[55] are all listed. Experimental results are also provided for comparison, which correspond to works of adhesion, (W_{ad}), for the $M/(Al_2O_3)_{All}$ interfaces as discussed in the paragraph below Eq. (2). [†]This work. [‡]Chatain et al.[17] [§]Saiz et al.[14] [¶]Zhang et al.[35] [‖]Li.[56] [††]Feng et al.[36] [‡‡]Lipkin et al.[56] [§§]Zhang and Smith.[33] [¶¶]Batyrev and Kleinman.[30] [‖‖]Siegel et al.[29] [†††]Batyrev et al.[32] [‡‡‡]Zhang and Smith.[34]

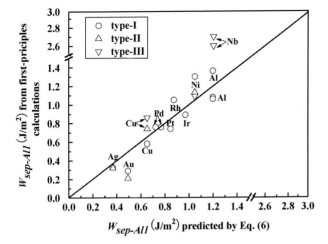

Fig. 3. $W_{\text{sep-All}}$ from first-principles calculations versus $W_{\text{sep-All}}$ predicted by Eq. (6) for the stoichiometric metal/alumina interfaces. All the data are listed in Table IV.

in which the empirical constant c was estimated to be 0.22, N is Avogadro's constant, and V_{Me} is the molar volume of the metal Me. $\Delta \bar{H}^{\infty}_{\text{O(Me)}}$ and $\Delta \bar{H}^{\infty}_{\text{Al(Me)}}$ are the enthalpies of mixing at infinite dilution of oxygen and aluminum atoms in the bulk metal Me, respectively.

Li[55] regarded the $W_{\text{sep-All}}$ as a function of the electron density parameter n_{ws}, although no straightforward expression was given. From Table IV, it is observed that most of the results obtained here and Chatain and colleagues' model predictions agree well with experimental data, while there are several large discrepancies for Li's model, such as the Pd-, Pt-, Cu-, Au-, Ag-, and Al-related systems. Considering the mixed ionic-covalent and metallic character of chemical bonding at the $M/(\text{Al}_2\text{O}_3)_{\text{All}}$ interfaces, the electron density parameter n_{ws} is not sufficient to describe the adhesion of the stoichiometric interface. The contribution from the metal–oxygen ionic–covalent bonding has to be considered, even though it is not as large.

(4) All Stoichiometries of M/α-Al₂O₃ Interfaces

Combining the results for the interfaces of different terminations, a simple, semiempirical relationship ensues for the work

of separation, applicable to all the metals and interfacial stoichiometries:

$$W_{\text{sep}} = a n_{\text{ws}} + b|EOF| + c \qquad (8)$$

The nature of the chemical bonding of metal/α-Al₂O₃ interfaces underlying Eq. (8) is now clear. The coefficients a–c are different for the different stoichiometries. However, the above comparisons with results of experiments and first-principles computations demonstrate clearly that there does exist a group of $(a$–$c)$ parameters to describe the adhesion of metals with α-Al₂O₃ for stoichiometric, aluminum-rich, and oxygen-rich interfacial stoichiometries. It is reasonable to expect that the model has some applicability to other interfaces between metals and oxides.

IV. Conclusions

Through analyses of interfacial stoichiometry, atomic relaxation, electronic structure, chemical bonding, and adhesion between α-Al₂O₃ and a series of metals including Al, Ni, Cu, Au, Ag, Rh, Ir, Pd, Pt, Nb, and β-NiAl, one can conclude the following. Chemical bonding is found to be metallic for the aluminum-rich interfacial stoichiometry, primarily ionic with some covalent contributions for the oxygen-rich stoichiometry, and of mixed ionic-covalent and metallic for the stoichiometric interfaces. Simple empirical relationships were identified. It was found that the metallic component of bonding scales with $(B/V)^{1/2}$, where B is the bulk modulus and V is molar volume of the metal. Similarly, it was found that the ionic-covalent component of bonding is equal to a constant plus a term that scales with the enthalpy of oxide formation $|EOF|$ of the metal. A linear combination of a constant, $(B/V)^{1/2}$, and the enthalpy of oxide formation, both of the latter of the metal, provides a unified relationship to evaluate the works of separation for all the stoichiometries: aluminum-rich, oxygen-rich, and stoichiometric metal/α-Al₂O₃ interfaces. This relationship is consistent with the results of first-principles calculations and experimental measurements for all three potential stoichiometries of the eleven metal/α-Al₂O₃ interfaces. These analytic functions of a constant, $(B/V)^{1/2}$, and $|EOF|$ suggest also that Al-metal–O interactions are important contributors to stoichiometric effects in bonding at these interfaces. In general, higher surface energies and lower interfacial energies lead to the higher works of separation found for O-rich and Al-rich interfaces.

Acknowledgment

The authors would like to acknowledge Prof. Dominique Chatain for providing the useful information of his model and experimental data. H. Li and W. Zhang thank the SCCAS for time on the Shengteng7000 supercomputer. J. R. Smith is supported for this work by the U.S. Office of Naval Research under Grant N00014-08-1-1164.

References

[1]A. Rabiei and A. G. Evans, "Failure Mechanisms Associated with the Thermally Grown Oxide in Plasma-Sprayed Thermal Barrier Coatings," *Acta Mater.*, **48**, 3963–76 (2000).
[2]N. P. Padture, M. Gell, and E. H. Jordan, "Materials Science-Thermal Barrier Coatings for Gas-Turbine Engine Applications," *Science*, **296**, 280–4 (2002).
[3]D. W. Goodman, "Model Studies in Catalysis Using Surface Science Probes," *Chem. Rev.*, **95**, 523–36 (1995).
[4]R. C. Santana, S. Jongpatiwut, W. E. Alvarez, and D. E. Resasco, "Gas-phase Kinetic Studies of Tetralin Hydrogenation on Pt/Alumina," *Ind. Eng. Chem. Res.*, **44**, 7928–34 (2005).
[5]F. Ahmed, M. K. Alam, A. Suzuki, M. Koyama, H. Tsuboi, N. Hatakeyama, A. Endou, H. Takaba, C. A. Del Carpio, M. Kubo, and A. Miyamoto, "Dynamics of Hydrogen Spillover on Pt/Gamma-Al₂O₃ Catalyst Surface: A Quantum Chemical Molecular Dynamics Study," *J. Phys. Chem. C*, **113**, 15676–83 (2009).
[6]C. C. Young, J. G. Duh, and C. S. Huang, "Improved Characteristics of Electroless Cu Deposition on Pt-Ag Metallized Al₂O₃ Substrates in Microelectronics Packaging," *Surf. Coat. Technol.*, **145**, 215–2 (2001).
[7]H. J. Grabke, D. Wiemer, and H. Viefhaus, "Segregation of Sulfur During Growth of Oxide Scales," *Appl. Surf. Sci.*, **47**, 243–50 (1991).

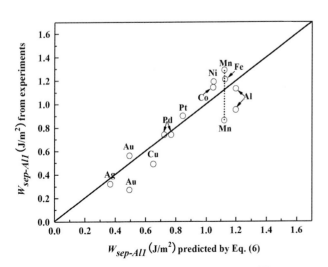

Fig. 4. $W_{\text{sep-All}}$ by experimental measurements versus $W_{\text{sep-All}}$ predicted by Eq. (6) for the stoichiometric metal/alumina interfaces. The slope of solid line is unity. All the data are listed in Table IV, and the short vertical dotted line designates the range of experimental data[17] for the $W_{\text{sep-All}}$ of $Mn/(Al_2O_3)_{All}$. *Experimental data correspond to the work of adhesion (W_{ad}) for the M/(Al₂O₃)$_{\text{All}}$ interfaces as discussed in the paragraph below Eq. (2).*

[8]H. J. Grabke, G. Kurbatov, and H. J. Schmutzler, "Segregation Beneath Oxide Scales," *Oxid. Met.*, **43**, 97–114 (1995).

[9]J. A. Haynes, B. A. Pint, K. L. More, Y. Zhang, and I. G. Wright, "Influence of Sulfur, Platinum, and Hafnium on the Oxidation Behavior of CVD NiAl Bond Coatings," *Oxid. Met.*, **58**, 513–44 (2002).

[10]P. Y. Hou, "Segregation Phenomena at Thermally Grown Al₂O₃/Alloy Interfaces," *Annu. Rev. Mater. Res.*, **38**, 275–98 (2008).

[11]F. Ernst, "Metal–Oxide Interfaces," *Mater. Sci. Eng.*, **R 14**, 97–156 (1995).

[12]M. W. Finnis, "The Theory of Metal–Ceramic Interfaces," *J. Phys.: Condens. Matter*, **8**, 5811–36 (1996).

[13]J. M. Howe, "Bonding, Structure, and Properties of Metal–Ceramic Interfaces .1. Chemical Bonding, Chemical-Reaction, and Interfacial Structure," *Int. Mater. Rev.*, **38**, 233–56 (1993).

[14]E. Saiz, A. P. Tomsia, and R. M. Cannon, "Wetting and Work of Adhesion in Oxide/Metal Systems"; pp. 65–82 in *Ceramic Microstructure: Control at the Atomic Level*, Edited by A.P Tomsia, and A. M. Glaeser. Plenum Press, New York, 1998.

[15]E. Saiz, R. M. Cannon, and A. P. Tomsia, "Energetics and Atomic Transport at Liquid Metal/Al₂O₃ Interfaces," *Acta Mater.*, **47**, 4209–20 (1999).

[16]E. Saiz, R. M. Cannon, and A. P. Tomsia, "High-Temperature Wetting and the Work of Adhesion in Metal/Oxide Systems," *Annu. Rev. Mater. Res.*, **38**, 197–226 (2008).

[17]D. Chatain, I. Rivollet, and N. Eustathopoulos, "Thermodynamic Adhesion in Non-Reactive Liquid Metal–Alumina Systems," *J. Chim. Phys.*, **83**, 561–7 (1986).

[18]D. Chatain, I. Rivollet, and N. Eustathopoulos, "Estimation of The Thermodynamic Adhesion and the Contact-Angle in the Nonreactive metal-Ionocovalent Oxide Systems," *J. Chim. Phys.*, **84**, 201–3 (1987).

[19]D. Chatain, L. Coudurier, and N. Eustathopoulos, "Wetting and Interfacial Bonding in Ionocovalent Oxide-liquid Metal Systems," *Rev. Phys. Appl.*, **23**, 1055–64 (1988).

[20]M. Degraef, B. J. Dalgleish, M. R. Turner, and A. G. Evans, "Interfaces Between Alumina and Platinum: Structure, Bonding and Fracture Resistance," *Acta Metal. Mater.*, **40**, S333–S344 (1992).

[21]M. Rühle, "Structure and Composition of Metal/Ceramic Interfaces," *J. Eur. Ceram. Soc.*, **16**, 353–65 (1996).

[22]N. Eustathopoulos and B. Drevet, "Determination of the Nature of Metal-Oxide Interfacial Interactions from Sessile Drop Data," *Mater. Sci. Eng. A*, **249**, 176–83 (1998).

[23]N. Eustathopoulos, B. Drevet, and M. L. Muolo, "The Oxygen-Wetting Transition in Metal/Oxide Systems," *Mater. Sci. Eng. A*, **300**, 34–40 (2001).

[24]N. Eustathopoulos, "Progress in Understanding and Modeling Reactive Wetting of Metals on Ceramics," *Curr. Opin. Solid State Mater. Sci.*, **9**, 152–60 (2005).

[25]A. Bogicevic and D. R. Jennison, "Variations in the Nature of Metal Adsorption on Ultrathin Al₂O₃ Films," *Phys. Rev. Lett.*, **82**, 4050–3 (1999).

[26]C. Verdozzi, D. R. Jennison, P. A. Schultz, and M. P. Sears, "Sapphire(0001) Surface, Clean and with D-Metal Overlayers," *Phys. Rev. Lett.*, **82**, 799–802 (1999).

[27]J. R. Smith and W. Zhang, "Stoichiometric Interfaces of Al and Ag with Al₂O₃," *Acta Mater.*, **48**, 4395–403 (2000).

[28]Z. Łodziana and J. K. Norskov, "Adsorption of Cu and Pd on Alpha-Al₂O₃(0001) Surfaces with Different Stoichiometries," *J. Chem. Phys.*, **115**, 11261–7 (2001).

[29]D. J. Siegel, L. G. Hector, and J. B. Adams, "Adhesion, Atomic Structure, and Bonding at the Al(111)/Alpha-Al₂O₃(0001) Interface: A First Principles Study," *Phys. Rev. B*, **65**, 085415, 19pp (2002).

[30]I. G. Batyrev and L. Kleinman, "In-plane Relaxation of Cu(111) and Al(111)/Alpha-Al₂O₃ (0001) Interfaces," *Phys. Rev. B*, **64**, 033410, 4pp (2001).

[31]I. G. Batirev, A. Alavi, M. W. Finnis, and T. Deutsch, "First-Principles Calculations of the Ideal Cleavage Energy of Bulk Niobium(111)/Alpha-Alumina(0001) Interfaces," *Phys. Rev. Lett.*, **82**, 1510–3 (1999).

[32]I. G. Batyrev, A. Alavi, and M. W. Finnis, "Equilibrium and Adhesion of Nb/Sapphire: The Effect of Oxygen Partial Pressure," *Phys. Rev. B*, **62**, 4698–706 (2000).

[33]W. Zhang and J. R. Smith, "Nonstoichiometric Interfaces and Al₂O₃ Adhesion with Al and Ag," *Phys. Rev. Lett.*, **85**, 3225–8 (2000).

[34]W. Zhang and J. R Smith, "Stoichiometry and Adhesion of Nb/Al₂O₃," *Phys. Rev B*, **61**, 16883–9 (2000).

[35]W. Zhang, J. R. Smith, and A. G. Evans, "The Connection between Ab Initio Calculations and Interface Adhesion Measurements on Metal/Oxide Systems: Ni/Al₂O₃ and Cu/Al₂O₃," *Acta Mater.*, **50**, 3803–16 (2002).

[36]J. W. Feng, W. Q. Zhang, and W. Jiang, "Ab Initio Study of Ag/Al₂O₃ and Au/Al₂O₃ Interfaces," *Phys. Rev. B*, **72** [1–11] 115423 (2005).

[37]W. Zhang, J. R. Smith, and X. G. Wang, "Thermodynamics from Ab Initio Computations," *Phys. Rev. B*, **70**, 024103, 8pp (2004).

[38]J. Bruley, R. Brydson, H. Mülleejans, J. Mayer, G. Gutekunst, W. Mader, D. Knauss, and M. Rühle, "Investigations of the Chemistry and Bonding at Niobiumsapphire Interfaces," *J. Mater. Res.*, **9**, 2574–83 (1994).

[39]G. Dehm, M. Ruhle, G. Ding, and R. Raj, "Growth and Structure of Copper Thin-Films Deposited on (0001) Sapphire by Molecular-Beam Epitaxy," *Philos. Mag. B*, **71**, 1111–24 (1995).

[40]C. Scheu, G. Dehm, M. Rühle, and R. Brydson, "Electron-Energy-Loss Spectroscopy Studies of Cu-Alpha-Al₂O₃ Interfaces Grown by Molecular Beam Epitaxy," *Philos. Mag.A*, **78**, 439–65 (1998).

[41]J. P. Perdew, K. Burke, and M. Ernzerhof, "Generalized Gradient Approximation Made Simple," *Phys. Rev. Lett.*, **77**, 3865–8 (1996).

[42]G. Kresse and J. Hafner, "Ab Initio Molecular Dynamics for Liquid Metals," *Phys. Rev. B*, **47**, 558–61 (1993).

[43]G. Kresse and J. Furthmuller, "Efficient Iterative Schemes for Ab Initio Total-Energy Calculations using a Plane-Wave Basis Set," *Phys. Rev. B*, **54**, 11169–86 (1996).

[44]P. E. Blöchl, "Projector Augmented-Wave Method," *Phys. Rev. B*, **50**, 17953–79 (1994).

[45]G. Kresse and D. Joubert, "From Ultrasoft Pseudopotentials to the Projector Augmented-Wave Method," *Phys. Rev. B*, **59**, 1758–75 (1999).

[46]H. T. Li, J. W. Feng, W. Q. Zhang, W. Jiang, H. Gu, and J. R. Smith, "Ab Initio Thermodynamic Study of the Structure and Chemical Bonding of a Beta-Ni₁₋ₓAlₓ/Alpha-Al₂O₃ Interface," *Phys. Rev. B*, **80**, 205422, 12pp (2009).

[47]I. Batyrev, A. Alavi, and M. W. Finnis, "Ab Initio Calculations on the Al₂O₃ (0001) Surface," *Faraday Discuss.*, **114**, 33–43 (1999).

[48]X. G. Wang, A. Chaka, and M. Scheffler, "Effect of the Environment on Alpha-Al₂O₃ (0001) Surface Structures," *Phys. Rev. Lett.*, **84**, 3650–3 (2000).

[49]F. R. D. Boer, R. Boom, W. C. M. Mattens, A. R. Miedema, and A. K. Niessen, *Cohesion in Metals*, North-Holland, New York, 1988.

[50]A. K. Niessen and A.R Miedema, "The 'Macroscopic Atom' Model: An Easy Tool to Predict Thermodynamic Quantities"; pp. 29–54 in *Thermochemistry of Alloys*, edited by H. Brodowsky, and H. J. Schaller. Kluwer Academic Publishers, Boston, 1987.

[51]A. R. Miedema, "Surface Energies of Solid Metals," *Z. Metallkd.*, **69**, 287–92 (1978).

[52]R. S. Mulliken, "Electronic Population Analysis on Lcao-Mo Molecular Wave Functions .1," *J. Chem. Phys.*, **23**, 1833–40 (1955).

[53]O. Kubaschewski, C. B. Alcock, and P. J. Spencer, *Materials Thermochemistry*. Pergamon Press, Oxford, 1993.

[54]D. Lide, *CRC Handbook of Chemistry and Physics.* CRC Press, Boca Raton, 1997.

[55]J. G. Li, "Wetting and Interfacial Bonding of Metals with Ionocovalent Oxides," *J. Am. Ceram. Soc.*, **75**, 3118–26 (1992).

[56]D. M. Lipkin, J. N. Israelachvili, and D. R. Clarke, "Estimating the Metal-Ceramic Van der Waals Adhesion Energy," *Philos. Mag. A*, **76**, 715–28 (1997).

[57]C. Kittel, *Introduction to Solid State Physics.* Wiley, New York, 1971.

[58]J. E. McDonald and J. G. Eberhart, "Adhesion in Aluminum Oxide-Metal Systems," *Trans. Metall. Soc., AIME*, **233**, 512–7 (1965).

[59]Y. V. Naidich, "The Wettability of Solids by Liquid Metals," *Prog. Surf. Membr. Sci.*, **14**, 353–484 (1981).

[60]A. M. Stoneham and P. W. Tasker, "Metal Non-Metal and Other Interfaces—The Role of Image Interactions," *J. Phys. C: Solid State Phys.*, **18**, L543–L548 (1985).

[61]M. W. Finnis, "Metal Ceramic Cohesion and the Image Interaction," *Acta Metall. Mater.*, **40**, S25–37 (1992).　□

J. Am. Ceram. Soc., **94** [S1] S160–S167 (2011)
DOI: 10.1111/j.1551-2916.2011.04459.x
© 2011 The American Ceramic Society

journal _____

Elastodynamic Erosion of Thermal Barrier Coatings

Man Wang,[‡] Norman A. Fleck,[†,‡] and Anthony G. Evans[*,§]

[‡]Department of Engineering, Cambridge University, Cambridge CB2 1PZ, U.K.

[§]Materials Department, University of California Santa Barbara, California 93106

The finite element method is used to analyze the elastodynamic response of a columnar thermal barrier coating due to normal impact and oblique impact by an erosive particle. An assessment is made of the erosion by crack growth from preexisting flaws at the edge of each column: it is demonstrated that particle impacts can be sufficiently severe to give rise to columnar cracking. First, the transient stress state induced by the normal impact of a circular cylinder or a sphere is calculated in order to assess whether a 2D calculation adequately captures the more realistic 3D behavior. It is found that the transient stress states for the plane strain and axisymmetric models are similar. The sensitivity of response to particle diameter and to impact velocity is determined for both the cylinder and the sphere. Second, the transient stress state is explored for 2D oblique impact by a circular cylindrical particle and by an angular cylindrical particle. The sensitivity of transient tensile stress within the columns to particle shape (circular and angular), impact angle, impact location, orientation of the angular particle, and to the level of friction is explored in turn. The paper concludes with an evaluation of the effect of inclining the thermal barrier coating columns upon their erosion resistance.

I. Introduction

THERMAL barrier coatings (TBCs) are widely used in gas turbines for aerospace propulsion and power generation. The current coating system consists of three layers.[1] The outer layer is the thermal barrier (TB) against hot combustion gases, and typically comprises yttria-stabilized zirconia (YSZ). Oxidation resistance is achieved by an underlying thermally grown oxide (TGO) layer of alumina. The TGO progressively thickens during service by the oxidation of an underlying aluminum-rich bond coat. Two distinct topologies of the TB layer exist. One comprises layers of splats deposited by an air plasma spray technique and the other consists of a columnar microstructure produced by electron beam physical vapor deposition (EB-PVD). There is continued development on the topology of EB-PVD coatings in order to optimize their TB properties, ideally without the loss of durability. For example, inclined TBC columns have increased thermal resistance[2,3] but the recent experimental evidence of Nicholls and colleagues[4–6] reveals that they have a reduced resistance to erosion. In this paper, we limit attention to the erosion of columnar TBCs prepared by the EB-PVD route and explore the sensitivity of the elastodynamic response (and thereby the anticipated erosion response) to the topology of coating, level of contact friction, size and shape of impacting particle, and to the angle and site of impact of the incoming particle.

Two distinct erosion mechanisms have been identified for EB-PVD TBCs.[7] One involves plastic indentation and the other is elastodynamic in nature. The plastic damage, including plastic densification and microbuckling of the columnar TBCs, accompanies impact by large particles with high momentum and at high temperature. The deformation zones develop over a millisecond timescale as the impacting particle decelerates.[8]

In contrast, elastodynamic failure is caused by the impact of small particles of low momentum and at low temperature. Elastic waves emanate from the contact and lead to trans-columnar cracks beneath the surface. Zisis and Fleck[9,10] have studied the elastodynamic response of the columns to normal impact by a spherical particle using the finite element method. The columns act as wave-guides, and the transient tensile stress generated by the impact event is sufficient to crack the columns. Typical stress histories are reported in detail in Chen and colleagues[8–10] but attention is limited to normal impact by spherical particles at a limited range of speeds. The predicted site of column cracking is in good agreement with the experimental observations of[11]: the columns crack at a depth somewhat less than their diameter.

In the practical case of erosion of TBC coatings by foreign particles, there are a number of complicating features that have hitherto been ignored in the idealized finite element calculations. The particles may not be smooth and spherical, they may impinge the TBC at an angle that deviates from the normal direction, and the TBC columns themselves can be inclined. We address the significance of these complicating features in the current study. The focus of this paper is on the elastodynamic response of EB-PVD TBCs to normal and oblique impact by a foreign particle. The ABAQUS/Explicit finite element program (Version 6.5)[12] is used to analyze the impact event for both circular and angular erosive particles. The primary objective is to explore the sensitivity of the transient stress state to the wide range of geometric and kinematic variables that define the impact event, including the size of erosive particle, the initial velocity of the particle, angle of incidence, impact location at the top of a TBC column, inclination of the angular particle, and the level of Coulomb friction between columns and at the particle–column contact. The transient stress state is used to determine the potential for cracking: the static stress intensity factor is calculated for a putative crack in the columns. Recently, it has been suggested that the thermal resistance of TBC columns is increased by depositing them at an inclination to the normal direction.[2,3] We end the paper with an assessment of the degree to which this influences the erosion resistance under elastodynamic loading.

A full 3D finite study of a columnar TBC impacted by a 3D particle is prohibitively expensive in computer resources, and so the study focuses on the 2D, plane strain problem of the oblique impact of an array of strip-like TBC columns by a cylindrical projectile. To explore the significance of 2D vs 3D particles, the case of normal impact is considered for both axisymmetric and plane strain cases, and it is demonstrated that the two idealizations yield similar predictions.

II. Normal Impact of the Columnar TBC Layer

In order to study the normal impact event, a plane strain finite element model and an axisymmetric finite element model are each established, as sketched in Figs. 1(a) and (b), respectively (the direction of normal impact is defined by $\phi = 0°$ in Fig. 1(a)).

R. McMeeking—contributing editor

Manuscript No. 28789. Received October 15, 2010; approved January 21, 2011.
[†]Author to whom correspondence should be addressed. e-mail: nafl@eng.cam.ac.uk
[*]Deceased.

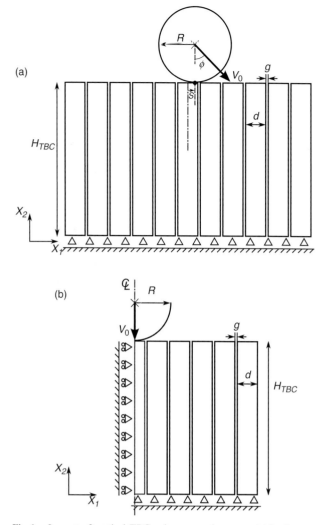

Fig. 1. Impact of vertical TBC columns, resting on a rigid substrate. (a) plane strain case, circular cylindrical particle obliquely impacting an array of rectangular strips; (b) axisymmetric case, normal impact of a concentric array of annuli by a spherical particle.

The EB-PVD columns are perfectly bonded to a rigid substrate for both models. In the plane strain model, the columnar TBC layer is represented by a set of vertical aligned rectangular strips, and the foreign particle is represented by a circular cylinder. In contrast, for the axisymmetric model, the TBC layer is simulated by a nested annular array of circular cylinders, and the particle is a sphere of radius R. A Cartesian reference frame is adopted for both models.

(1) Material and Geometric Parameters

The columns in both the plane strain model of Fig. 1(a) and the axisymmetric model of Fig. 1(b) are of width d, height $H_{TBC} = 20\,d$, and intercolumnar gap $g = 0.01\,d$. The columnar structure is ascribed a linear elastic response with Young's modulus E_{TBC}, Poisson's ratio v, and density ρ_{TBC}. The 1D elastic wave speed is $c = \sqrt{E_{TBC}/\rho_{TBC}}$, and is used below for normalization of the impact velocity. The incident particle is taken to be rigid and of density $\rho_P = 0.34\rho_{TBC}$, to represent the relative density values for a silica particle and a YSZ TBC. Its radius ranges from $R = 0.5$ to $10\,d$, and its initial velocity ranges from $V_0 = 0.01$ to $0.3c$. The particle is allowed to decelerate after impact due to the transient contact force exerted upon it by the TBC columns.

(2) The Finite Element Simulations

A mesh sensitivity study reveals that the plane strain column model of Fig. 1(a) achieves adequate accuracy by using 240 697

four-noded quadrilateral plane strain elements (CPF4R in ABAQUS); likewise, the axisymmetric model has adequate accuracy by using 164 126 four-noded quadrilateral axisymmetric elements (CAX4R in ABAQUS).

The displacement boundary conditions for the two models are $u_1 = u_2 = 0$ along the bottom. Additionally, the displacement boundary condition $u_1 = 0$ is prescribed along the center line in the axisymmetric model. Traction-free side boundaries are assumed. Numerical experiments reveal that axial stress waves in the central column propagate to the substrate in about 50 ns after impact; as our study is limited to the initial elastodynamic response near the tops of the columns, the total calculation time is limited to 50 ns. Thus, the choice of boundary conditions on the side boundary is not significant as the stress waves do not propagate to the sides within 50 ns.

Small deformations are assumed in all calculations. Automatic time stepping is performed, with the Courant condition automatically satisfied. The default value of material viscosity in ABAQUS is used throughout. The contact pair algorithm in ABAQUS is chosen to calculate the contact between the incoming particle and the TBC, and the default options of mechanical constraint formulation and sliding formulation are used. Frictionless conditions are assumed between columns, and between the particle and columns.

(3) Typical Elastodynamic Response

Immediately following impact, an elastic wavetrain emanates from the contact and travels down the TBC columns, as outlined in Zisis and Fleck[9] for the axisymmetric case. Wave propagation is by a combination of shear waves, longitudinal waves, and surface waves. The stress wave reinforcement is largely a consequence of the columnar geometry. The degree to which the columns behave as independent waveguides or as the elements of a half-space depends on the column diameter d and the gap size g between columns in relation to the radius R of the particle. A typical elastodynamic response for the axisymmetric problem is reported in Zisis and Fleck[9]; here, we summarize the equivalent plane-strain case to show that the response is qualitatively similar to the axisymmetric response. Snapshots of the transient stress distribution are given in Fig. 2(a) for normal impact; for later discussion, predictions are shown in Fig. 2(b) for impact at an angle of $\phi = 45°$. (The results are given for $E_{TBC} = 140$ GPa, $v = 0.3$, $\rho_{TBC} = 5900$ kgm^{-3}, $d = 10$ μm, $g = 0.1$ μm, $H_{TBC} = 200$ μm, $V_0 = 300$ ms^{-1}, $R = 25$ μm, $\rho_P = 2000$ kgm^{-3}, and $\mu_S = \mu_C = 0$ in order to allow for a comparison with the axisymmetric results given in Fig. 3 of Zisis and Fleck.[9])

Contours of tensile axial stress σ_{22} are shown, with the color white indicating axial compression. Immediately following impact, a combination of longitudinal, shear, and Rayleigh waves emanates from the contact into the central column. After a time $t = 1$ ns, a local maximum stress of 1.3 GPa is attained at the side surface of the central column and at a depth of 1 μm below the surface. The central column then comes into contact with the adjacent column, and the ensuing response is similar to that already experienced by the central column. A peak tensile stress of 3.7 GPa is attained in both space and time (at $t = 6$ ns), compared with an overall peak value of 3.9 GPa for the axisymmetric case reported in Zisis and Fleck.[9] The transient stress travels down the columns as depicted in the snapshot at 8 ns and then 15 ns. It is clear that the wavetrain from the impact site disperses throughout the TBC coating by the collision of neighboring columns, leading to a complex and highly nonuniform transient stress state. In addition to the tensile waves shown in Fig. 2(a), compression waves travel down the central columns but further consideration of these is discounted here; the present study concerns itself with the more damaging tensile waves near the surface of the coating that lead to surface erosion.

Our main concern is the possibility of cracking at the edge of the columns due to the transient tensile axial stress σ_{22}. Write $(\sigma_{22})_{max}$ as the spatial and temporal maximum value of tensile stress within the columns over the initial impact period of 50 ns.

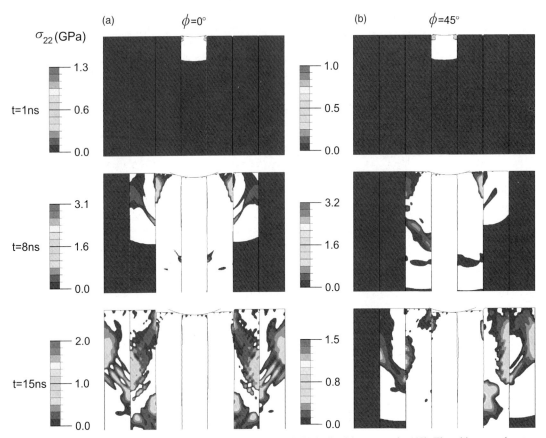

Fig. 2. Time sequences of axial tensile stress for (a) normal impact $\phi = 0°$ and (b) inclined impact at $\phi = 45°$. The white zone denotes compression.

Then, define the normalized maximum tensile stress $\bar{\sigma}_{22}$ as

$$\bar{\sigma}_{22} = \frac{(\sigma_{22})_{\max}}{V_0 \sqrt{E_{TBC} \rho_{TBC}}} \quad (1)$$

The denominator of Eq. (1) is a reference tensile stress: it is the uniaxial tensile stress in an elastic rod (made from TBC material) when subjected to a step-wise jump in the axial velocity of V_0 at one end.[13] An assessment is now made of the value of $\bar{\sigma}_{22}$ that leads to cracking from the edge of the columns. Then, in the remainder of this section, we shall explore the sensitivity of $\bar{\sigma}_{22}$ to the size and initial velocity of the impacting particle for both the plane strain and axisymmetric problems.

(A) Assessment of Crack Initiation Under the Transient Stress Field: It is instructive to assess whether the stress level $\bar{\sigma}_{22}$ from the elastodynamic simulations is sufficiently high to lead to tensile failure from the edge of the feathery TBC columns. Here, we perform a preliminary fracture mechanics assessment and calculate the maximum mode I stress intensity factor for incipient flaws at the edge of each TBC column. The role of material inertia is ignored in the calculation of the mode I stress intensity factor from the stress state $\bar{\sigma}_{22}$: it is appreciated that this is only an approximation, but our intent is to provide an overall assessment of whether the elastodynamic stress field is sufficiently intense to lead to erosion of the coating.

Microstructural observations of TBC coatings suggest that the feathery edges of the EB-PVD columns possess crack-like flaws of length a on the order of 1 μm along the sides of each column. Further, it is observed from the finite element simulations of the present study that the transient tensile stress distribution occurs to a depth on the order of 1 μm from the free edge; this stress state is approximated by a triangular waveform from a maximum value at the edge of the column to zero at the crack tip. Now, the stress intensity factor K_I for an edge crack of length a in a semiinfinite plate,[¶] subjected to a linear distribution of traction from σ at the

crack mouth to zero at the crack tip, is given by $K_I = 1.208$ $(1 - 2/\pi)\sigma\sqrt{\pi a}$.[14] Assume a representative value of fracture toughness of the TBC layer to be $K_{IC} = 1$ MPa $\sqrt{m}$ and, with $a = 1$ μm, this yields a critical value of stress $\sigma_C = 1.65$ GPa. For an impact velocity $V_0 = 300$ ms^{-1}, along with $E_{TBC} = 140$ GPa and $\rho_{TBC} = 5900$ kg/m^3, relation (1) implies that the critical value of $\bar{\sigma}_{22}$ is given by $\bar{\sigma}_{22} = \bar{\sigma}_C = 0.19$. We shall show below that the level of peak axial stress $\bar{\sigma}_{22}$ in many of the elastodynamic simulations exceeds this threshold value and consequently crack growth from the edge flaws is anticipated. It is recognized that the loading duration by wave propagation is on the order of nanoseconds. However, the loading period is much greater than the time period for atomic vibration, and is thereby sufficiently long to allow for the possibility of cleavage fracture.

(4) The Effect of Particle Size and Initial Impact Velocity on the Transient Tensile Stress

The FE simulations reveal that the maximum transient tensile stress occurs at the edge of the column and at a depth of 0.2–2.2 d. The nondimensional peak stress $\bar{\sigma}_{22}$ is evaluated for selected values of V_0/c and is plotted as a function of d/R in Fig. 3(a) for the plane strain problem, and in Fig. 3(b) for the axisymmetric case. For both problems, $\bar{\sigma}_{22}$ decreases somewhat with increasing d/R. The mild but nonmonotonic dependence of $\bar{\sigma}_{22}$ upon V_0/c is associated with the details of contact evolution between columns. In broad terms, the plane strain model adequately mimics the elastodynamic response of the axisymmetric model but predicts somewhat higher stress levels. For both types of models, the impacting particle has retarded to a minor degree (5%–20% depending on the value of d/R) over the time interval for the attainment of maximum tensile stress. An analytical model for the deceleration of the particle has been given by Fleck and Zisis,[10] including validation by finite element simulation. They show that the contact pressure exerted by the columnar TBC on the particle scales with the instantaneous particle velocity V according to

$$p = V \sqrt{E_{TBC} \rho_{TBC}} \quad (2)$$

[¶]It is appreciated that a more realistic estimate of K_I can be obtained by considering a single edge-notched specimen rather than an edge crack in a semi-infinite plate. The K_I value will be increased by about 10% for the finite width plate.

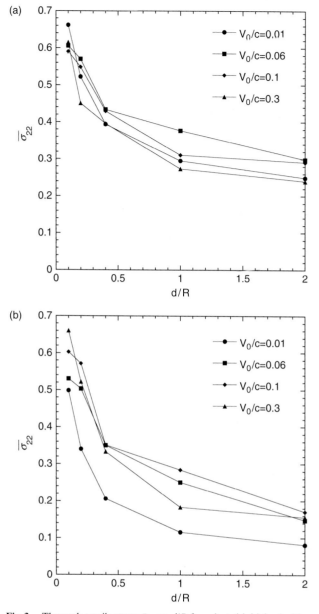

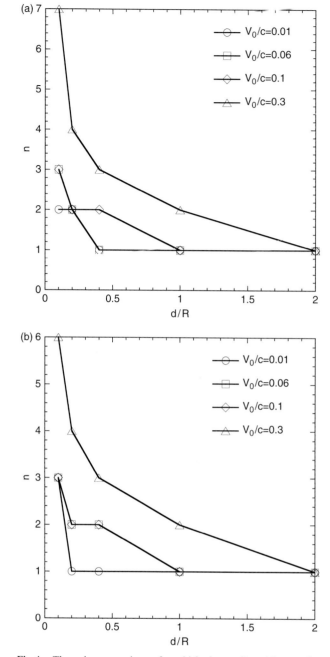

Fig. 3. The peak tensile stress $\bar{\sigma}_{22}$ vs d/R for selected initial velocities. (a) plane strain case, (b) axisymmetric case.

Fig. 4. The column number n for which the tensile axial stress has a global and temporal maximum value versus d/R. (a) plane strain and (b) axisymmetric case.

The column number n for which the maximum tensile stress occurs, counted from the initial contact point, is plotted against the normalized particle size d/R in Fig. 4(a) for the plane strain model and in Fig. 4(b) for the axisymmetric case. Both models show similar trends: the column that suffers the highest transient stress is located further from the center line with increasing V_0/c and decreasing d/R. The surprising result that the peak stress is attained in a column several diameters away from the impacted column is a consequence of the highly nonlinear contact interactions along with the ensuing complex pattern of wave motion.

III. Oblique Impact of the Columnar TBC Layer

It is of interest to determine the sensitivity of axial stress to the obliquity of impact and to the impact site at the top of the columns. We limit attention to the 2D plane strain case, and assume that the cylindrical particle has an incoming velocity V_0 at an angle ϕ to the normal, and impacts a column at a distance s from its midplane, as defined in Fig. 1(a). The same FE mesh, boundary conditions and time-stepping conditions were used for oblique impact as for normal impact. Again, the maximum value of transient tensile stress along the axis of the

columns was determined over the first 50 ns required for stress wave propagation to the bottom of the TBC columns.

The impact problem now involves a larger number of geometric parameters. The sensitivity of elastodynamic response to the intercolumnar gap size g and height of column H_{TBC} has been addressed previously[10] for the case of normal impact. Accordingly, in the remainder of this study, the values for (g, H_{TBC}) are fixed at $g = 0.01\ d$ and $H_{TBC} = 20\ d$. We also limit attention to a circular cylinder of radius $R = 2.5\ d$ and of density $\rho_p = 0.34\rho_{TBC}$. The particle impinges the underlying TBC at an initial velocity $V_0 = 0.062\sqrt{E_{TBC}/\rho_{TBC}}$ and, for the case of a TBC made from YSZ, this corresponds to an impact velocity of the tip of the turbine blades, $V_0 = 300\ \text{ms}^{-1}$.

The sensitivity of the maximum axial stress to the impact angle ϕ for three selected values of impact location $s = -d/4$, 0, and $d/4$ is given in Fig. 5 for the frictionless case $\mu_s = \mu_c = 0$. The peak stress declines by a factor of about four when the angle of obliquity ϕ is increased from 0° to 75°, and increases slightly

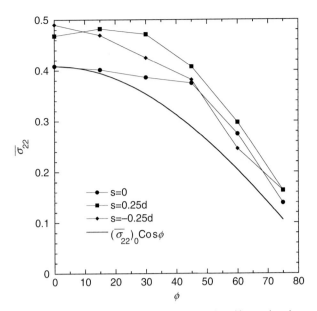

Fig. 5. Peak stress $\bar{\sigma}_{22}$ versus impact angle ϕ for selected impact locations.

when the impact site is located off-center $s \neq 0$. In order to assess whether the peak axial stress is dictated simply by the *vertical* component of particle velocity, an additional stress trajectory has been added to Fig. 5, as follows: the normal component of impact velocity is given by $V_0 \cos \phi$ and if this velocity component were to dictate the peak stress attained then the peak stress $\bar{\sigma}_{22}$ at an incidence of ϕ would be related to the peak stress $(\bar{\sigma}_{22})_0$ for normal impact according to $\bar{\sigma}_{22} = (\bar{\sigma}_{22})_0 \cos \phi$. This function is plotted in Fig. 5 for the choice $s = 0$; it is an adequate approximation for $\phi < 15°$ but underestimates the peak stress by about 20% at large ϕ.

The transient stress history for oblique impact is asymmetrical with respect to the center line of the column, but is otherwise qualitatively similar. An example of the axial stress distribution at the instant of peak local stress for $s = \mu_s = \mu_c = 0$ and $\phi = 45°$ is shown in Fig. 6(c). The location of peak stress is almost identical to that for $s = \mu_s = \mu_c = 0$ and $\phi = 0°$; recall Fig. 6(a). For completeness, the axial stress distribution at peak local stress is

included in Fig. 6 for normal impact $\phi = 0°$ and inclined impact $\phi = 45°$, with $\mu_s = \mu_c = 1$ and $s = 0$. This high value of friction coefficients leads to significant changes in the location and value of peak stress (an increase of 20% for $\phi = 0°$, and of 80% for $\phi = 45°$).

(1) The Effect of Particle Shape on the Transient Tensile Stress

Foreign object debris is usually angular in shape rather than circular, and it is of interest to determine the sensitivity of the elastodynamic response to the details of the particle shape. We consider a prototypical angular particle in the form of a square cylinder of side length $a = 4.4\ d$ with a rounded-off corner of radius $r = 0.2\ d$; see Fig. 7. The angle of attack ψ of the square particle is also defined in Fig. 7; it has the selected values of $0°$, $22.5°$, $45°$, and $67.5°$ in our study.

The nondimensional tensile stress $\bar{\sigma}_{22}$ depends on several factors such as the particle shape, the impact angle ϕ, the friction coefficient, the initial impact location, and the orientation of the angular particle. The sensitivity of $\bar{\sigma}_{22}$ to ϕ is shown in Fig. 8 for the circular and square particles, for (a) $\mu_s = \mu_c = 0$ and (b) $\mu_s = \mu_c = 1$. Both particles impact the midpoint of the TBC columns, $s = 0$, and the square particle is oriented so that it has angle of attack, $\psi = 45°$. It is clear from Fig. 8 that the dependence of peak stress $\bar{\sigma}_{22}$ on the angle of incidence ϕ is sensitive both to the particle shape and to the level of friction. The circular particle contacts and bends two underlying columns at the same time, whereas the square particle initially touches and bends the first column by its corner and then bends the second column by its side. Consequently, the stress transients are different. For the circular particle and frictionless contacts, $\bar{\sigma}_{22}$ gradually reduces from 0.41 to 0.14 when ϕ increases from $0°$ to $75°$. With high friction present, $\bar{\sigma}_{22}$ first increases with increasing ϕ and then attains a peak value at ϕ equal to approximately $30°$.

For the square particle and frictionless conditions, $\bar{\sigma}_{22}$ is on the order of 0.3 for ϕ in the range $0°–45°$, then increases rapidly to 0.44 as ϕ is increased to $\phi = 60°$ and then declines to 0.35 at $\phi = 75°$. The maximum tensile stresses occur at initial 2 to 3 ns and at a depth of $0.2\ d$ at small impact angles ($\phi = 0°–30°$). However, at large impact angles ($\phi = 45°–75°$), the maximum tensile stresses occur after 30–40 ns and at a depth of $3–6\ d$. An increase in the friction level causes $\bar{\sigma}_{22}$ to attain a higher peak

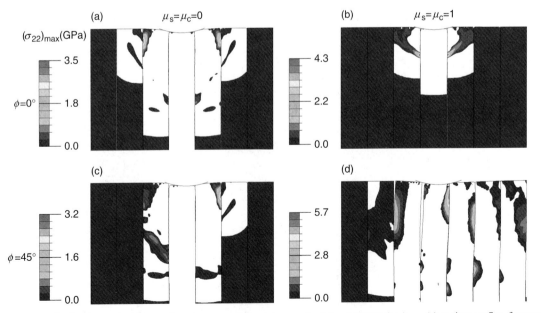

Fig. 6. The axial stress distribution at the instant of maximum spatial and temporal axial component $(\sigma_{22})_{max}$. (a) at time $t = 7$ns, for normal impact with frictionless conditions, (b) $t = 3.5$ ns, under normal impact with $\mu_S = \mu_C = 1$, (c) $t = 8$ ns under oblique impact $\phi = 45°$ with frictionless conditions, and (d) $t = 14.5$ ns under oblique impact $\phi = 45°$ with $\mu_S = \mu_C = 1$.

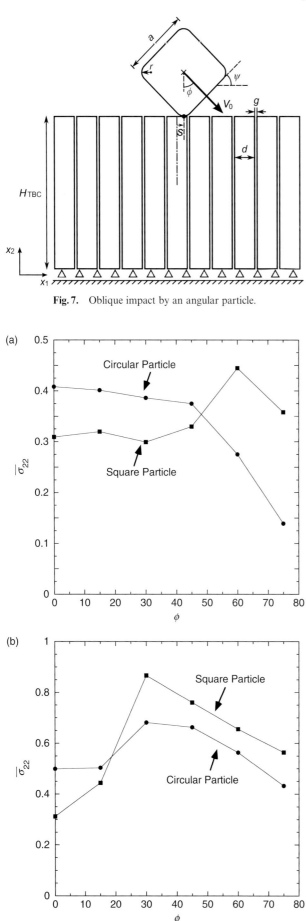

Fig. 7. Oblique impact by an angular particle.

value of about 0.85 but at a reduced value of impact angle, $\phi = 30°$; see Fig. 8(b).

In broad terms, $\bar{\sigma}_{22}$ varies over a wide range, from 0.15 to 0.85, depending on the particle shape, friction level, and angle of incidence. This is consistent with the feature that particle erosion of TBC systems is a highly stochastic event: there is a wide variation in impact damage from one particle to the next.[6]

(2) Square Particles: The Effect of Angle of Attack and Impact Location on Stress State

The effect of impact location s and angle of attack ψ of the square particle on the $\bar{\sigma}_{22}$ vs ϕ response is shown in Fig. 9, for frictionless contacts. Three initial impact locations are chosen, $s = 0.0, 0.25\,d$, and $-0.25\,d$, in Fig. 9(a), with $\psi = 45°$. The initial impact location has a minor effect on the magnitude of the nondimensional maximum tensile stress, as for the circular particle; recall Fig. 5. The angle of attack ψ has a more dramatic effect on $\bar{\sigma}_{22}$, particularly at oblique impacts with ϕ on the order of 60°; see Fig. 9(b) For this value of ϕ, $\bar{\sigma}_{22}$ increases from 0.2 to 0.6 as ψ is increased from 0° to 67.5°.

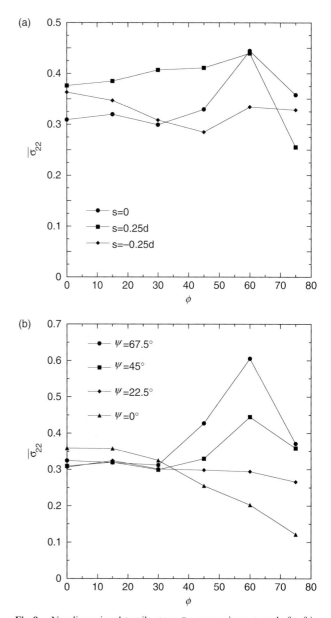

Fig. 8. Nondimensional tensile stress $\bar{\sigma}_{22}$ versus impact angle for the circular particle, and square particle with $\psi = 45°$ and initial location $s = 0$. (a) $\mu_S = \mu_C = 0$ and (b) $\mu_S = \mu_C = 1$.

Fig. 9. Nondimensional tensile stress $\bar{\sigma}_{22}$ versus impact angle for frictionless impact by the square particle. (a) impact locations $s = 0, 0.25\,d$ and $-0.25\,d$, with $\psi = 45°$; (b) initial location $s = 0$ but particle oriented at $\psi = 67.5°, 45°, 22.5°$, and 0°.

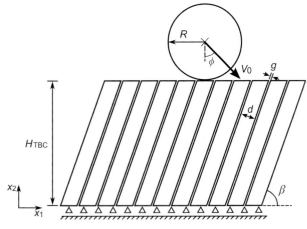

Fig. 10. Sketch of the plane strain model for the impact of inclined columns.

IV. The Elastodynamic Response of Inclined TBC Columns

Inclined TBC columns provide greater thermal insulation than vertical columns, but have a reduced erosion resistance. In this section, the elastodynamic response of inclined columns is compared with that of vertical columns, first for normal impact and then for inclined impact by circular cylindrical particles.

(1) Normal Impact of Inclined Columns

The impact response has been explored for normal impact ($\phi = 0°$) of an array of inclined columns, as defined by the angle β in Fig. 10. Otherwise, the geometry and material parameters are the same as those used for the vertical columns as defined above. The rigid particle is a circular cylinder of radius $R = 2.5\, d$ and density $\rho_p = 0.34\rho_{TBC}$, and it impacts the underlying columnar coating perpendicularly at an initial velocity $V_0 = 0.062\sqrt{E_{TBC}/\rho_{TBC}}$ (corresponding to 300 ms^{-1} in a practical system). For the present calculations, all the contacts are assumed to be frictionless.

The commercial finite element software ABAQUS/Explicit was again used, with the implementation details and boundary conditions as described in previous sections of this paper. A typical elastodynamic response for columns inclined at $\beta = 60°$ is shown in Fig. 11: contours of the tensile stress σ_a coaxial with the columns are shown after 2 ns and after 8 ns. It is instructive

to compare the response with that shown in Fig. 2 for normal impact ($\phi = 0°$) of vertical columns. For the case of inclined columns ($\beta = 60°$), the peak axial stress (of 10.2 GPa) occurs shortly after impact at $t = 2$ ns, along the edge of the first column and at a depth of 0.15 d. Stress wave propagation follows into neighboring inclined columns via contact from one column to the next, and a complex distribution of dispersive bending waves is established in the array of columns. After $t = 8$ ns, the peak axial stress is of magnitude 3.9 GPa at the outer edge of the second column. This resembles the peak state of stress in the vertical array of columns that occurred after $t = 8$ ns; recall Fig. 2.

The sensitivity of peak axial stress to the inclination β of inclined columns has also been determined by a series of FE simulations for normal impact. The nondimensional maximum tensile stress $\bar{\sigma}_a$ in the axial direction is related to the temporal and spatial maximum tensile stress $(\sigma_a)_{max}$ within the columns according to

$$\bar{\sigma}_a = \frac{(\sigma_a)_{max}}{V_0\sqrt{E_{TBC}\rho_{TBC}}} \tag{3}$$

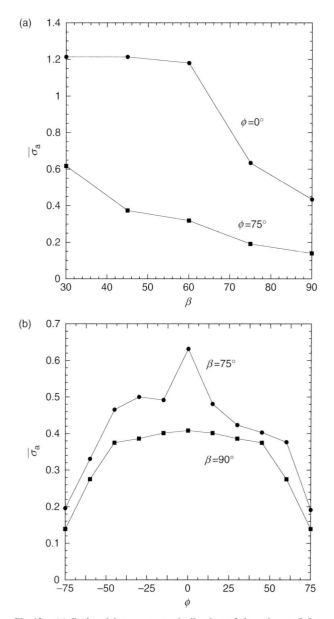

Fig. 12. (a) Peak axial stress versus inclination of the columns β for normal impact $\phi = 0°$ and oblique impact $\phi = 75°$; (b) peak axial stress versus impact angle ϕ for $\beta = 75°$ and 90°.

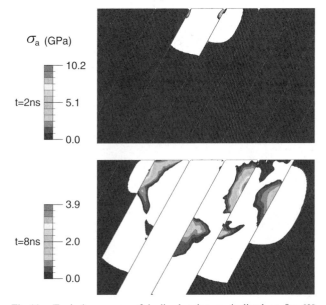

Fig. 11. Typical response of inclined columns, inclined at $\beta = 60°$, showing the axial tensile stress following normal impact. The white color denotes a state of compression.

The dependence of $\bar{\sigma}_a$ on β is plotted in Fig. 12(a) for normal impact. The tensile stress $\bar{\sigma}_a$ equals 1.2 for β in the range $30°$–$60°$, and then decreases to 0.43 as β is increased to the limit of vertical columns, $\beta = 90°$. The maximum transient tensile stress $(\sigma_a)_{max}$ occurs within 5 ns of impact and at $<0.2\ d$ from the top surface of the columns. The sensitivity of $\bar{\sigma}_a$ to β is consistent with the observations of Nicholls[5]: vertical columns are more erosion resistant than inclined columns under normal impact.

(2) Oblique Impact of Inclined Columns

The dependence of $\bar{\sigma}_a$ on β is plotted for a particular impact angle $\phi = 75°$ in Fig. 12(a). The tensile stress $\bar{\sigma}_a$ gradually decreases from 0.62 to 0.14 when β is increased from $30°$ to $60°$. The stress levels for oblique impact are significantly below those for normal impact $\phi = 0°$. However, the location of peak stresses and the time to attain it are comparable for the two values of ϕ.

Recall from Fig. 5 that the peak axial stress within frictionless, vertical columns decreases with increasing angle of obliquity of impact ϕ. Does this conclusion still hold for the case of inclined columns? To investigate this, a limited number of impact simulations have been performed on an array of frictionless columns inclined at $\beta = 75°$. Otherwise, the finite element model is the same as that described in the previous section. The observed dependence of $\bar{\sigma}_a$ on ϕ is plotted is Fig. 12(b) for $\beta = 75°$ and for $\beta = 90°$ (from Fig. 5). For both values of β, the highest axial stresses are generated by normal impact, $\phi = 0°$. For $\beta = 75°$, the normalized tensile stress $\bar{\sigma}_a$ declines from 0.63 to 0.20 as the magnitude of ϕ increases from $0°$ to $75°$. Again, we conclude that vertical columns ($\beta = 90°$) are more erosion resistant than inclined columns, regardless of the impact angle.

V. Concluding Remarks

The explicit finite element method has been used to explore the elastodynamic response of the EB-PVD columnar structure to normal and oblique impact by a foreign particle. A complex transient stress pattern emerges from the contact, and contact between the columns can lead to the most intense tensile stresses occurring in columns adjacent to the impacted column. The level of peak axial stress is sufficiently high to induce cracking from the edge of the column and thereby erosion. It is found that the transient tensile stress is sensitive to the size, shape, and approach direction of the incoming particle and to the details of the TBC geometry. Moreover, the peak tensile stress is sensitive

to the level of friction between columns and between the particle and the top of TBC coating. Additionally, the maximum tensile stress significantly depends on the inclination of the columns: vertical columns have a greater erosion resistance than inclined columns for the elastodynamic mechanisms. Consequently, the erosion of TBC coatings is highly stochastic in nature.

Acknowledgments

This research was initiated when Prof. Evans was on sabbatical from University of California, Santa Barbara, at Cambridge University. The two other authors are indebted to him for his contributions to this article, and for his wise counsel and friendship over the years.

References

[1]A. G. Evans, D. R. Mumm, J. W. Hutchinson, G. H. Meier, and F. S. Pettit, "Mechanisms Controlling the Durability of Thermal Barrier Coatings," *Prog. Mater. Sci.*, **46** [5] 505–53 (2001).

[2]D. D. Hass, A. J. Slifka, and H. N. G. Wadley, "Low Thermal Conductivity Vapour Deposited Zirconia Microstructures," *Acta Mater.*, **49** [6] 973–8 (2001).

[3]S. Gu, T. J. Lu, D. D. Hass, and H. N. G. Wadley, "Thermal Conductivity of Zirconia Coatings with Zig-Zag Pore Microstructures," *Acta Mater.*, **49** [13] 2539–47 (2001).

[4]J. R. Nicholls, M. J. Deakin, and D. S. Rickerby, "A Comparison Between the Erosion Behaviour of Thermal Spray and Electron Beam Vapour Deposition Thermal Barrier Coatings," *Wear*, **233**, 352–61 (1999).

[5]R. G. Wellman, M. J. Deakin, and J. R. Nicholls, "The Effect of TBC Morphology on the Erosion Rate of EB-PVD TBCs," *Wear*, **258** [1–4] 349–56 (2005).

[6]R. G. Wellman, M. J. Deakin, and J. R. Nicholls, "The Effect of TBC Morphology and Aging on the Erosion Rate of EB-PVD TBCs," *Tribol. Int.*, **38** [9] 798–804 (2005).

[7]A. G. Evans, N. A. Fleck, S. Faulhaber, N. Vermaak, M. Maloney, and R. Darolia, "Scaling Laws Governing the Erosion and Impact Resistance of Thermal Barrier Coatings," *Wear*, **260** [7–8] 886–94 (2006).

[8]X. Chen, M. Y. He, I. Spitsberg, N. A. Fleck, J. W. Hutchinson, and A. G. Evans, "Mechanisms Governing the High Temperature Erosion of Thermal Barrier Coatings," *Wear*, **256** [7–8] 735–46 (2004).

[9]Th. Zisis and N. A. Fleck, "Mechanisms of Elastodynamic Erosion of Electron-Beam Thermal Barrier Coatings," *Int. J. Mater. Res.*, **98** [12] 1196–202 (2007).

[10]N. A. Fleck and Th. Zisis, "The Erosion of EB-PVD Thermal Barrier Coatings: The Competition Between Mechanisms," *Wear*, **268** [11–12] 1214–2 (2010).

[11]R. G. Wellman and J. R. Nicholls, "Some Observations on Erosion Mechanisms of EB PVD TBCs," *Wear*, **242** [1–2] 89–96 (2000).

[12]Hibbitt, Karlsson and Sorensen Inc. *ABAQUS version 6.5 User's Manual.* Hibbitt, Karlsson and Sorensen Inc., Pawtucket, 2005.

[13]K. L. Johnson, *Contact Mechanics*. Cambridge Press, Cambridge, U.K., 1985.

[14]H. Tada, P. Paris, and G. Irwin, *The Stress Analysis of Cracks Handbook*. Del Research Corporation, St. Louis, 1973. □

J. Am. Ceram. Soc., **94** [S1] S168–S177 (2011)
DOI: 10.1111/j.1551-2916.2011.04531.x
© 2011 The American Ceramic Society

journal

Phase Stability of t′-Zirconia-Based Thermal Barrier Coatings: Mechanistic Insights

Jessica A. Krogstad,[†,‡] Stephan Krämer,[‡] Don M. Lipkin,[§] Curtis A. Johnson,[§] David R. G. Mitchell,[¶] Julie M. Cairney,[¶] and Carlos G. Levi[‡]

[‡]Materials Department, University of California Santa Barbara, Santa Barbara, California 93106 5050

[§]GE Global Research, One Research Circle, Niskayuna, New York 12309

[¶]Australian Center for Microscopy and Microanalysis, The University of Sydney, NSW 2006 Australia

The temperature capability of yttria-stabilized zirconia thermal barrier coatings (TBCs) is ultimately tied to the rate of evolution of the "nontransformable" t′ phase into a depleted tetragonal form predisposed to the monoclinic transformation on cooling. The t′ phase, however, has been shown to decompose in a small fraction of the time necessary to form the monoclinic phase. Instead, a modulated microstructure consisting of a coherent array of Y-rich and Y-lean lamellar phases develops early in the process, with mechanistic features suggestive of spinodal decomposition. Coarsening of this microstructure leads to loss of coherency and ultimately transformation into the monoclinic form, making the kinetics of this process, and not the initial decomposition, the critical factor in determining the phase stability of TBCs. Transmission electron microscopy is shown to be essential not only for characterizing the microstructure but also for proper interpretation of X-ray diffraction analysis.

I. Introduction

METASTABLE zirconia-based materials and their mechanical behavior were a subject of interest to Professor Evans for over 30 years, first in the context of transformation toughening of ceramics,[1,2] and later in their role as thermal barrier coatings (TBCs) for gas turbine components,[3] where toughness is postulated to derive from ferroelectric domain switching.[4] The classical system is ZrO_2–$YO_{1.5}$, Fig. 1, and the enabling phase, in both cases, is the tetragonal solid solution (t). For transformation toughening,[5] t domains must be retained well into the monoclinic (m) stability range delineated by the T_0 (t/m) curve, until the t→m transformation is triggered by interaction with a propagating crack.[2] This mechanism is not useful in TBCs because the transformation is thermodynamically forbidden above T_0 (t/m) and thermal cycling through T_0 (t/m) degrades the integrity of the coating owing to the associated changes in molar volume.

Ideally, TBCs require "nontransformable" (t′) compositions that remain tetragonal at all temperatures within the thermal cycle.[6,7] These are epitomized by the industry standard, ZrO_2+7–8 wt% (7.6–8.7 mol%)$YO_{1.5}$,[8] commonly referred to as 7YSZ—see Fig. 1. Because the t′ phase is supersaturated at all temperatures of interest, it is driven to decompose into a mixture of Y-lean tetragonal (t) and Y-rich cubic (c) phases. The depleted tetragonal phase would then be susceptible to the deleterious t↔m transformation on cycling.[6] The lifetime of current TBCs (T≤1250°C) is not limited by destabilization because other mechanisms usually intervene before the structure becomes transformable.[4] With prospective TBC surface temperatures approaching 1482°C (2700°F) in future turbines, the loss of t′ stability becomes more critical. Enhanced understanding of this problem is thus required to guide material design and improve life prediction models, motivating the present work.

TBCs are typically fabricated by air plasma spray (APS) or electron-beam physical vapor deposition (EBPVD), both yielding nominally single-phase t′ solid solutions. Whereas t′-YSZ grows directly from the vapor in EBPVD, in APS it evolves from the displacive transformation of the melt-crystallized cubic phase as it cools across the T_0 (c/t) curve in Fig. 1. The microstructures of EBPVD and APS t′ phases are thus distinctly different, with potential implications to their decomposition kinetics. Prior studies have shown, nonetheless, that the broader features of the phase evolution on aging of t′-YSZ are common to both types of TBCs; upon isothermal annealing the t′ phase evolves gradually into t and c phases, as expected, but the depleted tetragonal is not immediately transformable to the monoclinic symmetry.[6,9–14] The t→m transformation is martensitic[15] but nucleation controlled,[16] therefore suppressible upon cooling in sufficiently small particles[17] or fine polycrystalline aggregates.[18,19] The nucleation hindrance in particles arises primarily from the substantial difference in surface energy between the parent and product phases,[20] and in dense bodies from the mechanical constraint of the surrounding material. Nucleation can be activated by residual stresses, which scale with the grain size,[16,21] or by environmental factors.[22–25]

The above discussion suggests that the relative volume fractions, compositions, morphologies and size of the tetragonal, and cubic phases evolving from the decomposition of t′-YSZ play a role in the transformability of the structure and hence may influence the durability of the TBC. This has been demonstrated in other forms of ZrO_2–$YO_{1.5}$ materials, notably in the "colonies" of tetragonal phase evolving by precipitation from supersaturated cubic (c′) YSZ.[26,27] The colonies are comprised of twin related tetragonal domains with {110}-type habit planes. Prior TEM work in other systems has shown lamellar structures resulting from the decomposition of t′-YSZ,[11,28–30] but clearly different from the "colony" structure. There is also a long-standing debate on whether the c′/t′ → t+c reaction occurs by

A. Heuer—contributing editor

Manuscript No. 29064. Received December 19, 2010; approved February 25, 2011.
This paper is dedicated to the memory of Professor Anthony G. Evans: inspiring leader, generous colleague and good friend.
This investigation was financially supported under Grant DMR-0605700 from the National Science Foundation. The research made use of the UCSB-MRL Central Facilities supported by NSF under Grant DMR-0080034. Use of the Advanced Photon Source at Argonne National Laboratory was supported by the U. S. Department of Energy, Office of Basic Energy Sciences, under Contract No. DE-AC02-06CH11357. The scienctic input and technical support of the AMMRF node at the University of Sydney, Australia, is gratefully acknowledged. The authors are also grateful to Drs. Y. Gao (GE-GRC) for illuminating discussions and to Dr. R.M. Leckie (UCSB) for assistance during coating fabrication.
[†]Author to whom correspondence should be addressed. e-mail: jkoschmeder@engineering.ucsb.edu

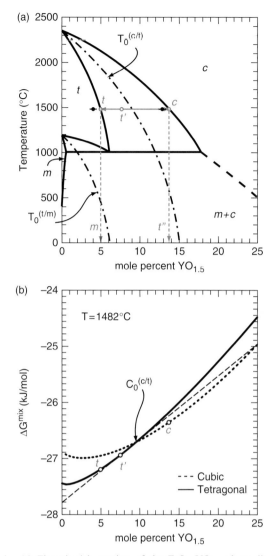

Table I. Evolving Phase Fractions Extracted from the Rietveld Refinement as a Function of Aging Time at 1482°C and Corresponding Hollomon–Jaffee Parameter[†]

Conditions		Phase fraction					Statistics	
HJP	h	t'	t	t''	c	m	R_p	χ^2
	0	100	0	0	0	0		
45 000	0.26	91.0	5.1	4.0	0	0	0.085	1.82
48 000	1.4	68.0	20.6	9.8	0	1.7	0.121	10.29
51 000	7.8	40.1	37.7	15.5	5.2	1.7	0.149	8.71
54 000	43.2	12.3	49.2	25.2	7.7	5.6	0.150	4.54
57 000	238.5	0	13.9	25.5	17.2	43.4	0.095	7.37

[†]The Hollomon-Jafee parameter is defined as HJP = $T[27+\ln(t)]$ where T is in K and t in h.

that both techniques are necessary to develop a consistent picture of the mechanisms underlying the phase evolution processes in EBPVD 7YSZ TBCs.

(1) Sample Preparation

7YSZ coatings ~ 200 μm thick were deposited on high-purity alumina substrates (99.5% purity, CoorsTek, Golden, CO) at ~ 1000°C using an in-house EBPVD facility and procedures outlined elsewhere.[35] Thermal exposures involved (i) inserting the specimen(s) into a furnace preheated to 1482°C, (ii) holding for the prescribed time, and (iii) withdrawing rapidly to cool in air. Dwell times from 0.26 to 238.5 h, listed in Table I, were selected to match specific values of a Hollomon-Jaffe type parameter[36] derived from aging studies of APS coatings.[37]

(2) Electron Microscopy

TEM was performed using FEI Tecnai G2 Sphera (Hillsboro, OR) and FEI Titan 300 kV FEG TEM/STEM systems, both equipped with energy dispersive X-ray spectroscopy (EDS). A JEOL 2100 microscope (Tokyo, Japan) with a Gatan 652 double-tilt heating stage (Pleasanton, CA) was used for *in situ* high-temperature TEM analysis. Most specimens were extracted from polished cross sections using an FEI DB235 dual-beam focused ion beam (FIB) system; those with microstructural features < 10 nm, where the higher voltage ion beam artifacts could interfere, were prepared either via traditional wedge polishing or using an FEI Helios 600 FIB at 8 kV. Microstructures were imaged using both conventional bright and dark field conditions, and in selected cases with high-angle annular dark field (HAADF). The {112} reflections on the $\langle 111 \rangle$ zone axes were selected for dark field imaging.

(3) X-Ray Characterization and Refinement

Before heat treatment, the coatings for X-ray characterization were carefully removed from the substrate using 30 μm lapping films and lightly separated to randomize the textured columnar structure typical of EBPVD TBCs. High-resolution synchrotron powder diffraction data were collected using beamline 11-BM at the Advanced Photon Source (APS), Argonne National Laboratory, with an average wavelength of 0.458643 Å. Discrete detectors were scanned at 0.15°/s over the range $-6 \leq 2\theta \leq 65°$, with data points collected at $\Delta 2\theta \approx 0.002°$ intervals.

The XRD data were refined via Rietveld's method[38] using the GSAS software[39] with the EXPGUI graphical interface.[40] The background was fit manually and the line profiles were fit using a convolution of pseudo-Voigt and asymmetry function, while also accounting for microstrain broadening, Γ_s^2, as described by Stephens.[41]

$$\Gamma_s^2 = S_{400}\left(h^4 + k^4\right) + S_{004}l^4 + 3S_{220}h^2k^2 + 3S_{202}(h^2l^2 + k^2l^2) \tag{1}$$

where h, k, l are the indices of the relevant peak and S is a corresponding scale factor. LaB$_6$ was used to calculate the

Fig. 1. (a) Zirconia-rich portion of the ZrO$_2$–YO$_{1.5}$ phase diagram, adapted from.[60] The hypothetical decomposition of t' 7YSZ and subsequent transformations upon cooling are illustrated. The solid circles represent the compositions estimated from X-ray diffraction tetragonality measurements after 43.2 h at 1482°C. (b) Calculated free energy curves at $T = 1482$°C for the tetragonal and cubic phases based on the assessed thermodynamic model.[51]

nucleation and growth[14,28,31,32] or by spinodal decomposition.[29,33,34] The inference is that the understanding of the mechanisms underpinning the evolution of these $t+c$ microstructures is inadequate.

This paper presents a systematic study of the phase evolution in EBPVD coatings aged at 1482°C (2700°F) for times up to ~ 240 h. The goal is not to assess the durability of YSZ, which is known to be inadequate at this temperature, but to understand the early stages of the decomposition process and their potential influence on the onset of transformability. The insight would ideally help identify whether there are any realistic paths to use alternate zirconia compositions with the metastable t' structure in future TBCs. EBPVD was selected because of the greater simplicity and homogeneity of the microstructure relative to APS coatings, with the idea of providing a baseline for ongoing studies on APS microstructures to be reported later.

II. Experimental Procedure

The approach involves characterization of pristine and aged columnar TBCs via high-resolution X-ray diffraction (XRD) and transmission electron microscopy (TEM). It will be shown

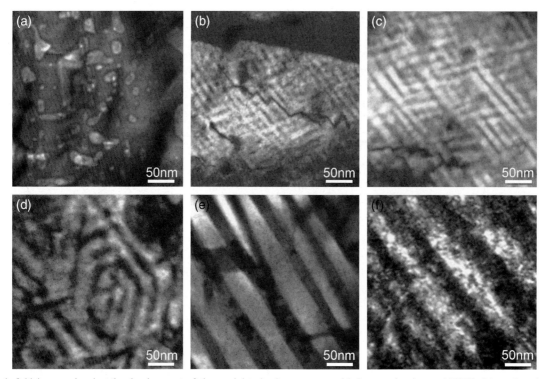

Fig. 2. Dark field images showing the development of the modulated microstructure with increased aging at 1482°C: (a) as deposited and after (b) 0.26 h, (c) 1.4 h, (d) 7.8 h, (e) 43.2 h, (f) 86 h. The lighter features in (a) are pores. The contrast in (b–f) reflects modulation in Y concentration. The column growth axis is parallel to the scale bar.

instrumental contribution to the peak width, and the line profiles for the t and t'' phases were assumed to be similar and thus refined together, allowing for the lattice parameter of each phase to vary. A maximum of 30 parameters were refined; the total depending on the number of phases present and the number of coefficients used to describe the background.

Five phases are possible, not including the internal alumina standard: the cubic (c) and monoclinic (m) phases, plus three tetragonal phases of different composition, t', t, and t'', the latter resulting from the partitionless transformation of the high-temperature c phase upon cooling (Fig. 1). The same structural model was used for all tetragonal phases, based on the primitive tetragonal cell, $P4_2/nmc$.[42] However, for the purposes of illustrating the relationship with the cubic form, the tetragonal phases will be discussed using a pseudo-cubic unit cell,[43] which is effectively a distorted fluorite with twice the volume of the primitive cell. The cubic and monoclinic structural models, $Fm\bar{3}m$ and $P2_1/c$, respectively, were adopted from Howard et al.[44]

(4) Lattice Mismatch and Coherency Strain
The lattice misfit, δ, between the t and t'' phases was calculated using the expression:

$$\delta = (d_{Y\text{-lean}}/d_{Y\text{-rich}}) - 1 \tag{2}$$

where d is the lattice spacing along the relevant crystallographic direction and the subscripts denote the Y-lean phase (t) and Y-rich (t'') phases. The spacings were determined from the XRD profile; because of coherency strains, these may differ from those of a stress-free crystal but little variation was observed after long annealing times. The XRD spacings were consistent with values calculated from the equilibrium compositions in Fig. 1 using established relationships between lattice parameters and Y-content (J. A. Krogstad et al., unpublished data).

Misfit values were calculated at ambient and 1482°C using recently determined coefficients of thermal expansion, dependent on composition, x(mol% $YO_{1.5}$), for the high-temperature

lattice parameters (Y. Gao et al., unpublished data):

$$\alpha_a = [10.00017 + 0.19413x] \times 10^{-6}\,K^{-1}$$
$$\alpha_c = [12.239 + 0.32576x - 0.5776x^2] \times 10^{-6}\,K^{-1} \tag{3}$$

III. Results

(1) TEM
Key features of the microstructure evolution up to the onset of monoclinic formation are summarized in Fig. 2. In the as-deposited condition, Fig. 2(a), each column is a porous tetragonal single crystal with uniform Y concentration and a $\langle100]$ growth axis, with the c-axis normal to the rotation direction.[45] Dark field imaging reveals contrast modulation normal to {111}-type planes after only 0.26 h (15.4 min) at 1482°C, Fig. 2(b). The diffraction pattern (not shown) is essentially identical to that of the as-fabricated coating, revealing that the initial single tetragonal orientation persists: the contrast arises from varying degrees of tetragonality and, by inference, composition. The darker cuboids‖ are Y-rich (lower tetragonality) whereas the lighter matrix is Y-lean (higher tetragonality). The cuboids are fully coherent with the matrix, as illustrated in Fig. 3(a). Because the specimen thickness was considerably greater than the observed domain size, it was not possible to accurately quantify the concentration of the cuboids or matrix. However, the correlation between composition and HAADF contrast can be established after some coarsening has occurred, as in Fig. 4. Furthermore, the compositions at the core of the lamellae are close to those expected in equilibrium, cf. Fig. 1.

On further aging, the cuboids merge and coarsen into lamellae with habit planes and interfaces along the {111} tetragonal planes, as illustrated in Figs. 2(c)–(f). Coarsening is considerably faster along the habit plane than normal to it, as shown in Fig. 5. The thickness evolves with the cube root of time in accordance with classical models.[46,47] However, the lengthening of the

---

‖The term cuboid is used loosely. Based on subsequent evolution, these are likely to be octahedra viewed along a [110] zone axis and bound by {111} interfaces, e.g Fig. 6.

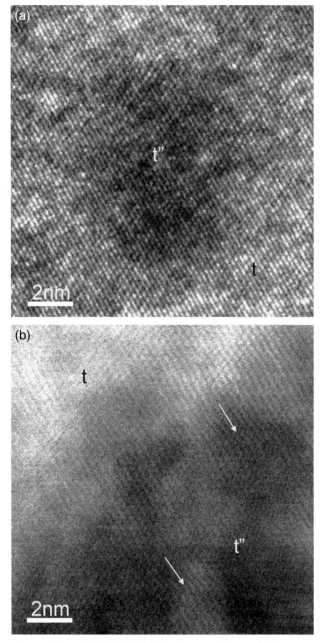

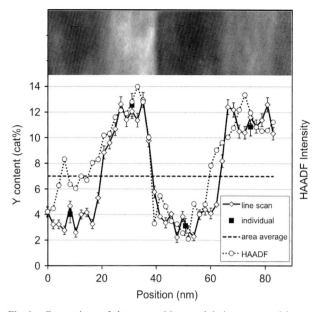

Fig. 4. Comparison of the composition modulation measured by energy dispersive X-ray spectroscopy (EDS) and the high-angle annular dark field (HAADF) contrast variation across a series of lamellae for a coating aged for 7.8 h. The measured compositions are consistent with those estimated from X-ray diffraction tetragonality values and marked in Fig. 1.

Fig. 3. High-resolutiontransmission electron microscopic (HRTEM) images of the modulated microstructure after (a) 0.26 h (15.4 min) and (b) 7.8 h at 1482°C. The dark cuboid of t'' phase in (a) is fully coherent with the surrounding t phase. The image in (b) reveals lattice bending near the diffuse interface between t and t'', examples of which are marked by arrows. No dislocations were identified by TEM over a much larger area of the latter interface.

lamellae appears to conform initially to a nearly linear relationship. The lamellae appear to remain fully coherent at least through 7.8 h, e.g. Fig. 3(b). Distortion of lattice fringes across the interface is evident (see arrows), but no dislocations were found in the areas examined.

In principle, the lamellar microstructure can adopt any of the orientations corresponding to the four possible {111} habit planes, e.g. Figs. 2(b)–(d) and 6. It is notable that all of these orientations maintain the same c-axis as the parent columnar crystal. For longer times, coarsening appears to favor one of the orientations over a broader area, whereupon the lamellae can extend across the entire width of the columns, as shown for the sample aged 43.2 h in Fig. 7. Pairs of branches and terminations, reminiscent of the fault migration coarsening mechanism in eutectics,[48] are also identifiable and denoted by arrows in Fig. 7. Closer examination of this figure reveals that pores do

not seem to significantly affect the lamellar pattern. Additionally, slight rotation of the convergent beam electron diffraction pattern between neighboring lamellae has also been observed (not shown), consistent with the expected degradation of interfacial coherency.

The extent of coherency loss is exacerbated in the more heavily aged coatings, as evident in the dark field image for the sample aged 86 h in Fig. 8. The modulated microstructure is still present in some regions (position A) but it is no longer the dominant feature. A second type of lamellar structure is observed (position B), but the correlation between contrast and compositional variation (cf. Fig. 4) is lost and the spatial orientation is distinct from that in the initial modulated structure. Fully incoherent, globular grains with compositions reasonably consistent with the equilibrium t and c boundaries in

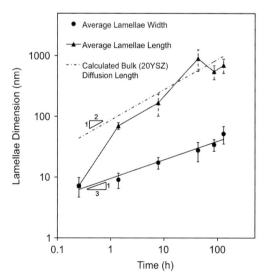

Fig. 5. Coarsening behavior of the modulated microstructure in terms of lamellar thickness and length, which follow different trends. The apparent drop in length after 43.2 h reflects the onset of breakdown of the lamellar structure. The calculated bulk diffusion distance in 20YSZ using the diffusion coefficient reported in Kilo *et al.*[60] is also plotted for comparison.

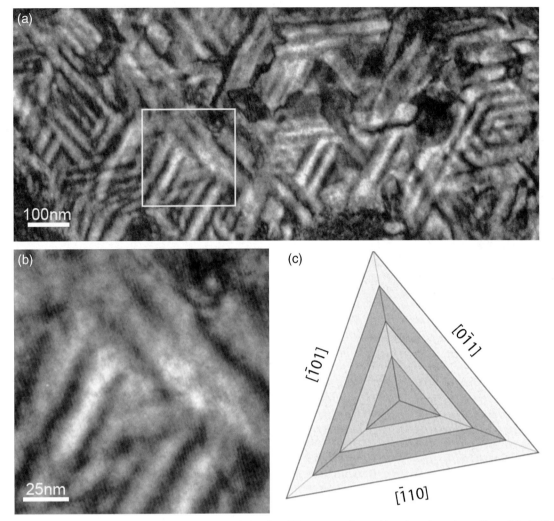

Fig. 6. (a) Lamellar packets in the sample aged for 7.8 h at 1482°C showing different crystallographic orientations along all possible {111} planes, but with a common [001] direction. (b) Shows a detail of lamellae converging in a tetrahedral-like pattern, schematically depicted in (c). The mismatch between neighboring lamellae is similar along the [$\bar{1}$01] and [0$\bar{1}$1] directions, but different along [$\bar{1}$10].

Fig. 1 are also observed (position C). However, the orientation of the *c*-axis within these grains is no longer in agreement with the initial column orientation. Moreover, the Y-rich regions (such as the grain-labeled C) are not cubic but exhibit three tetragonal variants in a "mottled" pattern, similar to those reported elsewhere.[14] Hence, the microstructure at this point is quite complex and not uniform across the entire sample. Additional observation after 129 h confirms further evolution of the modulated lamellae toward the more globular microstructure.

Direct observation of the *t→m* transformation was made on a coating aged for 238.5 h using an *in situ* TEM heating stage, see Fig. 9. Vestiges of lamellar composition modulation, albeit irregular, were still observed, as illustrated in the EDX map. Within the Y-lean regions, the monoclinic symmetry was confirmed from diffraction and clearly distinguishable transformation twins. Complete reversal of the transformation was observed above 700°C and, if properly quenched, the tetragonal structure was maintained at room temperature. However,

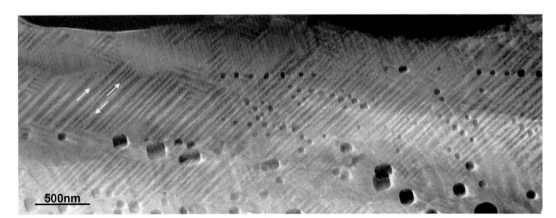

Fig. 7. The lamellar structure after 43.2 h of heat treatment is dominated by one of the original {111} orientations that has extended across much of the column width. Arrows denote lamellae ends suggestive of a fault migration coarsening mechanism. (high-angle annular dark field contrast in a ⟨110⟩ zone axis.)

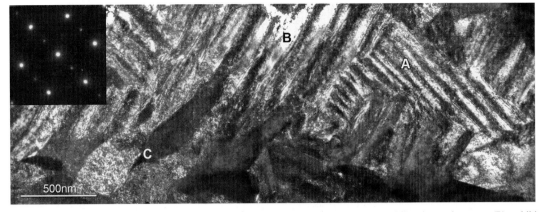

Fig. 8. Evidence of the original modulated microstructure is still observed after 86 h in location (A), while other regions, e.g. (B), exhibit composition variation no longer parallel to any of the original {111} planes. However, (A) and (B) still share the same cation lattice. Conversely, area (C) shows globular domains with crystallographic orientations that do not share the same [001] axis with their neighbors. The diffraction pattern is taken from area (C), multiple {112}-type reflections are present.

the tetragonal phase was not stable upon reheating and the transformation to the monoclinic phase proceeded rapidly above 150°C.

(2) XRD

The (004)/{400} region of the diffraction patterns for as-fabricated and aged coatings is presented in Fig. 10. Note that the entire profile was refined although only the (004)/{400} region is shown for clarity. The profile from the as-fabricated coating confirms the specimen is single-phase t'-YSZ. For a coating heat treated just 0.26 h at 1482°C, the analysis indicates that the majority of the coating has retained the t' structure (Table I); however, the microstructure of this coating, Fig. 2(b), clearly shows two fully coherent phases. After 1.4 h, the presence of the t and t'' phases is clearly evident in Fig. 10 but the pattern is still dominated by the t' reflection with no evident shift in lattice parameter. Over the subsequent heat treatments, the apparent t' phase fraction gradually decreases while the fractions of the t and t'' phases continue to increase. By 43.2 h, the t'-phase is no longer detected in the refinement. This is at variance with the TEM observations, which show only two phases over the same aging time (Fig. 2). A cubic phase distinct from t''-YSZ is observable by XRD after 7.8 h; however, this contribution is minor until much of the t phase has transformed into monoclinic symmetry after 238.5 h.

The refined lattice parameters for all phases in question remain essentially constant (maximum deviation of 0.15%, with no apparent trends) over the course of the heat treatment. However, the integral peak breadth is neither in agreement with the expected linear relationship to 2θ nor consistent from one heat treatment to another. Stephens' microstrain model (Eq. 1) has been applied to account for this. These values are reported in Fig. 11(a). Because the t and t'' phases were refined assuming the same profile description during the refinement, the microstrain in the t'' phase is omitted for clarity. The (004) and {400} have been chosen on the basis that these planes best characterize the anion sublattice and are also observed to deviate considerably from the expected linear behavior. The calculated microstrain for the t' phase is found to increase upon aging to 43.2 h, while the opposite is true for the t and t'' phases. This trend is concurrent with the observed longitudinal coarsening of the lamellae in Fig. 5 up to the same aging time, followed by an apparent drop after the structure starts breaking down. The accelerated increase in microstrain calculated in the coating after 43.2 h, Fig. 11, is ascribed to the formation of the monoclinic phase and associated transformation strains.

IV. Discussion

The observations above highlight a number of issues that merit further discussion. These will be broadly classified into those pertaining to the nature of the microstructural constituents and those related to the mechanisms of evolution. An overarching theme is the role of coherency in understanding (i) the apparent discrepancies between XRD and TEM observations, (ii) the

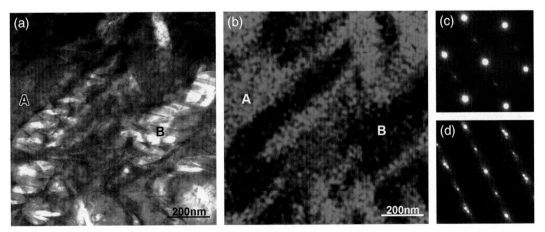

Fig. 9. After 238.5 h at 1482°C coarsened lenticular features denoted by (B) show transformation twins in (a) suggestive of the martensitic $t \rightarrow m$ transformation (high-angle annular dark field in a $\langle 110 \rangle$ zone axis). The EDX Y map in (b) reveals these regions to be lean in Y, and the diffraction pattern in (d) is indexed to monoclinic ZrO₂. Regions rich in Y, e.g. (A), exhibit the tetragonal pattern (c) consistent with t''. Note the EDX map is reminiscent of the modulated composition pattern evolving earlier in the process.

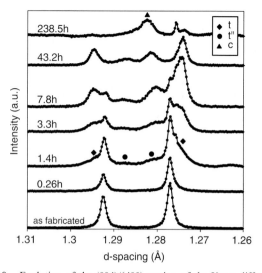

Fig. 10. Evolution of the (004)/{400} section of the X-ray diffraction profile upon aging at 1482°C highlighting the changes in phase constitution with aging.

initial decomposition stage, and (iii) the coarsening behavior preceding formation of the monoclinic phase.

(1) Microstructure and Phase Constitution

Comparison of the TEM and XRD observations clearly shows that the former are essential for proper interpretation of the latter. The phase analysis by XRD is broadly consistent with prior studies[6,9–11,13,14,32] and raises issues that had not been resolved satisfactorily. Notably, three distinct tetragonal phases with different Y content are identified in all aged structures before the appearance of the monoclinic phase, in addition to a cubic phase at times $\geq$ 7.8 h (Table I).

Two of the tetragonal phases, t and t''-YSZ are readily explained. The Y-lean t phase $(3.98 \pm 0.35 YO_{1.5})$ should form monoclinic on cooling (Fig. 1), but its retention is ascribed to kinetic limitations resulting from a combination of coherency, domain size, and mechanical constraint by the surrounding matrix.[16] The Y-rich t'' phase $(12.76 \pm 0.49 YO_{1.5})$ is also expected, stemming from the high-temperature cubic phase crossing the T_0 (c/t) curve upon cooling (Fig. 1). The concurrent presence of a cubic phase of similar composition, however, is more difficult to rationalize. The partitionless $c \rightarrow t$ transformation is reported to be displacive and nonsuppressible on cooling for <14% $YO_{1.5}$.[48] Hence, all t'' and c compositions—estimated from XRD lattice parameters—should be transformable.[49] Indeed, no cubic phase was detected in any of the TEM specimens, wherein all Y-rich grains examined showed one or more {112} reflections forbidden for cubic symmetry. Similar observations were made on aged compacts of precursor-derived powders[14] where c and t'' phases were identified by XRD but all grains showed tetragonality in the TEM. As in that study, no significant evidence of twins associated with the displacive transformation is observed. The coarsened Y-rich domains found in the later stages of the process, e.g. Fig. 8 (area C), showed a "mottled" structure with tetragonal domains of order $\sim$ 10 nm. These microstructures have been ascribed to an alternate short-range anion ordering mechanism that operates when the driving force for the shear transformation is diminished, i.e. when T_0 (c/t) approaches ambient.[14] In principle, these grains are also t''-YSZ but the degree of tetragonality is smaller and the collective interaction between the variants yields an apparent "cubic" peak in XRD. The reasons for the dual population of t'' domains in the XRD profile are further discussed later.

The apparent persistence of the original t' phase in the XRD profile has been previously noted[11–13] but is neither expected nor consistent with the present TEM observations. The current-free energy model for the tetragonal phase, Fig. 1(b), allows for nucleation of a cubic phase from t'-YSZ, but not of a separate t phase, so the coexistence of a Y-lean t-YSZ and a compositionally invariant t'-YSZ is problematic. In APS coatings, one might argue that the transformation is not spatially uniform because of local variations in composition, but such an argument is not applicable to these EBPVD coatings and no evidence was found of untransformed regions in the TEM, even at the shortest times examined. Indeed, at 0.26 h the structure is clearly modulated in composition but the XRD is largely dominated by t'-YSZ (Table I). Near the coherent interfaces between the Y-rich and Y-lean lamellae, the composition and lattice parameters are still nominally those of t'-YSZ, thus the t'-signal in XRD is hypothesized to originate from this diffraction volume. This putative t' "phase" is termed *pseudo-t'* for the purposes of the present discussion.

Clarification of the true nature of the persistent t' phase is particularly relevant to the fundamental understanding of phase stability because it has been assumed that its disappearance is linked to the onset of monoclinic formation.[6,10] It is evident from the TEM results that the true t'-YSZ disappears very early

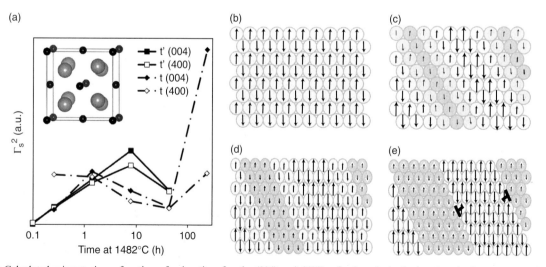

Fig. 11. (a) Calculated microstrain as function of aging time for the (004) and {400} reflections in both t' and t, the inset tetragonal structure is represented in the pseudo-cubic setting. (b–e) Schematic depictions of the anion lattice projected on a {111} plane. The magnitude of the arrows is intended to reflect the extent of tetragonality and hence Y content. (a) As-fabricated single-phase t' coating. (b) Lightly aged coating showing composition modulation. (c) Coarsened domains with only a small remaining *pseudo-t'* interfacial region. (d) Structure after loss of coherency with lattice relaxation enabled by interfacial misfit dislocations.

in the process although there is still a connection between the vanishing of the *pseudo-t′* peaks and the transformability of the Y-lean phase (Table I). This is further discussed below.

It is also worth noting that the lamellar microstructures in this study exhibit similar resistance to the $t \to m$ transformation as the "colony" structure resulting from decomposition of a saturated cubic phase above the T_0 (c/t) curve in Fig. 1.[26,27] However, the two structures are distinctly different. The colony structure consists of multiple alternating twin variants of the Y-lean tetragonal phase with each "packet" surrounded by the cubic matrix. Conversely, the lamellar structure here comprises alternating coherent Y-rich (t'') and Y-lean (t) phases of a single variant with a common c-axis orientation inherited from the original $t′$ crystal. Moreover, the habit plane of the "colony" structure is {110} whereas that of the modulated lamellae is {111}. It is not clear why the diffusional decomposition of $t′$- and $c′$-YSZ should give rise to such distinct microstructures, especially when the volume changes involved are so small, but the implication is that the mechanisms are different. Earlier work on precursor-derived powders showed indeed that the kinetics of decomposition at a given temperature are different for compositions lower and higher than that corresponding to T_0 (c/t).[51]

(2) Microstructure Evolution: Decomposition Stage

Whether the $t′ \to t+c$ transformation occurs by nucleation and growth or spinodal decomposition is still a subject of debate three decades after the phase stability problem in TBCs came to light. Rigorously resolving the issue is beyond the scope of this paper, but the present results provide substantial new insight into the mechanism. At early stages, the crystal structure is continuous and compositionally modulated on a nanometer scale, e.g. 5–7 nm in Fig. 2(b), with fully coherent interfaces, Fig. 3, and an inferred gradient in tetragonality across those interfaces. These observations are all consistent with spinodal decomposition but difficult to reconcile with classical nucleation and growth. However, the thermodynamic model for the tetragonal phase derived from the assessment of the phase diagram (Fig. 1)[52] does not contain a chemical spinodal, whereupon cubic precipitation can only occur by nucleation and growth.[26] There is an alternate thermodynamic approach that allows for a spinodal in the energy landscape by introducing an order parameter reflecting the relative anion displacements, which vary with composition.[33,34,52] While the latter models qualitatively replicate some aspects of the experimental observations,[29] the physical connection between the order parameter and the free energy is tenuous. Notably, the energy contribution of coherency strains is not always explicitly incorporated. To date, no model exists that can account for the all of the experimental observations, including, but not limited to, the alternate shear and ordering mechanisms in the diffusionless $c′ \to t′$ transformation, for the evolution of the "colony" structure when $c′$ is the parent phase, or for the absence of any signs of decomposition in the as-deposited structure, which involves substantial times at temperatures of order 1000°C.

Hence, while the present results are more consistent with a spinodal decomposition scenario, further refinements of the model are needed to account for the role of coherency in the transformation. One possible approach where the free energy is formulated as a function of composition and the tetragonal strain, referred to the cubic reference structure, has been applied by Van der Ven to similar problems in other systems[53,54] and will be investigated in the next stage of this project.

(3) Evolution of the Lamellar Structure

The coherency at interfaces plays a prominent role in the initial decomposition and the subsequent evolution of the morphology from the original "cuboids" to well-defined lamellae, as well as in the rate of coarsening. At the earliest times, the domains are so small that the constraint applied to the interface extends across the entire structure, preventing the Y-rich and Y-lean phases from relaxing to their equilibrium bulk lattice parame-

ters, effectively appearing as single phase $(t′)$ in XRD. Comparison of the domain size with the characteristic diffusion distance (Fig. 5) suggests that solute redistribution is initially controlled by the chemical and strain contributions to the gradient term in the free energy.[55] This excess energy also drives the structure to coarsen, allowing for progressive relaxation of the individual regions toward their equilibrium compositions and lattice parameters.

It is hypothesized that as long as the interfaces remain fully coherent, the structure is tetragonal everywhere, with c/a varying from the lowest to the highest Y contents. This is illustrated schematically in Fig. 11, where the arrows indicate both the degree and direction of the tetragonal shift of the anions, neglecting local distortions introduced by the presence of oxygen vacancies. For a pristine $t′$ coating the array is uniform, Fig. 11(b), but even light aging at 1482°C induces dopant segregation, with the anion shifts diverging from the $t′$ value according to the local Y content, Fig. 11(c). At this stage, the majority of the lattice is constrained by coherency to an average lattice parameter equivalent to that of the original $t′$ phase. Domain coarsening leads to progressive relaxation of the lattice parameters away from the coherent interface, Fig. 11(d), whereupon the Y-rich and Y-lean phases become evident in the diffraction pattern. Continued aging allows further relaxation of the lattice until coherency can no longer be maintained, Fig. 11(e).

The microstrain analysis in Fig. 11(a) reflects the evolving order/disorder within the t, t'', and *pseudo-t′* regions upon aging. While phenomenological, interpretation of the evolving microstrain provides insight on potential physical scenarios, such as interfacial coherency strain, which lead to the discrepancies between TEM and XRD observations. The *pseudo-t′* volume is initially dominant with nearly uniform lattice parameters, i.e. minimal microstrain, whereas the incipient t and t'' phases are highly disordered owing to a wider distribution of lattice parameters. As coarsening proceeds, the volume fraction of relaxed material within the t and t'' layers increases, while that constrained by the interfacial coherency (*pseudo-t′*) decreases. This sharpens the strain gradient at the interface and broadens the distribution of lattice parameters that are associated with the *pseudo-t′* region. Thus the microstrain drops steadily for the t and t'' phases and increases for *pseudo-t′*. When misfit dislocations are introduced to relieve the coherency strain at the interface, Fig. 11(e), the microstrain associated with the diminishing *pseudo-t′* volume drops slightly before vanishing. At that point, the microstrain in Fig. 11(a) increases rapidly because of the transformation stresses.

The coherency strains also influence the increase in aspect ratio of the second phase during coarsening, as noted in Fig. 5. The morphological evolution in Fig. 2 is reminiscent of those predicted by Wang *et al.*[56] The transformation starts by the development of concentration waves along all soft directions of the crystal, assumed $\langle 111 \rangle$ as in cubic YSZ.[29] Clusters form at the intersection of these waves in a pattern reflecting the symmetry of the lattice, cf. Figs. 3(a) and 2(b), respectively. Over time and under the mismatch strain, these clusters coalesce into plates normal to the soft directions, e.g. by anisotropic aggregation of material at edges following a so-called "macrodislocation climb" mechanism.[56] At intermediate times, all orientations are equally likely, occasionally converging in tetrahedral patterns such as that in Fig. 6(b). With further coarsening, however, one of the orientations tends to predominate over a scale much larger than that of the packets in Fig. 6(b), e.g. Fig. 7. Because the tetragonal lattice parameters exhibit opposite trends with Y content (J. A. Krogstad *et al.*, unpublished data), the mismatch of t and t'' plates along $[110]/[\bar{1}10]$ is larger than that along $[\bar{1}01]/[0\bar{1}1]$, Fig. 6(c), and also larger at the aging temperature than at ambient (Table II). Every initial interface contains two $[\bar{1}01]/[0\bar{1}1]$ directions and one $[110]/[\bar{1}10]$; hence, one might expect that extension of the plates be favored along the better-matched directions. Alternatively, if the planes extended isotropically, coherency would be lost first along the $[110]/[\bar{1}10]$

Table II. Lattice Mismatch and Dislocation Spacings for Neighboring Tetragonal Lamellae with Low and High Stabilizer Contents

	25°C	1482°C
a (t, nm)	0.5094	0.5172
c (t, nm)	0.5183	0.5279
a (t'', nm)	0.5132	0.5226
c (t'', nm)	0.5149	0.5193
δ [101] (%)	−0.0365	0.3126
δ [110] (%)	−0.7399	−1.0268
s_D [101] (nm)	996	119
s_D [110] (nm)	48	35

directions, with a misfit dislocation spacing (s_D) of 35–48 planes (Table II).

(4) Loss of Coherency and Impact on Coarsening

Regardless of preferred coarsening direction, it is evident from the length of the lamellae in Fig. 7 ($\geq 3\ \mu m$) that at this stage the structure cannot remain fully coherent and mismatch dislocations must be present at the interfaces. These dislocations have Burgers vectors of $a/2 \langle 110 \rangle$ and lie on $\{111\}$ planes, making them glissile. Because the stacking fault energy is low, the dislocations can also dissociate with a spacing between partials of ~ 100 nm,[57] although that would not be achievable along the direction of greater mismatch because the nominal spacing of the total dislocations is smaller. Figure. 7 also shows faults in the form of lamellar terminations (denoted by arrows) that can play a significant role in coarsening by "climbing," essentially reversing the process by which clusters coalesced into lamellae at earlier stages. Misfit dislocations also offer faster diffusion paths along the interfaces, regardless of their state of dissociation. The net result is that loss of coherency should be accompanied by an acceleration of the coarsening kinetics.

Soon after the loss of full interfacial coherency, the lamellar structure starts breaking down, eventually leading to globularization of the phases. The structure at this point becomes heterogeneous, with some areas retaining the partially coherent lamellar modulation, others with modified lamellar orientations, and still others with incoherent globules, e.g. Fig. 8. The Y-lean domains are now transformable to the monoclinic form, with kinetics depending on the degree of coherency, while the Y-rich t'' domains may exhibit clear tetragonality if partially constrained, or appear cubic in the XRD pattern if incoherent.

The inference from the discussion above is that prolonged stability of the coatings depends on the preservation of the coherent, modulated microstructure. This is particularly evident when comparing the EBPVD microstructure to APS coatings or materials produced from precursors, both of which have an abundance of boundaries, preexisting faults, and in some cases inhomogeneous solute distribution. The latter generally experience destabilization at significantly shorter times, even if the modulated structure is present in the early stages of the transformation (Krogstad et al.; Lipkin et al., unpublished data). One might hypothesize that manipulating the coherency strains through codoping may be an avenue to improve the durability of the t' structures, but studies with alternate trivalent stabilizers in the 1350–1450°C range[14,58] suggest the potential improvements are likely to be modest.

V. Conclusions

Key factors and the underlying mechanisms influencing the phase stability of t'-YSZ-based TBCs have been identified. The true t' phase has been shown to be rapidly replaced by a modulated microstructure comprising a coherent, ordered array of Y-rich and Y-lean lamellar domains. The features of the transformation in the early stages are largely consistent with a

spinodal decomposition mechanism, followed by morphological evolution influenced by the coherency strain. As the structure ages, full interfacial coherency is lost, leading to a second stage of coarsening and eventual transformability to the monoclinic phase. This establishes that it is the rate of coarsening that dictates the phase stability of a conventional TBC, rather than the preservation of a single t' phase.

Proper interpretation of the evolving microstructure was achieved by relying on both TEM and XRD characterization techniques. This parallel diagnostic approach clarified several outstanding issues. The observation of the t' phase in XRD profiles of aged coatings is shown to be an artifact of the coherency between the t and t'' elements of the structure. Moreover, the cubic phase reported by XRD upon moderate aging is actually a form of t''-YSZ that exhibits the characteristic "mottled" structure and lower tetragonality. The phase evolution described in this study for TBCs is clearly a complex process that will require additional investigation, particularly to address the role of strain as a motivation for spinodal decomposition.

References

[1]D. L. Porter, A. G. Evans, and A. H. Heuer, "Transformation Toughening in Partially Stabilized Zirconia (PSZ)," *Acta Metall.*, 27 [10] 1649–54 (1979).

[2]A. G. Evans and R. M. Cannon, "Toughening of Brittle Solids by Martensitic Transformations," *Acta Metall.*, 34 [5] 761–800 (1986).

[3]A. G. Evans, D. R. Mumm, J. W. Hutchinson, G. H. Meier, and F. S. Pettit, "Mechanisms Controlling the Durability of Thermal Barrier Coatings," *Prog. Mater. Sci.*, 46 [5] 505–3 (2001).

[4]A. Evans, D. Clarke, and C. Levi, "The Influence of Oxides on the Performance of Advanced Gas Turbines," *J. Eur. Ceram. Soc.*, 28 [7] 1405–19 (2008).

[5]R. H. J. Hannink, P. M. Kelly, and B. C. Muddle, "Transformation Toughening in Zirconia-Containing Ceramics," *J. Am. Ceram. Soc.*, 83 [3] 461–87 (2000).

[6]R. A. Miller, "Phase Stability in Plasma Sprayed, Partially Stabilized Zirconia-Yttria"; pp. 241–53 in *Science and Technology of Zirconia*, Vol. I, Edited by A. H. Heuer and L. W. Hobbs. The American Ceramics Society Inc., Columbus, OH, 1981.

[7]T. A. Schaedler, R. M. Leckie, S. Kramer, A. G. Evans, and C. G. Levi, "Toughening of Nontransformable t′-YSZ by Addition of Titania," *J. Am. Ceram. Soc.*, 90 [12] 3896–901 (2007).

[8]NASA Technical Memorandum. NASA TM-78976, National Aeronautics Space Administration, 1978.

[9]J. R. Brandon and R. Taylor, "Phase-Stability of Zirconia-Based Thermal Barrier Coatings. 1. Zirconia Yttria Alloys," *Surf. Coat. Tech.*, 46 [1] 75–90 (1991).

[10]U. Schulz, "Phase Transformation in EB-PVD Yttria Partially Stabilized Zirconia Thermal Barrier Coatings During Annealing," *J. Am. Ceram. Soc.*, 83 [4] 904–10 (2000).

[11]A. Azzopardi, R. Mevrel, B. Saint-Ramond, E. Olson, and K. Stiller, "Influence of Aging on Structure and Thermal Conductivity of Y–PSZ and Y–FSZ EB-PVD Coatings," *Surf. Coat. Tech.*, 177, 131–9 (2004).

[12]V. Lughi and D. R. Clarke, "Transformation of Electron-Beam Physical Vapor-Deposited 8 wt% Yttria-Stabilized Zirconia Thermal Barrier Coatings," *J. Am. Ceram. Soc.*, 88 [9] 2552–8 (2005).

[13]G. Witz, V. Shklover, W. Steurer, S. Bachegowda, and H. P. Bossmann, "Phase Evolution in Yttria-Stabilized Zirconia Thermal Barrier Coatings Studied by Rietveld Refinement of X-Ray Powder Diffraction Patterns," *J. Am. Ceram. Soc.*, 90 [9] 2935–40 (2007).

[14]J. M. Cairney, N. R. Rebollo, M. Rühle, and C. G. Levi, "Phase Stability of Thermal Barrier Oxides: A Comparative Study of Y and Yb Additions," *Int. J. Mater. Res.*, 98 [12] 1177–87 (2007).

[15]G. K. Bansal and A. H. Heuer, "On a Martensitic Phase Transformation in Zirconia (ZrO₂)—1. Metallographic Evidence," *Acta Metall.*, 20 [11] 1281–9 (1972).

[16]A. H. Heuer and M. Rühle, "On the Nucleation of the Martensitic Transformation in Zirconia (ZrO₂)," *Acta Metall.*, 33 [12] 2101–12 (1985).

[17]R. C. Garvie, "Stabilization of Tetragonal Structure in Zirconia Microcrystals," *J. Phys. Chem.*, 82 [2] 218–24 (1978).

[18]W. Z. Zhu, "Grain Size Dependence of the Transformation Temperature of Tetragonal to Monoclinic Phase in ZrO₂ (Y₂O₃) Ceramics," *Ceram. Int.*, 22, 389–95 (1996).

[19]A. Suresh, M. J. Mayo, W. D. Porter, and C. J. Rawn, "Crystallite and Grain-Size-Dependent Phase Transformation in Yttria-Doped Zirconia," *J. Am. Ceram. Soc.*, 86 [2] 360–2 (2003).

[20]M. W. Pitcher, S. V. Ushakov, A. Navrotsky, B. F. Woodfield, G. Li, J. Boerio-Goates, and B. M. Tssue, "Energy Crossovers in Nanocrystalline Zirconia," *J. Am. Ceram. Soc.*, 88 [1] 160–7 (2005).

[21]M. Rühle and A. G. Evans, "High Toughness Ceramics and Ceramic Composites," *Prog. Mater. Sci.*, 33, 85–167 (1989).

[22]J. Chevalier, B. Cales, and J. M. Drouin, "Low-Temperature Aging of Y–TZP Ceramics," *J. Am. Ceram. Soc.*, 82 [8] 2150–4 (1999).

[23]K. Yasuda, Y. Goto, and H. Takeda, "Influence of Tetragonality on Tetragonal-to-Monoclinic Phase Transformation During Hydrothermal Aging in

Plasma-Sprayed Yttria-Stabilized Zirconia Coatings," *J. Am. Ceram. Soc.*, **84** [5] 1037–42 (2001).

[24]V. Lughi and D. R. Clarke, "Low-Temperature Transformation Kinetics of Electron-Beam Deposited 5 wt.% Yttria-Stabilized Zirconia," *Acta Mater.*, **55** [6] 2049–55 (2007).

[25]J. Chevalier, L. Gremillard, A. V. Virkar, and D. R. Clarke, "The Tetragonal-Monoclinic Transformation in Zirconia: Lessons Learned and Future Trends," *J. Am. Ceram. Soc.*, **92** [9] 1901–20 (2009).

[26]A. H. Heuer and M. Rühle, "Phase Transformations in ZrO_2-Containing Ceramics: I, The Instability of c-ZrO_2 and the Resulting Diffusion-Controlled Reactions"; pp. 1–13 in *Science and Technology of Zirconia II*, Edited by N. Claussen, M. Rühle, and A. H. Heuer. The American Ceramics Society, Inc., Columbus, OH, 1984.

[27]V. Lanteri, T. E. Mitchell, and A. H. Heuer, "Morphology of Tetragonal Precipitates in Partially Stabilized ZrO_2," *J. Am. Ceram. Soc.*, **49** [7] 564–9 (1986).

[28]R. H. J. Hannink, "Growth Morphology of the Tetragonal Phase in Partially Stabilized Zirconia," *J. Mater. Sci.*, **13**, 2487–96 (1978).

[29]T. Sakuma, Y. I. Yoshizawa, and H. Suto, "The Modulated Structure Formed by Isothermal Agining in ZrO_2–5.2 mol%Y_2O_3 Alloy," *J. Mater. Sci.*, **20** [3] 1085–92 (1985).

[30]M. Doi and T. Miyazaki, "On the Spinodal Decomposition in Zirconia–Yttria (ZrO_2–Y_2O_3) Alloys," *Philos. Mag. B—Phys. Condens. Matter. Stat. Mech. Electron. Opt. Magn. Prop.*, **68** [3] 305–15 (1993).

[31]V. Lanteri, R. Chaim, and A. H. Heuer, "On the Microstructures Resulting from the Diffusionless Cubic–Tetragonal Transformation in ZrO_2–Y_2O_3 Alloys," *J. Am. Ceram. Soc.*, **69** [10] C258–61 (1986).

[32]V. Lughi and D. R. Clarke, "High Temperature Aging of YSZ Coatings and Subsequent Transformation at Room Temperature," *Surf. Coat. Tech.*, **200** [5–6] 1287–91 (2005).

[33]M. Hillert and T. Sakuma, "Thermodynamic Modeling of the $c \rightarrow t$ Transformation in ZrO_2 Alloys," *Acta. Metall. Mater.*, **39** [6] 1111–5 (1991).

[34]J. Katamura and T. Sakuma, "Computer Simulation of the Microstructural Evolution During the Diffusionless Cubic-to-Tetragonal Transition in the System ZrO_2–Y_2O_3," *Acta Mater.*, **46** [5] 1569–75 (1998).

[35]S. Krämer, J. Yang, C. G. Levi, and C. A. Johnson, "Thermochemical Interaction of Thermal Barrier Coatings with Molten CaO–MgO–Al_2O_3–SiO_2 (CMAS) Deposits," *J. Am. Ceram. Soc.*, **89** [10] 3167–75 (2006).

[36]J. H. Hollomon and L. D. Jaffe, "Time-Temperature Relations in Tempering Steel," *Trans. AIME*, **162**, 223–49 (1945).

[37]C. A. Johnson, "Phase Stability of Yttria-Stabilized Zirconia after Isothermal and Gradient Aging, *in International Conference & Exposition on Advanced Ceramics and Composites*, 2009, Daytona Beach, FL.

[38]H. M. Rietveld, "A Profile Refinement Method for Nuclear and Magnetic Structures," *J. Appl. Crystallogr.*, **2**, 65–71 (1969).

[39]A. C. Larson and R. B. Von Dreele, "General Structure Analysis System (GSAS)," *Los Alamos Natl. Lab. Rep. LAUR.*, 86–748 (2004).

[40]B. H. Toby, "EXPGUI, a Graphical User Interface for GSAS," *J. Appl. Crystallogr.*, **34**, 210–3 (2001).

[41]P. W. Stephens, "Phenomenological Model of Anisotropic Peak Broadening in Powder Diffraction," *J. Appl. Crystallogr.*, **32**, 281–9 (1999).

[42]G. Teufer, "Crystal Structure of Tetragonal ZrO_2," *Acta. Crystallogr.*, **15**, 1187 (1962).

[43]J. Lefevre, "De Differenetes Modifications Structurales des Phases de Type Fluorine Dans les Systemses a Base de Zircone ou D'Oxyde de Hafnium," *Ann. Chim.*, **8**, 117–49 (1963).

[44]C. J. Howard, R. J. Hill, and B. E. Reichert, "Structures of the ZrO_2 Polymorphs at Room-Temperature by High-Resolution Neutron Powder Diffraction," *Acta Crystallogr.*, *Sect. B: Struct. Sci.*, **44**, 116–20 (1988).

[45]U. Schulz, S. G. Terry, and C. G. Levi, "Microstructure and Texture of EB-PVD TBCs Grown Under Different Rotation Modes," *Mater. Sci. Eng. A.*, **A360**, 319–29 (2003).

[46]I. M. Lifshitz and V. V. Slyozov, "The Kinetics of Precipitation from Supersaturated Solid Solutions," *J. Phys. Chem. Solids.*, **19** [1–2] 35–50 (1961).

[47]C. Wagner, "Theorie Der Alterung von Niederschlagen durch Umlosen (Ostwald-Reifung)," *Z. Elektrochem.*, **65** [7–8] 581–91 (1961).

[48]J. Llorca and V. M. Orera, "Directionally Solidified Eutectic Ceramic Oxides," *Prog. Mater. Sci.*, **51** [6] 711–809 (2006).

[49]A. H. Heuer, R. Chaim, and V. Lanteri, "The Displacive Cubic–Tetragonal Transformation in ZrO_2 Alloys," *Acta Metall.*, **35** [3] 661–6 (1987).

[50]N. R. Rebollo, "Phase Stability of Thermal Barrier Oxides Based on t'-Zirconia with Trivalent Oxide Additions"; pp. 161 in *Materials Department*. University of California, Santa Barbara, Santa Barbara, CA, 2005.

[51]O. Fabrichnaya, M. Zinkevich, and F. Aldinger, "Thermodynamic Modelling in the ZrO_2–La_2O_3–Y_2O_3–Al_2O_3 System," *Int. J. Mater. Res.*, **98** [9] 838–46 (2007).

[52]D. Fan and L. Q. Chen, "Possibility of Spinodal Decomposition in ZrO_2–Y_2O_3 Alloys: A Theoretical Investigation," *J. Am. Ceram. Soc.*, **78** [6] 1680–6 (1995).

[53]J. Bhattacharya and A. Van der Ven, "Mechanical Instabilities and Structural Phase Transitions: The Cubic to Tetragonal Transformation," *Acta Mater.*, **56** [16] 4226–32 (2008).

[54]A. Van der Ven, K. Garikipati, S. Kim, and M. Wagemaker, "The Role of Coherency Strains on Phase Stability in Li_xFePO_4: Needle Crystallites Minimize Coherency Strain and Overpotential," *J. Electrochem. Soc.*, **156** [11] A949–57 (2009).

[55]J. W. Cahn and J. E. Hilliard, "Free Energy of a Non-Uniform System. 1. Interfacial Free Energy," *J. Chem. Phys.*, **28** [2] 258–67 (1958).

[56]Y. Wang, L. Q. Chen, and A. G. Khachaturyan, "Kinetics of Strain-Induced Morphological Transformation in Cubic Alloys with a Miscibility Gap," *Acta. Metall. Mater.*, **41** [1] 279–96 (1993).

[57]U. Messerschmidt, D. Baither, B. Baufeld, and M. Bartsch, "Plastic Deformation of Zirconia Single Crystals: A Review," *Mater. Sci. Eng., A.*, **233** [1–2] 61–74 (1997).

[58]C. G. Levi, "Emerging Materials and Processes for Thermal Barrier Systems," *Curr. Opin. Solid State Mater. Sci.*, **8**, 77–91 (2004).

[59]O. Fabrichnaya, C. Wang, M. Zinkevich, C. G. Levi, and F. Aldinger, "Phase Equilibria and Thermodynamic Properties of the ZrO_2–$GdO_{1.5}$–$YO_{1.5}$ System," *J. Phase Equilib.*, **26** [6] 591–604 (2005).

[60]M. Kilo, M. A. Taylor, C. Argirusis, G. Borchardt, B. Lesage, S. Weber, S. Scherrer, H. Scherrer, M. Schroeder, and M. Martin, "Cation Self-Diffusion of 44Ca, 88Y, 96Zr in Single-Crystalline Calcia and Yttria-Doped Zirconia," *J. Appl. Phys.*, **94** [12] 7547–52 (2003). □

J. Am. Ceram. Soc., **94** [S1] S178–S185 (2011)
DOI: 10.1111/j.1551-2916.2011.04448.x
© 2011 The American Ceramic Society

Journal

Chemical and Mechanical Consequences of Environmental Barrier Coating Exposure to Calcium–Magnesium–Aluminosilicate

Bryan J. Harder,[‡,||] Joaquin Ramírez-Rico,[‡,††] Jonathan D. Almer,[§] Kang N. Lee,[¶] and Katherine T. Faber[†,‡]

[‡]Department of Materials Science and Engineering, Northwestern University, Evanston, Illinois 60208

[§]Argonne National Laboratory, Advanced Photon Source, Argonne, Illinois 60439

[¶]Rolls-Royce Corporation, Materials, Processes and Repair Technologies, Indianapolis, Indiana 46206

The success of Si-based ceramics as high-temperature structural materials for gas turbine applications relies on the use of environmental barrier coatings (EBCs) with low silica activity, such as $Ba_{1-x}Sr_xAl_2Si_2O_8$ (BSAS), which protect the underlying components from oxidation and corrosion in combustion environments containing water vapor. One of the current challenges concerning EBC lifetime is the effect of sandy deposits of calcium–magnesium–aluminosilicate (CMAS) glass that melt during engine operation and react with the EBC, changing both its composition and stress state. In this work, we study the effect of CMAS exposure at 1300°C on the residual stress state and composition in BSAS–mullite–Si–SiC multilayers. Residual stresses were measured in BSAS multilayers exposed to CMAS for different times using high-energy X-ray diffraction. Their microstructure was studied using a combination of scanning electron microscopy and transmission electron microscopy techniques. Our results show that CMAS dissolves the BSAS topcoat preferentially through the grain boundaries, dislodging the grains and changing the residual stress state in the topcoat to a nonuniform and increasingly compressive stress state with increasing exposure time. The presence of CMAS accelerates the hexacelsian-to-celsian phase transformation kinetics in BSAS, which reacts with the glass by a solution–reprecipitation mechanism. Precipitates have crystallographic structures consistent with Ca-doped celsian and Ba-doped anorthite.

I. Introduction

SILICON-BASED ceramics, such as SiC or Si_3N_4, are proposed candidates for high-temperature structural applications such as gas turbine components due to their high thermomechanical stability and low density.[1,2] One of the main drawbacks of these materials is their low oxidation resistance in combustion environments, as the protective SiO_2 that forms on the surface of Si-based ceramics reacts with water vapor at high temperatures to form a volatile silicon hydroxide, leading to surface recession and limiting component lifetime.[3–7] For this reason, environmental barrier coatings (EBCs) were developed as means of protecting Si-based ceramics from corrosion and recession in combustion environments.[2,8–16]

EBCs based on $Ba_{1-x}Sr_xAl_2Si_2O_8$ (BSAS) constitute the baseline performance of current EBCs, although other compositions such as rare-earth silicates are currently being explored.[2,11,17] BSAS EBCs are air plasma sprayed at high temperature ($\approx 1200°C$) on SiC/SiC substrates as the topcoat of multilayer coating systems. Typically, a bond coat of silicon ($\approx 125 \mu m$) is deposited on the ceramic substrate to enhance adhesion of the subsequent layers, as well as to provide additional oxidation resistance.[2] A thicker ($\approx 200 \mu m$) intermediate layer consisting of mullite ($3Al_2O_3 \cdot 2SiO_2$) or mullite plus $SrAl_2Si_2O_8$ (SAS) is then deposited and serves both to prevent the topcoat from reaction with the silica scale on the substrate and to better match the thermoelastic properties of the substrate with those of the topcoat to minimize residual stresses. Finally, a $\approx 200 \mu m$ layer of BSAS ($x \le 0.25$), constituting the EBC, is deposited.

In room-temperature plasma-sprayed systems, BSAS is deposited as an amorphous phase and crystallizes in the metastable hexacelsian phase (hexagonal) when heated. Heating the substrate to 1200°C during the plasma-spray process is sufficient to deposit BSAS in the hexacelsian phase. This hexacelsian phase then undergoes a phase transformation at temperatures over 1200°C into the stable celsian phase (monoclinic).[18–23] This transformation is accompanied by a 0.5% volume reduction and drastically changes the elastic modulus and coefficient of thermal expansion (CTE) of the topcoat (see Table I). Harder and Faber[24] showed that the transformation rate of plasma-sprayed BSAS was faster as a coating than when crushed into a powder due to the presence of unmelted celsian particles from the plasma-spray powder feed. The unmelted particles act as seeds, and thereby effectively reduce the activation energy of the hexacelsian-to-celsian transformation of freestanding plasma-sprayed BSAS bars compared with BSAS powders obtained by crushing and grinding plasma-sprayed material.

The CTEs and compliances of the different layers within the EBC system must be matched to minimize residual stresses that arise upon thermal cycling of the multilayer. The celsian/hexacelsian phase fraction in BSAS was shown to have a high impact on the residual stresses due to CTE differences.[9] The high CTE of the hexacelsian phase induced large tensile stresses in the topcoat that lead to through-thickness cracking upon cooling. The stress drop due to the formation of these cracks was detected *in situ* using a high-energy diffraction technique,[10] which showed that when the topcoat was completely transformed to the celsian phase by a static heat treatment, the resulting compressive residual stress inhibited crack formation.

A challenge for EBC operation is the deleterious effect of reactive glasses that deposit onto hot surfaces within the engine, which was identified by Evans and colleagues as one of the *extrinsic* factors governing thermal barrier coating (TBC) lifetimes.[25,26] These glasses can form from dust and sand deposits in engines operated at temperatures over 1200°C[27–31] and can infiltrate pores and small cracks to subsequently react with the coating, limiting its lifespan by causing premature failure. The composition of deposits on turbine blades coated with a

C. Levi—contributing editor

Manuscript No. 28705. Received September 30, 2010; approved January 14, 2011.
This work was supported by the Department of Energy, Office of Basic Energy Science, under contract number DE-AC02-06CH11357.
[†]Author to whom correspondence should be addressed. e-mail: k-faber@northwestern.edu
[||]Present address: NASA Glenn Research Center, Cleveland, Ohio.
[††]Present address: University of Seville, Spain.

Table I. Thicknesses, Elastic Coefficients for the Crystallographic Planes Measured by Transmission Diffraction and Coefficient of Thermal Expansions (CTEs) of the Multilayer System[9]

Phase	Thickness (μm)	hkl	E_{hkl} (GPa)	ν_{hkl}	α ($10^{-6}\,°C^{-1}$)
BSAS (Hexacelsian)	200	$(1\bar{1}0)$	129	0.35	8.37
BSAS (Celsian)	200	$(\bar{1}12)$	81	0.30	4.28
Mullite	200	(220)	145	0.16	5.50
SAS	200	(041)	97	0.37	4.28
Silicon	125	(113)	187	0.15	4.44
Silicon carbide	≈ 3000	(220)	422	0.17	5.06

yttria-stabilized zirconia (YSZ) TBCs is mostly independent of geography, operating conditions or degree of exposure and consists of a calcium–magnesium aluminosilicate (CMAS) eutectic ($33CaO–9MgO–13AlO_{1.5}–45SiO_2$) with a low-melting point ($T_m = 1250°C$).[31] The effects of CMAS attack on YSZ TBCs have been extensively studied and shown to be highly detrimental on TBC performance and lifetime,[32] as it can induce failure in multiple ways.[26] At high temperatures, CMAS glass can infiltrate pores and cracks and the CTE mismatch between the solid CMAS glass ($\approx 8.1 \times 10^{-6}\,°C^{-1}$)[33] and the YSZ ($\approx 11.1 \times 10^{-6}\,°C^{-1}$)[34] introduces large compressive residual stresses that can lead to delamination and cold-shock fracture.[33] Additionally, chemical interaction between CMAS and the TBC has been shown to reduce the coating's lifespan, as it destabilizes tetragonal YSZ.[35,36] The mechanism of this transformation involves the dissolution of the metastable tetragonal phase and the reprecipitation of ZrO_2 in the stable monoclinic phase. This mechanism is accompanied by a change in microstructure and a CTE mismatch with the substrate, resulting in a loss of strain tolerance and limiting service lifetimes.

BSAS EBCs showed similar CMAS corrosion problems to TBCs.[37] In freestanding plasma-sprayed specimens with low glass loading (≈ 10 mg/cm^2) and short hold times (1 h) at 1300°C, no substantial reaction of the glass with the coating was observed, but was evident at longer holding times (4 h). Higher loading (60 mg/cm^2) resulted in increased penetration and greater dissolution of BSAS into the molten glass. CMAS penetrated the grain boundaries of the EBC coating and dislodged the BSAS crystallites, which were dissolved into the glass and reprecipitated as a calcium-containing modified celsian phase. At low CMAS loading, precipitates of anorthite ($Ca_2Al_2Si_2O_8$) with Ba and Sr in solid solution and diopside ($CaMgSi_2O_6$) were also detected. The penetration depths observed after 4 h were on the order of typical plasma-sprayed EBC coating thicknesses, making CMAS corrosion a concern regarding lifetimes. However, it is important to note that although the CMAS was shown to penetrate over 200 μm, the loadings used to reach that depth (60 mg/cm^2) were extremely large, and may not be representative of CMAS exposure in typical engine environments.

In addition to altering the EBC composition, CMAS interaction is expected to have an effect on mechanical performance of these multilayer systems due to CTE mismatch effects. Although extensively studied for TBCs, no data exist in the literature for BSAS EBC systems. The aims of the present work were twofold. Firstly, we studied the effect of CMAS on the stress state of BSAS EBC multilayers by means of high-energy X-ray diffraction using a synchrotron source, a technique that has been shown to be well suited for this purpose as it allows stress measurement in multilayers as a function of depth with a spatial resolution of $< 50\ \mu$m.[38] Not only can the average stress state in the different layers that compose the system be measured with this method but also local stress variations due to partial infiltration and reaction of CMAS can be elucidated. Also, because X-rays provide compositional information, this technique allows the quantification of hexacelsian/celsian fractions within

the BSAS topcoat with spatial resolution, to determine whether the presence of aliovalent cations from the glass affects the hexacelsian-to-celsian transformation kinetics. Secondly, we explored the effect of CMAS exposure on the microstructure and composition of BSAS for glass loadings closer to those commonly found in field tests[36] and for longer exposure times than those reported previously.

II. Experimental Methods

(1) Plasma-Spray Processing

Multilayer systems were deposited using a procedure described previously.[9] In brief, doped BSAS coatings were air plasma sprayed onto coupons (25.4 mm × 6.4 mm × 3.5 mm) of melt-infiltrated SiC/SiC ceramic matrix composite that were grit blasted with alumina to improve bonding to the substrate. A bond coat of silicon (Atlantic Equipment Engineers, Bergenfield, NJ) was applied first, followed by a layer of mullite+SAS from a mechanical mixture of 80 wt% mullite and 20 wt% SAS. In a final step, a topcoat of BSAS (HC Stark, Goslar, Germany) was applied. Details of the plasma spraying are proprietary, but are similar to those used by Lee in previous works.[12] The coatings were sprayed into a furnace where the substrate was heated to 1200°C in order to avoid depositing amorphous material that would cause catastrophic failure upon crystallization.[12,13]

Free-standing BSAS bars were plasma sprayed at room temperature for CMAS reaction studies using the procedure described in Harder and Faber.[24] BSAS powders (HC Stark) were sprayed onto grit-blasted steel substrates (55 mm × 25 mm × 3 mm), which were ultrasonically cleaned in isopropanol for approximately 5 min before deposition. The deposited material had a final thickness of ≈ 2.5 mm. After the deposition, the coated substrates were cooled at room temperature. Within 2–3 min, the BSAS delaminated cleanly from the substrate due to CTE mismatch between the steel and coating. The freestanding material was cut into 5 mm × 12 mm bars.

Following a procedure described previously,[24] the freestanding BSAS bars were heat treated at 990°C for 10 h to crystallize the amorphous, as-deposited BSAS into the metastable hexacelsian form. Subsequently, both the multilayer system and the freestanding samples were heat treated at 1300°C for 20 h to promote partial transformation of hexacelsian BSAS into the stable, low-CTE celsian phase, resembling the baseline or Gen 1 EBC as defined by Eaton *et al.*[39] and Spitsberg and Steibel.[40] The resulting coatings and freestanding bars were approximately 50 wt% celsian.

(2) CMAS Interaction Studies

The interaction of CMAS glass was investigated on both multilayer EBC coatings as well as freestanding BSAS bars. The glass was generated by mixing CaO (Aldrich, St. Louis, MO), MgO (Alfa Aesar, Ward Hill, MA), Al_2O_3 (Praxair, Danbury, CT), and SiO_2 (Alfa Aesar) in ratios consistent with the $33CaO–9MgO–13AlO_{1.5}–45SiO_2$ formula. The mixed powders were ball milled using Al_2O_3 media and water for 24 h. The mixture was dried and heated to 1550°C for 4 h in a Pt crucible to form a glass. The glass was ground with a mortar and pestle and fired a second time at 1550°C for 4 h to ensure a homogeneous mixture. Finally, the glass was removed from the crucible and ground to a fine powder ($\leq 50\ \mu$m particle size).

Multilayer EBC samples (BSAS/mullite/Si/SiC) with a topcoat composition of ≈ 50 wt% celsian were exposed to the CMAS by mixing 0.5 g of glass with 0.75 g of ethanol to create a mixture that could be subsequently applied to the surface of the coatings until there was a dried loading of 35 mg/cm^2 on the sample, which is close to values usually found in field tests and follows previous studies by Aygun *et al.*[36] Upon application of the glass, samples were heated to 1300°C for 4, 12, 24, or 48 h with a temperature ramp rate of 10°C/min.

In addition to multilayer EBCs, freestanding BSAS coating material was exposed to CMAS glass. Wells were bored into the material using an ultrasonic drill press (Branson Ultrasonics, Danbury, CT) with a radius of 1.56 mm to a depth of ≈ 1.2 mm. This diameter and depth of the trough were chosen to expose the surface area of the BSAS to 35 mg/cm^2 CMAS when completely filled. After drilling, the freestanding material was also heated to 1300°C for 20 h to promote the hexacelsian-to-celsian transformation. The wells were filled with ≈ 10 mg of CMAS glass, and heat treated at 1300°C for 4, 12, 24, and 48 h with a 10°C/min ramp rate. This approach ensured a constant CMAS loading, in contrast with the multilayer samples where variations in CMAS layer thickness were unavoidable due to coating roughness.

(3) High-Energy X-Ray Diffraction

High-energy X-ray diffraction was used to characterize the crystal structures, phases, and stresses within the multilayer coatings. Experiments were conducted at the 1-ID beam line of the Advanced Photon Source at Argonne National Laboratory. The general setup for transmission diffraction is shown in Fig. 1. High-brilliance, high-energy (80–86 keV) X-rays were produced by a Laue monochromator and were vertically focused using Si refractive lenses. A Mar345 image plate (345 mm diameter, 150 μm pixel size) was set at a distance of 1.2 m from the sample, and a CeO$_2$ standard was used to determine the sample-to-detector distance, beam tilt, and beam center.

Strains were measured in the multilayer EBCs as a function of depth at room temperature. The X-ray beam size was 50 μm × 100 μm, and during each exposure the sample was translated horizontally 2 mm to increase the number of sampled grains, resulting in a total diffracted volume of 0.35 mm.^{3} Samples were moved vertically with respect to the beam in 25 μm steps to probe the layers and the substrate.

To assess internal and residual stresses, strains were determined from the distortion of the Debye rings using a series of MATLAB programs developed at Sector 1 of the Advanced Photon Source at Argonne National Laboratory. The radii of diffraction planes for the materials were determined using the unit cell dimensions specified by powder diffraction files 00-026-0182 (hexacelsian BSAS), 38-1451 (celsian BSAS), 00-015-0776 (mullite), 38-1454 (celsian SAS), 00-027-1402 (silicon), and 00-029-1129 (β-silicon carbide).[41] Planes were chosen so as not to overlap with any other phases or materials. The reflections used to calculate strains present in the coating materials are shown in Table I. Diffraction peaks were fit to pseudo-Voigt functions to determine the peak centers, which were mapped as a function of azimuth to determine the strain calculated by the equation:

$$\varepsilon_\eta = \frac{r_0 - r_\eta}{r_0} \tag{1}$$

where r_η is the radius at azimuthal angle η, and r_0 is the strain-free radius, which is the measured radius at the strain-free angle (η^*). The latter is calculated from the X-ray elastic constants included in Table I using the following expression, valid under the assumption of a purely biaxial stress state[38]:

$$\sin^2 \eta^* = \frac{2\nu_{hkl}}{1 + \nu_{hkl}} \tag{2}$$

Error in the strain measurements were computed based on statistical error from peak fitting and standard errors from the biaxial fit of lattice parameter versus azimuth. The strains in the in-plane (ε_{11}) and the out-of-plane (ε_{33}) directions, corresponding to $\eta = 0°$ and $\eta = 90°$, were used to determine the in-plane equibiaxial stress ($\sigma_\parallel$) by[42]

$$\sigma_\parallel = \left(\frac{E}{1+\nu}\right)\varepsilon_{11} + \frac{\nu E}{(1-2\nu)(1+\nu)}(2\varepsilon_{11} + \varepsilon_{33}) \tag{3}$$

The X-ray elastic constants, E_{hkl}, ν_{hkl}, and the CTE, α, included in Table I were measured *in situ* at the synchrotron beam line; details of the measurement technique were published in a previous work.[9] In brief, α was determined by measuring changes in d-spacing as a function of temperature in crushed, plasma-sprayed BSAS, SAS, silicon, and mullite, as well as in the SiC/SiC coupon. Elastic constants E_{hkl} and ν_{hkl} were determined by *in situ* loading experiments on freestanding plasma-sprayed bars in compression, where Eq. (1) was used to determine strains parallel and perpendicular to the loading direction, as a function of applied stress. Elastic constants for all materials comprising the multilayers were determined in this manner, except for the SiC/SiC substrate, where an average of the Voigt and Reuss models from single crystal elastic constants[43] was used.

The ratio of celsian-to-hexacelsian phases in wt% in the EBCs was determined from the acquired X-ray diffraction patterns as a function of depth and exposure time to CMAS. The 2D patterns were radially integrated to obtain profiles of diffracted intensity versus 2θ that were then fit using the Rietveld method as implemented in the EXPGUI/GSAS package.[44,45] Starting structures were obtained from the ICSD database (references 97411–hexacelsian BSAS, 40530–celsian BSAS, 78793–SAS[46]). Initial values for diffractometer geometry and peak profiles were obtained by refining a pattern from the CeO$_2$ standard. These parameters were later refined along with lattice parameters and phase fractions in patterns from the BSAS topcoat.

(4) Microstructure Characterization

Representative cross sections of the multilayers before and after CMAS interaction at 1300°C for different times were prepared using conventional metallographic procedures for observation in the scanning electron microscope (SEM S3400, Hitachi High Technologies, Tokyo, Japan) using backscattered electrons (BSE) for compositional contrast. Sections of the freestanding bars were also prepared in this manner. From these samples, thin lamellae in areas representative of the observed features were prepared using a dual-beam focused ion beam/SEM (Helios NanoLab, FEI, Hillsboro, OR) and mounted onto copper grids using a micromanipulator (Autoprobe 200, Omniprobe, Dallas, TX). These lamellae were then observed under the transmission electron microscope (TEM, Jeol 2100F) in both conventional and scanning modes. X-ray energy-dispersive spectra and maps were acquired, and selected area diffraction (SAD) patterns were obtained for identification of precipitated phases in the BSAS/CMAS interaction zone. ICSD structure references used for indexing of SAD patterns were those mentioned in Section III (3) and 78793 (anorthite).

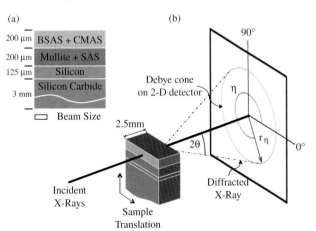

Fig. 1. Schematic for the transmission high-energy X-ray diffraction experiments. (a) To-scale sketch of the multilayer environmental barrier coating studied, showing typical coating thicknesses and diffraction beam size. (b) Experimental setup at beamline 1-ID showing the diffraction geometry used.

3. Results and Discussion

(1) Microstructure of As-Deposited Multilayer EBCs

A representative cross section of the microstructure of plasma-sprayed EBCs multilayers studied in this work is shown in Fig. 2, for a sample heat treated at 1300°C for 20 h. The roughness of the interfaces, due to the plasma-spray process, is easily seen. The average thickness of the layers were close to those reported in Table I although, due to the mentioned roughness, thickness varied as much as ± 50 μm in some parts of the coating. The topcoat layer corresponds to BSAS, which contains a significant amount of porosity. The contrast that can be seen is due to an inhomogeneous distribution of Ba and Sr, changing the local Ba/Sr ratio in adjacent grains. The intermediate layer consists of a mixture of mullite (dark contrast) and 20 wt% SAS. The silicon bondcoat serves to enhance adhesion between the intermediate layer and the SiC/SiC substrate, which can be seen in the bottom part of Fig. 2. In the composite, the SiC fibers are woven and run parallel to the layer interfaces both along and normal to the micrograph.

(2) CMAS Exposure Effects on Multilayer BSAS EBCs

Stress as a function of depth, determined by X-ray diffraction, is shown in Fig. 3 for a multilayer sample that had been exposed to 1300°C for 20 h. The stresses observed in all layers are consistent with those reported previously in this system.[9] The variation of stress was likely due to lack of uniformity in the coating thicknesses as well as the presence of pores.

For samples exposed to CMAS, stresses were measured in each layer of the coating as well as the SiC substrate, but the differences in the stress magnitude of these layers were nearly negligible over the course of the CMAS exposure. The BSAS topcoat in the current study was most affected by the exposure

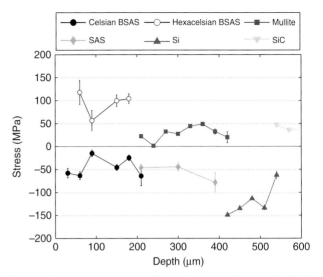

Fig. 3. Biaxial residual stress profile as a function of depth in the BSAS multilayer EBC system after heat treatment at 1300°C for 20 h, measured by transmission X-ray diffraction.

to CMAS. In Fig. 4, the stress in the BSAS topcoat is plotted as a function of depth in each of the samples. The stress at the surface of the BSAS became increasingly compressive with CMAS exposure time. The elevated stress was a result of the interaction of the glass with the BSAS and not the heat treatment, as it was reported that samples of this EBC system that were thermally cycled to 1300°C for 100 h exhibited a constant compressive stress of ≈30 MPa across the coating layer.[9] The elevated compressive stress at the surface of samples exposed to CMAS was likely caused by the CTE of the residual glass (or byproducts that were formed from the interaction) to be larger than the topcoat. The CTE of amorphous CMAS was estimated by Chen to be $8.1 \times 10^{-6} °C^{-1}$.[33] This is similar to the CTE of hexacelsian BSAS ($8.37 \times 10^{-6} °C^{-1}$), but much larger than the celsian BSAS ($4.28 \times 10^{-6} °C^{-1}$).

The microstructure of the BSAS topcoat also changed with exposure to the CMAS. The surface was roughened considerably during the hexacelsian-to-celsian transformation due to the more faceted nature of the precipitated celsian grains. Additionally, the glass incorporated some of the BSAS material in the form of precipitates, as shown in Fig. 5. At 4 h, a significant number of elongated precipitates appear to grow from the BSAS/glass interface with varied levels of BSE contrast, suggesting a nonhomogeneous composition. A different, smaller type of precipitate can be seen with dark BSE contrast at the

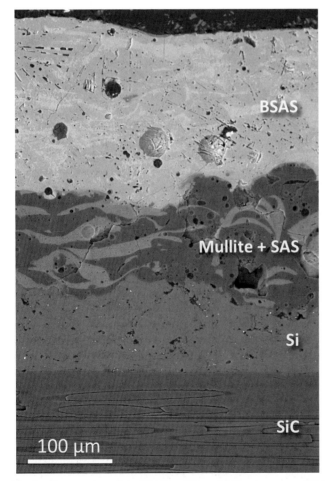

Fig. 2. Cross section of the BSAS multilayer EBC system on an SiC/SiC melt infiltrated composite, after heat treatment at 1300°C for 20 h.

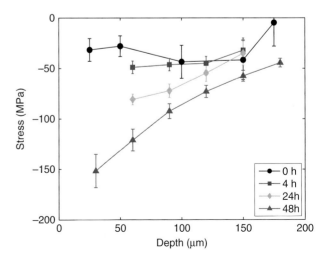

Fig. 4. Biaxial residual stress profiles as a function of depth in the BSAS topcoat after exposure to CMAS at 1300°C for 0, 4, 24, and 48 h, measured by transmission X-ray diffraction.

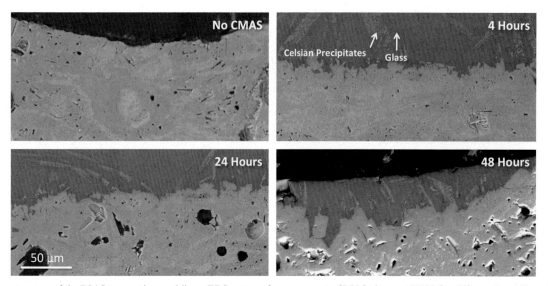

Fig. 5. Microstructure of the BSAS topcoat in a multilayer EBC system after exposure to CMAS glass at 1300°C for different times. From left to right and top to bottom: 0, 4, 24, and 48 h.

BSAS/glass interface. EDS was used to determine that these precipitates contain Ba, Sr, Al, Ca, and trace amounts of Mg. One explanation for their presence is that the glass attacks the BSAS coating and incorporates it into the molten glass, which crystallizes from the melt as it cools. Similar precipitates were observed in TEM studies by Grant *et al.*[37] who identified them as celsian, anorthite, or diopside. The BSAS coating also showed a "reaction zone" at the interface with the glass. This region exhibited a more faceted grain and pore structure, in addition to a more homogeneous contrast in the BSE images. Reaction time did not affect the morphology of these precipitates, but longer exposure times resulted in fewer and smaller precipitates, due to the small amount of unincorporated glass remaining and longer holding times. Indeed, the 48 h sample exhibited much less residual glass on the surface, suggesting that the CMAS was incorporated into the coating during the exposure.

To investigate the effect of the CMAS glass on the crystal structure of BSAS, the weight percents of celsian and hexacelsian phases in the topcoat the multilayer samples were determined by Rietveld fitting of 1D patterns calculated by integration of Debye ring radii from $\eta = 0$–360°. The weight percent celsian was determined as a function of depth in approximately 25 μm increments from the glass on the surface to the interface with mullite layer and shown in Fig. 6. The results of these phase fraction measurements were compared with published values for samples that underwent identical heat treatments without CMAS exposure,[24] represented as horizontal lines. The celsian content is increased at the top of the coating and at the BSAS/mullite interface, and lower in the center of the layer, suggesting that this transformation is favored at the surfaces and interfaces. This was also observed in literature results, as BAS and BSAS were shown to transform at free surfaces and at interfaces before the center of the bulk.[47,48] BSAS material that was exposed to CMAS demonstrated an accelerated transformation rate. The effect was pronounced near the surface of the sample exposed for 4 h. The weight percent celsian at the surface was ≈25% higher than the expected average value, and was much higher than the remainder of the sample, which was close to the condition without the glass. The microstructure results in Fig. 5 also support this, as faceted grain and pore structure in the reaction zone more closely resemble the celsian BSAS. In the sample exposed for 24 h, the celsian content is elevated over a depth of ≈125 μm into the sample, which suggested deeper CMAS infiltration than the 4 h sample. After 48 h, the topcoat shows a constant celsian content close to 100%, which was ≈10% higher than the heat treating alone would have provided. In these CMAS exposures, the data suggest that the exposure of the glass accelerated the hexacelsian-to-celsian

transformation in BSAS. Several mechanisms or combinations thereof could be responsible for the enhancement of this transformation rate. In YSZ-based TBC coatings, CMAS glass attacks by a dissolution–reprecipitation mechanism. Metastable tetragonal YSZ is dissolved into solution, and upon cooling, is reprecipitated as the stable monoclinic phase with an equiaxed microstructure.[35,36] If BSAS is attacked by CMAS in a similar way, the metastable (hexacelsian) and stable (celsian) material would be dissolved into the glass, and would reprecipitate as the stable celsian crystal structure. This mechanism was suggested by Grant *et al.*[37] from results from TEM studies on CMAS-reacted BSAS, and corroborated here. BSAS was dissolved into the CMAS glass and reprecipitated as a modified phase with the incorporation of Ca and a different Ba: Sr ratio.

Another mechanism that could accelerate the transformation rate of hexacelsian to celsian is the incorporation of Mg and Ca cations into the hexacelsian lattice. There is evidence in the literature that the integration of some additives to BAS material accelerates the conversion from hexacelsian to celsian.[49,50] The incorporation of these smaller cations, for which CMAS may act as a source, may reduce the transformation energy by sub-

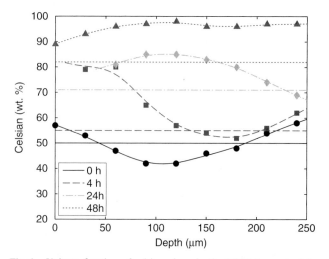

Fig. 6. Volume fraction of celsian phase in the BSAS topcoat of the studied multilayer EBC system after high-temperature exposure to CMAS at 1300°C for 0, 4, 24, and 48 h. Symbols represent experimental values determined from Rietveld refinements of diffraction patterns obtained by high-energy X-ray diffraction. Horizontal lines correspond to the average measured fraction of celsian phase in BSAS samples that underwent identical heat treatments without CMAS exposure.[24]

stitutionally doping for the Ba and Sr ions in the BSAS. The doped lattice would lower the activation energy and enhance the transformation rate. In any case, such a mineralizing effect will be convoluted with the main solution–reprecipitation mechanism.

(3) CMAS Exposure Effects on Freestanding BSAS

Freestanding BSAS samples were exposed to CMAS at 1300°C for 4, 12, 24, and 48 h, and the microstructures of the bottom of the wells are shown in Fig. 7. The sample exposed for 4 h contained a large amount of residual glass on the surface, but heating for 12 h reduced the remaining glass considerably, with almost none remaining after 24 or 48 h. The BSE images show a reaction zone that extends into the BSAS similar to what was seen in the coatings in Fig. 5, but the penetration depth was greater due to the increased crack density of the bulk BSAS samples, which had a greater number of pathways than the multilayer coatings for the glass to infiltrate. The grain and pore structures in the reaction zone were also more faceted than the underlying material. In the 4 h sample, the reaction zone was small, on the order of $\approx 10\ \mu m$ from the CMAS/BSAS interface. After 12 h, the reaction zone was $\approx 30\ \mu m$, and after 24 h the reaction zone grew to $\approx 50\ \mu m$. Heating for 48 h did not extend the zone any further than the 24 h sample due to the complete incorporation of the glass after 24 h.

TEM observations confirmed that the precipitates observed at the early stages of CMAS/BSAS interaction had a modified celsian structure that differed slightly in lattice parameters but showed essentially the same SAD patterns as the celsian present in the as-sprayed bars. Figure 8 shows an STEM image of a representative precipitate growing into the glass after 4 h of CMAS exposure, accompanied by EDS compositional maps and an SAD pattern. The diffraction pattern could only be indexed as a celsian structure. Although interplanar angles and spacings calculated from SAD patterns differed slightly from the BSAS celsian phase, they were not compatible with other possible structures such as anorthite or diopside. The EDS maps additionally show the presence of Ba- and Ca-rich regions in the precipitate, incorporating little or no Mg. Strontium is found not only in the precipitate but also in the glass, suggesting that Ca could be substituting Sr in the precipitates. Barium was not observed in the glass near the precipitate, suggesting that BSAS

is dissolved into the glass and reprecipitated as a Ca-doped BAS phase while the Sr remains in the glass.

Precipitates with SAD patterns consistent with anorthite were found to be present in very small amounts at the BSAS/CMAS interface (seen in Fig. 7 as dark, faceted phases), and EDS measurements showed that some Ba was present in the structure. This observation is again consistent with precipitates observed in previous works at higher CMAS loads,[37] although, contrary to that case, no diopside was found. The absence of diopside could be related to the amount of CMAS available for reaction. Combining our results with the observations of Grant and colleagues, we can conclude that anorthite, diopside and celsian precipitate from molten CMAS glass at loadings between 10 to 35 mg/cm^2. At loadings between 35 and 60 mg/cm^2, celsian and anorthite are present as precipitates, while for CMAS loadings of 60 mg/cm^2 and higher, only the celsian phase is found.

IV. Summary

We have studied exposure effects of CMAS deposits at 1300°C on both the composition and residual stresses of plasma-sprayed BSAS multilayer EBCs, for different exposure times. CMAS reacts with the topcoat dissolving the BSAS topcoat preferentially along the grain boundaries and dislodging the crystallites, roughening the BSAS–CMAS interface. Owing to the higher CTE of the CMAS glass, the result is a composite layer where the residual stress changes from uniform compression to an increasingly compressive stress gradient reaching values as high as ≈ -160 MPa after 48 h of exposure to 35 mg/cm^2 of CMAS. The residual stresses in the underlying layers and SiC/SiC substrates were unaffected by CMAS exposure. The average composition of the topcoat was evaluated by Rietveld fitting, and it was found that the presence of aliovalent atoms such as Ca and Mg accelerated the hexacelsian-to-celsian phase transformation in BSAS. CMAS reacts with BSAS following a solution–reprecipitation mechanism by which BSAS grains are dissolved into the molten glass. Ba and Sr atoms are diffused away from the CMAS/BSAS interface where they precipitate as a modified celsian phase that contains Ca in solid solution. No diopside precipitates were found for loadings of 35 mg/cm^2, but a small number of anorthite grains were identified using electron diffraction in the TEM.

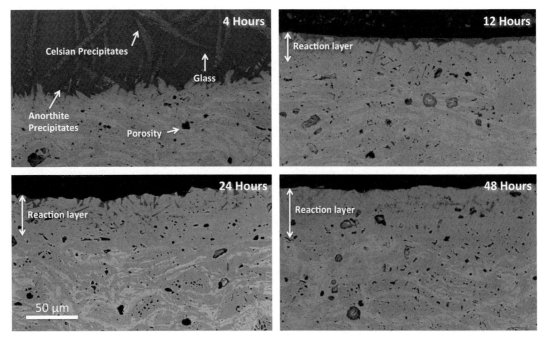

Fig. 7. Microstructure in a freestanding plasma-sprayed BSAS bar after exposure to CMAS glass at 1300°C for different times. From left to right and top to bottom: 4, 12, 24, and 48 h.

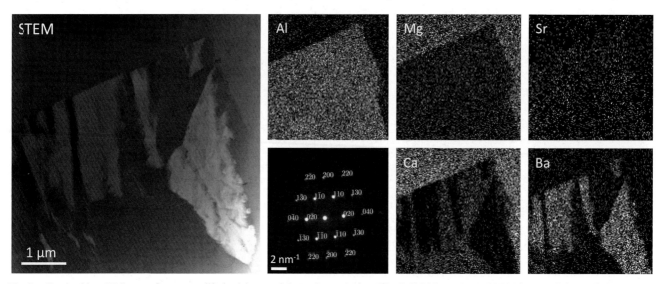

Fig. 8. Composition EDS maps from a modified celsian precipitate observed after 4 h of CMAS exposure, STEM image of the studied area, and selected area electron diffraction pattern from the precipitate, with indexing consistent with a celsian structure.

Acknowledgments

Microscopy images were collected at the NUANCE Center at Northwestern University, which is supported by the MRSEC program of the National Science Foundation (DMR-0520513). J. Ramirez-Rico was partially funded by the Spanish Ministry of Science and Technology under grant MAT2007-30141-E.

References

[1]N. S. Jacobson, "Corrosion of Silicon-Based Ceramics in Combustion Environments," *J. Am. Ceram. Soc.*, **76** [1] 3–28 (1993).

[2]K. N. Lee, "Current Status of Environmental Barrier Coatings for Si-Based Ceramics," *Surf. Coat. Technol.*, **133**, 1–7 (2000).

[3]E. J. Opila and R. E. Hann, "Paralinear Oxidation of CVD SiC in Water Vapor," *J. Am. Ceram. Soc.*, **80** [1] 197–205 (1997).

[4]E. J. Opila, "Variation of the Oxidation Rate of Silicon Carbide with Water-Vapor Pressure," *J. Am. Ceram. Soc.*, **82** [3] 625–36 (1999).

[5]E. J. Opila, J. L. Smialek, R. C. Robinson, D. S. Fox, and N. S. Jacobson, "SiC Recession Caused by SiO₂ Scale Volatility Under Combustion Conditions: Ii, Thermodynamics and Gaseous-Diffusion Model," *J. Am. Ceram. Soc.*, **82** [7] 1826–34 (1999).

[6]R. C. Robinson and J. L. Smialek, "SiC Recession Caused by SiO Scale Volatility Under Combustion Conditions: I, Experimental Results and Empirical Model," *J. Am. Ceram. Soc.*, **82** [7] 1817–25 (1999).

[7]J. L. Smialek, R. C. Robinson, E. J. Opila, D. S. Fox, and N. S. Jacobson, "SiC and Si₃N₄ Recession Due to SiO₂ Scale Volatility Under Combustor Conditions," *Adv. Compos. Mater.*, **8** [1] 33–45 (1999).

[8]K. T. Faber, C. M. Weyant, B. Harder, J. Almer, and K. Lee, "Internal Stresses and Phase Stability in Multiphase Environmental Barrier Coatings," *Int. J. Mater. Res.*, **98** [12] 1188–95 (2007).

[9]B. J. Harder, J. Almer, C. M. Weyant, K. N. Lee, and K. T. Faber, "Residual Stress Analysis of Multilayer Environmental Barrier Coatings," *J. Am. Ceram. Soc.*, **92** [2] 452–9 (2009).

[10]B. J. Harder, J. Almer, K. N. Lee, and K. T. Faber, "*In Situ* Stress Analysis of Multilayer Environmental Barrier Coatings," *Powder Diffr.*, **24** [2] 94–8 (2009).

[11]K. N. Lee, D. S. Fox, and N. P. Bansal, "Rare Earth Silicate Environmental Barrier Coatings for SiC/SiC Composites and Si₃N₄ Ceramics," *J. Eur. Ceram. Soc.*, **25** [10] 1705–15 (2005).

[12]K. N. Lee, R. A. Miller, and N. S. Jacobson, "New-Generation of Plasma-Sprayed Mullite Coatings on Silicon-Carbide," *J. Am. Ceram. Soc.*, **78** [3] 705–10 (1995).

[13]K. N. Lee and R. A. Miller, "Development and Environmental Durability of Mullite and Mullite/YSZ Dual Layer Coatings for SiC and Si₃N₄ Ceramics," *Surf. Coat. Technol.*, **86** [1–3] 142–8 (1996).

[14]M. Moldovan, C. M. Weyant, D. L. Johnson, and K. T. Faber, "Tantalum Oxide Coatings as Candidate Environmental Barriers," *J. Therm. Spray Technol.*, **13** [1] 51–6 (2004).

[15]C. M. Weyant, K. T. Faber, J. D. Almer, and J. V. Guiheen, "Residual Stress and Microstructural Evolution in Tantalum Oxide Coatings on Silicon Nitride," *J. Am. Ceram. Soc.*, **88** [8] 2169–76 (2005).

[16]C. M. Weyant, K. T. Faber, J. D. Almer, and J. V. Guiheen, "Residual Stress and Microstructural Evolution in Environmental Barrier Coatings of Tantalum Oxide Alloyed with Aluminum Oxide and Lanthanum Oxide," *J. Am. Ceram. Soc.*, **89** [3] 971–8 (2006).

[17]K. N. Lee, J. I. Eldridge, and R. C. Robinson, "Residual Stresses and Their Effects on the Durability of Environmental Barrier Coatings for SiC Ceramics," *J. Am. Ceram. Soc.*, **88** [12] 3483–8 (2005).

[18]D. Bahat, "Kinetic Study of Hexacelsian–Celsian Phase Transformation," *J. Mater. Sci.*, **5** [9] 805–10 (1970).

[19]N. P. Bansal and M. J. Hyatt, "Crystallization Kinetics of BaO–Al₂O₃–SiO₂ Glasses," *J. Mater. Res.*, **4** [5] 1257–65 (1989).

[20]N. P. Bansal, "Solid State Synthesis and Properties of Monoclinic Celsian," *J. Mater. Sci.*, **33** [19] 4711–5 (1998).

[21]N. P. Bansal and C. H. Drummond, "Kinetic-Study on the Hexacelsian–Celsian Phase-Transformation—Comment," *J. Mater. Sci. Lett.*, **13** [6] 423–4 (1994).

[22]N. P. Bansal and C. H. Drummond, "Kinetics of Hexacelsian-To-Celsian Phase Transformation in SrAl₂Si₂O₈," *J. Am. Ceram. Soc.*, **76** [5] 1321–4 (1993).

[23]K. T. Lee and P. B. Aswath, "Synthesis of Hexacelsian Barium Aluminosilicate by a Solid-State Process," *J. Am. Ceram. Soc.*, **83** [12] 2907–12 (2000).

[24]B. J. Harder and K. T. Faber, "Transformation Kinetics in Plasma-Sprayed Barium- and Strontium-Doped Aluminosilicate (BSAS)," *Scr. Mater.*, **62** [5] 282–5 (2010).

[25]A. G. Evans, D. R. Clarke, and C. G. Levi, "The Influence of Oxides on the Performance of Advanced Gas Turbines," *J. Eur. Ceram. Soc.*, **28** [7] 1405–19 (2008).

[26]S. Krämer, S. Faulhaber, M. Chambers, D. R. Clarke, C. G. Levi, J. W. Hutchinson, and A. G. Evans, "Mechanisms of Cracking and Delamination within Thick Thermal Barrier Systems in Aero-Engines Subject to Calcium–Magnesium–Alumino-Silicate (CMAS) Penetration," *Mater. Sci. Eng. A—Struct.*, **490** [1–2] 26–35 (2008).

[27]J. Kim, M. G. Dunn, A. J. Baran, D. P. Wade, and E. L. Tremba, "Deposition of Volcanic Materials in the Hot Sections of Two Gas-Turbine Engines," *J. Eng. Gas. Turb. Power*, **115** [3] 641–5 (1993).

[28]J. L. Smialek, F. A. Archer, and R. G. Garlick, "Turbine Airfoil Degradation in the Persian-Gulf-War," *J. Min. Met. Mater. S*, **46** [12] 39–41 (1994).

[29]D. J. Dewet, R. Taylor, and F. H. Stott, "Corrosion Mechanisms of ZrO₂–Y₂O₃ Thermal Barrier Coatings in the Presence of Molten Middle-East Sand," *J. Phys. IV*, **3** [C9] 655–63 (1993).

[30]F. Stott, D. J. Dewet, and R. Taylor, "Degradation of Thermal-Barrier Coatings at Very High-Temperatures," *MRS Bull.*, **19** [10] 46–9 (1994).

[31]M. P. Borom, C. A. Johnson, and L. A. Peluso, "Role of Environmental Deposits and Operating Surface Temperature in Spallation of Air Plasma Sprayed Thermal Barrier Coatings," *Surf. Coat. Technol.*, **86** [1–3] 116–26 (1996).

[32]C. Mercer, S. Faulhaber, A. G. Evans, and R. Darolia, "A Delamination Mechanism for Thermal Barrier Coatings Subject to Calcium–Magnesium–Alumino-Silicate (CMAS) Infiltration," *Acta Mater.*, **53** [4] 1029–39 (2005).

[33]X. Chen, "Calcium–Magnesium–Alumina–Silicate (CMAS) Delamination Mechanisms in EB-PVD Thermal Barrier Coatings," *Surf. Coat. Technol.*, **200** [11] 3418–27 (2006).

[34]A. G. Evans, D. R. Mumm, J. W. Hutchinson, G. H. Meier, and F. S. Pettit, "Mechanisms Controlling the Durability of Thermal Barrier Coatings," *Prog. Mater. Sci.*, **46** [5] 505–53 (2001).

[35]S. Krämer, J. Yang, C. G. Levi, and C. A. Johnson, "Thermochemical Interaction of Thermal Barrier Coatings with Molten CaO–MgO–Al₂O₃–SiO₂ (CMAS) Deposits," *J. Am. Ceram. Soc.*, **89** [10] 3167–75 (2006).

[36]A. Aygun, A. L. Vasiliev, N. P. Padture, and X. Ma, "Novel Thermal Barrier Coatings that are Resistant to High-Temperature Attack by Glassy Deposits," *Acta Mater.*, **55** [20] 6734–45 (2007).

[37]K. M. Grant, S. Krämer, J. P. A. Lofvander, and C. G. Levi, "CMAS Degradation of Environmental Barrier Coatings," *Surf. Coat. Technol.*, **202** [4–7] 653–7 (2007).

[38]J. Almer, U. Lienert, R. L. Peng, C. Schlauer, and M. Oden, "Strain and Texture Analysis of Coatings Using High-Energy X-Rays," *J. Appl. Phys.*, **94** [1] 697–702 (2003).

[39]H. E. Eaton, W. P. Allen, N. S. Jacobson, N. P. Bansal, E. J. Opila, J. L. Smialek, K. N. Lee, I. T. Spitsberg, H. Wang, P. J. Meschter, and K. L. Luthra, "Article Having Silicon-Containing Substrate and Barrier Layer and Production Thereof"; US Patent 6,387,456, 2001.

[40]I. Spitsberg and J. Steibel, "Thermal and Environmental Barrier Coatings for SiC/SiC CMCs in Aircraft Engine Applications," *Int. J. Appl. Ceram. Technol.*, **1** [4] 291–301 (2004).

[41]International Centre for Diffraction Data, Newton Square, PA, JCPDS, 2008.

[42]I. C. Noyan and J. B. Cohen, *Residual Stress: Measurement by Diffraction and Interpretation*. Springer-Verlag, New York, 1986.

[43]K. H. Hellwege (ed.), *Elastic, Piezoelectric and Related Constants of Crystals*, Vol. 11. Springer-Verlag, Berlin, 1979.

[44]A. C. Larson and R. B. Von-Dreele, "General Structure Analysis System (GSAS)"; Los Alamos National Laboratory Report, LAUR 86-748, Los Alamos, NM, 2000.

[45]B. H. Toby, "EXPGUI, a Graphical User Interface for GSAS," *J. Appl. Crystallogr.*, **34**, 210–3 (2001).

[46] International Centre for Structure Data, Karlsruhe, ICSD, 2010.

[47]J. I. Eldridge and K. N. Lee, "Phase Evolution of BSAS in Environmental Barrier Coatings," *Cer. Eng. Sci. Proc.*, **22** [4] 383–90 (2001).

[48]M. J. Hyatt and N. P. Bansal, "Crystal Growth Kinetics in $BaO \cdot Al_2O_3 \cdot 2SiO_2$ and $SrO \cdot Al_2O_3 \cdot 2SiO_2$ Glasses," *J. Mater. Sci.*, **31** [1] 172–84 (1996).

[49]K. T. Lee and P. B. Aswath, "Kinetics of the Hexacelsian to Celsian Transformation in Barium Aluminosilicates Doped with CaO," *Int. J. Inorg. Mater.*, **3** [7] 687–92 (2001).

[50]K. T. Lee and P. B. Aswath, "Role of Mineralizers on the Hexacelsian to Celsian Transformation in the Barium Aluminosilicate (BAS) System," *Mat. Sci. Eng. A—Struct.*, **352** [1–2] 1–7 (2003).   □

J. Am. Ceram. Soc., **94** [S1] S186–S195 (2011)
DOI: 10.1111/j.1551-2916.2011.04556.x
© 2011 The American Ceramic Society

journal

A Method for Assessing Reactions of Water Vapor with Materials in High-Speed, High-Temperature Flow

Sergio L. dos Santos e Lucato, Olivier H. Sudre, and David B. Marshall[†]

Teledyne Scientific Company, Thousand Oaks, California 91360

A simple method is described for measuring material erosion by reaction with water vapor under high-speed flow conditions, with H_2O partial pressures, velocities, temperatures, and erosion rates representative of those experienced in gas turbine engines. A water vapor jet is formed by the feeding water at a controlled rate into a capillary inside a tube furnace, where the large expansion of vaporization within the confines of the capillary accelerates the jet. With modest flow rates of liquid water, steam jets with temperatures up to $\sim 1400°C$ and velocities in the range 100–300 m/s have been achieved. The partial pressure of water vapor in the 100% steam jet is the same as in an industrial turbine operating at 10 atm total pressure with 10% water vapor. In preliminary experiments with SiC, erosion rates of the order of 1 μm/h have been observed.

I. Introduction

HIGH-TEMPERATURE metallic and nonoxide structural materials in gas turbine engines rely on protection from oxidation by oxide coatings. In some cases, these form naturally from oxidation of the material itself, as in the SiO_2 layer that forms on SiC and Si_3N_4, or the thermally grown oxide layers on superalloys, which are based on Al_2O_3. In other cases, oxide coatings are applied for additional protection, often in multiple layers, either as environmental barrier layers or as thermal barrier layers.

In dry oxidizing environments, continuous adherent layers of SiO_2 and Al_2O_3 are effective for the protection of the base material. The oxidation rate is limited by diffusion of oxygen through the oxide layer, resulting in parabolic kinetics; as the layer thickness increases the oxidation rate decreases. In most practical systems, the oxidation rate becomes negligible when the oxide layers reach thicknesses of several micrometers to tens of micrometers.

However, in combustion environments, which contain a large fraction of water vapor, the protection is dramatically reduced by reaction of the water vapor with the oxide to produce volatile oxy-hydroxide species.[1–18] In the case of silica, the following reactions occur, with relative importance being dependent on the temperature/pressure ranges[19]:

$$SiO_2(s) + 1/2H_2O(g) \rightarrow SiO(OH)(g) + 1/4O_2(g) \qquad (1)$$

$$SiO_2 + H_2O(g) \rightarrow SiO(OH)_2(g) \qquad (2)$$

$$SiO_2(s) + 2H_2O(g) \rightarrow Si(OH)_4(g) \qquad (3)$$

F. Zok—contributing editor

Manuscript No. 28937. Received November 18, 2010; approved March 11, 2011.
This work was financially supported by the U. S. Office of Naval Research under contract N00014-02-C-0025, the NASA/AFOSR National Hypersonic Science Center for Materials and Structures (AFOSR Contract No. FA9550-09-1-0477).
[†]Author to whom correspondence should be addressed. e-mail: dmarshall@teledyne.com

$$2SiO_2 + 3H_2O(g) \rightarrow Si_2O(OH)_6(g) \qquad (4)$$

Several studies have shown that reaction (3) is dominant at temperatures up to $\sim 1200°–1400°C$, with reaction (2) becoming more significant at higher temperatures.

Similar reactions occur for all oxides, albeit with reaction rates differing by many orders of magnitude.[19–21] This leads to erosion of the outer surface of the oxide coating and a change from parabolic oxidation kinetics with weight gain to linear kinetics with weight loss. After an initial period of increasing oxide thickness, a steady-state condition of constant oxide thickness is approached, in which rates of material loss at the outer surface and oxidation at the interface of the oxide and the base material are equal.[3]

The time to reach steady state, the steady-state recession rate, and the limiting oxide thickness are all very sensitive to flow conditions, both pressure and gas velocity.[8] With the especially strong dependence on gas velocity, conditions typically achieved in laboratory experiments give erosion rates many orders of magnitude slower than in real gas turbine engines (and other engines including rocket and combined cycle engines). It is perhaps for this reason that the role of water vapor erosion as the life-limiting mechanism in SiC and Si_3N_4 engine components was for many years overlooked in favor of other mechanisms such as creep and oxidative strength degradation that dominate in less aggressive laboratory conditions.[22,23] However, from the references mentioned above,[1–17] which include burner rig and long-term engine testing, the dominant role of water vapor in combustion environments is clear.

This realization has motivated efforts to develop environmental barrier coatings consisting of complex oxides that are more resistant than SiO_2 to water vapor corrosion.[24–32] However, the data on corrosion rates for other oxide compounds are very limited. The difficulty of measuring the response of developmental coating materials without going to extremely costly engine tests remains a barrier to development. Moreover, without well-controlled comparative data from complex coatings and from pure forms of the compounds that constitute the coatings, it is often difficult to distinguish effects of the compounds themselves from the role of the microstructures of the coatings.

The purpose of this paper was to demonstrate a simple approach for overcoming this barrier, using a jet of high-temperature water vapor to generate flow conditions with H_2O partial pressures, velocities, temperatures, and erosion rates representative of those experienced in turbine engines. The jet is formed by feeding water at a controlled rate into a capillary tube inside a tube furnace, where the large expansion of vaporization within the confines of the capillary accelerates the jet. With very modest flow rates of liquid water, steam jet velocities typical of flow rates in industrial turbines can be achieved (e.g., flow of water at rates in the range 1–2 mL/min into a 1-mm-diameter capillary creates jets with temperatures between 1100°C and 1300°C and 1 atm pressure with a velocity in the range 160–300 m/s). The partial pressure of water vapor in the 100% steam jet is the same as in an industrial turbine operating at 10 atm total pressure with 10% water vapor. The test set-up builds on work reported earlier by Ferber and Lin[33] who used a capillary to produce a steam jet, although with gas velocities an order of magnitude lower (35 m/s). More recently, Sudhir and Raj[34] used a steam jet in a different configuration to measure oxidation weight changes in silicon nitride at velocities up to 0.35 m/s.

The area of the test sample exposed to high-velocity steam in these experiments is small (a few millimeters diameter). Whereas this would be a disadvantage for measuring degradation in the conventional manner through weight changes, the presence of an adjacent unexposed reference surface makes possible a more direct and informative approach by measurement of surface recession.

II. Experimental Procedure

The test set-up, shown in Fig. 1, uses a tube furnace with a mullite tube (length 1 m) and flanges that accommodate several inlet and outlet attachments for gas and water supply, as well as an attachment to hold an alumina tube inside the furnace. The alumina tube extends to the center of the hot zone of the furnace and serves as specimen holder. A quartz glass capillary with a hole of diameter 1 mm is mounted along the center of the alumina tube and connects through the inlet flange to a peristaltic pump, which feeds ultrahigh purified liquid water into the capillary. As the water enters the hot zone of the furnace, it evaporates and accelerates to form a jet of high-temperature steam. The surrounding atmosphere (between the emerging jet and the mullite furnace tube) is flowing argon, which enters through a second inlet in the same flange at a rate of 4 L/min, giving a gas velocity ~ 0.1 m/s in hot zone. The argon and the water vapor are extracted through a single outlet and passed through a condenser before being released into the atmosphere.

The temperature of the steam jet emerging from the end of the capillary was measured using a thermocouple (type R), routed through the alumina tube and held in front of the capillary tip. A set of calibration tests were carried out with the thermocouple bead placed about 1 mm from the tip of the capillary, at the same position the test specimens were mounted during the experiments. The calibration test matrix included water flow rates between 1 and 6 mL/min and furnace temperatures between 1300°C and 1500°C. The temperatures were

recorded after reaching steady state before and after the water was turned on, as well as after the water was turned off, to ensure that the thermocouple did not degrade during the experiments.

The average velocity, v_{av}, of the steam jet was estimated using the ideal gas law

$$v_{av} = \frac{\dot{m} R T}{A \, MW \, P} \tag{5}$$

where $\dot{m}$ is the mass flow rate, R the universal gas constant, T the temperature of the jet, A the cross-section area of the capillary, MW the molecular weight of water, and P the pressure at the exit of the capillary. The exit pressure was taken to be approximately equal to the ambient pressure (1 atm), an assumption that was confirmed to be accurate to within $<0.2\%$ from CFD analysis (Appendix A). The velocity along the centerline of the jet decreases beyond the capillary exit, although the decrease is expected to be small over a distance of ~ 5 times the capillary diameter.

The water vapor jet was directed onto the leading edge of a sharp wedge-shaped test specimen as shown in Fig. 1. The test material was polycrystalline SiC produced by chemical vapor deposition (Morton International Inc., Woburn, MA). The test specimens were prepared by grinding and polishing blocks of dimensions 10 mm × 10 mm × 1 mm to form sharp wedges with an included angle of 6° (Fig. 1(b)). In this configuration, the sharp edge of the specimen was aligned with the center of the jet, so that the jet was divided into two equal streams flowing almost parallel to the two wedge faces. During initial set-up of the test, another configuration was investigated with the jet incident on a flat polished surface inclined at an angle of 45° to the flow. While this was also convenient for producing measurable recession, the flow conditions adjacent to the specimen surface are more complex.

The specimens were mounted in an alumina holder and locked in place in front of the capillary, with an alumina tube being used to hold and align all components. Photographs taken before and after each test were used to confirm that no parts shifted during the experiment. Three combinations of water flow rate and duration were chosen from the initial set of calibration data for furnace temperature of 1500°C: 1.87 mL/min for 20 h;

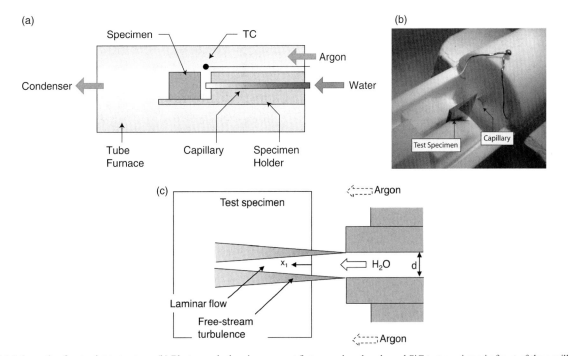

Fig. 1. (a) Schematic of water jet test set-up. (b) Photograph showing support fixture and wedge-shaped SiC test specimen in front of the capillary tube. (c) Schematic showing flow of water vapor jet over wedge-shaped test specimen (flow rate of surrounding argon gas is low, ~ 0.1 m/s, compared with ~ 100 m/s for water jet).

1.87 mL/min for 12 h; and 1 mL/min for 20 h. After the experiments, the depths of erosion were measured by profilometry and optical interferometry.

III. Results

(1) Temperature Calibration

Results of the temperature calibration are summarized in Fig. 2. In the set-up used in these experiments, the length of capillary tube in the hot zone of the furnace is not sufficient to allow equilibration of the temperatures of the water vapor and the furnace atmosphere, so that the temperature of the vapor jet is lower than the furnace temperature by an amount that is sensitive to the flow rate. Moreover, because this temperature difference increases with increasing flow rate, the velocity of the vapor jet at given furnace temperature does not increase monotonically with the water flow rate (Fig. 2(b)).

An estimate of the position at which the liquid water vaporized could usually be obtained from a slight discoloration visible on the inside of the capillary tube after the test run (from trace amounts of contaminants in the water). The distances between this position and the exit of the capillary were approximately 12 and 17 cm for flow rates of 1.0 and 1.87 mL/min, corresponding to a residence time of the accelerating gas in the capillary of the order of 10^{-3} s.

(2) Recession Measurements and Observations

A sharp wedge-shaped SiC test sample was observed as shown in Fig. 3(a), after exposure for 20 h in a jet of mean velocity 260 m/s and temperature 1160°C (furnace temperature 1500°C). Severe recession occurred both in the position of the leading edge

(originally a straight line) and in grooves along the initially flat polished faces of the wedge. The positions of the deepest grooves coincide with flow at the periphery of the water vapor jet.

From the profilometer data shown in Fig. 3(b), the recession depth in the central region of the jet flow is ~9 μm, and is roughly constant within a distance of ~4 mm from the leading edge. The grooves at the edge of the flow are deeper by a factor of 3 or more near the leading edge (~27 and 30 μm) and become shallower with increasing distance downstream. The formation of grooves of this depth on both sides of the wedge-shaped test samples would cause the leading edge of the wedge to recede by ~500 μm. This is close to the observed recession of the leading edge in Fig. 3(c), thus indicating that material removal in this region is predominantly from the flat faces of the wedge rather than from the stagnation point at the leading edge itself. This conclusion is further supported by the observation that the leading edge remains sharp in the grooves.

Examination by SEM and EDS analysis indicated that there was a continuous SiO_2 layer outside the region of jet flow. However, there was no detectable evidence for the oxide layer within the deep grooves, while in the central region of the jet flow, discontinuous patches of thin oxide were detected. Elemental maps of silicon, oxygen, carbon, and aluminum from the region of Fig. 3(c) are shown in Fig. 4 (note the absence of oxygen along the grooves, reduced oxygen signal at the central region of jet flow, and complementary regions of high and low oxygen and carbon signals). The presence of a small concentration of aluminum outside the region of the jet flow is indicative of formation of volatile aluminum oxy-hydroxides[19-21] by the reaction of water vapor in the surrounding gas with the alumina furnace hardware. There was no evidence of Al within the region exposed to the jet.

Independent measurements of recession depths and oxide thickness were obtained using optical interference microscopy. Three sets of fringes are visible in the white light interference micrograph shown in Fig. 3(d): one set is from interference between reflections from the microscope reference mirror and the top surface (oxide) of the test specimen; a second set from interference between the reference mirror and the SiO_2–SiC interface; and the third set from interference between the top of the oxide layer and the oxide–SiC interface. For a flat surface (i.e., before exposure to the water vapor jet), the first two sets of fringes would be straight and horizontal in Fig. 3(d). The downward displacement of these fringes, visible in Fig. 3(d), is proportional to the recession depth of the surface, while the separation of the two sets of fringes is proportional to the thickness of the oxide layer. The near-vertical fringes from interference between the top and bottom of the oxide layer also give a measure of the layer thickness. Measurements from these fringes as well as measurements from interference micrographs obtained using monochromatic light indicated that the thickness of the oxide layer is ~4 μm outside the region impacted by the jet (right side of the micrograph) and the thickness decreases continuously toward the center of the groove, where the oxide thickness is below the detection limit of the interference measurements (~30 nm). In the central region of the jet flow, a thin oxide layer (thickness <200 nm) was detected in patches separated by regions where no oxide was detected.

Higher magnification views of the specimen in Fig. 3 are shown in Fig. 5. Large differences in the surface texture are evident in various regions: the region outside the vapor jet, which is covered with a layer of glassy SiO_2, is relatively smooth, whereas the region within the grooves is much rougher and pitted, with no evidence of glassy SiO_2 detected either from the appearance of the SEM image or from EDS analysis. In the central region of the vapor jet, a mixture of surface textures are evident, circular depressions with rough pitted surfaces surrounded by a smoother thin film of glassy SiO_2. Many of the circular depressions showed radial texture that might suggest an association with cristobalite spherulites.[35,36] However, no evidence of cristobalite crystals was seen in the pitted surfaces.

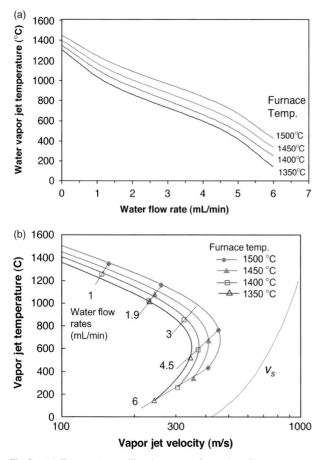

Fig. 2. (a) Temperature calibration curves for various furnace temperatures and water flow rates. (b) Vapor jet temperatures and velocities for various furnace temperatures and water flow rates. Speed of sound in water vapor, v_s, indicated for comparison.

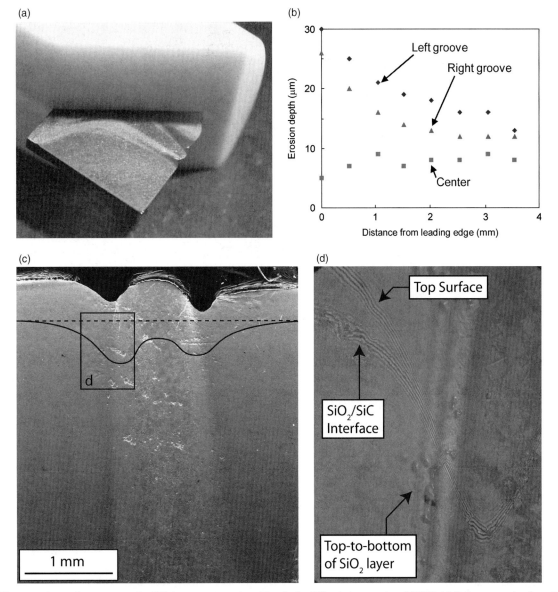

Fig. 3. SiC test specimen after exposure for 20 h in water vapor jet with velocity 260 m/s, temperature 1160°C. (a) Before removing from holder used during test: grooves coincide with flow of the edge of the jet. (b) Erosion depths measured from profilometer data. (c) SEM image with superimposed profilometer trace along broken line. (d) Optical interference micrograph (white light) from area indicated in (c) showing three sets of interference fringes.

Erosion patterns similar to Fig. 3 were observed under several conditions of jet velocity and temperature, although the relative depths of the central region and the edge grooves varied with conditions. At lower velocity (160 m/s) and higher temperature (1350°C), the side grooves were not as distinct, while at high velocity their depth increased with velocity and exposure time. However, in all cases, regions similar to Fig. 5(a) were observed in the side grooves, with rough, pitted surfaces and no evidence of glassy SiO_2 detectable by SEM imaging or EDS analysis. Similar rough pitted surfaces were also observed in the tests in which the jet was incident on a flat polished surface inclined at an angle of 45° to the flow. An example is shown in Fig. 6(a).

IV. Discussion

(1) Flow Conditions in Water Vapor Jet
The main characteristics of the flow conditions in the water vapor jet can be deduced from standard fluid dynamics relations. Within the capillary, the flow is characterized by the Reynolds number[37]:

$$Re_d = \frac{\rho \, d \, v}{\mu} \qquad (6)$$

where v, ρ, and μ are the velocity, density, and viscosity of the fluid, and d is the capillary diameter. The flow is laminar for low values of Re_d, with the onset of turbulence occurring at $Re_d = 2300$ and fully developed turbulent flow at $Re_d > 10\,000$. For the range of conditions of the water vapor jet at temperatures up to ~ 1400°C ($d = 1$ mm, $\rho \sim 1.3 \times 10^{-4}$ g/cc, $\mu \sim 5 \times 10^{-4}$ g/cm·s, and $v < 300$ m/s), Eq. (6) gives $Re_d < 1000$. Therefore, the flow at the exit of the capillary is laminar in all of the experiments.

The fluid velocity for laminar flow in a tube varies across the diameter of the tube, approaching a steady-state parabolic profile after flowing a distance greater than the hydrodynamic entry length, x_o. For a fluid injected into the tube with uniform velocity profile and flowing with uniform mean velocity, x_o is given by [32]

$$\frac{x_o}{d} \approx 0.05 Re_d \qquad (7)$$

With $d = 1$ mm and $Re_d < 1000$, as in the present experiments, Eq. (7) gives $x_o < 50$ mm. The observations in Section III(1) indicate that evaporation occurred at a distance of ~ 120–170 mm from the exit of the capillary. At the evaporation position,

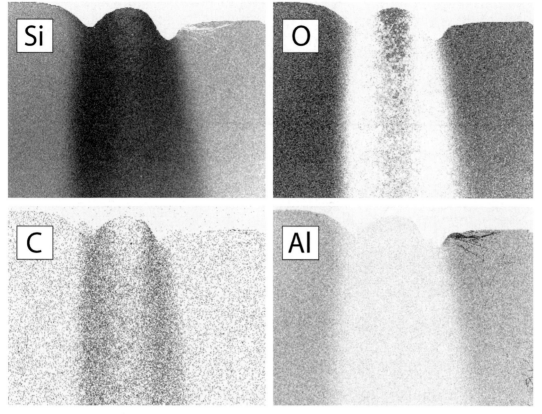

Fig. 4. EDS maps from area shown in Fig. 3(b) (dark indicating high-intensity signal).

the velocity profile would be expected to be uniform. Therefore, if flow between the evaporation position and the exit were to occur under isothermal conditions, the steady-state parabolic profile would be approached, for which the maximum velocity (at the center of the jet) is double the average velocity. However, in these experiments, the temperature of the water vapor increases by a factor ~ 4 between the evaporation point and the exit, resulting in a corresponding increase in average velocity. The velocity profile at the exit in this case is dependent on the variation of gas temperature along the capillary and would be expected to be much more uniform than for the steady-state case. In the absence of a detailed analysis of the velocity profile, the average velocity will be used in the analysis here.

As the water vapor jet exits the capillary, shear forces from interaction with the surrounding atmosphere (slowly flowing argon gas, velocity <0.1 m/s) generate turbulence around the periphery of the jet as depicted in Fig. 1(c), while the central region remains laminar. The intersection of this free-stream turbulence with surface of the test specimen corresponds with the location of the eroded grooves in Fig. 3. This observation is consistent with the expectation that mass transfer (of the volatile reaction products) would be more rapid through the boundary layer in a region of turbulent flow than in the laminar flow region in the central area of the jet.

As the laminar region in the center of the jet flows along the specimen surface, a boundary layer develops adjacent to the surface, characterized by the Reynolds number[32]:

$$Re_x = \frac{\rho\, x_1\, v_0}{\mu} \qquad (8)$$

where x_1 is the distance from the leading edge (Fig. 1(c)). The flow remains laminar for $Re_x < 5 \times 10^5$. Within the region of interest ($x_1 < 4$ mm), Eq. (8) gives $Re_x < 1000$, indicating that the flow remains laminar within this region. Under these conditions, the thickness of the boundary layer that controls diffusion of

volatile reaction products (and hence erosion rate) is given by[5]

$$\delta = \frac{1.5\, x_1}{Re_x^{1/2}\, Sc^{1/3}} \qquad (9)$$

where Sc is the Schmidt number (defined in Appendix B). Under the range of conditions of these experiments, the boundary layer thickness within 4 mm of the leading edge is always <0.2 mm (see Appendix B). Therefore, the boundary layer on the specimen surface along the centerline of the jet remains confined within the jet (radius 0.5 mm).

(2) Comparison of Water Vapor Jet with Hydrocarbon Combustion: The Silica System

Opila *et al.*[5,8] have provided a comprehensive analysis of the behavior of silica-forming materials (SiC and Si_3N_4) in flowing gas containing water vapor, under conditions where reaction to form $Si(OH)_4$ dominates at the outer surface and the reaction rate is limited by diffusion of the reaction product through a laminar boundary layer. This analysis involved use of a correlation function for the Sherwood number for flow over a flat plate, which relates the mass flow rate of the reaction product to the Reynolds number and Schmidt number (hence to velocity, pressure, and viscosity of the flowing gas), the diffusion coefficient, and the equilibrium reaction product concentration at the outer surface. From these results, combined with analysis of the oxidation reaction at the SiC–SiO_2 interface, the following expressions were derived for the steady-state erosion rate, $\dot{y}_L$, the limiting oxide thickness, x_L, and the time, t_L, to achieve steady state

$$\dot{y}_L = C_1 \frac{v^{1/2} P_{H_2O}^2}{P^{1/2}} \qquad (10)$$

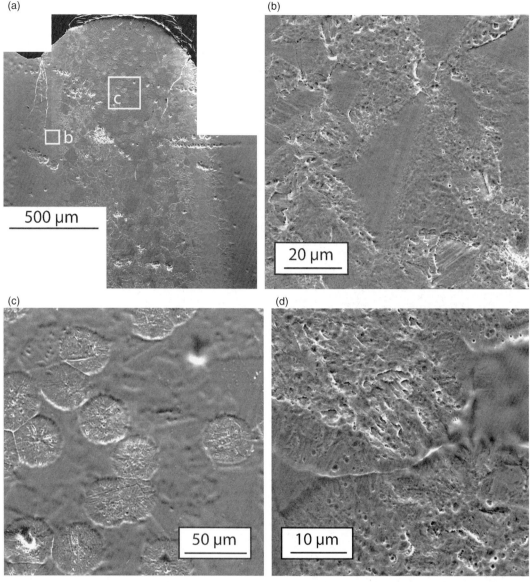

Fig. 5. SEM images from area of Fig. 3: (b), (c) and (d) higher magnification from areas indicated in (a).

$$x_L = C_2 \frac{P^{1/2}}{P_{H_2O}^3 v} \tag{11}$$

$$t_L = C_3 \frac{P}{P_{H_2O}^3 v} \tag{12}$$

where v is the gas velocity, P_{H_2O} is the partial pressure of water vapor, P is the total pressure, and C_1, C_2, and C_3 are constants dependent on gas properties, reaction enthalpy, and temperature. Test data for recession of SiC from several facilities (high-pressure burner rig (HPBR) and synthetic combustion gas furnace) were correlated with these expressions and used to calibrate the constants C_1, C_2, and C_3 at a temperature of 1316°C. The values thus obtained were $C_1 = 0.18$, $C_2 = 0.615$, $C_3 = 1.71$, with units of pressure in atm, v in m/s, t_L in hours, x_L in μm, and $\dot{y}$ in μm/h. The results from Eqs (10–12) with these calibrated constants are plotted in Fig. 7 in terms of gas velocity for several combinations of total and partial pressures.

Industrial turbines typically operate at 10 atm total pressure, with combustion gases containing ~10% water vapor. Therefore, the partial pressure of water vapor is equivalent to that in a 100% steam jet at 1 atm pressure. However, the erosion rate at given gas velocity depends also on the total pressure (Eqs (10–12)), at least within the regime where the erosion is controlled by

the transport of the reaction product through a laminar boundary layer. This dependence on total pressure, which enters via its influence on the diffusion coefficient for the reaction product through the boundary layer as well as the Reynolds number, makes the water vapor jet more severe than the combustion environment at given gas velocity and equivalent partial pressure of H_2O.

Also shown in Fig. 7 are operating conditions for various test configurations and real gas turbine engines taken from Opila[8] (who plotted these equations in a different form, as contour maps for hydrocarbon combustion in terms of velocity and total pressure). It is apparent that the water jet provides erosion conditions (steady-state recession rates, limiting oxide thicknesses, and time constant) close to those of the turbine engine, whereas even sophisticated burner rig facilities (Mach 0.3 burner rig and HPBR) do not achieve the conditions in an industrial gas turbine, the steady-state erosion rates are smaller by a factor of 10–100, the limiting oxide thicknesses are larger by a factor of 10, and the times taken to achieve steady state are longer by a factor of 100. The conditions in laboratory furnaces, either with flowing gas or in a special high-pressure furnace, are many orders of magnitude further from the turbine conditions.

Closer examination of Fig. 7 suggests limits to the ranges of pressure and temperature over which this erosion mechanism applies. At low gas velocities and pressures typical of laboratory

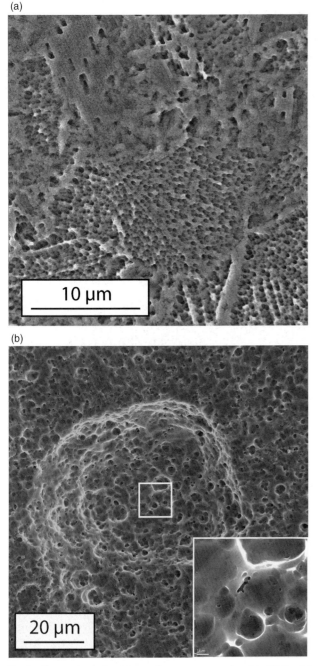

(a)

(b)

Fig. 6. Comparison of erosion surface textures produced in SiC by (a) turbulent region of water vapor jet incident at an angle of 45° to specimen surface; and (b) H_2/O_2 combustion plume in rocket nozzle conditions (surface temperature $\sim 1600°C$); inset showing higher magnification image from area indicated.

water vapor jet in Figs. 5(b) and 6(a)) and a change in the rate-controlling mechanism from volatility of SiO_2 to the direct reaction of SiC or Si_3N_4 with water vapor. Several possibilities were proposed by Opila[8]: (i) rate controlled by reaction of SiC (or Si_3N_4) with water to give SiO_2, with the oxide being swept away as soon as it forms, (ii) "active" volatilization similar to the active oxidation that occurs for SiO_2 formers at low partial pressures of O_2, where SiO(g) is formed rather than SiO_2. In this case, an analogous reaction would be SiC reacting directly with water vapor to give $Si(OH)_4(g)$, without the intermediate SiO_2 formation step. Evidence for a transition in behavior to a direct reaction was reported[8,33] for Si_3N_4 vanes after exposure in industrial turbines. The observations in Figs. 5(b) and 6(a) of deeply pitted surfaces in regions of turbulent flow of the water vapor jet are suggestive of a transition in mechanism. Also shown for comparison in Fig. 6(b) is erosion damage on the surface of an SiC test specimen exposed to high-speed H_2/O_2 combustion products from a rocket nozzle.[‡] A similar rough surface is evident with pitting at multiple scales. Even if a very thin oxide layer exists in these cases, with thickness below the detection level of SEM analysis, it is clear that the erosion process is more complex than a simple model of diffusion through a laminar planar boundary layer.

Because these conditions cannot be readily obtained in a laboratory environment, there is very little understanding of the details of the mechanism. However, there is evidence for highly accelerated erosion rates in this regime.[8,33] The upper limit for erosion rate in this case is given by the Langmuir equation for free evaporation into a vacuum.[3] This limit is indicated in Fig. 7(a). It is evident from Fig. 7(a) that extrapolation of Eq. (10) to conditions of H_2/O_2 rocket engines, where the water vapor pressure is substantially >1 atm, gives erosion rates approaching the Langmuir limit. Extrapolation for conditions of the water vapor jet ($p H_2O = 1$ atm) with laminar flow indicates that a significantly higher jet velocity ($\sim 10^5$ m/s) would be required to reach the Langmuir limit.

(3) Comparison with Measured Recession Rates

Along the line x_1 in Fig. 1(c), where the flow is laminar, the measured recession rates can be compared with the analysis summarized in the previous sections. Two conditions of velocity and temperature were used in the experiments: (A) velocity of 260 m/s at temperature of 1160°C; and (B) velocity of 160 m/s at temperature of 1350°C.

The temperature dependence (Eq. (10)) enters through the parameter C_1, which can be written in the form

$$C_1 = \beta \, T^n \exp\left(\frac{-\Delta H}{R\,T}\right) \tag{13}$$

where the exponent n is close to $-1/4$, ΔH is the reaction enthalpy of the volatile species (in this case $Si(OH)_4$), and β is a temperature-independent function of molecular weights, collision cross sections, viscosity, and the length dimension that appears in the Reynolds number for flow over the specimen surface (see Appendix B). The temperature variation due to the term T^n is negligible compared with the influence of uncertainties in the exponential term. Thermodynamic measurements by Jacobson et al.[38] over the temperature range 1000–1400 K, within which $Si(OH)_4$ is the dominant reaction product, give $\Delta H = 54.6$ kJ/mol, whereas the HPBR erosion data used in the calibration of the constant C_1 in Eq. (10)[4,8] gave an activation energy of 108 kJ/mol, a result that was interpreted as implying the presence of appreciable amounts of $SiO(OH)_2$ as well the predominant $Si(OH)_4$. Steady-state recession rates predicted from Eqs (10) and (13) with the latter value of activation energy and with the calibration at $T = 1316°C$ mentioned above are shown in Fig. 8 at the temperatures of the water jet experiments.

furnaces, the time required to approach steady state is $>10\,000$ h. In any practical experiment, the erosion rates are negligible and the kinetics is dominated by the parabolic growth of the oxide layer. Moreover, the limiting oxide thickness becomes sufficiently large that spalling can occur.[8]

At the other extreme of high velocity and pressure, as experienced in industrial gas turbines, the time to reach steady state is only a few minutes, the erosion rates are high (~ 1 μm/h) and the oxide thickness is very small (<0.1 μm). Similar erosion rates and oxide thicknesses were observed in the central laminar region of the water vapor jet in Section III. Conditions are even more severe in H_2/O_2 rocket combustion environments, which experience higher pressures and velocities of water vapor (and higher temperatures), as indicated in Fig. 7. In these cases, the small oxide thickness (<10 nm in Fig. 7(b)) may lead to breakup of the oxide layer (as possibly seen in the turbulent region of the

[‡]O. Sudre, unpublished work on SiC composite exposed to testing at Cell 22 facility at NASA Glenn Research Center.

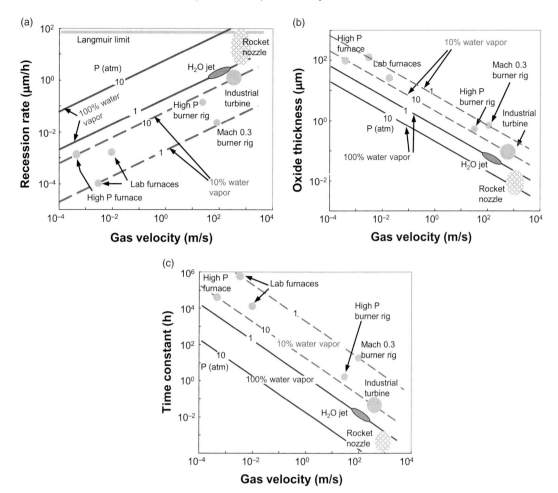

Fig. 7. Plots from Eqs (10–12) comparing erosion of SiC in flowing water vapor (blue solid lines) and in hydrocarbon combustion (red broken lines, 10% water vapor) at temperature of 1316°C, assuming laminar flow over a flat plate and volatilization of $Si(OH)_4$: (a) steady-state recession rate, (b) limiting oxide thickness, and (c) time taken to reach steady state. Conditions corresponding to several test facilities, including the high-speed water vapor jet, are indicated and compared with typical conditions for industrial turbine engines and rocket nozzles (note temperatures are generally higher in rocket nozzles).

Also shown in Fig. 8 are the measured recession rates. The recession rates on the line x_1 in Fig. 1(c) ($\sim$0.5–0.8 µm/h for condition A and 0.6–0.7 µm/h for condition B) were smaller by a factor of 2 to 3 than the predicted values, while the recession

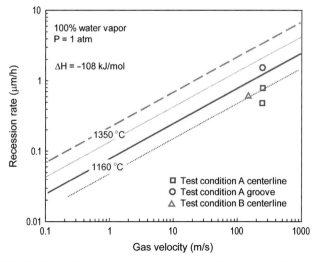

Fig. 8. Comparison of measured recession rates at two test conditions A (v_{av} = 260 m/s, T = 1160°C) and B (v_{av} = 160 m/s, T = 1350°C) with predictions from Eq. (10) using thermodynamic data ($\Delta H = -108$ kJ/mol) from burner rig testing.[4,8] Dotted lines indicate approximate shift of predicted curves that would result from referencing the calibration of the constant C_1 in Robinson and Smialek[4] to the measured gas temperature (1416°C) rather than the measured test specimen temperature (1316°C).

rates in the grooves at the turbulent periphery of the jet were slightly higher than the predicted value for laminar flow ($\sim$1.5 µm/h). These differences likely relate to different flow conditions in the HPBR tests used to calibrate the constant C_1 in Eq. (10)[4,8]: (i) the test surfaces in the HPBR tests were oriented at an angle $\sim$45° to the flow, whereas those in the present experiments were parallel to the flow; and (ii) the gas temperatures in the HPBR tests were higher by $\sim$100°C than the measured specimen temperatures used in the calibration, whereas the temperatures referenced in the present tests were measured values for the water vapor jet. If the calibration from the HPBR tests were to be referenced to the gas temperature rather than the test specimen temperature (i.e., 1416°C rather than 1316°C), the predicted erosion rates at the temperatures of the water jet tests would be lowered by approximately 40% (dotted lines in Fig. 8), giving close agreement with the measured erosion rates for condition A and reducing the discrepancy for condition B. Additional uncertainty is introduced through differences in the length parameter in the Reynolds number and Sherwood number in Eqs (B-1) and (B-2) and the likelihood that the water vapor jet contains some $Si(OH)_4$ from reaction with the capillary wall, which would act to retard the volatilization reaction on the test specimen surface. Given these uncertainties, the measured recession rates appear to be reasonably consistent with the analysis of Opila *et al.*[5,8]

V. Conclusions

The method demonstrated here provides a means for efficient laboratory scale studies of corrosion of materials and coatings by water vapor, under conditions that are representative of the gas velocity, partial pressure, and temperatures of gas turbine engines. The configuration with a sharp wedge-shaped test specimen appears to be suitable for quantitative analysis of mass transport through a well-controlled boundary layer. Along the centerline of the vapor jet, the conditions are laminar and analytical correlation solutions are available. Around the periphery of the jet, free stream turbulence creates much higher mass transfer conditions (and hence erosion rates), and possible transition in erosion mechanism. The presence of two regions with well-defined sets of conditions in the one test is a useful feature. Further quantitative analysis of both regions would be possible if the experiments were to be combined with detailed computational fluid dynamics analysis to assess the variation of velocity within the jet (radial and axial) and diffusion of reactants away from the surface in the turbulent region.

Acknowledgment

The authors acknowledge helpful discussions with Prof. Beth Opila.

Appendix A: CFD Analysis of Water Vapor Jet

A preliminary axisymmetric CFD analysis (using commercial software, ANSYS FLUENT (ANSYS Inc., Canonsburg, PA)) was used to investigate several characteristics of the water vapor jet under typical experimental conditions described in Section III(2), namely, whether the pressure at the exit of the capillary is significantly higher than ambient pressure and whether the composition of the jet changed significantly due to radial diffusion of water out of the jet and argon into the jet. The jet in the analysis was created by feeding water vapor into one end of a capillary (1 mm diameter, 150 mm length) at constant mass flow rate ($\dot{m} = 0.03$ g/s) and uniform velocity (300 m/s), corresponding to gas temperature of 1500°C. The jet exited the other end of the capillary into surrounding static argon gas at pressure of 1 atm and temperature 1500°C. Although in the experiments the surrounding gas was moving, the velocity (0.1 m/s) was negligible compared with the jet velocity and the assumption of static surrounding gas is accurate. The analysis did not account for the presence of the test specimen. Nor did it account for the fact that the water vapor temperature and hence velocity increases continuously between the location of injection (vaporization) and the exit. Therefore, the analysis was not used to assess velocity distributions (radial or axial), which would be expected to be sensitive to this temperature variation.

The pressure at the exit of the capillary is 1.002 atm; and is thus close to the ambient pressure, as assumed in using Eq. (5) to calculate the average jet velocity, v_{av}. The analysis also confirms that the gas composition at the center of the jet remains pure water vapor (i.e., mixing and diffusion of argon into the center of the jet is negligible), at least within 5 mm of the capillary exit, as might be intuitively expected.

Appendix B: Diffusion Equations and Boundary Layer Thickness

Along the line x_1 in Fig. 1(c), where the flow is laminar, the steady-state recession rate in the transport-limited regime can be calculated following the analysis of Opila,[8] which begins with a dimensionless correlation function for the Sherwood number, Sh, in terms of the Reynolds number, Re, and Schmidt number, Sc:

$$Sh = \frac{k_1 L}{D \rho_v} = 0.664 Re^{1/2} Sc^{1/3} \tag{B-1}$$

which, with Re given by Eq. (8) and $Sc = \eta/\rho_B D$, becomes

$$k_1 = 0.664 \left(\frac{L v \rho_B}{\eta}\right)^{1/2} \left(\frac{\eta}{\rho_B D}\right)^{1/3} \frac{D \rho_v}{L} \tag{B-2}$$

where k_1 is the average mass flux, v is the gas velocity, ρ_v the equilibrium concentration of the volatile species (Si(OH)$_4$), ρ_B the density of the boundary layer gas (H$_2$O), D the diffusion coefficient of the volatile species in the boundary layer, η the gas viscosity, and L is a characteristic specimen length, equal to x_1 for the wedge-shaped specimen here. Equation (B-2) can be written in the form of Eqs (10) and (13) with the following substitutions:

$$\dot{y}_L = k_1 \frac{\rho_{SiO_2}}{MW_{SiO_2}} \frac{MW_{SiC}}{\rho_{SiC}} \tag{B-3}$$

$$\rho_B = P_B \frac{MW_B}{RT} \tag{B-4}$$

$$\rho_v = P_{Si(OH)_4} \frac{MW_{Si(OH)_4}}{RT} = k_{eq} \frac{MW_{Si(OH)_4}}{RT} P_{H_2O}^2 \tag{B-5}$$

$$k_{eq} \propto \exp\left(\frac{-\Delta H}{RT}\right) \tag{B-6}$$

$$D = \left[\frac{0.0018 \sqrt{\frac{1}{MW_v} + \frac{1}{MW_B}}}{\sigma_{vB}^2 \Omega}\right] \frac{T^{3/2}}{P_B} \tag{B-7}$$

where the subscripts B and v refer to the boundary layer gas and the volatile species, P is pressure, ρ is density, MW is molecular weight, k_{eq} and ΔH are the equilibrium constant and enthalpy for reaction (3), σ_{vB} is the mean of the collision diameters for the species v and B (i.e., Si(OH)$_4$ and H$_2$O in our case), and Ω is a dimensionless collision integral.[39] With the temperature dependence introduced through the parameters ρ_v, ρ_B, and D, the exponent n in Eq. (13) is $-1/6$. If in addition we account for the temperature dependence of viscosity, given approximately by $\eta \propto T^{1/2}$, the exponent becomes $n \approx -1/4$.

The thickness of the diffusion boundary layer may be evaluated by substituting the expressions above into Eq. (9). The boundary layer thickness increases with increasing x, increasing T and decreasing v. Therefore, the largest thickness in the current experiments will occur for the higher temperature/lower velocity condition ($T = 1350°C$, $v = 160$ m/s). The parameters needed to estimate the diffusion coefficient are $MW_v = 96$ g/mol (Si(OH)$_4$), $MW_B = 18$ g/mol (H$_2$O), $\sigma_{vB} \approx 3$ Å, $\Omega \approx 1$.[39] With $P_B = 1$ atm, the diffusion coefficient from Eq. (B-7) is D = 3.6 cm^2/s (substitution of these parameters with the given units into Eq. (B-7) gives D in cm^2/s). With $\eta = 5.7 \times 10^{-4}$ g/cm·s, $\rho_B = 0.00013$ g/cc, and $x_1 = 4$ mm, we get $Re \approx 1500$, $Sc \approx 1.2$, and maximum thickness $\delta \approx 0.15$ mm.

References

[1] C. S. Tedmon Jr., "The Effect of Oxide Volatilization on the Oxidation Kinetics of Cr and Fe–Cr Alloys," *J. Electrochem. Soc.*, **113** [8] 766–8 (1967).

[2] E. J. Opila, "Oxidation Kinetics of Chemically Vapor-Deposited Silicon Carbide in Wet Oxygen," *J. Am. Ceram. Soc.*, **77** [3] 730–6 (1994).

[3] E. J. Opila and R. E. Hann, "Paralinear Oxidation of CVD SiC in Water Vapor," *J. Am. Ceram. Soc.*, **80** [1] 197–205 (1997).

[4] R. C. Robinson and J. L. Smialek, "SiC Recession Caused by SiO$_2$ Scale Volatility under Combustion Conditions: I, Experimental Results and Empirical Model," *J. Am. Ceram. Soc.*, **82** [7] 1817–25 (1999).

[5] E. J. Opila, J. L. Smialek, R. C. Robinson, D. S. Fox, and N. S. Jacobson, "SiC Recession Caused by SiO$_2$ Scale Volatility under Combustion Conditions: II, Thermodynamics and Gaseous Diffusion Model," *J. Am. Ceram. Soc.*, **82** [7] 1826–34 (1999).

[6] K. L. More, P. F. Tortorelli, M. K. Ferber, and J. R. Keiser, "Observations of Accelerated Silicon Carbide Recession by Oxidation at High Water-Vapor Pressures," *J. Am. Ceram. Soc.*, **83** [1] 211–3 (2000).

[7] M. K. Ferber, H. T. Lin, V. Parthasarathy, and R. A. Wenglarz, "Degradation of Silicon Nitrides in High-Pressure, Moisture-Rich Environments, Paper No. 2000-GT-0661"; pp. 1–17 in *Proceedings of IGTI Conference*, Munich, Germany, May 8–11, 2000. ASME, New York, 2000.

[8] E. J. Opila, "Oxidation and Volatilization of Silica Formers in Water Vapor," *J. Am. Ceram. Soc.*, **86** [8] 1238–48 (2003).

[9] E. J. Opila, "Variation of the Oxidation Rate of Silicon Carbide with Water-Vapor Pressure," *J. Am. Ceram. Soc.*, **82** [3] 625–36 (1999).

[10]H. T. Lin, M. K. Ferber, and M. van Roode, "Evaluation of Mechanical Reliability of Si₃N₄ Nozzles after Exposure in an Industrial Gas Turbine"; pp. 97–102 in *7th International Symposium of Ceramic Materials and Components for Engines, June 19–21*, Edited by J. G. Heinrich, and F. Aldinger. Wiley-VCH, Goslar, Germany, 2000.

[11]H. T. Lin and M. K. Ferber, "Mechanical Reliability Evaluation of Silicon Nitride Ceramic Components After Exposure in Industrial Gas Turbines," *J. Eur. Ceram. Soc.*, **22** 2789–97 (2002).

[12]M. K. Ferber, H. T. Lin, and J. Keiser, "Oxidation Behavior of Non-Oxide Ceramics in a High-Pressure, High-Temperature Steam Environment"; pp. 201–15 in *Mechanical, Thermal and Environmental Testing and Performance of Ceramic Composites and Components, ASTM STP 1392*, Edited by M. G. Jenkins, E. Lara-Curzio, and S. T. Gonczy. American Society for Testing and Materials, West Conshohocken, PA, 2000.

[13]H. T. Lin, M. K. Ferber, W. Westphal, and F. Macri, "Evaluation of Mechanical Reliability of Silicon Nitride Vanes After Field Tests in an Industrial Gas Turbine, ASME 2002-GT-30629"; pp. 147–54 in the Proceedings of at Turbo Expo Land Sea, and Air, Amsterdam, the Netherlands, June 3–6, 2002 ASME New York, 2002.

[14]K. L. More, P. F. Tortorelli, L. R. Walker, N. Miriyala, J. R. Price, and M. VanRoode, "High Temperature Stability of SiC-Based Composites in High Water-Vapor-Pressure Environments," *J. Am. Ceram. Soc.*, **86**, 1272–81 (2003).

[15]P. F. Tortorelli and K. L. More, "Effects of High Water-Vapor Pressure on Oxidation of Silicon Carbide at 1200°C," *J. Am. Ceram. Soc.*, **86** [8] 1249–55 (2003).

[16]D. S. Fox, E. J. Opila, Q. N. Nguyen, D. L. Humphrey, and S. M. Lewton, "Paralinear Oxidation of Silicon Nitride in a Water-Vapor/Oxygen Environment," *J. Am. Ceram. Soc.*, **86** [8] 1256–61 (2003).

[17]M. van Roode and M. K. Ferber, "Degradation of Ceramics for Gas Turbine Applications"; pp. 305–321 in ASME Conference Proceedings, Vol. 1, Turbo Expo 2007: Power for Land, Sea, and Air, Paper no. GT2007-27956 ASME, New York, 2007.

[18]M. van Roode, J. Price, J. Kimmel, N. Miriyala, D. Leroux, A. Fahme, and K. Smith, "Ceramic Matrix Composite Combustor Liners: A Summary of Field Evaluations," *J. Eng. Gas Turbines Power*, **129** [1] 21–31 (2007).

[19]N. Jacobson, D. Myers, E. Opila, and E. Copland, "Interactions of Water Vapor with Oxides at Elevated Temperatures," *J. Phys. Chem. Sol.*, **66**, 471–8 (2005).

[20]E. J. Opila and N. S. Jacobson, "Volatile Hydroxide Species of Common Protective Oxides and their Role in High Temperature Corrosion"; pp. 269–78 in *Proceedings of the Symposium on Fundamental Aspects of High Temperature Corrosion, Vol. 96-26*, Edited by D. A. Shores, R. A. Rapp, and P. Y. Hou. The Electrochemical Society, Pennington, NJ, 1997.

[21]E. J. Opila and D. L. Myers, "Alumina Volatility in Water Vapor at Elevated Temperatures," *J. Am. Ceram. Soc.*, **87** [9] 1701–5 (2004).

[22]N. S. Jacobson, "Corrosion of Silicon-Based Ceramics in Combustion Environments," *J. Am. Ceram. Soc.*, **76** [1] 3–28 (1993).

[23]R. E. Tressler, "Environmental Effects on Long-Term Reliability of SiC and Si₃N₄ Ceramics"; pp. 99–123 in *Ceramic Transactions, Vol. 10, Corrosion and Corrosive Degradation of Ceramics*, Edited by R. E. Tressler, and M. McNallan. American Ceramic Society, Westerville, OH, 1990.

[24]K. N. Lee, R. A. Miller, and N. S. Jacobson, "New Generation of Plasma-Sprayed Mullite Coatings on Silicon-Carbide," *J. Am. Ceram. Soc.*, **78** [3] 705–10 (1995).

[25]K. N. Lee, D. S. Fox, J. I. Eldridge, D. Zhu, R. C. Robinson, and N. P. Bansal *et al.*, "Upper Temperature Limit of Environmental Barrier Coatings Based on Mullite and BSAS," *J. Am. Ceram. Soc.*, **86** [8] 1299–306 (2003).

[26]J. R. Price, M. van Roode, and C. Stala, "Ceramic Oxide-Coated Silicon Carbide for High Temperature Corrosive Environments," *Key Eng. Mater.*, **72–74**, 71–84 (1992).

[27]K. N. Lee, "Current Status of Environmental Barrier Coatings for Si-Based Ceramics," *Surf. Coat. Technol.*, **1–7**, 133–4 (2000).

[28]J. Kimmel, N. Miriyala, J. Price, K. More, P. Tortorelli, H. Eaton, G. Linsey, and E. Sun, "Evaluation of CFCC Liners with EBC After Field Testing in a Gas Turbine," *J. Eur. Ceram. Soc.* **22** [14–15] 2769–75 (2002).

[29]K. L. More, P. F. Tortorelli, L. R. Walker, J. B. Kimmel, N. Miriyala, and J. R. Price, *et al.*, "Evaluating Environmental Barrier Coatings on Ceramic Matrix Composites after Engine and Laboratory Exposures. ASME paper 2002-GT-30630"; pp. 155–62 in *Proceedings of ASME Turbo Expo 2002*, Amsterdam, the Netherlands, June 3–6. ASME, New York, 2002.

[30]S. Ueno, T. Ohji, and H-T Lin, "Recession Behavior of Yb₂Si₂O₇ Phase Under High Speed Steam Jet at High Temperatures," *Corros Sci*, **50**, 178–82 (2008).

[31]S. Ueno, T. Ohji, and H.-T. Lin, "Recession Behavior of a Silicon Nitride with Multi-Layered Environmental Barrier Coating System," *Ceram Int.*, **33**, 859–62 (2007).

[32]K. N. Lee, D. S. Fox, and N. P. Bansal, "Rare Earth Silicate Environmental Barrier Coatings for SiC/SiC Composites and Si₃N₄ Ceramics," *J. Eur. Ceram. Soc.*, **25**, 1705–15 (2005).

[33]M.-K. Ferber and H.-T. Lin, "Environmental Characterization of Monolithic Ceramics for Gas Turbine Applications," *Key Eng Mater*, **287**, 368–80 (2005).

[34]B. Sudhir and R. Raj, "Effect of Steam Velocity on the Hydrothermal Oxidation/Volatilization of Silicon Nitride," *J. Am. Ceram. Soc.*, **89** [4] 1380–7 (2006).

[35]V. Presser, A. Loges, Y. Hemberger, and K. G. Nickel, "Microstructural Evolution of Silica on Single-Crystal Silicon Carbide. Part I: Devitrification and Oxidation Rates," *J. Am. Ceram. Soc.*, **92** [3] 724–31 (2009).

[36]V. Presser, A. Loges, R. Wirth, and K. G. Nickel, "Microstructural Evolution of Silica on Single-Crystal Silicon Carbide. Part II: Influence of Impurities and Defects," *J. Am. Ceram. Soc.*, **92** [8] 1796–805 (2009).

[37]F. P. Incropera and D. P. Dewitt, *Introduction to Heat Transfer*. John Wiley & Sons, New York, 1996.

[38]N. S. Jacobson, E. J. Opila, D. L. Myers, and E. H. Copland, "Thermodynamics of Gas Phase Species in the Si–O–H System," *J. Chem. Thermodyn.*, **37**, 1130–7 (2005).

[39]H. C. Graham and H. H. Davis, "Oxidation/Vaporization Kinetics of Cr₂O₃," *J. Am. Ceram. Soc.*, **54** [2] 89–93 (1971). □

J. Am. Ceram. Soc., **94** [S1] S196–S203 (2011)

DOI: 10.1111/j.1551-2916.2011.04530.x

© 2011 The American Ceramic Society

journal

Effect of Water Penetration on the Strength and Toughness of Silica Glass

Sheldon M. Wiederhorn,[†,‡] Theo Fett,[§] Gabriele Rizzi,[¶] Stefan Fünfschilling,[§] Michael J. Hoffmann,[§] and Jean-Pierre Guin[‖]

[‡]Materials Science and Engineering Laboratory, National Institute of Standards and Technology, Gaithersburg, Maryland

[§]Karlsruhe Institut für Technologie, Institut für Keramik im Maschinenbau, 76131 Karlsruhe, Germany

[¶]Karlsruhe Institut für Technologie, Institut für Materialforschung II, 76344 Karlsruhe, Germany

[‖]LARMAUR ERL-CNRS 6274, University of Rennes 1, Rennes, France

In this paper, we discuss the effect of water on the strength and static fatigue of silica glass. When a crack is formed in silica glass, the surrounding environment rushes into the crack; water then diffuses from the environment into the newly formed fracture surfaces to generate a zone of swelling around the crack tip. Because the swollen material is constrained from expanding by the surrounding glass, a zone of compressive stress is generated at the fracture surface around the crack tip. The results are similar to those found for transformation toughened zirconium oxide, with the exception that the transformation zone in silica glass grows with time, so that the effect gets progressively stronger. Using diffusion data from the literature, we show that the diffusion of water into silica glass can explain several significant experimental observations: the reported strengthening of silica glass by soaking in water at 88°C; an increase in the slope of dynamic fatigue curve by prior exposure to water at 88°C; the observation of a static fatigue limit in silica glass at very low values of the applied stress-intensity factor; and the observation of crack face displacements caused by water penetration into the glass at the crack tip.

I. Introduction

T HIS paper discusses the role played by water in determining the strength of silica glass. The detrimental effect of water on the strength of glass is well known, but not everyone agrees as to the causes of its effect. It is, however, universally believed that the strength of silica glass is controlled by cracks contained in the surface of the glass. Removal of these cracks enhances the strength of the glass, and if the glass can be made completely free of surface cracks then the strength approaches the theoretical strength of glass, ≈ 14 GPa, as it does in optical fiber glass.[1] Even slight mechanical contact with the glass surface reduces its strength substantially.[2,3]

Water is found to be deleterious to the strength of glass. With water present, the breaking stress under load is time dependent, the higher the load applied to the glass, the more rapidly it breaks. For very low loads, silica glass can support the load for a very long time before breaking[4]; some glass scientists believe that silica glass possesses a static fatigue limit, a stress below which the silica glass will not break no matter how long the load

is applied. A fatigue limit of this sort has been measured on silica glass by Sglavo and Green.[5]

This effect of time on the strength of glass is believed by many to be caused by subcritical crack growth, resulting from a stress-enhanced interaction between water and glass. Crack growth in silica glass has been very well characterized by fracture mechanics techniques, in which the crack velocity is measured as a function of the applied stress-intensity factor, relative humidity, and temperature.[6–10] Chemical reaction rate theory provides a sound basis for understanding crack growth in glass.[11,12] Furthermore, a chemical reaction path for crack growth has been identified, as has the molecular structure of the water that is responsible for crack growth.[13] Based on this discovery, other chemicals that cause crack growth in glass have also been identified, for example, ammonia.[13] The growth of cracks in large fracture mechanics specimens and in glass fibers is similar enough that the belief that the strength of glass is controlled by subcritical crack growth is almost unquestioned.

Areas where scientists disagree on the causes of strength degradation and static fatigue are centered on the nature of the crack tip and how the crack tip controls the growth of cracks in silica glass. Some glass scientists believe that cracks in glass are basically elastic, i.e., the stresses, strains, and displacements within the silica glass follow the elastic solution almost to the crack tip, differing from the elastic solution only where the molecular structure of the glass becomes apparent. Their view is supported by studies on the atomic force microscope (AFM), which demonstrate that displacements near the crack tip follow the elastic solution to within 10 nm of the crack tip.[14] Experiments using spatially and spectrally resolved cathode-luminescence show that stresses follow the predicted stress level to within 6 nm of the crack tip.[15] Finally, transmission electron micrographs demonstrated that the crack tip "radius of curvature" in silica glass was <1.5 nm.[16] These measurements limit the size of any possible nonlinear zone near the crack tip.

In apparent contradiction to the idea that cracks are "sharp" and that strength and lifetime are limited by crack length, some glass scientists argue that crack tip blunting, either by precipitation of silica at the crack tip, or by plastic flow, is crucial to the measured time to failure.[17–20] Blunted cracks cannot propagate until they become atomically sharp; hence, the kinetics of blunting and sharpening is important in establishing the strength and dynamic fatigue behavior of silica glass.[17–20] In support of their argument, Hirao and Tomozawa[20] show that the strength of silica glass can be increased by annealing the silica glass at high temperature, 910°C, while Ito and Tomozawa show that strength can be increased by soaking the glass in water at 88°C.[18] In either case the strength of silica glass increases by

A. Heuer—contributing editor

Manuscript No. 29002. Received December 3, 2010; approved February 25, 2011.
[†]Author to whom correspondence should be addressed. e-mail: smwied@aol.com

about 10%–20%, the increase being attributed to blunting of the crack by the annealing process.

Other strengthening mechanisms have been suggested to explain changes in the strength of glass as a consequence of ageing or annealing without invoking crack tip blunting as a part of the mechanism. For example, most cracks in glass are formed as a result of mechanical impact that also leaves residual damage behind in the form of plastic deformation and densification in the surface of the glass. Such damage raises the stresses at the tips of cracks and is the main driving force for crack growth.[21] Annealing the glass at high temperature relieves these stresses and thus increases the strength of the glass. Alternatively, aging glass in water results in the formation of an interfacial layer that acts as a "glue" to bind the two fracture surfaces together, thus, increasing the strength of the glass.[22]

In this paper, we discuss an alternative explanation of the observations of the strength increase reported by Hirao and Tomozawa.[19,20] We assume that the cracks are always sharp, and that the observed strengthening occurs by the penetration of water into the glass surrounding the crack tip and into the adjacent walls of the crack. Such penetration induces a compressive stress within the walls of the crack, which in turn results in a negative stress-intensity factor at the crack tip. A mechanism such as this has been reported for soda lime silicate glass as a consequence of ion exchange near the crack tip, hydronium ions for alkali ions in the glass.[23,24] The stresses set up by the ion exchange process, ≈ 2000 MPa,[25] can result in a substantial time delay in restarting crack growth in soda lime silicate glass.[23,26,27] Similar levels of stress can be achieved when water penetrates into silica glass.

II. Water Diffusion and Volume Expansion in Silica

A substantial literature exists documenting the diffusion of water into silica glass.[28–31] Water diffuses by a diffusion-reaction process,[28] in which the water molecules can react with the Si–O bonds to form –SiOH groups. At low temperatures, $<250°C$, the –SiOH groups are immobile, whereas, at high temperatures, $\approx 1000°C$, they exhibit a limited mobility and can take part in the equilibrium reaction that establishes the relationship between the concentration of –SiOH and the molecular water. Part of the water is present in molecular form and part is present in reacted form and both parts have to be taken into account when water diffuses though silica glass. The transport of water through the glass structure, however, is by motion of the molecular water alone.

The diffusion coefficient, D_w, of the molecular water is given by the following equation:

$$D_w = D_{w0} \exp(-Q_w/R\Theta) \qquad (1)$$

where Q_w is the activation energy, Θ is the absolute temperature, and R is the gas constant. As reported in Zouine *et al.*,[30] $Q_w = 71.2 \pm 1$ kJ/mol and $\log_{10} D_o = -7.78 \pm 0.06$, where D_o is in m^2/s.

For a constant diffusion coefficient the water concentration profile, $C(z,t)$, as a function of depth z and time t is given by

$$C_w(z,t) = C_{w0} \cdot \mathrm{erfc}\left(\frac{z}{2\sqrt{Dt}}\right) \qquad (2)$$

where C_{w0} is the surface concentration of the water.[32] As shown by Nogami and Tomozawa,[32] diffusion is enhanced in the presence of stresses.

The diffusion coefficient as a function of the hydrostatic stress, σ_h, can be written

$$D = D(0) \exp(\Delta V_w \sigma_h/R\Theta) \qquad (3)$$

where, ΔV_w is the activation volume for water diffusion in glass and $D(0)$ is the diffusivity at zero pressure, i.e., $\sigma_h = 0$ atm. The

value of the diffusivity at 1 atm is almost the same as at 0 atm, so for practical purposes the diffusivity at 1 atm is used in this paper. At room temperature, $D(0) \cong 10^{-21}$ m^2/s.[30] The activation volume for water at room temperature has not yet been evaluated, but is believed to be approximately equal to the molar volume of water, 18×10^{-6} m^3/mol (M. Tomozawa, private communication). This paper suggests a new method of calculating the activation volume of water. The value obtained, 10×10^{-6} m^3/mol, is reasonably close to the molar volume of water. Clear evidence has been reported in the literature for volume swelling, ε_0, measured directly via dimension changes in the silica glass, and indirectly via stress generation within the glass. The swelling behavior has been determined by length[33] and curvature measurements on silica glass bars[34] and by X-ray diffraction strain measurements[35] on silica glass. Gorbacheva and Zaoints[33] soaked fused silica prisms ($25 \times 5 \times 5$ mm) in water at 80°C for 20 months. They found the measured length to have increased by 0.17% by this treatment. Because only a thin surface layer of water penetrated glass, ≈ 5 μm,[††] could have been generated during the water storage, the suppressed linear strain in the surface must be clearly larger than 0.17%.

An estimate of the surface stress caused by water penetration is relevant to the present study, because it is this surface stress that affects the stress-intensity factor driving the crack growth. Knowing the effect of water penetration into the glass on the volume of the glass, permits us to determine that surface stress by first estimating the volumetric strain of the glass as a consequence of the water penetration. The volumetric strain, $\varepsilon(z,t)$, is a function of time, t, and distance, z, from the free surface:

$$\varepsilon(z,t) = \varepsilon_0 \cdot \mathrm{erfc}\left(\frac{z}{2\sqrt{Dt}}\right) \qquad (4)$$

with ε_0 being the time-dependent volumetric strain at the surface. The stresses caused by swelling are equi-biaxial ($\sigma_z = 0$) and given by

$$\sigma_x = \sigma_y = -\frac{\varepsilon(z,t)E}{3(1-\nu)} \qquad (5)$$

where E is Young's modulus and ν is Poisson's ratio. Consequently, the hydrostatic stress is

$$\sigma_h = \frac{1}{3}(\sigma_x + \sigma_y) = -\frac{2\varepsilon(z,t)E}{9(1-\nu)} \qquad (6)$$

Because the diffusion coefficient depends on the hydrostatic stress component according to Eq. (3), the hydrostatic stress due to swelling must affect the concentration profile. Hence, Eqs. (2) and (4) are no longer correct with the consequence that the diffusion differential equation has to be solved numerically. For reasons of clarity and simplicity, such higher-order effects on concentration profiles will be neglected in the following discussion.

III. An Approximate One-Dimensional Treatment of the Swelling Zones

A static crack undergoing water diffusion shows a diffusion zone such as that represented in Fig. 1(a). Water entering the glass through the plane of the crack can be described by a one-dimensional diffusion problem, the solution of which is given by Eq. (4). Near the crack tip and ahead of the crack a much more complicated two-dimensional nonaxial solution to the diffusion equations is required.

The following considerations were made to allow the representation of the general trends by approximate analytical expressions from the one-dimensional analysis. First, we ignore

---

[††]This distance is calculated from the known diffusivity[30] and the time of exposure using the definition of the diffusion distance, $x = \sqrt{(Dt)}$.

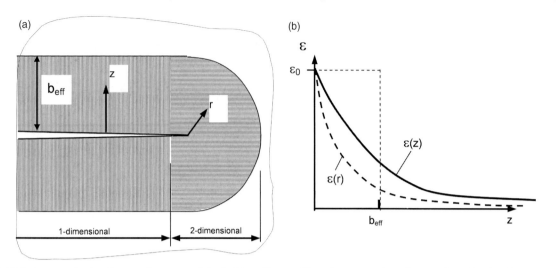

Fig. 1. (a) Diffusion zone in the absence of an external load, (b) swelling profile under assumption of a one-dimensional (solid curve) and axial-symmetric two-dimensional water diffusion (dashed curve); effective zone thickness b_{eff} in the one-dimensional case.

the dilution of water concentration due to the increase of the area elements $dA \propto r\,dr$ in the two-dimensional problem and use $dA \propto dz$ from the one-dimensional model. Also, the continuously varying swelling strain $\varepsilon(r)$ of Eq. (4) is replaced by a step-shaped dependency with a change at the characteristic contour height b_{eff} (Fig. 1(b)) resulting in the same shielding stress-intensity factor.

(1) Swelling Zone for Unloaded Cracks

The volume expansion at crack surfaces must result in an intrinsic shielding stress-intensity factor K_{sh}. In the case of an unloaded crack, the hydrostatic stress term in Eq. (3) disappears. Because of the Arrhenius dependence of the diffusivity on temperature, Eq. (1), the diffusion process becomes more important as the temperature is increased, even though no stress-enhancement takes place. The shielding stress-intensity factor, K_{sh}, after a time, t, of water contact is given by[36]

$$K_{sh} = -0.25 \frac{\varepsilon_0 E}{1-\nu} \sqrt{b_{eff}} \tag{7}$$

with an effective water layer thickness, b_{eff}, of

$$b_{eff} \cong \sqrt{D(0)t} \tag{8}$$

which is roughly the distance from the surface at which the strain has dropped to one-half of the surface value (Fig. 1(b)). Finally, it holds that

$$K_{sh} \cong -0.25 \frac{\varepsilon_0 E}{1-\nu} [D(0)t]^{1/4} \tag{9}$$

(2) Swelling Zone Near the Tip of a Nonpropagating Crack Under Load

The hydrostatic stress is given by the trace of the stress tensor,

$$\sigma_h = \frac{1}{3}(\sigma_{rr} + \sigma_{\varphi\varphi} + \sigma_{zz}) \tag{10}$$

Because very high tensile stresses occur in the vicinity of a crack tip, a water containing zone must rapidly extend from the tip to the surrounding material during crack growth. As the crack moves, the zone extends along the sides of the crack at the same velocity relative to the crack tip as the crack is moving in the test material.

If K_I denotes the stress-intensity factor, the near-tip stress field for the stationary crack in the absence of a T-stress term[36] reads[‡‡]

$$\sigma_{ii} = \frac{K_{tip}}{\sqrt{2\pi r}} g_{ii}(\varphi) \tag{11}$$

where r is the crack-tip distance. In Eq. (11), g_{ii} are well-known geometric functions depending on the polar angle φ.[37] The hydrostatic stress under plane strain conditions is given by

$$\sigma_h = \frac{2}{3}(1+\nu) \frac{K_{tip}}{\sqrt{2\pi r}} \cos(\varphi/2) \tag{12}$$

Introducing Eq. (12) into Eq. (3) yields location-dependent diffusion coefficients. The introduction of such diffusion coefficients into the problem makes the resulting equations highly intractable, because Eqs. (2) and (4) are correct only for constant diffusion coefficients. The authors are unaware of an exact analytical solution for this diffusion problem. Therefore, to simplify the problem, we assumed that the two-dimensional diffusion problem at the crack tip can be represented by the one-dimensional problem. Therefore, the erfc-relation, Eq. (4), was used for the near-tip zones in the presence of stresses.

Using Eqs. (12), (4), (3) and (1), an equation for the swelling strain can be derived:

$$\varepsilon(r, \varphi, t) = \varepsilon_0 \cdot \mathrm{erfc}\left[\frac{r}{2\sqrt{D(0)t}} \times \exp\left(-\frac{1}{3}(1+\nu)K_{tip}\frac{\cos(\varphi/2)}{\sqrt{2\pi r}}\frac{\Delta V_w}{R\Theta} \right) \right] \tag{13}$$

The contour for an effective swelling strain of $\varepsilon(r, \varphi, t) = 0.5\varepsilon_0$ results in

$$\frac{r}{2\sqrt{D(0)t}} \exp\left(-\frac{1}{3}(1+\nu)K_{tip}\frac{\cos(\varphi/2)}{\sqrt{2\pi r}}\frac{\Delta V_w}{R\Theta} \right) = 0.477 \cong 0.5 \tag{14}$$

with a solution of

[‡‡]The stress distribution around a crack tip is usually expressed in the form of a power series in $r^{n/2}$, where r is the distance from the crack tip. The series starts with $n = -1$. The coefficient of the first term is the stress-intensity factor; the coefficient of the second term of the series is the T-stress. See Fett[36] for further discussion of this subject.

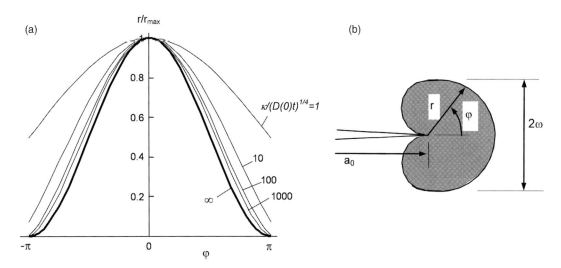

Fig. 2. (a) Diffusion zone radius normalized on the maximum value as a function of the polar angle φ, (b) shape of the swelling zone.

$$r = \frac{\kappa^2 \cos^2(\varphi/2)}{4\left(\text{PLog}\left[\frac{\kappa \cos(\varphi/2)}{2[D(0)t]^{1/4}}\right]\right)^2} \qquad (15)$$

with the abbreviation for κ:

$$\kappa = \frac{(1+\nu)}{3} \frac{K_{\text{tip}}}{\sqrt{2\pi}} \frac{\Delta V_{\text{w}}}{R\Theta} \qquad (16)$$

"PLog" stands for the Lambert W function or *product log function*, i.e., the solution $x = f(y)$ of the equation $y = x \exp(x)$ (in *Mathematica*[38] called ProductLog).[‡] Contour radii for constant swelling are plotted in Fig. 2(a) with $\kappa/(D(0)t)^{1/4}$ as the parameter.

In the case of strongly stress-enhanced diffusion, i.e., $\kappa/(D(0)t)^{1/4} \gg 1$, Eq. (15) tends to

$$\frac{r}{r_{\text{max}}} = \cos^2(\varphi/2), \; r_{\text{max}}$$

$$= \kappa^2 \left(2\text{Plog}\left[\frac{\kappa}{2[D(0)t]^{1/4}}\right]\right)^{-2} \qquad (17)$$

This can be seen from Fig. 2(a) where the thick curve represents the limit case for $\kappa/(D(0)t)^{1/4} \to \infty$ in K_{sh}. The shape of the related swelling zone is shown in Fig. 2(b). The result is similar to that for the phase-transformation zones in zirconia ceramics. But in contrast to those zones, the size of the swelling zones is time-dependent.

The height of the swelling contours for an arbitrary $\kappa/(D(0)t)^{1/4}$ results in very complicated expressions. For reasons of simplicity, the contour radius for $\varphi = 0$, r_{max}, may therefore be used as a characteristic size parameter. According to the analysis by McMeeking and Evans[39] this type of zone must result for $\kappa/(D(0)t)^{1/4} \to \infty$ in $K_{\text{sh}} = 0$.

IV. Shielding Term for Growing Cracks

A crack growing at a constant subcritical velocity, v, develops an extended zone of biaxial compression on the newly formed crack faces. The zone forms from the crack tip and as the crack moves forward, it passes through the compressive zone leaving half the compressive zone along each fracture surface. Figure 3(a) shows the swelling zone after a crack increment of Δa. For a very sim-

plified analysis, the continuously varying swelling strain $\varepsilon(r)$ may be replaced by a step-shaped dependency with a change at the characteristic contour height $\omega = b$ (Fig. 3(b)), defined by the same shielding stress-intensity factor.

In this normalized representation, $K_{\text{sh,max}}$ denotes the shielding for an infinitely long zone, $\Delta a/\omega \to \infty$. Without placing great demands on accuracy we can represent the $K_{\text{sh}}-\Delta a$ dependency, numerically given by McMeeking and Evans,[39] by the approximation:

$$\frac{K_{\text{sh}}}{K_{\text{sh,max}}} \cong \tanh\left[0.757\left(\frac{\Delta a}{\omega_{\text{eff}}}\right)^{3/4}\right]^{2/3} \qquad (18)$$

with an effective zone height $\omega_{\text{eff}} \approx r_{\text{max}}$, Eq. (17) (Fig. 3(b)).

The time available for the formation of the swelling zone is roughly $t = r_{\text{max}}/v$. Instead of Eq. (17), this now results in the implicit equation:

$$r_{\text{max}} = \kappa^2 \left(2\text{Plog}\left[\frac{\kappa}{2[D(0)r_{\text{max}}/v]^{1/4}}\right]\right)^{-2} \qquad (19)$$

The solution in terms of the power law of subcritical crack growth, which holds far from the thermodynamic threshold for crack growth[12]

$$v = A K_{\text{tip}}^n \qquad (20)$$

is then

$$r_{\text{max}} = \frac{\kappa^2}{\text{PLog}^2\left[\kappa\sqrt{\frac{v}{D(0)}}\right]},$$

$$\kappa = \frac{(1+\nu)}{3} \frac{(v/A)^{1/n}}{\sqrt{2\pi}} \frac{\Delta V_{\text{w}}}{R\Theta} \qquad (21)$$

It has to be reemphasized that this maximum zone size is only reached if a crack has grown a sufficiently large increment of $\Delta a > r_{\text{eff}}$ at a constant rate.

The maximum shielding stress-intensity factor increases with decreasing crack growth rate

$$K_{\text{sh,max}} \cong -0.22 \frac{\varepsilon_0 E}{1-\nu} \frac{\kappa}{\text{PLog}\left[\kappa\sqrt{\frac{v}{D(0)}}\right]} \qquad (22)$$

[‡]The use of commercial names is for identification only and does not imply endorsement by the National Institute of Standards and Technology.

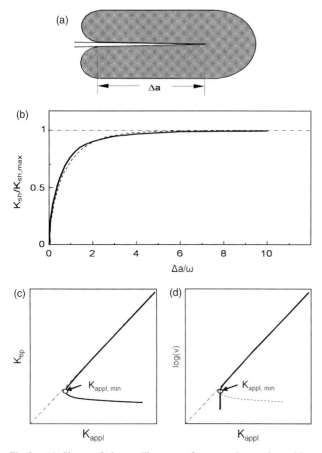

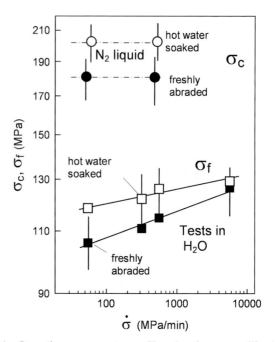

Fig. 4. Strength measurements on silica glass in two modifications; solid symbols: freshly abraded glass rods, open symbols: measurement after hot-water storage for 112 h at 88°C (circles: inert strength σ_c measured in liquid nitrogen, squares: strength σ_f affected by subcritical crack growth in water) from Hirao and Tomozawa.[19]

Fig. 3. (a) Shape of the swelling zone for a crack growing with a constant crack rate v, (b) shielding stress-intensity factor K computed with the method proposed by McMeeking and Evans[39] normalized on the limit value for $\Delta a/\omega \rightarrow \infty$ (solid curve), approximation by Eq. (18) (dashed curve), (c) related K_{tip} versus K_{appl} curve (solid curve); the straight dashed curve is an extension of the linear portion of the curve, extended from high K_{tip}, (d) related log (v) versus K_{appl} curve. Below $K_{appl, min}$ the crack stops growing; the dashed curve represents a continuation of the log (v) versus K_{appl} relationship; this portion of the curve is not accessible to real crack growth.

with the coefficient 0.22 derived in.[39] This equation implies that $K_{sh,max} \rightarrow -\infty$ for $v \rightarrow 0$. Consequently, the K_{tip} versus K_{appl} curve is given by

$$K_{tip} = K_{appl} + K_{sh}, \quad K_{sh} < 0 \tag{23}$$

For fully developed zones according to Fig. 3(a), the crack tip stress-intensity factor K_{tip} is plotted in Fig. 3(c) as a function of the applied stress-intensity factor K_{appl}. The minimum applied stress-intensity factor, $K_{appl,min}$, is indicated by the circle. This curve together with Eq. (23) establishes a threshold value for subcritical crack growth as shown in Fig. 3(d). The experimentally nonaccessible region of the $v(K_{appl})$-relation is indicated by the dashed curve in Fig. 3(d).

V. Interpretation of Experimental Results from Literature

(1) Inert Strength

The parameter ε_0 can be estimated from strength measurements by Hirao and Tomozawa[19] on hot-water soaked and freshly abraded silica glass specimens. Hirao and Tomozawa determined strength of high-silica glass for freshly abraded samples and specimens that were hot-water soaked for 112 h at 88°C. The results of strength tests in liquid N_2 as the inert medium under two loading rates are shown in Fig. 4 by the circles; closed circles for freshly abraded glass; open circles for abraded glass

aged in hot water. Whereas the freshly produced samples had strength of $\sigma_c = 180$ MPa, an increase of $\approx 12\%$ was caused by the water soaking procedure. Subcritical crack growth was absent in liquid N_2 as can be seen from the independence of strength on loading rate.

Because all specimens were prepared in the same way, their initial flaw population must have been the same. This makes it clear that the apparent toughness must have been increased during soaking by about 12% according to

$$K_{appl} = \sigma_c Y \sqrt{a} = K_{Ic} + |K_{sh}| = 1.12 K_{Ic} \tag{24}$$

(Y = geometric function). As a consequence of these inert strength results, the difference in stress-intensity factor $|K_{sh}| = 0.12 K_{Ic}$ must have been generated during water storage with $K_{Ic} \cong 0.8$ MPa $\cdot \sqrt{m}$, $K_{sh} \cong -0.096$ MPa $\cdot \sqrt{m}$ results.

The intrinsic stress-intensity factor, K_{sh}, caused by the compressive stresses in this layer is given by Eq. (9) where the surface strain was assumed to be a constant value ε_0, independent of time. Because K_{sh} is negative for volume expansion, the compressive zone of expansion partially shields the crack tip from the applied load.

With $E = 73.3$ GPa and $v = 0.17$ from[40] it follows from Eq. (7) that

$$\varepsilon_0 \sqrt{b_{eff}} = 4 \times 10^{-6} \sqrt{m} \tag{25}$$

The diffusion coefficient for the interesting temperature of 88°C could be taken from Zouine et al.[30] to be $D_0 \cong 3 \times 10^{-19}$ m²/s. With $E = 73$ GPa and $v = 0.17$ from[40] and $K_{sh} = -0.096$ MPa$\sqrt{m}$; from Eq. (8) we obtain,

$$b_{eff} \cong 0.3 \, \mu m \tag{26}$$

and

$$\varepsilon_0 \approx 0.7\% \tag{27}$$

Introducing this value in (5) gives at the surface (where $\varepsilon = \varepsilon_0$) a biaxial compression stress of $\sigma_x = \sigma_y = -200$ MPa.

In context with the result of (27), it has to be mentioned that volumetric strains may not be fully developed in the case of the strength measurements. In the first part of the soaking time, the crack is free of stresses and water can freely diffuse into the crack faces. With increasing time, a negative stress-intensity factor develops, so that the crack faces are pressed on each other and water diffusion is suppressed, resulting in a thinner zone compared with that in specimens under a superimposed positive stress-intensity factor. Nevertheless, we expect the general conclusions of this section to be correct: the differences in strength of the two types of specimens, freshly abraded and aged, occur as a consequence of the zones of swelling around the tips of cracks contained in the aged glass, but not in the freshly abraded glass.

(2)　Static Lifetime Tests

Figure 5 shows a schematic v–K curve for tests on fresh cracks of initial length a_0. Under constant load, the curve increases monotonically with $K_{appl} \propto (a_0 + \Delta a)^{1/2}$. The circle at the origin of the dashed curve applies for the cardioid shape of the swelling zone developing at $\Delta a = 0$. Because the shielding term for this zone is very small, $K_{sh} \to 0$, this point is also located on the v–K_{tip} curve represented by the light dash-dotted line. Experimentally, this curve is difficult to determine, because all crack growth data are affected more or less by a shielding stress-intensity factor term. The solid curve represents the v–K_{appl} curve for a test with a fully developed zone length, $K_{sh} = K_{sh,max}$. The upper limit curve for $K_{appl} = K_{tip} + |K_{sh,max}|$ will be reached for a swelling zone length of $\Delta a \geq 3\omega$ (Fig. 3(a)). The presence of a minimum in K_{appl}, $dK_{appl}/d(\log(v)) = 0$, suggests a static fatigue limit for the v–K_{appl} curve, see Fig. 3(d).

The dashed curve in Fig. 5(a) describes the v–K_{appl} curve for a lifetime test, which uses fresh cracks previously exposed to an aqueous environment, but not yet propagated in the presence of water. These cracks have no swelling along the fracture surface that would result in toughening. The water that penetrates into the tensile stress region around the crack tip has a cardioid shape indicated in the figure. On propagation in a moist environment, a zone of swelling sweeps from the crack tip to the surface of the crack and develops into a shielding zone that toughens the crack, Section IV. The crack growth curve starts at ($\Delta a = 0$, $K_{sh} = 0$) and asymptotically approaches the solid curve as the length of the swelling zone increases. The zone thickness is less for more rapidly moving cracks, and the slope of the v–K_{appl}

curve increases as the curve approaches the solid curve. These results suggest that different slopes can be obtained for v–K_{appl} curves depending on how the data are collected. Curves obtained on fresh cracks, are expected to be shallower than curves from cracks with well-developed zones of swelling. Data that distinguishes between these two kinds of behaviour are discussed in the next section.

(3)　A Possible Explanation of Results from Dynamic Strength Tests

In dynamic bending tests carried out in water, the slope of log (σ_f) versus the log $(\dot{\sigma})$ was found by Hirao and Tomozawa[19] to depend on test history. Curves obtained on freshly abraded glass were steeper than those obtained on glass soaked in hot water (Fig. 4). In the power-law representation of subcritical crack growth data,[12,37] the value of n for the data in Fig. 4 was found to be greater for data collected on hot water soaked specimens than for data collected on freshly abraded specimens. These results can be explained by the ideas presented in the previous section. For freshly abraded specimens, shielding zones develop during the test and resistance to crack growth gradually builds up as the crack progresses, as illustrated in Fig. 6 by the thick dashed curve. By contrast, specimens that have been exposed to water at elevated temperatures will contain cracks that have fully developed shielding zones and for each value of log(v), crack growth will require a higher value of K_{appl}, solid line in Fig. 6. The slope of the v–K_{appl} curves from such data will be steeper than curves from freshly abraded specimens. As the slope on log–log representation of a v–K_{appl} curve is n, the value of n for the specimens exposed to hot water is expected to be greater than the value of n obtained on fresh specimens, as is observed experimentally[19] (Fig. 4).

We expect as the general conclusions of this section: the differences in slopes of the two types of specimens, freshly abraded and aged, occur as a consequence of differences in the zones of swelling around the tips of cracks contained in their surfaces (preexisting from hot-water soaking or to be generated in the dynamic strength test).

(4)　Crack-Face Measurements by AFM

A discussion of crack-surface inspections by atomic force microscopy (AFM) is given in Wiederhorn *et al.*[41] In particular, data was obtained on a crack in silica glass that was first arrested

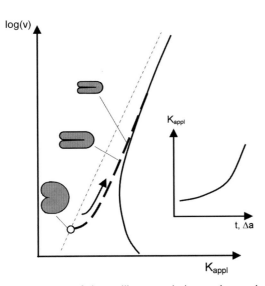

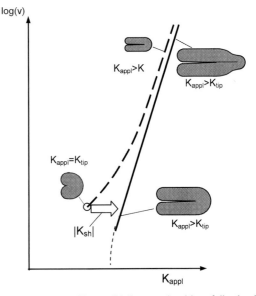

Fig. 5. Development of the swelling zone during crack growth. The solid curve is expected for tests carried out with v = constant over long crack extensions Δa (e.g., in a DCB-test) reaching saturation in the shielding term. The heavy dashed curve is for a crack that develops a zone of swelling along the fracture surface as it propagates. The light dash-dotted line represents the v–K_{tip} curve.

Fig. 6. v–K curve for silica; solid line: crack with a fully developed shielding zone present already at the beginning of the test, dashed line: test with a shielding zone to be created during a test. The thin dashed curve indicates the possibility of fatigue limit at low values of K_{appl}, as in Figs. 3 and 5.

and held under load and then repropagated under a much higher load. The stress-intensity factor applied to the crack tip at the point of arrest was 0.245 MPa·$\sqrt{m}$; the load was maintained for 80 days in water at room temperature, then increased to 0.48 MPa·$\sqrt{m}$ to restart the crack. Crack motion was immediate upon increasing the load on the specimen.

Examination of a crack profile by comparing the two fracture surfaces after completing the fracture (Fig. 7(a)) indicates that with the exception of a small area near the crack tip, a, the two surfaces matched over the length of the crack that had been opened and exposed to the water. An interesting observation in Fig. 7(a) is the overlap of the two fracture surfaces near the crack tip, a. One possible explanation of the overlap might be the occurrence of silica precipitate from the crack tip solution and its deposition at the crack tip before final fracture.[16] This was the interpretation given for the observation in Wiederhorn et al..[41] We now discount this possibility for two reasons. First, the experiment was carried out in water, and the specimen had to be dried before any studies could be carried out on the fracture surface. In drying the fracture surface, it was wiped with high-quality tissue paper, which would have removed any precipitation deposited on the surface during the experiment. In our previous experience, gels that precipitate on the fracture surface are easily removed by such a procedure.

Second, an AFM scan of the apparent excess material at the crack arrest point using the tapping mode™ indicated no shift in the phase image, which suggests that the viscoelastic/elastic properties of the excess material were close to that of the original glass. If the excess material represented a dried precipitate, the elastic constants would have been lower because of internal porosity and a phase shift would have been observed in the phase image.

Based on these considerations, we interpret the excess material ahead of the actual crack tip as a water diffusion zone where swelling by the volumetric strain, ε_0, took place. After complete cracking of the specimen, each half of this region can expand in the direction normal to the new crack face. This results in a volume expansion and in an overlapping of the surface profiles. Figure 7(b) gives measurements of the overlapping displacements with a maximum value of about 6 nm.

In order to study this effect, a finite element (FE) modelling of the heart-shaped zone according to Fig. 2(b) was performed. The case of a half-space was realized by a plate 60ω wide and 30ω high including 1400 elements and 4300 nodes. Solid continuum elements (8-node bi-quadratic) in plane strain were chosen and the computations carried out with ABAQUS version 6.8. The volume strain was replaced by the equivalent thermal problem by heating the inner of the zone with a temperature of ΔT keeping the temperature outside the zone constant (e.g., at

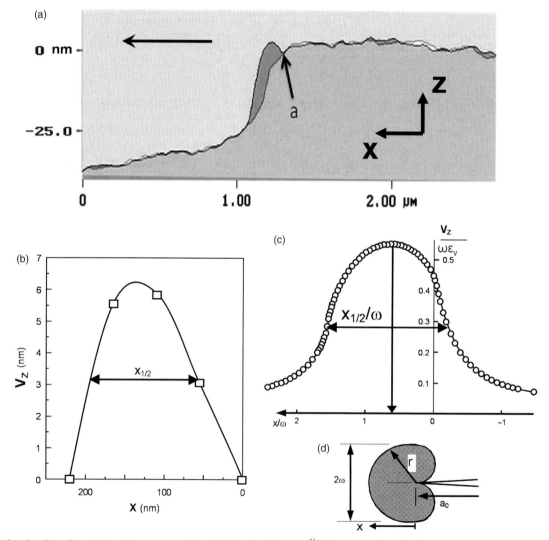

Fig. 7. (a) Overlapping of crack faces for an arrested crack after final fracture.[41] The arrow gives the direction of crack growth, (b) crack–face displacement, V_z, from (a), (c) finite element result for the surface displacements due to a volumetric swelling strain ε_v, (d) the cardioid shape of the region of the glass penetrated by water. The width of the cardioid is 2ω and the distance from the crack tip in front of the crack is x (geometric data are defined in the bottom schematic).

$T = 0$). This results in the volume strain

$$\varepsilon_v = 3\alpha\Delta T \tag{28}$$

The displacements are shown in normalized representation in Fig. 7(c). The maximum displacement from FE-modelling was found at 0.6ω ahead of the crack tip to be $v_y/(\omega\varepsilon_v) \cong 0.55$. From Figs. 7(b) and (c) we can conclude that $\omega\varepsilon_0 \cong 11$ nm.

The width at half of maximum displacement, $x_{1/2}/\omega$, is (see Fig. 7(c)) $x_{1/2}/\omega = 1.8$. The corresponding measured value of $x_{1/2}$ is $x_{1/2} \approx 140$ nm (see Fig. 7(b)). From these two results we obtain: $\omega \cong 80$ nm $\Rightarrow \varepsilon_0 = 13.7\%$, or, a linear strain of about 4.6%, which is equivalent to a biaxial compression stress of $\sigma_x = \sigma_y = -4000$ MPa, Eq. (5).

The height of the zone (identical with $r \times \sin\varphi$ for $\varphi \cong \pi/3$) and an estimate of ΔV_w can be determined from Eq. (17). Inserting the values of $D_0 = 10^{-21}$ m^2/s^{25} into Eq. (17) yields $\omega \cong 140$ nm for $\Delta V_w = 30 \times 10^{-6}$ m^3/mol, $\omega \cong 110$ nm for $\Delta V_w = 20 \times 10^{-6}$ m^3/mol and $\omega \cong 80$ nm for $\Delta V_w = 10 \times 10^{-6}$ m^3/mol. The latter value of ω is in good agreement with the value resulting from the evaluation of Fig. 7(b). This calculation suggests that $\Delta V_w = 10 \times 10^{-6}$ m^3/mol, a value not that far from the molar volume of water, $\Delta V_m = 18 \times 10^{-6}$ m^3/mol. Thus, a systematic investigation of the zone size at different holding loads, K, holding times t, and temperatures, Θ, provides an independent tool for the determination of the activation volume, ΔV_w, for the diffusion of water in silica glass.

VI. Summary/Conclusions

In this paper, we investigated the effect of water diffusion into silica glass fracture surfaces on the strength and static fatigue of this glass. The approach used is identical to that used by McMeeking and Evans[39] in their calculation of the effect of a phase transformation on the strength and toughening of zirconium oxide. The toughness of zirconium oxide was enhanced substantially by a volume expansion of the material around the crack tip, caused by a stress-induced phase transformation in the zirconium oxide. In the case of silica glass, the volume expansion is the result of swelling when water diffuses into the fracture surface of the glass and the glass around the crack tip. The swelling is time-dependent and so the toughening of the glass increases with time. The theory is able to predict an increase of strength when testing is carried out in liquid nitrogen; a small change in the slope of the static fatigue curve as a result of the zone of swelling around the crack tip, and an overlapping of the fracture surfaces by about 6 nm in the near vicinity of the crack tip due to water diffusion into the highly strained region surrounding the crack tip. A further benefit of this study was the development of a new method of measuring the activation volume for the diffusion of water through silica glass.

References

[1]C. R. Kurkjian, P. K. Gupta, R. K. Brow, and N. Lower, "The Intrinsic Strength and Fatigue of Oxide Glasses," *J. Noncrystal. Solids*, **316**, 114–24 (2003).

[2]F. W. Preston, "The Mechanical Properties of Glass," *J. Appl. Phys.*, **13**, 623–34 (1942).

[3]R. E. Mould, "The Strength of Inorganic Glasses"; pp. 119–49 in *Fundamental Phenomena in the Materials Sciences, Vol. 4, Fracture of Metals, Polymers and Glasses*, Edited by L. J. Bonis, J. J. Duga, and J. J. Gilman. Plenum Press, New York, 1967.

[4]B. A. Proctor, I. Whitney, and J. W. Johnson, "The Strength of Fused Silica," *Proc. R. Soc. A*, **297** [1451] 534–57 (1967).

[5]V. M. Sglavo and D. J. Green, "Fatigue Limit in Fused Silica," *J. Eur. Ceram. Soc.*, **21**, 561–7 (2001).

[6]S. M. Wiederhorn and L. H. Bolz, "Stress Corrosion and Static Fatigue of Glass," *J. Am. Ceram. Soc.*, **53** [10] 543–8 (1970).

[7]S. M. Wiederhorn, H. Johnson, A. M. Diness, and A. H. Heuer, "Fracture of Glass in Vacuum," *J. Am. Ceram. Soc.*, **57** [8] 336–41 (1974).

[8]S. Sakaguchi and Y. Hibino, "Fatigue in Low Strength Silica Optical Fibres," *J. Mater. Sci.*, **19**, 3416–20 (1984).

[9]M. Muraoka and H. Abé, "Subcritical Crack Growth in silica Optical Fibers in a Wide Range of Crack Velocities," *J. Am. Ceram. Soc.*, **79** [1] 51–7 (1996).

[10]T. I. Suratwala and R. A. Steele, "Anomalous Temperature Dependence of Sub-Critical Crack Growth in Silica Glass," *J. Non-Crystalline Sol.*, **316**, 174–82 (2003).

[11]S. M. Wiederhorn, S. W. Freiman, E. R. Jr. Fuller, and C. J. Simmons, "Effect of Water and Other Dielectrics on Crack Growth," *J. Mater. Sci.*, **17**, 3460–78 (1982).

[12]B. R. Lawn, *Fracture of Brittle Solids*, 2nd edition, Cambridge University Press, Cambridge, 1993.

[13]T. A. Michalske and S. W. Freiman, "A Molecular Mechanism for Stress Corrosion in Vitreous Silica," *J. Am. Ceram. Soc.*, **66** [4] 284–8 (1983).

[14]K. Han, M. Ciccotti, and S. Roux, "Measuring Nanoscale Stress Intensity Factors with an Atomic Force Microscope," *EPL*, **89**, 66003 (2010).

[15]G. Pezzotti and A. Leto, "Contribution of Spatially and Spectrally Resolved Cathodo-luminescence to Study Crack-Tip Phenomena in Silica Glass," *Phys. Rev. Lett.*, **103**, 175501 (2009).

[16]Y. Bando, S. Ito, and M. Tomozawa, "Direct Observation of Crack Tip Geometry of SiO$_2$ Glass by High-Resolution Electron-Microscopy," *J. Am. Ceram. Soc.*, **67**, C36–7 (1984).

[17]R. J. Charles and W. B. Hillig, *Symposium on Mechanical Strength of Glass and Ways of Improving It*. Florence, Italy, September 25–29, 1961, pp. 511–27. Union Scientifique Continentale du Verre, Charleroi, Belgium, 1962.

[18]S. Ito and M. Tomozawa, "Crack Blunting of High-Silica Glass," *J. Am. Ceram. Soc.*, **65** [8] 368–71 (1982).

[19]K. Hirao and M. Tomozawa, "Kinetics of Crack Tip Blunting of Glasses," *J. Am. Ceram. Soc.*, **70** [1] 43–8 (1987).

[20]K. Hirao and M. Tomozawa, "Dynamic Fatigue of Treated High-Silica Glass: Explanation by Crack Tip Blunting," *J. Am. Ceram. Soc.*, **70** [6] 377–82 (1987).

[21]B. R. Lawn, K. Jakus, and A. C. Gonzalez, "Sharp vs Blunt Crack Hypotheses in the Strength of Glass: A Critical Study Using Indentation Flaws," *J. Am. Ceram. Soc.*, **68** [1] 25–34 (1985).

[22]D. H. Roach, S. Lathabai, and B. R. Lawn, "Interfacial Layers in Brittle Cracks," *J. Am. Ceram. Soc.*, **71** [2] 97–105 (1988).

[23]T. A. Michalske and B. C. Bunker, "Effect of Surface Stress on Stress Corrosion of Silicate Glass"; pp. 3689–701 in *Advances in Fracture Research*, Vol. 6, Edited by K. Salama, K. Ravi-Chandler, D. M. R. Tablin, and P. Ramao Rao. Pergamon Press, New York, 1989.

[24]J.-P. Guin, S. M. Wiederhorn, and T. Fett, "Crack-Tip Structure in Soda-Lime–Silicate Glass," *J. Am. Ceram. Soc.*, **88** [3] 652–9 (2005).

[25]T. Fett, J. P. Guin, and S. M. Wiederhorn, "Stresses in Ion-Exchange Layers of Soda-Lime–Silicate Glass," *Fatigue Fract. Eng. Mater. Struct.*, **28**, 507–14 (2005).

[26]T. A. Michalske, "The Stress Corrosion Limit: Its Measurement and Implications"; pp. 277–89 in *Fracture Mechanics of Ceramics, Vol. 5, Surface Flaws, Statistics, and Microcracking*, Edited by R. C. Bradt, A. G. Evans, D. P. H. Hasselman, and F. F. Lange. Plenum Press, New York, 1977.

[27]E. Gehrke, Ch. Ullner, and M. Hähnert, "Fatigue Limit and Crack Arrest in Alkali-Containing Silicate Glasses," *J. Mater. Sci.*, **26**, 5445–55 (1991).

[28]R. H. Doremus, "Diffusion of Water in Silica Glass," *J. Mater. Res.*, **10** [9] 2379–89 (1995).

[29]K. M. Davis and M. Tomozawa, "Water Diffusion into Silica Glass: Structural Changes in Silica Glass and their Effect on Water Solubility and Diffusivity," *J. Non-Cryst. Solids*, **185**, 203–20 (1995).

[30]A. Zouine, O. Dersch, G Walter, and F. Rauch, "Diffusivity and Solubility of Water in Silica Glass in the Temperature Range 23–200°C," *Phys Chem. Glasses: Eur. J. Glass Sci Technol. B*, **48** [2] 85–91 (2007).

[31]A. Oehler and M. Tomozawa, "Water Diffusion into Silica Glass at a Low Temperature Under High Water Vapor Pressure," *J. Non-Cryst. Glasses*, **347**, 211–9 (2004).

[32]M. Nogami and M. Tomozawa, "Effect of Stress on Water Diffusion in Silica Glass," *J. Am. Ceram. Soc.*, **67**, 151–4 (1984).

[33]M. I. Gorbacheva and R. M. Zaoints, "Moisture Expansion of Various Crystalline and Amorphous Phases," *Steklo Keram.*, **11**, 18–9 (1974).

[34]J. Thurn, "Water Diffusion Coefficient Measurements in Deposited Silica Coatings by the Substrate Curvature Method," *J. Non-Cryst. Solids*, **354**, 5459–65 (2008).

[35]W. M. Kuschke, H. D. Carstanjen, N. Pazarkas, D. W. Plachke, and E. Arzt, "Influence of Water Absorption by Silicate Glass on the Strains in Passivated Al Conductor Lines," *J. Electron. Mater.*, **27**, 853–7 (1998).

[36]T. Fett, *Stress intensity factors, T-stresses, Weight functions, IKM 50*. Universitätsverlag Karlsruhe, Karlsruhe, Germany, 2008.

[37]D. Munz and T. Fett, *Ceramics, Mechanical Properties, Failure Behaviour, Materials Selection*. Springer, Berlin, 1999.

[38]Wolfram Research. *Mathematica*. Wolfram Research, Champaign, IL, USA.

[39]R. M. McMeeking and A. G. Evans, "Mechanics of Transformation-Toughening in Brittle Materials," *J. Am. Ceram. Soc.*, **65**, 242–6 (1982).

[40]D. B. Fraser, "Acoustic Properties of Vitreous Silica," *J. Appl. Phys.*, **39** [13] 5868–78 (1968).

[41]S. M. Wiederhorn, J.-P. Guin, and T. Fett, "The Use of Atomic Force Microscopy to Study Crack Tips in Glass," *Met. Mater. Trans.*, **42A**, 267–78 (2011). □

J. Am. Ceram. Soc., **94** [S1] S204–S214 (2011)
DOI: 10.1111/j.1551-2916.2011.04516.x
© 2011 The American Ceramic Society

journal

A Constitutive Description of the Inelastic Response of Ceramics

Vikram S. Deshpande[†]

Department of Engineering, University of Cambridge, Cambridge CB2 1PZ, U.K.

E. A. Nell Gamble, Brett G. Compton, Robert M. McMeeking, Anthony G. Evans, and Frank W. Zok

Materials Department, University of California, Santa Barbara, 93106 California

The objective of the article is to present a unified model for the dynamic mechanical response of ceramics under compressive stress states. The model incorporates three principal deformation mechanisms: (i) lattice plasticity due to dislocation glide or twinning; (ii) microcrack extension; and (iii) granular flow of densely packed comminuted particles. In addition to analytical descriptions of each mechanism, prescriptions are provided for their implementation into a finite element code as well as schemes for mechanism transitions. The utility of the code in addressing issues pertaining to deep penetration is demonstrated through a series of calculations of dynamic cavity expansion in an infinite medium. The results reveal two limiting behavioral regimes, dictated largely by the ratio of the cavity pressure p to the material yield strength σ_Y. At low values of p/σ_Y, cavity expansion occurs by lattice plasticity and hence its rate diminishes with increasing σ_Y. In contrast, at high values, expansion occurs by microcracking followed by granular plasticity and is therefore independent of σ_Y. In the intermediate regime, the cavity expansion rate is governed by the interplay between microcracking and lattice plasticity. That is, when lattice plasticity is activated ahead of the expanding cavity, the stress triaxiality decreases (toward more negative values) which, in turn, reduces the propensity for microcracking and the rate of granular flow. The implications for penetration resistance to high-velocity projectiles are discussed. Finally, the constitutive model is used to simulate the quasi-static and dynamic indentation response of a typical engineering ceramic (alumina) and the results compared to experimental measurements. Some of the pertinent observations are shown to be captured by the present model whereas others require alternative approaches (such as those based on fracture mechanics) for complete characterization.

I. Introduction

CERAMICS have been used extensively for armor protection. When confined within a metallic medium, their exceptional dynamic strength in compression causes impacting projectiles to deform and erode. Absent the confinement, the tensile stresses induced in the ceramic cause extensive cracking that eliminates the benefit. The practical challenge is to conceive designs that maintain the confinement for a sufficient duration to realize the full benefit. Current design practice has been developed through extensive testing but has limited scope given the large design space; the most efficient approach would incorporate numerical simulations.

Various computational tools have been developed and dynamic constitutive models devised that characterize the response of ceramics to extreme loads.[1–6] In general, two approaches have been used:

(i) Phenomenological damage mechanics, wherein the ceramic is regarded as an elastic–plastic material subject to a notional damage process that reduces the strength as the deformation proceeds.

(ii) Micromechanically motivated models that incorporate aspects of the physics governing compressive damage and plastic deformation.

Each of these approaches has its benefits and draw-backs. The micromechanical models give insight into the mechanisms at play but are computationally too expensive to be used in large-scale structural simulations. On the other hand, while the phenomenological approaches are suited to large-scale computations, they require extensive calibration and typically have a regime of applicability that is limited to scenarios resembling the calibration schemes. Among the phenomenological models, the one that is most complete is that devised by Johnson and Holmquist.[5] The model incorporates a phenomenological law inspired by the response of ceramics impacted at high velocity under highly confined conditions. It embodies a Drucker–Prager yield surface[7] that evolves with damage through the effective plastic strain (analogous to that in the Johnson–Cook model[8] for metals). The representation has multiple coefficients and exponents that require calibration through dynamic measurements. It does not incorporate any microstructural characteristics (such as grain size) or normative material properties (such as toughness and hardness). Nevertheless, when a material has been calibrated in accordance with the proposed protocol, projectile penetration can be simulated with adequate fidelity.[9] The evident limitation is that, for each candidate ceramic, the coefficients and exponents must be recalibrated, because they are not connected to basic microstructure/property relationships.

Recently, Deshpande and Evans[10] (subsequently referred to as DE) devised a micromechanically motivated model that is amenable to large-scale dynamic structural computations. Two specific inelastic phenomena have been included in this model. The first is lattice plasticity due to dislocations and twins, characterized by the von Mises stress relative to the (rate-dependent) flow strength of the material. This property can be probed using hardness measurements conducted with a spherical indenter. The second is damage in the form of microcracks that emanate from preexisting flaws. They evolve subject to a combination of local deviatoric and hydrostatic stresses. The microcracks have dimensions and separations that scale with the grain size. They extend in a manner dictated by the short crack toughness of the ceramic. The constitutive law embodying these mechanisms has been used to successfully simulate the influence of normative material properties, including grain size, hardness, and toughness, on the extents of plastic deformation and damage that occur in polycrystalline alumina upon impact of a hard spherical projectile[11] and upon quasi-static penetration by a hard sphere.[12] The limitation of the foregoing representation is that

A. Heuer—contributing editor

Manuscript No. 28828. Received October 27, 2010; approved February 19, 2011.
This work is financially supported by the Office of Naval Research through a Multidisciplinary University Research Initiative Program on "Cellular Materials Concepts for Force Protection," Prime Award No. N00014-07-1-0764.
[†]Author to whom correspondence should be addressed. e-mail: vsd@eng.cam.ac.uk

the mechanical response beyond the point at which the damage becomes saturated (namely, the microcracks have coalesced) has yet to be addressed. The present article establishes an approach for representing the behavior of a damage-saturated ceramic and combines with the foregoing lattice plasticity and damage models to produce a unified model that can be used to simulate deep penetration.

The fully comminuted ceramic created within the highly confined environment beneath an impacting projectile is in a unique state. It consists of grain-sized particles bounded by narrow separations between them (Fig. 1) and has a relative density very close to unity because the volume occupied by the separations is small. This state differs from that found in particulate materials such as soils and in powder compacts, which have relative densities around 60%. When subjected to a combination of compressive and shear stress, the particles can slide and rotate past one another: a phenomenon hereafter referred to as *granular plasticity*.

The article is organized as follows: it begins with a description of the unified model, its physical interpretation, and the associated constitutive law. Thereafter, preliminary features of the constitutive response are elucidated via the predicted stress/strain response of a prototypical hard ceramic (namely alumina) under varying levels of stress triaxiality. Subsequently, two specific loading scenarios are examined and, where possible, used to interpret phenomena found experimentally. The first is the expansion of a spherical cavity in an infinite medium. The objective is to characterize the development and propagation of elastic, plastic, and damage waves as well as understand the interplay between plasticity and damage. The second considers quasi-static and dynamic penetration. Here comparisons are made of the predictions of the original DE model and the current constitutive law, with emphasis on the damage evolution pattern and the load versus displacement response.

II. Development of the Constitutive Model

The inelastic deformation of ceramic polycrystals occurs in accordance with the foregoing three primary mechanisms (Fig. 1) proceeding partially in series and partially in parallel. We envisage the ceramic with a population of preexisting flaws or heterogeneities that, under stress, can extend into microcracks. Before microcracks form, plastic deformation can only occur by dislocation slip or twinning: termed *lattice plasticity*. Its onset is characterized by the von Mises yield criterion, depicted in effective stress (σ_e)–mean stress (σ_m) space as a horizontal line (Fig. 2). When microcracks grow and coalesce, the ceramic transitions into a granular medium comprising narrowly

separated granules with dimensions dictated typically by the grain size. The fully comminuted ceramic has a small shear yield strength governed by friction between the particles. Its yield criterion has a form similar to Drucker–Prager for granular media. The transition between these two yield surfaces with increasing microcrack density is depicted in Fig. 2. At each level of damage, D, yield is characterized by a sloping line in σ_e–σ_m space; as described below, we shall assume that all of these lines intercept the lattice plasticity line at the same location. Upon full comminution, inelastic deformation can occur by one of two mechanisms: granular plasticity of the comminuted particles or lattice plasticity within the particles themselves.

To devise a mechanism-inspired model from the foregoing physical picture, the following four elements must be incorporated:

(i) A lattice plasticity model.

(ii) A criterion for the evolution of damage due to microcracking.

(iii) The transition between the yield surfaces for lattice and granular plasticity.

(iv) The incorporation of elasticity (with lattice and granular plasticity) to give the overall constitutive description.

(1) The Constituent Models

(A) Lattice Plasticity: In structural ceramics, lattice plasticity occurs in accordance with two limits: at low strain rates, resistance from the lattice governed by the Peierls stress and, at high rates, phonon drag.[13] With this in mind, we approximate the effective plastic strain rate $\dot{\varepsilon}_e^{pl}$ versus the von Mises effective stress σ_e relationship by

$$\frac{\dot{\varepsilon}_e^{pl}}{\dot{\varepsilon}_0} = \begin{cases} \left(\dfrac{\dot{\varepsilon}_0}{\dot{\varepsilon}_t}\right)^{(1-n)/n}\left(\dfrac{2\sigma_e}{\sigma_0}-1\right) & 2\sigma_e > \sigma_0\left[\left(\dfrac{\dot{\varepsilon}_t}{\dot{\varepsilon}_0}\right)^{1/n}+1\right] \\ \left(\dfrac{2\sigma_e}{\sigma_0}-1\right)^n & \sigma_0 < 2\sigma_e \le \sigma_0\left[\left(\dfrac{\dot{\varepsilon}_t}{\dot{\varepsilon}_0}\right)^{1/n}+1\right] \\ 0 & 2\sigma_e \le \sigma_0 \end{cases}$$

$$(1)$$

where $\sigma_0(\varepsilon_e^{pl})$ is the flow stress at an equivalent plastic strain ε_e^{pl}, while $\dot{\varepsilon}_0$ and n are the reference strain rate and strain rate sensitivity exponent, respectively, and $\dot{\varepsilon}_t$ is the transition strain-rate above which plasticity becomes phonon drag limited. Note that the exponent n sets the strain rate sensitivity of the ceramic in the low strain rate regime. Strain hardening is characterized by a conventional power law, notably:

$$\sigma_0 = \frac{\sigma_Y}{2}\left[1 + (\varepsilon_e^{pl}/\varepsilon_Y)^M\right] \tag{2}$$

where σ_Y is the uniaxial yield strength, ε_Y the plastic strain at which $\sigma_0 = \sigma_Y$, and M the strain hardening exponent. Associated flow is assumed with a flow potential, $G_p \equiv \sigma_e$. The plastic strain rate is then

$$\dot{\varepsilon}_{ij}^{pl} = \dot{\varepsilon}_e^{pl}\frac{\partial G_p}{\partial \sigma_{ij}} \tag{3}$$

(B) Microcrack Evolution: The model envisages an array of f microcracks per unit volume, subject to principal stresses σ_1 and σ_3 and growing in an otherwise elastic medium (Fig. 3). Each microcrack develops from an initial flaw, radius a, by means of two wings, length ℓ, extending parallel to X_1 (Fig. 3). The inclined flaws are subject to Coulomb friction, with friction coefficient μ. The radius and separation of the flaws scale with the grain size as $a = g_1 d$ and $1/f^{1/3} = g_2 d$, respectively, where g_1 and g_2 are parameters. The initial and current levels of damage are expressed respectively as $D_0 = (4/3)\pi(\alpha a)^3 f$ and $D = (4/3)\pi(l+\alpha a)^3 f$, where $\alpha \equiv \tan\psi$ is a shape factor to account for the angle of the initial flaw as shown in Fig. 3. The

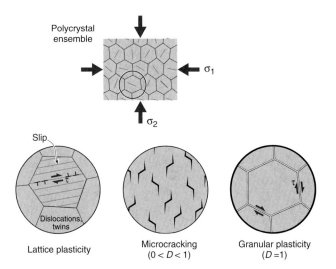

Polycrystal
ensemble

σ_1

σ_2

Slip

Dislocations,
twins

Lattice plasticity

Microcracking
$(0 < D < 1)$

Granular plasticity
$(D = 1)$

Fig. 1. The inelastic mechanisms included in the constitutive model for ceramics.

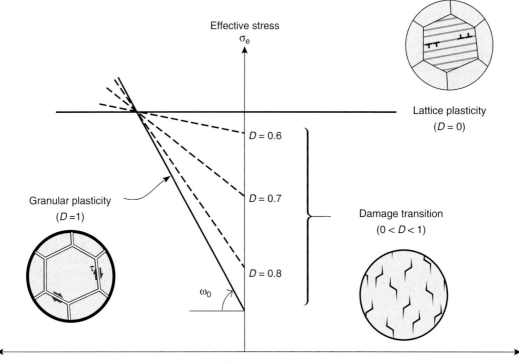

Fig. 2. Schematic of the yield surface transition from lattice to granular plasticity.

development of the microcracks is governed by the mode I stress intensity, K_I, at the tip of the wing cracks. In turn, K_I depends on the stress triaxiality, $\lambda \equiv \sigma_m/\sigma_e$, in accordance with three behavioral regimes (*I, II, III*) with four associated nonlinear functions of damage and friction (*A, B, C,* and *E,* defined in the Appendix A). In *Regime I* ($\lambda \leq -B/A$), the cracks are shut, with $K_I = 0$. In *Regime II*, frictional slip occurs along the initial flaw surfaces, with

$$K_I/\sqrt{\pi a} = A\sigma_m + B\sigma_e \qquad (4)$$

In *Regime III*, $\lambda \geq AB/[C^2 - A^2]$, contact along the faces of the initial flaw is lost, where upon

$$K_I/\sqrt{\pi a} = (C^2\sigma_m^2 + E^2\sigma_e^2)^{1/2} \qquad (5)$$

The crack growth rate, $\dot{\ell}$ is related to the stress intensity factor by

$$\dot{\ell} = \begin{cases} \min\left[\dot{\ell}_0(K_I/K_{IC})^m, \sqrt{G/\rho_0}\right] & D < 1 \\ 0 & \text{otherwise} \end{cases} \qquad (6)$$

where K_{IC} is the mode I (short-crack) fracture toughness, m is a rate exponent, and $\dot{\ell}_0$ the reference crack growth rate at $K_I = K_{IC}$. The crack speed is limited by the shear wave speed $\sqrt{G/\rho_0}$, where G and ρ_0 are the shear modulus and density of the intact ceramic, respectively. When $D \rightarrow 1$, the microcracks coalesce, causing $K_I \rightarrow \infty$ (in *Regimes II and III*), thus requiring that the microcracking model be supplanted by the ensuing model for granular plasticity. This is where the new model deviates from that originally proposed by Deshpande and Evans.[10]

(C) Granular Plasticity: The comminuted ceramic is modeled as a granular medium using a nonassociated, viscoplastic, Drucker–Prager type constitutive law, with an effective stress, $\hat{\sigma}$, defined as

$$\hat{\sigma} \equiv \frac{\sigma_e + (\tan\omega)\sigma_m}{1 - \tan\omega/3} \qquad (7)$$

where ω is the friction angle. The factor in the denominator ensures that, under a uniaxial compressive stress σ, the effective stress $\hat{\sigma} \equiv |\sigma|$.

The effective stress is used to define an effective strain rate $\dot{\varepsilon}_e^g$, motivated by the following considerations. Under compressive hydrostatic stress states ($\sigma_m < 0$), the flow of the comminuted ceramic is expected to follow a Bagnold-type granular flow law,[14] wherein the effective stress for the medium scales with the square of the effective strain rate. This relation arises because the stresses generated in the medium at high-strain rates are not associated with interparticle friction but, rather, are due to repeated collisions between particles. During these collisions, both the momentum change per collision and the number of collisions per unit time are proportional to the strain rate, resulting in the quadratic scaling. In contrast, for tensile hydrostatic stress states ($\sigma_m > 0$), particle interactions are negligible, violating the assumptions that underpin Bagnold flow. These features suggest an overstress model of the form:

$$\frac{\dot{\varepsilon}_e^g}{\dot{\varepsilon}_s} = \begin{cases} \left(\frac{\hat{\sigma}}{\Sigma_c} - 1\right)^s & \hat{\sigma} > \Sigma_c \\ 0 & \text{otherwise} \end{cases} \qquad (8a)$$

where Σ_c is the uniaxial compressive strength of the ceramic and $\dot{\varepsilon}_s$ a reference strain rate (specified subsequently in Section II(2)). Consistent with Bagnold scaling, the exponent is selected to be $s = 0.5$ when $\sigma_m \leq 0$: whereupon (8a) gives the Bagnold scaling $\dot{\varepsilon}_e^g \sim \sqrt{\hat{\sigma}}$ in the limit $\hat{\sigma} \gg \Sigma_c$. Conversely, when $\sigma_m > 0$, s needs to be large (typically $s = s_0 \geq 10$) to ensure that $\hat{\sigma}$ does not significantly exceed the uniaxial strength, Σ_c, thereby preventing the comminuted ceramic from attaining unreasonably high-tensile stresses. To ensure that the strain rate $\dot{\varepsilon}_e^g$ is continuous around $\sigma_m = 0$, a piecewise function for s is specified, with

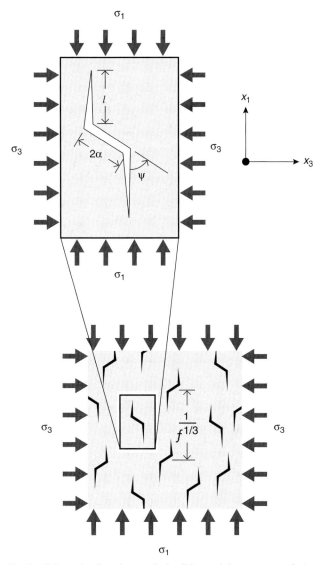

Fig. 3. Schematic of a microcracked solid containing an array of wing cracks.

the transition from $s = 0.5$ to s_0 occurring over the range $-\Sigma_c \leq \sigma_m \leq \Sigma_c$ and given by

$$
s = \begin{cases}
s_0 & \sigma_m \geq \Sigma_c \\
s_0 + \dfrac{1}{2}(s_0 - 0.5)\left(\dfrac{\sigma_m - \Sigma_c}{\Sigma_c}\right)^q & 0 \leq \sigma_m < \Sigma_c \\
0.5 + \dfrac{1}{2}(s_0 - 0.5)\left(\dfrac{\sigma_m + \Sigma_c}{\Sigma_c}\right)^q & -\Sigma_c \leq \sigma_m < 0 \\
0.5 & \text{otherwise}
\end{cases}
$$

(8b)

where the exponent q governs the transition rate.

To complete the characterization, a flow rule is specified. Experiments by Bagnold[14] and others suggest that, when deformation is governed by particle collisions, flow is nonassociated such that the flow potential is not determined by $\hat{\sigma}$. To ensure this generality, a flow surface G_g is defined as

$$
G_g \equiv \begin{cases}
\sigma_e + (\tan \delta)\sigma_m & \sigma_m < 0 \\
\sqrt{(\varepsilon \Sigma_t)^2 + \sigma_e^2} + \dfrac{\Sigma_c}{\Sigma_t}\sigma_m & \text{otherwise}
\end{cases}
$$

(9a)

where Σ_t is the hydrostatic tensile strength, given by setting $\hat{\sigma} = \Sigma_c$ and $\sigma_m/\sigma_e = 1/3$ in Eq. (7)

$$
\Sigma_t \equiv \frac{\Sigma_c(1 - \tan \omega/3)}{\tan \omega}
$$

(9b)

The quantity δ is the flow dilatation angle of the comminuted ceramic defined such that, when $\delta = 0$, the flow is volume conserving under compressive hydrostatic stress states. The parameter ε is a nondimensional regularization parameter (taken as 0.01) used to ensure that the flow is purely dilatational under hydrostatic tension. The granular strain rate follows from the flow surface G_g as

$$
\dot{\varepsilon}_{ij}^g = \frac{\dot{\varepsilon}_e^g}{\varsigma}\frac{\partial G_g}{\partial \sigma_{ij}}
$$

(10a)

where the normalization parameter is

$$
\varsigma \equiv \begin{cases}
\sqrt{\dfrac{3}{2} + \dfrac{1}{3}(\tan \delta)^2} & \sigma_m < 0 \\
\sqrt{\dfrac{(3/2)\sigma_e^2}{(\varepsilon \Sigma_t)^2 + \sigma_e^2} + \dfrac{1}{3}\left(\dfrac{\Sigma_c}{\Sigma_t}\right)^2} & \text{otherwise}
\end{cases}
$$

(10b)

The factor ς is included in the flow rule so that

$$
\left(\frac{1}{\varsigma}\frac{\partial G_g}{\partial \sigma_{ij}}\right)\left(\frac{1}{\varsigma}\frac{\partial G_g}{\partial \sigma_{ij}}\right) \equiv 1
$$

(11)

i.e., the magnitude of the effective granular plastic strain rate is set solely by $\dot{\varepsilon}_e^g$ with $\dot{\varepsilon}_e^g \equiv \sqrt{\dot{\varepsilon}_{ij}^g \dot{\varepsilon}_{ij}^g}$, while the direction of plastic straining is given by $\frac{\partial G_g}{\partial \sigma_{ij}}$.

(2) The Transition from Lattice to Granular Plasticity

Now that the basic models have been established, linkages must be provided. Namely, because the granular plasticity model is valid for all states of the comminuted ceramic (over the entire range $D_0 \leq D \leq 1$), evolution equations of the granular plasticity parameters from $D = D_0$ to $D = 1$ are required. These equations are specified through four parameters: (i) the uniaxial compressive strength Σ_c; (ii) the friction angle ω; (iii) the flow dilatation angle δ and (iv) a reference granular strain rate $\dot{\varepsilon}_s$. In its initial intact state (with $D = D_0$), the ceramic behaves as a von-Mises plastic material (with no contribution from granular plasticity). Accordingly, the granular plasticity parameters are chosen to ensure that, when $D = D_0$, the Drucker–Prager model reduces to the von-Mises model and that the granular strain rate is zero. The following evolution equations then ensue.

(i) *Uniaxial compressive strength* Σ_c. As D increases, Σ_c decreases from the plastic yield strength σ_0 to the fully comminuted strength σ_c, specified using

$$
\Sigma_c = \sigma_0 - (\sigma_0 - \sigma_c)\left(\frac{D - D_0}{1 - D_0}\right)^q
$$

(12)

The exponent q governs the transition rate with respect to D.

(ii) *The friction angle* ω. When the granular strength Σ_c is equal to the plastic yield strength σ_0, the friction angle $\omega = 0$. The angle increases with damage, reaching a value ω_0 upon complete comminution. Its evolution with the current material strength is specified by:

$$
\tan \omega = \frac{(\sigma_0 - \Sigma_c)\tan \omega_0}{\sigma_0 - \sigma_c(1 - \tan \omega_0/3) - \Sigma_c \tan \omega_0/3}
$$

(13a)

Thus, the granular yield surface evolves by pivoting about a coordinate in stress space defined by $\sigma_e = \sigma_0$ and

$$
\sigma_m = \frac{\sigma_0 - \sigma_c}{\tan \omega_0} + \frac{\sigma_c}{3}
$$

(13b)

This evolution is depicted in Fig. 2.

(iii) *The flow dilatation angle* δ. Because the intact ceramic behaves as an incompressible plastic medium, the flow dilatation angle should vary between $\delta = 0$ at $D = D_0$ to that for the fully comminuted ceramic, δ_0, at $D = 1$. Thus, analogous to Eq. (12), the angle varies as

$$\delta = \delta_0 \left(\frac{D - D_0}{1 - D_0} \right)^q \tag{14}$$

(iv) *Reference granular strain rate* $\dot{\varepsilon}_s$. Granular straining does not occur in the undamaged ceramic. Thus $\dot{\varepsilon}_s$ is varied between $\dot{\varepsilon}_s = 0$ at $D = D_0$ to $\dot{\varepsilon}_{so}$ for the fully comminuted ceramic at $D = 1$, in accordance with:

$$\dot{\varepsilon}_s = \dot{\varepsilon}_{so} \left(\frac{D - D_0}{1 - D_0} \right)^q \tag{15}$$

It remains to estimate the reference strain rate $\dot{\varepsilon}_{so}$. The Bagnold[14] analysis suggests that the strength σ_c of the fully comminuted material scales as

$$\sigma_c \sim \left[\frac{e(1 + e)\pi}{6} \left(\frac{\bar{\rho}^{1/3}}{1 - \bar{\rho}^{1/3}} \right)^2 \rho_0 d^2 \right] \dot{\varepsilon}_{so}^2 \tag{16}$$

where d is the particle size, e is the co-efficient of restitution between particles while the relative packing density $\bar{\rho}$ specifies the interparticle spacing. Given values of $\bar{\rho}$ and σ_c, Eq. (16) can be used to estimate $\dot{\varepsilon}_{so}$.

(3) The Overall Constitutive Model

The lattice and granular plasticity models are combined with elasticity to complete the overall model. The dynamic response is dominated by lattice plasticity and granular flow with negligible energy absorption involved in microcracking.[11] Moreover, the elastic strains are small compared to those for lattice plasticity and granular flow. Thus, unlike the model of Deshpande and Evans,[10] the elastic response upon microcracking is left unmodified, i.e. no stiffness reduction compared to the undamaged ceramic for all D. The total strain rate $\dot{\varepsilon}_{ij}$ is the sum of the contributions from elasticity and lattice and granular plasticity. Thus, the elastic strain rate is given by

$$\dot{\varepsilon}_{ij}^e = \dot{\varepsilon}_{ij} - (\dot{\varepsilon}_{ij}^{pl} + \dot{\varepsilon}_{ij}^g) \tag{17}$$

The deviatoric elastic response is specified by an isotropic Hooke's law, with strain rate

$$\dot{D}_{ij} = \dot{\varepsilon}_{ij}^e - \dot{\varepsilon}_{kk}^e \delta_{ij} \tag{18a}$$

and deviatoric stress rate $\dot{S}_{ij}$ related via

$$\dot{S}_{ij} = 2G\dot{D}_{ij} \tag{18b}$$

where G is the shear modulus.

Two alternate formulations are used for the pressure, providing the flexibility needed to examine the importance of shocks within the ceramic under high rates of loading. (i) The *linear formulation* uses the isotropic Hooke's law, with bulk modulus κ related to shear modulus G and Poisson's ratio v by

$$\kappa = \frac{2G(1 + 2v)}{3(1 - 2v)} \tag{19a}$$

with pressure then related to the logarithmic volumetric elastic strain by

$$p \equiv -\frac{\sigma_{kk}}{3} = -\kappa\varepsilon_{kk}^e \tag{19b}$$

where $\varepsilon_{kk}^e = \int \dot{\varepsilon}_{kk}^e \, dt$. (ii) In order to allow for shock formation at high pressures, an alternative formulation wherein pressure is

specified by an equation of state can be used. The *Mie–Grüneisen equation of state* expresses the pressure in terms of the nominal volumetric elastic strain η and the internal energy per unit mass E_m. The internal energy is evaluated using the evolution equation

$$\frac{\partial(\rho E_m)}{\partial t} = \sigma_{ij}\dot{\varepsilon}_{ij} \tag{20a}$$

where ρ is the current density of the damaged ceramic, while the nominal compressive elastic strain is

$$\eta = 1 - \exp(\varepsilon_{kk}^e) \tag{20b}$$

These two state variables are related to the pressure by

$$p = \frac{\rho_0 c_0^2 \eta}{(1 - \vartheta\eta)^2} \left(1 - \frac{\Gamma_0 \eta}{2} \right) + \Gamma_0 \rho_0 E_m \tag{21}$$

where ρ_0 is the initial density of the ceramic, c_0 the initial p-wave speed ($\rho_0 c_0^2$ being equivalent to the elastic bulk modulus at small nominal strains), while Γ_0 and ϑ are dimensionless material constants measured from shock experiments. This equation predicts a linear shock velocity (u_s) versus particle velocity (u_p) relation of the form

$$u_s = c_0 + \vartheta u_p \tag{22}$$

In a finite element (FE) implementation, an artificial viscosity is included in Eq. (21) to enable the numerical solution to capture the development and propagation of the shock. Thus Eq. (21) is modified as

$$p = \frac{\rho_0 c_0^2 \eta}{(1 - s\eta)^2} \left(1 - \frac{\Gamma_0 \eta}{2} \right) + \Gamma_0 \rho_0 E_m + p_{bv} \tag{23}$$

where p_{bv} is the contribution to the pressure from the viscosity. Following von Neumann, a quadratic dependence of the viscous pressure on volumetric strain rate is typically employed:

$$p_{bv} = \begin{cases} 0 & \dot{\varepsilon}_{kk} \geq 0 \\ \rho_0 (b_1 L_e \dot{\varepsilon}_{kk})^2 & \text{otherwise} \end{cases} \tag{24}$$

where $b_1 \sim 1.2$ is the damping constant and L_e the characteristic length of the element in the FE mesh. We note in passing that an artifical shear viscosity is usually not required to stabilize the numerical calculations as both granular and lattice plasticity endow the material with a shear rate dependence.

With the deviatoric and mean stresses specified, the total stress σ_{ij} follows directly as

$$\sigma_{ij} = S_{ij} - p\delta_{ij} \tag{25}$$

(4) Choice of Material Parameters

In the subsequent section, the main features of the constitutive model are demonstrated through computations of various loading scenarios, with emphasis on effects of stress triaxiality. Except where otherwise noted, all computations use material parameters representative of alumina with a grain size $d = 3$ µm. These parameters have been carefully calibrated via a series of experiments reported in the companion paper Gamble *et al.*[15] The ceramic is elastically isotropic with Young's modulus $E = 405$ GPa, Poisson's ratio $v = 0.22$, density $\rho_0 = 3810$ kg/m^3, and fracture toughness $K_{IC} = 3$ MPa m$^{1/2}$. Based on a representative Vickers hardness, the yield strength and strain are taken as $\sigma_Y = 5.75$ GPa and $\varepsilon_Y = 0.002$, respectively, with $M = 0.1$ chosen to give a mildly strain hardening response.

The strain rate sensitivity parameters are chosen as $n = 34$, $\dot{\varepsilon}_0 = 10^{-3} \text{ s}^{-1}$ and $\dot{\varepsilon}_t = 10^6 \text{ s}^{-1}$ while the crack growth parameters are taken to be $l_0 = 10 \text{ mms}^{-1}$ and $m = 30$. Following Ashby and Sammis,[16] the geometrical constants are taken as $\beta = 0.45$, $\alpha = 1/\sqrt{2}$ and $\gamma = 2.0$ (see Appendix A) with a friction coefficient $\mu = 0.75$. The assumed flaw size and spacing parameters are $g_2 = 6$ and $g_1 = 1/2$, which yield $f = 1.74 \times 10^5 \text{ mm}^{-3}$ and $D_0 = 8.57 \times 10^{-4}$. The granular plasticity parameters are: friction angle $\omega_0 = 70°$, flow dilatation angle $\delta_0 = 0°$, and uniaxial compressive strength of the comminuted ceramic $\sigma_c = 1 \text{ MPa}$. The transition exponent is set as $q = 5$ while the granular rate sensitivity exponent in the hydrostatic tensile regime is chosen to be $s_0 = 10$. Taking $e = 1$ and $\bar{\rho} = 0.99$ in Eq. (16) gives the reference granular strain rate as $\dot{\varepsilon}_{s0} = 2 \times 10^4 \text{ s}^{-1}$. The pressure is specified using the linear (Hookean) formulation (i.e., shock effects as modeled by the Mie–Grüneisen equation of state are neglected).

III. Effects of Triaxiality on Intrinsic Material Response

The intrinsic stress versus strain response of the material was computed by integrating the constitutive equations detailed above for a total imposed strain rate of $\dot{\varepsilon}_e \equiv \sqrt{(2/3)\dot{\varepsilon}_{ij}^d \dot{\varepsilon}_{ij}^d} = 100 \text{ s}^{-1}$, neglecting inertial or wave effects. The results are plotted in Fig. 4(a) for several selected values of stress triaxiality, representative of low ($\lambda = -0.1$) to high-confining pressure ($\lambda = -0.5$).

The predictions are qualitatively similar to those of the DE model.[10] For $\lambda = -0.1$ and $-1/3$ (uniaxial compression), the response is approximately linear elastic up to a peak stress, whereupon a sharp load drop occurs. The drop corresponds to a sudden increase in D from D_0 to unity. Thereafter, continued deformation in both cases occurs by granular plasticity at the much reduced stress value of σ_c. For higher levels of imposed

pressure, characterized, for instance, by $\lambda = -0.5$, cracking does not occur and inelastic deformation occurs by lattice plasticity alone. The key difference between the behavior of the current model and the DE model is that, after the onset of damage, the DE model predicts a dilatant response, dictated by the extent of crack opening. By contrast, the addition of granular plasticity in the current formulation means that the postcracking dilatation is controlled independently: in the present computations, with the selection $\delta_0 = 0°$, no volume change occurs under compressive hydrostatic stresses.

Another feature of both the original DE model and the current one is that, once fully comminuted, the material can sustain considerable stresses provided sufficiently high-confining pressures are imposed during subsequent loading. This feature is illustrated in Fig. 4(b). Here the loading occurs in two steps. In the first, the ceramic is loaded at a strain rate $\dot{\varepsilon}_e = 100 \text{ s}^{-1}$ while the triaxiality is fixed at $\lambda = -1/3$ (as in Fig. 4(a)). Following the load drop at $\sigma_e \approx 6$ GPa, continued straining (up to $\varepsilon_e = 0.012$) occurs at a negligible stress. In the second step, the confining pressure is increased, to $\lambda = -0.6$, and deformation continued at $\dot{\varepsilon}_e = 100 \text{ s}^{-1}$. Under these conditions, granular plasticity is suspended and the ceramic is able to sustain high stresses and deform by lattice plasticity within the comminuted particles.

IV. Dynamic Cavity Expansion

The dynamic elasto-plastic field induced by a pressurized spherical cavity expanding in an infinite medium is widely used as a protocol for ascertaining the key phenomena accompanying penetration. An extensive review of early work has been given by Hopkins[17] with emphasis on metals characterized by Mises plasticity. More recent interest in penetration into concrete and other geomaterials has instigated investigations of dynamic qcavity expansion in pressure-sensitive solids described by Mohr–Coloumb or Drucker–Prager constitutive laws.[18–20] More recently, analogous approaches have been used for ceramic targets.[21–22] However, in contrast to the present formulation, the ceramic models used in the aforementioned studies do not differentiate explicitly between lattice and granular plasticity. When the two are indeed decoupled from one another in the constitutive law, the computations of dynamic cavity expansion reveal important interactions between the two deformation mechanisms. This feature is highlighted here.

(1) Boundary Value Problem

A spherical cavity of initial radius a_0 in an infinite medium is pressurized by a constant and spatially uniform pressure p. Specifically, the pressure is zero at time $t = 0^-$ and equal to p_0 at $t \geq 0$ (Fig. 5). Neglecting symmetry-breaking modes of deformation, the active components of the Cauchy stress are the radial stress σ_r and the hoop stresses $\sigma_\theta = \sigma_\phi$, and the radial

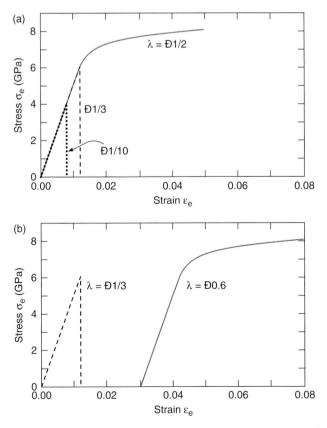

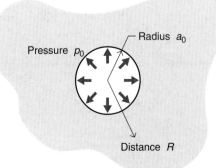

Fig. 4. (a) Effects of triaxiality on the predicted stress/strain response of alumina with reference properties at a strain rate of 100 s⁻¹. (b) Illustrative example showing effects of a change in stress triaxiality, from −1/3 to −0.6, during deformation on the flow response of alumina.

Fig. 5. Sketch of the boundary value problem for the expansion of a spherical cavity in an infinite medium.

equation of motion is given by

$$\frac{\partial}{\partial r}(r^2 \sigma_r) - 2r\sigma_\theta = \rho \ddot{u} r^2 \qquad (26)$$

where r denotes the radial position of a material point in the current configuration, u is the radial displacement of that point, the overdot denotes differentiation with respect to time, and ρ is the material density in the current configuration. Equation (26) is solved using an updated Lagrangian scheme, detailed in Deshpande and Evans.[10] In the FE calculations, the outer radius of the medium being analyzed is taken to be sufficiently large so as to ensure that the elastic wave does not reach the outer boundary over the time-span of the calculations reported here, thereby simulating an infinite medium. Unless otherwise specified, the computations use the property values for alumina (summarized in Section II(4)). The focus of subsequent numerical results and discussion is on the early time response, wherein dynamic effects dominate. For the expansion of a spherical cavity in an infinite *elastic* medium, Hopkins[17] has shown that the stress distributions near the vicinity of the cavity are well approximated by the corresponding quasi-static result at times $\bar{t} \equiv c_e t / a_0 > 4$, where c_e is the p-wave speed. Thus, all results presented here are for times $\bar{t} < 4$.

(2) Effect of Pressure

The predicted temporal variation of the radial displacement U/a_0 of the cavity surface is plotted against time $\bar{t}$ in Fig. 6 for three levels of the cavity pressure. The cavity displacement increases approximately linearly with time for the three pressure levels considered here. We proceed to investigate in further detail the cases in which $p_0 = 12$ and 24 GPa.

First consider the case where $p_0 = 12$ GPa. The spatial distributions of damage D, effective lattice plastic strain ε_e^{pl} and effective granular plastic strain ε_e^g are plotted in Fig. 7(a) at three select times. Note that ε_e^{pl} and ε_e^g are the time integrals of the strain rates given by Eqs. (1) and (8a), respectively. The distributions are plotted as a function of R/a_0, where R is the radial coordinate in the undeformed configuration. Full comminution of the ceramic (i.e., $D = 1$) first occurs at a finite distance from the cavity surface, viz. at $R/a_0 \approx 1.5$. The damage zone then expands radially inwards while simultaneously expanding outwards. However, while the damage zone continues to expand outwards for the duration of the computations, the progression of the damage zone toward the surface to the cavity halts at time $\bar{t} \approx 1.4$ when the inner boundary of the damage zone reaches $R/a_0 \approx 1.1$, i.e., the damage zone does not extend up to the cavity surface. We shall show subsequently that this feature is due to the fact that the large lattice plastic strains near the cavity

surface reduce the stress triaxiality and thereby reduce the propensity for microcracking. The spatial distributions of ε_e^{pl} included in Fig. 7(a) show that, for $\bar{t} > 1.4$, the lattice plastic strain rises sharply for $R/a_0 < 1.1$, i.e., the region over which damage does not occur. Correspondingly, the distribution of granular plastic strains clearly show that significant granular plasticity only occurs over the region where $D = 1$. Note that the total inelastic strains (i.e., lattice plus granular plastic strains) are continuous across the boundary of the damage zone near the surface of the cavity; while ε_e^g drops to zero across the boundary where the damage variable transitions from $D = 1$ to $D = D_o$, ε_e^{pl} rises sharply in order to keep the total inelastic strain approximately continuous.

At the higher applied cavity pressure of $p_0 = 24$ GPa, the qualitative picture discussed above remains unchanged. However, the higher pressure means that the stress triaxiality near the cavity surface is further reduced and hence microcracking initiates even further away from the cavity surface at $R/a_0 \approx 2.2$ and expands *only* away from the cavity surface, where the stress triaxiality increases. The discontinuity in the value of ε_e^{pl} across the damage zone interface near the surface of the cavity is now more obvious. However, similar to the $p_0 = 12$ GPa case, the total inelastic strains are approximately continuous across this interface, with granular plasticity inside the damage zone replaced by lattice plasticity.

Insights into the effects of lattice plasticity on stress triaxiality (which in turn affect microcracking) are gleaned from computations in which the damage is artificially suppressed, by setting $K_{IC} = \infty$. The results are plotted in Fig. 8 for three times and three values of yield strength, selected such that $p_0/\sigma_Y = 1.71$, 1.5, and 0 when $p_0 = 24$ GPa. For the purely elastic case ($p_0/\sigma_Y = 0$), the triaxiality is at its lowest (most negative) at the elastic wave front (as noted by Hopkins[17]). Thus, absent lattice plasticity, damage initiates at the cavity surface, where both the stresses and the stress triaxiality are greatest, and propagates radially outward, trailing the elastic wave front. When lattice plasticity is activated, as in the cases of $p_0/\sigma_Y = 1.71$ and 1.5, plasticity initiates on the cavity surface, where the deviatoric stress is highest, and subsequently spreads radially outward. This plasticity results in a reduction in the stress triaxiality in the plastic zone, as evident upon comparison of the results for $p_0/\sigma_Y = 1.71$ and 1.5 with those of the elastic case. The magnitude of this effect increases with increasing plastic strain. Moving radially outward from the cavity surface, the plastic strain diminishes and hence its effect on stress triaxiality decreases. Eventually, the triaxiality reaches the value corresponding to the elastic case at the elastic/plastic interface. Because of the opposing effects of elastic wave propagation and plasticity on stress triaxiality—the elastic wave reducing triaxiality with increasing R/a_0 and the near-surface plasticity reducing triaxiality as $R/a_0 \to 1$—the triaxiality attains a *maximum value at a finite distance* away from the cavity surface. For cases where damage is activated (such as that shown in Fig. 7), damage initiates at or near this location and then spreads radially both inward and outward. A corollary of the results presented in Fig. 8 is that the stress triaxiality decreases with increasing values of p_0/σ_Y, due to the increased levels of lattice plasticity. This is evident from comparisons of the results for $p_0/\sigma_Y = 1.71$ and 1.5 in Fig. 8. The reductions in the stress triaxiality for higher values of p_0/σ_Y result in damage initiating further away from the cavity surface as observed in Fig. 7(b).

(3) The Interplay Between Plasticity and Damage

The results in the preceding section demonstrate the interplay between lattice plasticity and stress triaxiality. Its connection to damage development and cavity expansion emerges from a series of computations for a cavity pressure $p_0 = 12$ GPa and three values of yield strength σ_Y (all other material properties being kept fixed at their reference values, including the finite fracture toughness). The variation in the displacement of the cavity surface U/a_0 with time $\bar{t}$ for each of these cases is plotted in

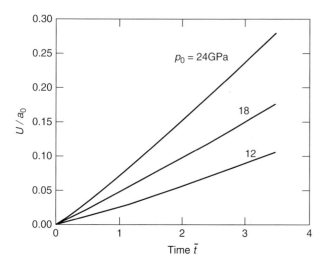

Fig. 6. The evolution of the radial displacement of the cavity surface with time. Results are for alumina with reference properties.

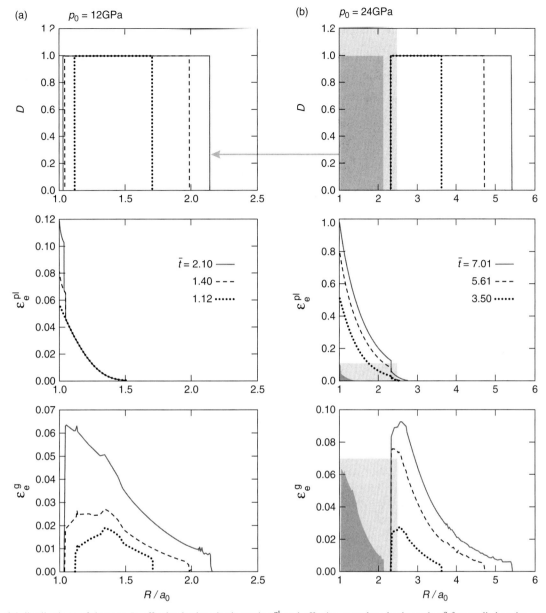

Fig. 7. Spatial distributions of damage D, effective lattice plastic strain ε_e^{pl} and effective granular plastic strain ε_e^g for applied cavity pressures, p_0, of (a) 12 GPa and (b) 24 GPa. Results are shown for alumina with reference properties at three select times $\bar{t}$.

Fig. 9(a). Interestingly, the cavity expansions are approximately equal for the $\sigma_Y = 5$ and 9 GPa cases (indeed, slightly lower for $\sigma_Y = 5$ GPa) but significantly higher for $\sigma_Y = 3$ GPa. In order to

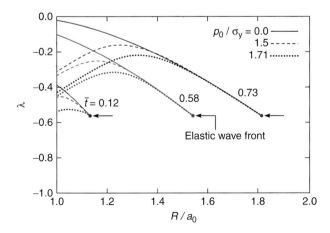

Fig. 8. Evolution of stress triaxiality associated with dynamic spherical cavity expansion in an elastic–plastic (nondamaging) alumina.

parameterise this information we plot the cavity displacement U_c/a_0 at a time $\bar{t} = 3$ against yield strength σ_Y in Fig. 9(b) for cavity pressures $p_0 = 12$ and 24 GPa. Three regimes emerge: one is dominated by damage, at high values of σ_Y, wherein U_c/a_0 is independent of σ_Y; a second dominated by lattice plasticity, wherein U_c/a_0 increases with decreasing σ_Y; and an intermediate regime in which lattice plasticity and damage occur simultaneously. Thus, there exists a critical value of σ_Y above which U_c does not decrease any further. This clearly shows that increasing the yield strength of the ceramic indefinitely will not bring any further performance benefits as the response becomes dominated by the microcracking deformation mode. However, the critical value of σ_Y increases with increasing applied pressure; for the parameters considered here, the critical yield strength increases from $\sigma_Y \approx 5$ to 8 GPa as p_0 increases from 12 to 24 GPa.

V. Indentation

The fidelity of the model has been assessed in two companion studies (B. Compton *et al.*, unpublished data).[15] Here we highlight some of their results in order to illustrate the applicability of the model. All tests were performed on an armor-grade

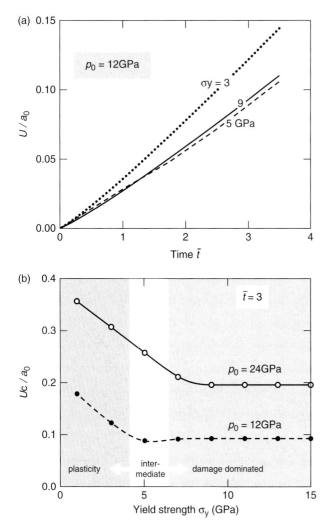

Fig. 9. (a) The evolution of radial displacement of the cavity surface with time for an applied cavity pressure $p_0 = 12$ GPa. (b) Effects of yield strength and pressure on the radial cavity displacement U_c at time $\bar{t} = 3$.

alumina (Corbit 98, produced by Industri Bitossi, Sovigliana vinci, Florence, Italy) with an average grain size $d = 3$ μm. It consists of 98% alumina and minor amounts of a glassy phase, the latter being situated predominantly at triple grain junctions. Details of the properties of this material and the model parameters (listed in Section II(4)) as well as the procedures used to obtain their values are in.[15]

(1) Quasi-static Indentation

The quasi-static penetration resistance was probed using a spheroconical diamond indenter with a 200 μm tip radius and 120° cone angle, mounted on a conventional servo-hydraulic mechanical test system.[15] Displacements were measured with submicrometer accuracy using a one-armed extensometer positioned on the alumina surface. Additionally, corrections for machine compliance and elastic (Hertzian) deformation of the indenter were used to remove displacements external to the indentation site. Corresponding FE computations were performed using the current model and the original DE model. (The parameter set of the DE model is a subset of the current model, i.e., all parameters of the current model expect those associated with granular plasticity. To make a fair comparison between the current and DE models we use the values parameters listed in Section II(4) in the calculations with the DE model as well.)

Figure 10 shows a comparison between predictions and measurements of both the load–displacement response and the damage immediately under the indenter at an applied load of 200 N. The current model is shown to accurately predict both the extent of the damage immediately under the indenter and the load versus penetration response. By contrast, the DE model predicts significantly greater damage and unrealistically large surface uplift around the indent periphery. This large damage manifests as large inelastic displacements during loading and hence the DE model also does not predict the measured load versus penetration response with an adequate level of accuracy. These discrepancies of the DE model are a consequence of the elastic dilatation associated with the cracks but not associated with granular plasticity.

(2) Dynamic Indentation

Constrained alumina tiles, each 50 mm × 50 mm × 12 mm thick, were impacted by 7.14 mm diameter steel spheres at velocities in the range 300–800 ms^{-1} (B. Compton *et al.*, unpublished data).

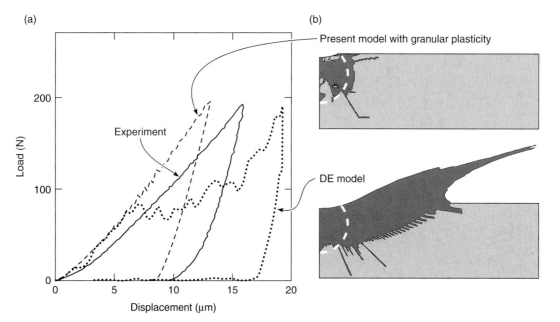

Fig. 10. (a) Comparisons between predicted and measured quasi-static indentation responses. (b) Corresponding subsurface damage at an applied load of 200 N. Predictions are shown for both the current and the DE models with the area shaded dark corresponding to fully microcracked ceramic. The dashed lines indicate the extent of the observed commutated zone (Adapted from[15]).

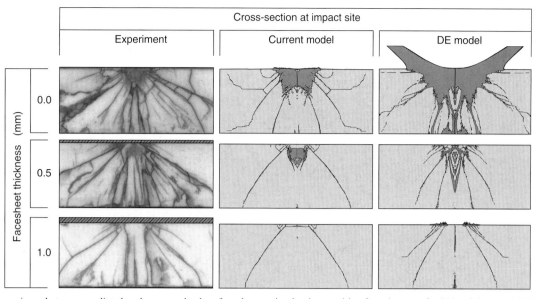

Fig. 11. Comparisons between predicted and measured subsurface damage in alumina resulting from impact of a 304 stainless-steel (SS) sphere at an incident velocity of 750 ms^{-1}. The dark shaded regions are fully damaged whereas the light shaded regions are undamaged. Predictions are shown for both the current and the DE models. Three cases are considered: no face-sheet, 0.5-mm-thick 304 SS face-sheet and 1-mm-thick 304 SS face-sheet.

Three situations were considered: (i) tiles with no face sheet; (ii) tiles with a 0.5 mm thick 304 stainless-steel face sheet; and (iii) tiles with a 1-mm-thick 304 stainless-steel face sheet. After the tests, the tiles were stained with a blue dye and impregnated with an epoxy resin in order to contain the comminuted ceramic. They were then sectioned in half in order to reveal the damage pattern directly beneath the impact site. Comparisons between the observations and the predictions (from both current and original DE models) for an impact velocity 750 ms^{-1} are presented in Fig. 11. The current model captures the observations for the three scenarios of present interest with reasonable fidelity, including the observation that the extent of damage is significantly reduced by the presence of the steel face sheets. By contrast, in line with the quasi-static indentation simulations, the DE model overestimates the damage significantly in all cases.

VI. Limitations of the Model

The micromechanical model for ceramics presented in this study has some inherent limitations that fall into two catagories: (i) limitations related to microstructural assumptions and (ii) limitations related to the neglecting of the statistical aspects of ceramic failure. Here we briefly discuss these limitations.

Three inelastic mechanisms have been accounted for in the approach presented here, viz. microcracking from preexisting flaws presumed to be present at grain boundaries, lattice plasticity and granular plasticity. Cracking in polycrystalline ceramics is however known to occur from numerous other sources of heterogeneities such as inclusions and voids,[23] glass present at grain boundaries[24] and surface flaws. The modeling framework presented here is sufficiently general that these mechanisms can be included in the constitutive model, albeit at the cost of greater complexity and additional material parameters that will inevitably require calibration. We thus expect to include these additional mechanisms as additional experimental data that suggests the need to incorporate these mechanisms becomes available.

It is well known that the failure strength of ceramics is both statistical in nature and size dependent. This is often well represented by the Weibull distribution.[25] The statistical effects give rise to the well-known observation that nominally identical ceramic specimens have different mechanical properties including sometime markedly different ballistic responses. To account for these statistical variations, Leavy *et al.*[26] included uncertainty into a deterministic model such as the Johnson and Holmquist[5]

model. This enabled them to not only capture the observed failure patterns in a Brazilian test but including a distribution of strengths also significantly alleviated the mesh dependency problems inherent in these ceramic models that are based on a damage mechanics approach. Inclusion of uncertainty into the present model is a necessary next step in developing a predictive ceramic model for armor applications.

VII. Conclusions

A unified model for the dynamic inelastic response of ceramics under compressive stress states has been developed. It incorporates three inelastic deformation mechanisms: (i) lattice plasticity due to dislocation glide or twinning; (ii) microcrack extension; and (iii) granular flow of densely packed comminuted particles. In addition to analytical descriptions of each of these mechanisms, prescriptions have been provided for their implementation into a FE code as well as schemes for transitioning between mechanisms.

The utility of the code in addressing issues related to deep penetration has been demonstrated through a series of calculations of dynamic cavity expansion in an infinite medium subject to a fixed internal pressure p. The results reveal two limiting behavioral regimes: one dictated by lattice plasticity and another by granular plasticity. The competition between these two deformation mechanisms along with the interplay between plasticity and stress triaxiality lead to an optimal value of yield strength at which the cavity expansion rate is minimized. This result suggests that, once the damage and fracture parameters are specified, an optimal yield strength exists for superior penetration resistance at a prescribed pressure. It remains to be established whether this effect can be exploited in designing ceramics to resist penetration over a targeted range of loadings.

Granular plasticity proves to be an essential ingredient in the constitutive law, as demonstrated by the quasi-static and dynamic indentation studies. When neglected (as in the original DE model), the spatial extent of damage and the magnitude of the strains are strongly overpredicted. Nevertheless, the model fails to capture all features in the indentation response. Most notably, the formation of lateral subsurface cracks during the unloading phase of indentation cannot be predicted. To capture this feature along with the ensuing spallation that commonly accompanies indentation, the model would need to be expanded to include a fracture mechanics-based component for modeling "macro" rather than "micro" cracking.

Journal of the American Ceramic Society—Deshpande et al. Vol. 94, No. S1

Appendix A: Summary of Microcrack Growth Coefficients

The coefficients used in Section II(*1*) are defined as follows. The parameters A and B are given as

$$A \equiv c_1(c_2 A_3 - c_2 A_1 + c_3) \tag{A-1}$$

and

$$B \equiv \frac{c_1}{\sqrt{3}}(c_2 A_3 + c_2 A_1 + c_3) \tag{A-2}$$

where

$$c_1 = \frac{1}{\frac{\pi^2}{2^{3/4}}\left[\left(\frac{D}{D_0}\right)^{1/3} - 1 + \beta\sqrt{2}\right]^{3/2}} \tag{A-3}$$

$$c_2 = 1 + 2\left[\left(\frac{D}{D_0}\right)^{1/3} - 1\right]^2 \left(\frac{D_0^{2/3}}{1 - D^{2/3}}\right) \tag{A-4}$$

and

$$c_3 = \pi^2 \left[\left(\frac{D}{D_0}\right)^{1/3} - 1\right]^2 \tag{A-5}$$

while

$$A_1 = \pi\sqrt{\frac{\beta}{3}}[(1 + \mu^2)^{1/2} - \mu] \tag{A-6}$$

and

$$A_3 = A_1 \left\{\frac{(1 + \mu^2)^{1/2} + \mu}{(1 + \mu^2)^{1/2} - \mu}\right\} \tag{A-7}$$

Here β is a coefficient introduced by Ashby and Sammis[16] to convert exact two-dimensional solutions to three-dimensional stress states. The coefficients E and C are defined as

$$E^2 = \frac{B^2 C^2}{C^2 - A^2} \tag{A-8}$$

and

$$C \equiv A + \gamma\sqrt{\frac{1}{\sqrt{2}}\left(\frac{D}{D_0}\right)^{1/3}} \tag{A-9}$$

with γ a constant used to match tensile data.

Acknowledgments

E. A. G. was supported by a National Defense Science and Engineering Fellowship. V. S. D. acknowledges ONR for support via award no. N00014-09-1-0573.

References

[1] A. M. Rajendran and D. J. Grove, "Modeling the Shock Response of Silicon Carbide, Boron Carbide and Titanium Diboride," *Int. J. Impact Eng.*, **18**, 611–31 (1996).

[2] G. T. Camacho and M. Ortiz, "Computational Modeling of Impact Damage in Brittle Materials," *Int. J. Solids Struct.*, **33**, 2899–938 (1996).

[3] P. J. Hazell and M. J. Iremonger, "Crack Softening Damage Model for Ceramic Impact and its Application with in a Hydrocode," *J. Appl. Phys.*, **82**, 1088–92 (1997).

[4] H. D. Espinosa, P. D. Zavattieri, and S. K. Dwivedi, "A Finite Deformation Continuum/Discrete Model for the Description of Fragmentation and Damage in Brittle Materials," *J. Mech. Phys. Solids*, **46**, 1909–42 (1998).

[5] G. R. Johnson and T. J. Holmquist, "Response of Boron Carbide Subjected to Large Strains, High Strain Rates, and High Pressures," *J. Appl. Phys.*, **85**, 8060–73 (1999).

[6] C. H. M. Simha, S. J. Bless, and A. Bedford, "Computational Modeling of the Penetration Response of a High-Purity Ceramic," *Int. J. Impact Eng.*, **27**, 65–861 (2002).

[7] D. C. Drucker and W. Prager, "Soil Mechanics and Plastic Analysis or Limit Design," *Quart. Appl. Math.*, **10**, 157–65 (1952).

[8] G. R. Johnson and W. H. Cook, "Fracture Characteristics of Three Metals Subjected to Various Strains, Strain Tatesm Temperatures and Pressures," *Eng. Fract. Mech.*, **21**, 31–48 (1985).

[9] D. W. Templeton, T. J. Holmquist, H. W. Meyer, D. J. Grove, and B. Leavy, "A comparison of ceramic material models"; pp. 299–308 in *Proceedings of the Ceramic Armor by Design Symposium*, Pacific Rim IV International Conference on Advanced Glass and Ceramics, Wailea, Maui, Hawaii, 2001.

[10] V. S. Deshpande and A. G. Evans, "Inelastic Deformation and Energy Dissipation in Ceramics: A Mechanism-Based Dynamic Constitutive Model," *J. Mech. Phys. Solids*, **56**, 3077–100 (2008).

[11] Z. Wei, V. S. Deshpande, and A. G. Evans, "The Influence of Material Properties and Confinement on the Dynamic Penetration of Alumina by Hard Spheres," *J. Appl. Mech.*, **76**, 051305, 8pp (2009).

[12] Z. Wei, A. G. Evans, and V. S. Deshpande, "Mechanisms and Mechanics Governing the Indentation of Polycrystalline Alumina," *J. Am. Ceram. Soc.*, **91**, 2987–96 (2008).

[13] H. J. Frost and M. F. Ashby, *Deformation Mechanism Maps*. Pergamon Press, Oxford, 1982.

[14] R. A. Bagnold, "Experiments on a Gravity-Free Dispersion of Large Solid Particles in a Newtonian Fluid Under Shear," *Proc. R. Soc. Lond.*, **A225**, 49–63 (1954).

[15] E. A. Gamble, B. Compton, V. S. Deshpande, A. G. Evans, and F. W. Zok, "Penetration resistance of an armor ceramic: I- Quasi-static loading," *J. Am. Ceram. Soc.*, (2010) (this issue).

[16] M. F. Ashby and C. G. Sammis, "The Damage Mechanics of Brittle Solids in Compression," *Pure Appl. Geophys.*, **133**, 489–521 (1990).

[17] H. G. Hopkins, "Dynamic Expansion of Spherical Cavities in Metal"; pp. 85–164 in *Progress in Solid Mechanics*, Vol. 1, Edited by I. N. Sneddon, and R. Hill. North-Holland, Amsterdam, 1960.

[18] M. J. Forrestal, D. Y. Tzou, E. Askari, and D. B. Longcope, "Penetration into Ductile Metal Targets with Rigid Spherical-Nose Rods," *Int. J. Impact Eng.*, **16**, 699–710 (1995).

[19] R. W. Macek and T. A. Duffey, "Finite Cavity Expansion Method for Near-Surface Effects and Layering During Earth Penetration," *Int. J. Impact Eng.*, **24**, 239–58 (2000).

[20] D. Durban and R. Masri, "Dynamic Spherical Cavity Expansion in a Pressure Sensitive Elastoplastic Medium," *Int. J. Solids Struct.*, **41**, 5697–716 (2004).

[21] S. Satapathy, "Dynamic Spherical Cavity Expansion in Brittle Ceramics," *Int. J. Solids Struct.*, **38**, 5833–45 (2001).

[22] J. D. Walker, "Analytic model for penetration of thick ceramic targets"; pp. 337–48 in *Proceedings of the ceramic armor by design symposium*, Pacific Rim IV International Conference on Advanced Glass and Ceramics, Wailea, Maui, Hawaii, 2001.

[23] C. G. Sammis and M. F. Ashby, "The Failure of Brittle Porous Solids Under Compressive Stress States," *Acta Metall.*, **34**, 511–26 (1986).

[24] D. R. Clarke, "Grain Boundaries in Polycrystalline Ceramics," *Annu. Rev. Mater. Sci.*, **17**, 57–74 (1987).

[25] W. Weibull, "A Statistical Distribution Function of Wide Applicability," *J. Appl. Mech.*, **18**, 293–7 (1951).

[26] R. B. Leavy, R. M. Brannon, and O. E. Strack, "The Use of Sphere Indentation Experiments to Characterize Ceramic Damage Models," *Appl. Ceram. Technol.*, **7**, 606–15 (2010). □

J. Am. Ceram. Soc., **94** [S1] S215–S225 (2011)
DOI: 10.1111/j.1551-2916.2011.04472.x
© 2011 The American Ceramic Society

Journal

Damage Development in an Armor Ceramic Under Quasi-Static Indentation

Eleanor A. Gamble,[‡] Brett G. Compton,[‡] Vikram S. Deshpande,[§] Anthony G. Evans,[‡] and Frank W. Zok[†,‡]

[‡]Materials Department, University of California, Santa Barbara, California 93106

[§]Engineering Department, Cambridge University, Cambridge CB2 1PZ, U.K.

The objective of the present study is to assess the capabilities of a recently developed mechanism-based model for inelastic deformation and damage in structural ceramics. In addition to conventional lattice plasticity, the model accounts for microcrack growth and coalescence as well as granular flow following comminution. The assessment is made through a coupled experimental/computational study of the indentation response of a commercial armor ceramic. The experiments include examinations of subsurface damage zones along with measurements of residual surface profiles and residual near-surface stresses. Extensive finite element computations are conducted in parallel. Comparisons between experiment and simulation indicate that the most discriminating metric in the assessment is the spatial extent of subsurface damage following indentation. Residual stresses provide additional validation. In contrast, surface profiles of indents are dictated largely by lattice plasticity and thus provide minimal additional insight into the inelastic deformation resulting from microcracking or granular flow. A satisfactory level of correlation is obtained using property values that are either measured directly or estimated from physically based arguments, without undue reliance on adjustable (nonphysical) parameters.

I. Introduction

CERAMIC materials are recognized for their effectiveness in protecting against ballistic threats. Their high hardness enables energy dissipation by plastic deformation and erosion of projectiles during the early stages of impact. But their brittleness precludes their use as primary load-bearing elements within structural systems. Furthermore, ceramics have limited utility in protecting against other dynamic threats such as blast loads. Consequently, in order to satisfy the full spectrum of structural and protective requirements, ceramics will need to be integrated into complex multimaterial systems with metallic alloys and fiber-reinforced polymer composites. With almost limitless configurational possibilities, the most rapid progress in the design of weight-efficient systems will be achieved through numerical simulations coupled with select validation tests. The success of this approach is predicated on the availability of reliable constitutive laws for the constituent materials.

The goal of the present article is to begin assessment of the capabilities of a recently developed mechanism-based model for ceramic deformation. The first version of the model, by Deshpande and Evans,[1] embodied two principal inelastic mechanisms: (i) lattice plasticity by dislocation glide or twinning (Fig. 1(a)) and (ii) extension of grain-sized microcracks (Fig. 1(b)). A recent extension of the model has addressed the material behavior after the damage has saturated, that is, once the microcracks have coalesced.[2] In this state, the material can deform by one of two mechanisms: lattice plasticity within the individual particles, or granular plasticity, wherein the particles flow past one another (Fig. 1(c)). The latter mechanism is closely analogous to that which occurs in soils, albeit with a much higher packing density of the constituent particles. The extended version of the Deshpande–Evans (DE) model is detailed in a companion paper in this issue.

The study focuses specifically on the response of a prototypical armor ceramic to *quasi-static indentation* by a hard spheroconical indenter. Such tests can reproduce the high contact pressures and level of confinement characteristic of the early stages of projectile impact. Examinations of the subsurface regions performed in this study and reported previously by Lawn and colleagues[3,4] reveal the presence of a heavily microcracked "quasi-plastic zone" beneath the indenter. The correlation between these microcracked regions and the Mescall zones formed during ballistic impact has been noted by LaSalvia and McCauley.[5] These correlations suggest that the quasi-static indentation tests are well suited to probing some of the pertinent inelastic deformation mechanisms: albeit lacking in the dynamic aspects of the loading. The principal advantage of the quasi-static tests (relative to impact) is their amenability to accurate instrumentation during loading. Secondary benefits include the lack of dynamic spallation or extraneous cracking caused by stress wave reflections in impact tests. Additionally, by using bonded interface specimens[4] (described below), the indentation tests allow for careful observations of subsurface damage, without the need for posttest sectioning and polishing and the associated artifacts that can ensue.

In addition to indentation tests and subsequent microstructural examinations, the experimental portion of the study includes measurement of the residual stresses in the region of the indents as well as the residual surface profiles. The experimental measurements are accompanied by extensive finite element computations using a range of material parameter values. A paper, comparing the predictions of the calibrated constitutive law with the damage and deformation caused by dynamic impact, is in preparation (B. G. Compton *et al.*, unpublished data).

The article is organized in the following way. Section II contains a description of the material and the experimental procedures employed in this study. Section III provides a synopsis of the extended DE model, with emphasis on the inelastic mechanisms and the associated material parameters required for calibration. Experimental measurements and observations are summarized in Section IV. This is followed by presentation of the computational results of the damage zones beneath the indents, in Section V, as well as an assessment of the residual stresses and surface profiles, in Section VI. Implications of the correlations and limitations of the extended DE model are addressed in the final section.

A. Heuer—contributing editor

Manuscript No. 28771. Received October 13, 2010; approved February 5, 2011.
This work was supported by the Office of Naval Research through a Multidisciplinary University Research Initiative Program on "Cellular Materials Concepts for Force Protection," Prime Award No. N00014-07-1-0764.
[†]Author to whom correspondence should be addressed. e-mail: zok@engineering.ucsb.edu

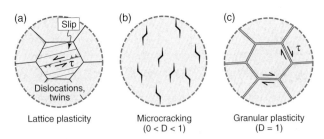

Fig. 1. Mechanisms of inelastic deformation and damage operative in polycrystalline ceramics under compressive stress states.

II. Material and Test Procedures

The material used for the experimental study was a fine-grained armor-grade ceramic (Corbit 98, produced by Industrie Bitossi, Sovigliana Vinci, Italy). It consists of 98% alumina and minor amounts of a glassy phase, the latter being situated predominantly at triple grain junctions. The glassy phase is evident on polished cross-sections in backscatter electron imaging mode in the scanning electron microscope (SEM) (Fig. 2(a)). It is also manifested as triple grain pores following either thermal etching (caused by volatilization of the glass, Fig. 2(b)) or chemical etching (Figs. 2(c) and (d)). The average grain diameter, d, is 3 μm.

A series of Vickers indentation tests was performed over a load range of 0.7–10 kg. The measurements yield a hardness $H = 15.9 \pm 1.8$ GPa and a (long crack) fracture toughness, $K_{IC} = 2.9 \pm 0.3$ MPa$\sqrt{m}$. The latter tests reveal that fracture occurs along an intergranular path, attributable to the presence of a thin grain-boundary glass layer.[6]

The Young's modulus, E, measured by instrumented nanoindentation, is 405 GPa. This value exceeds that reported by the manufacturer (384 GPa) by about 5%. The difference is attributable to the presence of both residual porosity and the compliant glassy phase at the grain boundaries. Neither of these features is probed by the nanoindentation tests. Consequently, the lower value is selected for use in subsequent computations.

Penetration resistance was probed using a spheroconical diamond indenter with a 200 μm tip radius and 120° cone angle. Two specimen geometries were employed. (i) Flat-polished plates were used to measure the plastic indentation stress–strain response. They were also used for measurement of surface profiles through the indents and the near surface residual stresses (described below). (ii) The nature and spatial extent of sub-surface damage were ascertained using bonded interface specimens.[4] To produce such specimens, pairs of polished alumina plates were bonded with a thin (~5 μm) layer of cyanoacrylate adhesive and clamped together with a pressure of about 10 MPa. One of the orthogonal surfaces was then polished. The clamping pressure was sufficient to maintain sample integrity during polishing and indentation but negligible in comparison to the those required to induce plastic deformation (on the order of 10 GPa). Indentation tests were performed on the latter surface with the indenter placed directly over the bonded interface (Fig. 3). Surface profiles around the indentations made on the bonded interface specimens were indistinguishable from those of corresponding indents on flat-polished samples. The inference is that the presence of the interface does not affect the overall deformation response. Whether some salient features of the deformation mechanisms directly on the bonded surfaces differ from those operating internally (say 10 μm below the bondline) remains to be critically addressed.

Following testing, the samples were separated and cleaned. The internal polished surfaces were coated with gold and examined using Nomarski interference optical microscopy and SEM. In some instances, the region directly beneath the indent was also ion-milled (to a depth of 200 nm) after indentation to remove surface texture resulting from plastic deformation. The latter surface treatment allows more definitive identification of grain-boundary cracks. Additionally, narrow grooves were cut perpendicular to the polished face using focused ion beam (FIB) machining. These grooves were cut both directly beneath the indent and at a remote location; the exposed surfaces were examined in the SEM.

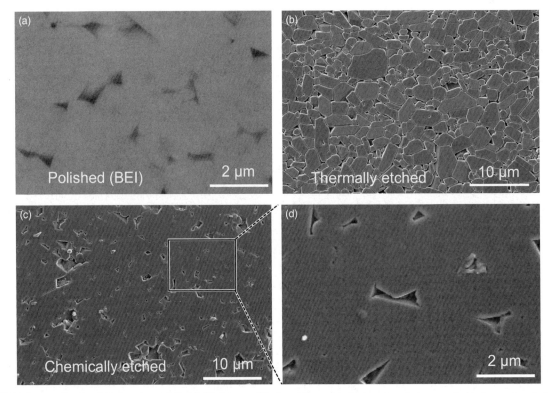

Fig. 2. SEM micrographs of polished sections of Corbit 98: (a) as-polished and imaged in backscatter electron mode (darker regions being the glassy phase); (b) polished and thermally etched at 1400°C for 1 h; and (c, d) polished and etched with phosphoric acid for 4 min at 130°C. Images in (b)–(d) taken in secondary electron mode.

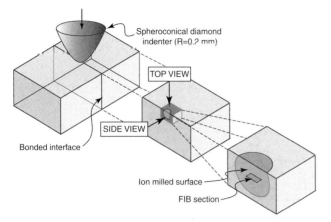

Fig. 3. Schematic of the bonded interface specimen.

Surface profiles on flat-polished specimens in the vicinity of the indents were measured using an optical profilometer (Wyko NT1100, Veeco, Tucson, AZ). Profilometry scans were also used to measure the diameter of the indent crater and, in turn, the nominal hardness (i.e., peak load/residual indent area). A series of such measurements to various peak loads was used to establish the plastic indentation stress–strain response.

Residual hydrostatic stresses in the near-surface vicinity of the indents were measured using a luminescence piezospectroscopy technique.[7,8] The technique involves use of a laser beam in an optical microscope to stimulate luminescence of trace Cr^{3+} ions and a spectrometer to measure the stress-induced frequency shift of the characteristic R_2 luminescence peak. The residual hydrostatic stress is proportional to the frequency shift through the piezospectroscopic coefficient (7.61 $(cm \cdot GPa)^{-1}$). The diameter of the beam employed in the present study was about 3 μm and its penetration depth 10 μm. Experiments performed to measure the effective beam diameter and the beam interactions with the subsurface regions of the material are detailed in Appendix A. The corresponding analysis (also in Appendix A) was subsequently used to establish a depth averaging protocol for the residual stresses computed by finite elements for comparison with the experimental measurements.

III. Constitutive Law and Finite-Element Analysis

The extended DE model[2] was used to represent the mechanical response of the alumina. A descriptive synopsis of the model follows and the key formulae are compiled in Appendix B. Graphical representations of the constitutive law are shown in Fig. 4.

(1) Lattice Plasticity

Plasticity is described by the effective stress σ_e and the von Mises yield criterion, with yield stress σ_y, and the subsequent hardening by a conventional power law, characterized by a hardening exponent M. (Strain rate sensitivity is also taken into account, through an overstress model, calibrated by fitting plasticity data for alumina.[9] However, such effects are exceedingly small in ceramics in the quasi-static loading regime at ambient temperature and, for all practical purposes, can be neglected.)

(2) Distributed Microcracking

Two non-dimensional parameters characterize the initial damage state: the normalized initial flaw size, $g_1 \equiv a/d$, and the normalized flaw spacing, $g_2 \equiv 1/(f^{1/3}d)$ (f being the number of microcracks per unit volume). Three behavioral regimes are obtained, delineated by the stress triaxiality, $\lambda \equiv \sigma_m/\sigma_e$ (σ_m being the mean stress) (Fig. 4(a)). In Regime I, preexisting cracks are closed and the shear stress is insufficient to overcome the frictional resistance of the contacting crack surfaces. Conse-

quently, the stress intensity is zero and crack growth cannot occur. This behavior is obtained when the stress triaxiality is less than a critical value, λ_C: dictated largely by the friction coefficient, μ, of the crack surfaces and the current level of microcrack damage. For friction coefficient values typical of engineering ceramics, $\lambda_C \approx -0.5$ to -1. At higher triaxialities, in Regime II, the crack surfaces remain in contact but are able to slide past one another. Sliding leads to a finite stress intensity which drives the formation of wing cracks at the tips of existing flaws. The cracks extend, initially stably, and eventually link to produce a fully comminuted ceramic. The crack growth rate is expected to increase with increasing stress intensity. Following Aeberli and Rawlings,[10] the crack growth rate is taken to scale with $(K_I/K_{IC})^m$ where $m \gg 1$. Furthermore, the growth rate is limited by the shear wave speed. In Regime III, the crack faces are not in contact with one another and thus the cracks can propagate unstably.

(3) Granular Plasticity

Once fully comminuted, the grains can rotate and slide past one another (analogous to deformation of soils). Granular flow is taken to obey the linear Drucker–Prager yield criterion, characterized by a critical strength σ_e^{cr} at $\sigma_m = 0$ and a friction angle ω (defined in Fig. 4(b)). The plastic potential is taken to be linear as well, with dilatational angle, ψ (Fig. 4(b)). Associated flow is obtained by setting $\psi = \omega$, whereupon the plastic strain exhibits significant dilatation. At the other extreme, where $\psi = 0$, the granular plastic strain is purely deviatoric.

Numerical simulations of penetration were performed using ABAQUS/Explicit V6.9-EF1. The use of the explicit analysis allows the same implementation of the constitutive law to be used for simulating both dynamic and quasi-static loading. The indenter was modeled as a rigid surface with a cone angle of 120° and a tip radius of 200 μm (matching that of the indenter used in the experiments). The alumina specimen being indented was represented by a biased mesh (2 μm edge length near the indentation surface and coarser toward its base and sides). Mass scaling was used to obtain a time step of 4 ns. The kinetic energy in the model was monitored to ensure that the stress due to inertial effects was minimal. Surface sliding was allowed to occur in accordance with Coulomb's law with a friction coefficient of 0.4.

To complete a full indentation simulation in a reasonable amount of time, the simulated indenter displacement rate was selected to be 125 mm/s: 750 000 times that used in the experiments (0.17 μm/s). The reference strain and crack growth rates were scaled by the same factor to reproduce the experimental stress-displacement relationship.

Calibration of the constitutive model was performed in the following way. First, the plastic indentation stress–strain response was computed using a range of parameter values, selected to be consistent with available data for alumina, and neglecting damage. The computed curve that best matched the experimental data was used to define σ_y and M. These parameter values were fixed for all subsequent calculations. Select calculations were performed to assess the sensitivity of the indentation stress–strain response to the choice of damage parameters. Additional calculations, both with and without damage, were performed to ascertain the residual indent profiles, the residual hydrostatic stresses and the spatial extent of sub-surface damage. Each of these metrics was compared with the corresponding experimental measurements. The pertinent material parameters and the range of values employed in the present computations are summarized in Table I.

IV. Indentation Deformation and Damage

The indentation stress–strain response is plotted in Fig. 5. A satisfactory fit of the data were obtained with the lattice plasticity model using a yield strength $\sigma_y = 5.75$ GPa (approximately one-third of the Vickers hardness) and a hardening exponent $M = 0.1$. The corresponding indent profiles are shown

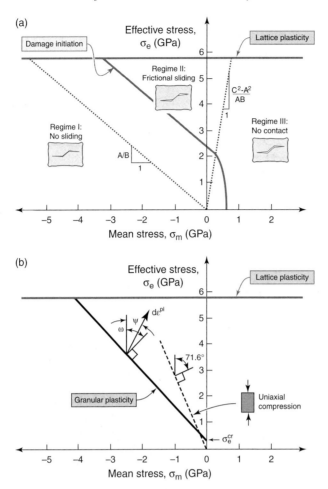

Fig. 4. (a) Conditions for microcracking and lattice plasticity in the alumina of present interest. The dotted lines represent the boundaries between the three microcracking regimes, each depicted by a corresponding schematic. The parameters A, B and C are defined in Appendix B. (b) Schematic of the linear Drucker-Prager yield criterion for granular plasticity. Also shown is the stress trajectory for uniaxial compression. Here $d\varepsilon^{pl}$ is the plastic strain increment.

Table I.　Summary of Property Data Used in the Extended DE Model

Parameter	Selected value	Range considered	Units	Description
ρ	3810		kg/m^3	Density
ν	0.22			Poisson's ratio
E	384		GPa	Young's modulus
d	3.0		μm	Grain diameter
σ_Y	5.75	3.0–8.5	GPa	Yield stress
K_{IC}	3.0	2.0–4.5	MPa√m	Fracture toughness
M	0.1	0.05–0.2		Strain hardening exponent
m	30			Crack growth rate sensitivity exponent
n	34			Strain rate dependence exponent
i_0	0.01[†]		m/s	Reference crack growth rate
ε_0	0.002			Reference strain
$\dot{\varepsilon}_0$	0.001[†]		1/s	Reference plastic strain rate
g_1	0.25	0.25–0.5		Normalized flaw size
g_2	6	2–12		Normalized flaw spacing
α	0.707			Crack orientation factor
β	0.45			Parameter to ensure 2-D compatibility
γ	2			Crack geometry factor
μ	0.75			Crack coefficient of friction
$\dot{\varepsilon}_t$	10^6		1/s	Transition shear strain rate
ω	70°	50°–70°		Soil friction angle
σ_c	10^6		Pa	Soil uniaxial compressive strength
x_d	5			Soil transition exponent
$\dot{\varepsilon}_{\text{cut-off}}$	2·10^6		1/s	Soil transition strain rate
ψ	0°	0°–70°		Dilatational angle

[†]Reference strain rate was scaled by the ratio of the indentation displacement rate to the displacement rate used in the FE computations (750 000), to ensure consistent quasi-static flow stress.

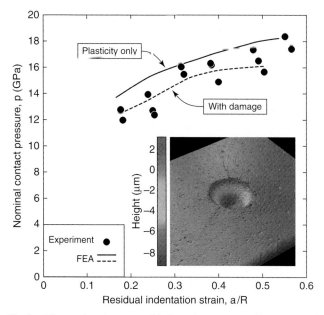

Fig. 5. Measured and computed indentation stress–strain response of alumina. Inset shows typical profilometry scan used to obtain residual diameter as well as indent profile (the latter plotted in Fig. 6). $\sigma_y = 5.75$ GPa and $M = 0.1$ for finite element computations. The key parameter values in the damage model are: $a/d = 1/2$, $K_{IC} = 3$ MPa$\sqrt{m}$, $\beta = 70°$, and $\psi = 0°$.

in Fig. 6. The plasticity model overestimates the maximum indent depths by only about 20% and slightly underestimates the surface uplift around the indent periphery. In light of the extensive subsurface damage (shown below), the agreement between the experimental and computed results is remarkably good. The implication is that the indent profile is dictated largely by the resistance to lattice plasticity and is not a discriminating metric for assessing the damage model.

Optical images of the bonded interface specimens reveal well-delineated subsurface damage zones with a truncated spherical shape (Fig. 7). The diameter of the damage zone scales with the residual indent diameter and coincides with this diameter at the indent surface. SEM examinations of the side surfaces (polished previously to a mirror finish) reveal numerous slip lines within each grain: a consequence of dislocation glide and/or twinning (Fig. 8). Because of the plasticity, the grains undergo a distortion that is manifested as surface rumpling at lower magnifications (such as those of the optical micrographs).

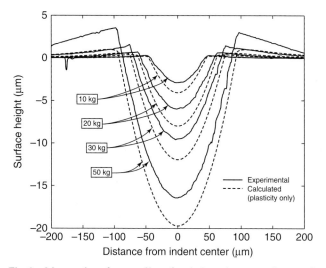

Fig. 6. Measured surface profiles after indentation to various peak loads. Also shown are the results of plasticity calculations.

The SEM images of the internal polished surfaces indicate the presence of grain-boundary microcracks, although the surface distortions preclude a conclusive assessment. They also reveal the presence of macroscopic lateral and radial cracks emanating from the periphery of the damage zone (Fig. 7(b)). Corroborating evidence of such cracks is found in the optical images as well as those obtained by optical profilometry (see, e.g., inset in Fig. 5).

Subsequent ion mill polishing of the polished surfaces yielded the most definitive evidence of microcracking. In the subsurface regions directly beneath an indentation site, the material was fully comminuted (Figs. 9(b) and (c)). That is, virtually *all* grain-boundary facets were cracked. Despite the comminution, the material strength was sufficient to prevent material loss from the indent site during normal specimen handling (including inversion and shaking). SEM images taken in a region remote from the indent revealed preferential removal of the glassy phase at the triple junctions but no relief at grain boundaries (Figs. 9(d) and (e)). Comparisons of images from the two regions reaffirmed that the boundary features near the indent surface are indeed due to microcracks and are not artifacts of the surface preparation procedure. Images of surfaces produced by FIB machining (shown in the lower parts of Figs. 9(c) and (e)) provided additional evidence that all grain boundaries in the near surface regions were cracked.

V. Damage Zone Simulations

An assessment of the extended DE model was made in the following way. First, for properties amenable to direct measurement, a narrow range of values was identified for the computations. For instance, the mode I long-crack fracture toughness of the alumina of present interest is 2.9 MPa$\sqrt{m}$. But, since the microcracks that form beneath the indent are inherently short and generally experience combined mode I/II loadings, the pertinent toughness may differ somewhat from this value. Furthermore, oxide ceramics are susceptible to subcritical cracking in the presence of water vapor, at stress intensities as low as about 2/3 of their fracture toughness. To cover the range of realistic possibilities, the fracture toughness values employed in the present computations span the range of 2.0–4.0 MPa$\sqrt{m}$. All property values are listed in Table I.

A few parameter values are less amenable to measurement, although realistic ranges can be identified. For example, the initial flaw size for intergranular fracture is expected to scale with the grain size. Assuming the existence of flaws on the grain-boundary facets, the flaw size would be $g_1 = a/d \approx 1/4$. Alternatively, if flaws are initiated by slip within the grain, the effective flaw size would be $g_1 \approx 1/2$. Thus the range of interest for the computations is expected to be bound approximately by $1/4 \leq g_1 \leq 1/2$. The flaw spacing parameter is more difficult to estimate *a priori*. But, as demonstrated below, the computed results are insensitive to its value in the domain $3 \leq g_2 \leq 12$.

Because of the high packing density of the comminuted material after microcrack coalescence, the friction angle ω is expected to be greater than that obtained for more loosely packed powders (typically 30°–50°).[11] However, to be consistent with experimental observations[12,13] an upper limit of $\omega = 71.6°$ is set by the requirement that microcracking and granular plasticity occur under uniaxial compression. Otherwise, for $\omega \geq 71.6°$, the only operative mechanism under uniaxial compression would be lattice plasticity: inconsistent with reported observations of compressive failure by microcrack coalescence at stresses below those needed for global lattice plasticity.[14] Thus, in subsequent calculations, ω was varied from 50° to 70°. A mechanistic basis for selecting the dilatational angle ψ is lacking and thus values over the entire parameter range ($0 \leq \psi \leq \omega$) were considered.

The spatial extent of the computed subsurface damage for a 20 kg indent is shown in Fig. 10 for the pertinent range of K_C and g_1, with the other parameters fixed ($\psi = 0°$, $\omega = 70°$, $g_2 = 6$). For comparison, an outline of the experimentally observed damage

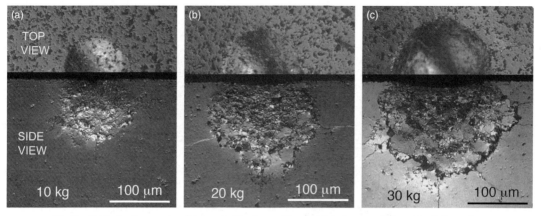

Fig. 7. Optical micrographs of top and side views of bonded interface specimens after indentation.

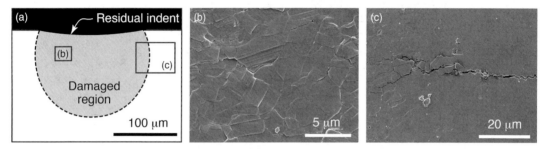

Fig. 8. (a, b) Slip lines and grain-boundary microcracks on polished surfaces of a bonded interface specimen directly beneath the indent (30 kg load). (c) Large lateral crack emanating from periphery of the damaged region.

region is superimposed on the computed results. Both the shape and extent of the damaged region were considered in assessing the degree of correlation between the experimental and computed results. The computed results exhibit an abrupt transition from damaged to undamaged material, closely analogous to that observed in experiments. The abrupt boundary is a manifestation of

the extremely small amount of crack extension that is predicted to occur under increasing stress. A closer examination of the present results indicate that the wing cracks are predicted to extend only very short distances (~ 200 nm, depending on the triaxiality) before the local stress reaches a maximum and further growth occurs under rapidly diminishing stress. Consequently, most wing

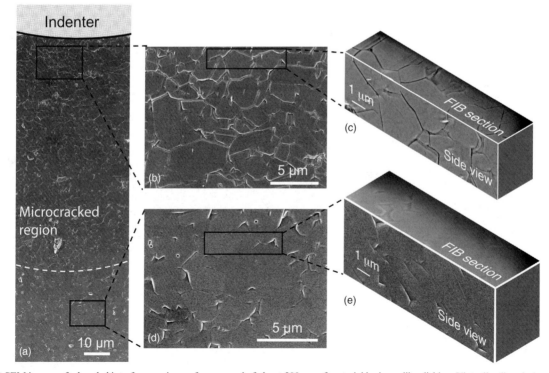

Fig. 9. (a) SEM images of a bonded interface specimen after removal of about 200 nm of material by ion mill polishing. Virtually all grain-boundary facets contain cracks in the regions directly beneath the indented surface (b, c). FIB machined surface perpendicular to the polished plane reveals similar grain-boundary cracks on all facets (bottom of (c)). In regions remote from the indented surface, image contrast is due solely to selective removal of the glassy phase at triple junctions (d, e). Similar features are evident on the corresponding FIB machined surface (bottom of (e)), without evidence of microcracking.

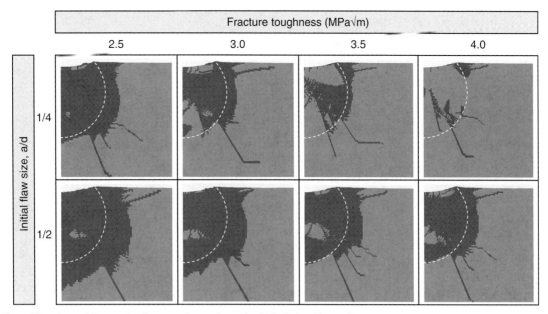

Fig. 10. Effects of flaw size and fracture toughness on damage beneath a 20 kg indent ($\psi = 0°$, $\beta = 70°$, $g_2 = 6$). Blue indicates intact material ($D = D_0$), while red indicates full comminution ($D = 1$). The white dashed lines correspond to the edge of the experimentally observed sub-surface damage region in Fig. 7(b).

cracks that extend even slightly will propagate unstably to full comminution. One implication is that partially extended wing cracks are unlikely to be observable in the experiments.

Satisfactory agreement between the computed results and experiments was obtained for the combination $K_C = 3.5$ MPa$\sqrt{m}$ and $g_1 = 1/2$. But the optimal combination is not unique. An equally good correlation was obtained for $K_C = 3$ MPa$\sqrt{m}$ and $g_1 = 1/4$. Nevertheless, the range of inferred property values is sufficiently narrow and well within the expected limits. Moreover, the computed damage pattern quickly deviates from the experimental observations when property values are selected outside this range. For instance, for $K_C = 3.5$ MPa$\sqrt{m}$ and $g_1 = 1/4$, anomalous undamaged regions appear both directly beneath the indent and further down along the centerline. Increasing the fracture toughness to 4 MPa$\sqrt{m}$ in combination with the same flaw size yields a prediction that damage is almost nonexistent.

The effects of flaw spacing, friction angle and dilatation angle on the damage pattern for $K_C = 3$ MPa$\sqrt{m}$ and $g_1 = 1/4$ for the same 20 kg indent are shown in Fig. 11. Among these, only the friction angle has an appreciable effect on the shape of the computed damage zone. Notably, for $\omega \leq 60°$, the boundary between damaged and undamaged material is highly irregular with multiple radial damage bands extending well beyond that observed in the experiments. Selecting $\omega = 70°$ yields acceptable results. This high friction angle (approaching the expected upper limit of 71.6°) is consistent with the high packing density (nearly 100%) of the comminuted grains. The flaw spacing and the dilatation angle play a decidedly secondary role over a realistic range of values. Consequently, further calculations presented here are based on the selections $g_2 = 6$ and $\psi = 0°$.

Finally, the effects of the indent load on both measured and computed damage patterns are shown in Fig. 12 for $K_C = 3$ MPa$\sqrt{m}$, $g_1 = 1/4$ or $1/2$, $\psi = 0°$, $\omega = 70°$, and $g_2 = 6$. The finite element results are in reasonable agreement with the

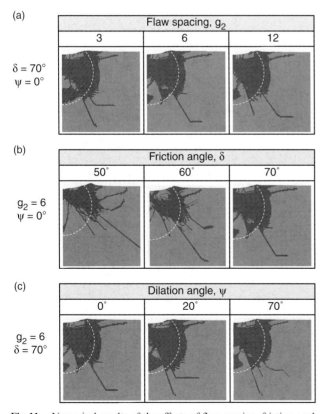

Fig. 11. Numerical results of the effects of flaw spacing, friction angle and dilatation angle on subsurface damage for a 20 kg indent ($a/d = 1/4$ and $K_C = 3.0$ MPa$\sqrt{m}$).

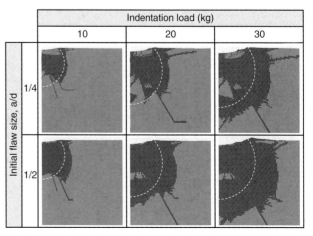

Fig. 12. Effect of indent load on subsurface damage. Dashed lines indicate the edge of the experimentally observed damage zones.

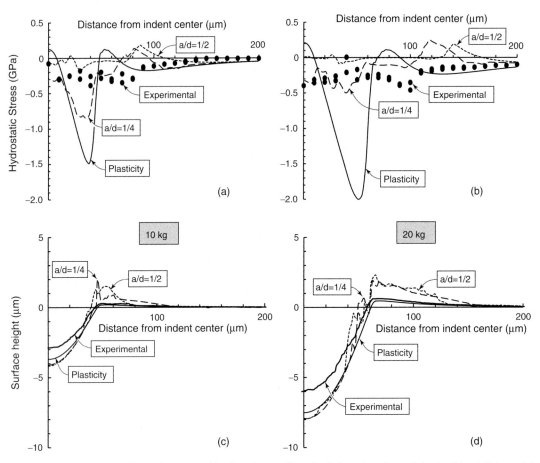

Fig. 13. Distributions of (a, b) residual hydrostatic stress and (c, d) surface profiles after indentation, for peak loads of (a, c) 10 kg and (b, d) 20 kg. In addition to the experimental measurements, computed curves are plotted for cases in which the ceramic deforms by lattice plasticity alone as well as plasticity accompanied by microcracking. The latter are based on a crack spacing $g_2 = 6$, fracture toughness $K_C = 3$ MPa$\sqrt{\text{m}}$, and two values of initial flaw size: $a/d = 1/4$ or $1/2$, as indicated.

experimental observations over the entire load range, with slight underestimation of the damage depth at 10 kg and slight overestimation at 30 kg.

VI. Residual Stresses and Indent Profiles

As a further assessment of the model, comparisons were made of the residual hydrostatic stress (the numerical results being averaged into the depth according to the method described in Appendix A). The comparisons are plotted on Fig. 13 for indent loads of 10 and 20 kg. Two sets of damage parameter values were considered, differentiated by the initial flaw size ($g_1 = 1/4$ or $1/2$). The other property values were fixed. Also shown are the finite element results assuming that the only inelastic mechanism is lattice plasticity.

The experimental measurements indicate that, within the indent, the residual stress is compressive and essentially uniform across the indent. Although significant scatter exists (attributable to the probe size being comparable to the grain diameter), the average values within the indents are roughly constant (-260 ± 100 MPa) for indent loads of 10 and 20 kg. These levels persist beyond the indent boundary, to a distance of about twice the indent radius. Thereafter, at greater distances from the indent, the stresses gradually approach zero.

The computed residual stress distributions for lattice plasticity alone are inconsistent with the measurements. For instance, the peak computed stresses (near the indent periphery) are about five times the measured values. This discrepancy is a direct consequence of the relaxation from granular plasticity that accompanies microcracking in the experiments but is not taken into account in the plasticity model. Nevertheless, in regions well away from the indent periphery ($\geq$ two indent radii), the plasticity calculations predict a gradual approach to zero stress at a rate comparable to

that of the experimental measurements. This correlation is consistent with the absence of damage in the latter regions.

The computations based on the damage model appear to adequately bind the residual stress measurements within the indent. That is, for $a/d = 1/2$ and an indent load of 10 kg, the magnitude of the average residual stress within the indent (≈ 250 MPa) is underestimated by the model; in contrast, for $a/d = 1/4$, it is overestimated. Both data sets exhibit a local tensile peak at a distance of about twice the indent size, before once again turning compressive and approaching the finite element results based on lattice plasticity alone. These tensile peaks coincide with the edge of the computed damage zone and are due to large tensile hoop stresses in the surrounding elastic material. In reality, these tensile stresses are at least partially relieved by the radial cracks emanating from the indent. Such cracks cannot form in the present computations because of the axisymmetric nature of the model and hence the residual stresses are overestimated. These unrelieved tensile hoop stresses may also lead to over-prediction of the spatial extent of the damage.

For completeness, comparisons were made of the residual surface profiles (Figs. 13(c) and (d)). Within the indents, the damage model yields results similar to those for lattice plasticity alone, reaffirming the notion that the profile is not appreciably affected by the damage. Outside the indent, additional pile-up associated with damage is evident, although the computed results overestimate the measured values.

VII. Concluding Remarks

The results of the extended DE model are in broad agreement with the experimental measurements and observations of quasi-static indentation tests. A particularly striking feature is the correlation obtained using property values that are either measured directly or estimated from physically based arguments, without undue reliance on adjustable (nonphysical) parameters. Specifically, for oxide ceramics, the effective fracture toughness pertinent to microcracking correlates well with the value measured by Vicker's indentations. Furthermore, the inferred initial flaw size falls in the range expected on the basis of grain-boundary facets or grain-sized slip-induced flaws. The inferred value of the friction angle that defines the flow stress for granular plasticity is near the upper limit required to yield damage under uniaxial compression. Such a high value is tentatively attributed to the high packing density of the comminuted material.

The dilatation angle and the flaw spacing have little effect on the predicted damage patterns. Loading conditions under which the these parameters play a significant role have yet to be ascertained. It is surmised that the dilatation angle may be more important in deep penetration experiments, wherein large amounts of confined granular flow are obtained, as described by Shockey *et al.*[15,16] In contrast, in the present calculations, the strains obtained via the granular plasticity mechanism were minimal.

The most discriminating metric in the assessment of the extended DE model is the spatial extent of subsurface damage. Residual hydrostatic stresses provide additional validation. In contrast, surface profiles of the indents are dictated largely by the plastic properties of the ceramic and thus provide minimal insight into the inelastic deformation resulting from microcracking or granular flow.

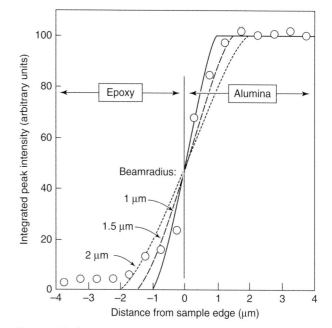

Fig. A1. Variation in peak intensity with position across a polished epoxy/alumina interface. The lines were calculated assuming a circular probe of uniform intensity, with the radii indicated.

Acknowledgments

E. A. G. was supported by a National Defense Science and Engineering Graduate Fellowship. The authors also acknowledge D.R. Clarke for fruitful discussions regarding residual stress measurement.

Appendix A: Measurement of Piezospectroscopic Interaction Volume

Experiments were conducted to ascertain the response function of the laser probe used for residual stress measurements. They were motivated by the recognition that the probe interrogates a finite volume of material. When stress gradients exist over length scales comparable to the probe dimensions, the true stress distributions can only be inferred through a deconvolution of a series of spectroscopic measurements. Alternatively, if the principal objective is to assess the fidelity of a model for the stress distribution, the computed results can be averaged accordingly for comparison with the experimental data. The latter approach was adopted in the present study.

The laser probe used to stimulate luminescence is focused, nominally to a point, within the area of interest in an optical microscope. In practice, the minimum probe diameter attainable depends on the system optics and the objective magnification and is typically in the range 1–2 μm. The probe also penetrates beneath the surface, to a depth controlled by the distribution in probe intensity with distance from the focal plane. The measurements are further influenced by the probe sensitivity: that is, the efficiency with which each volume element conveys information to the spectrometer. For instance, photons emitted deep below the surface are scattered by the material and yield a lower signal intensity than that emanating from photons emitted near the free surface. Combined, the probe size and probe sensitivity define the probe response function.[17,18]

To determine the size of the probe in the focal plane, a series of fluorescence measurements were taken across the interface between polished alumina and mounting epoxy. At each point, measurements were made of the integrated intensities of the R_1 and R_2 peaks. The translation increment between measurement points was 0.5 μm. The intensity distribution, shown in Fig. A1, was then modeled by assuming that the beam intensity is uniform in the focal plane and that the probe is circular

in shape. The fitting yields a probe diameter of 2–3 μm, virtually the same as the simulation element size in the contact region.

The coupled effects of the probe shape normal to the focal plane and the scattering of emitted photons were ascertained using a modification of the technique described by Lipkin and colleagues[14] and Wan and colleagues.[15] Specifically, spectroscopic measurements were made as the focal plane was translated normal to the sample surface. The variation in integrated intensity of the fluorescence spectrum with focal plane depth is plotted in Fig. A2. The probe intensity normal to the focal plane was assumed to be the sum of two components: a Lorentzian distribution, with a peak at the focal plane, and a constant value, associated with on-axis light. Furthermore, scattering effects are described by a conventional exponential decay function. The integrated intensity of a fluorescence spectrum obtained with the probe focused at depth c below the sample surface is thus given by

$$\frac{I}{I_0} = \int_0^\infty \left[B \cdot \exp(-2\alpha z) + (1-B) \cdot \exp(-2\alpha z) \left(\frac{p^2}{p^2 + (z-c)^2} \right) \right] dz$$

$$(A-1)$$

where B is the fraction of laser light that is on axis, z is depth below the surface, α is the absorption coefficient, p is a characteristic length describing the probe width along the z-axis and I_0 is a reference intensity. A least squares fit of this relation to the experimental data (Fig. A2) yields parameter values: $B = 0.138$, $\alpha = 0.059 \ \mu m^{-1}$ and $p = 5.25 \ \mu m$.

Appendix B: Constitutive Laws for Ceramic Deformation in the Extended Deshpande-Evans Model

(1) Lattice Plasticity

When the confining pressure is sufficiently high, plasticity occurs by slip and/or twinning. At room temperature, dislocation motion in oxide ceramics such as alumina is limited by the lattice resistance. Furthermore, according to Frost and Ashby,[9] the lattice resistance is largely rate independent until the dislocation velocity reaches a value limited by phonon drag. Here,

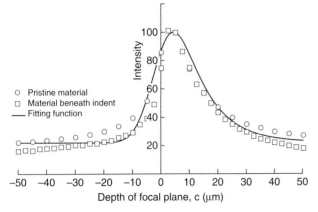

Fig. A2. Variation in peak intensity with focal plane position, for both pristine material and at the center of an indent. The solid line shows the least squares fit of Eq. (A-1).

the plastic strain rate can be described by

$$\frac{\dot{\varepsilon}_e^{pl}}{\dot{\varepsilon}_0} = \begin{cases} \min\left[\left(\frac{2\sigma_e}{\sigma_0}-1\right)^n, \left(\frac{\dot{\varepsilon}_0}{\dot{\varepsilon}_t}\right)^{(1-n)/n}\left(\frac{2\sigma_e}{\sigma_0}-1\right)\right] & 2\sigma_e > \sigma_0 \\ 0 & \text{otherwise} \end{cases}$$

(B-1)

where $\dot{\varepsilon}_t$ is the transition strain rate at which dislocation motion becomes limited by phonon drag. To ensure a rate-insensitive response, the strain rate sensitivity exponent, n, is selected to be $\gg 1$. In the present implementation, $n = 30$. The flow stress depends on the current plastic strain via a conventional power law

$$\sigma_0 = \frac{\sigma_y}{2}\left[1+\left(\frac{\varepsilon_e^{pl}}{\varepsilon_y}\right)^M\right]$$

(B-2)

where ε_y is the yield strain. Associated flow is assumed with a flow potential $G_p \equiv \sigma_e$.

(2) Microcrack Growth

Under compressive stress states, inelastic deformation occurs by stable extension of wing cracks from preexisting flaws (Fig. 1(b)). The stress intensity driving wing crack growth has been analyzed by Ashby and Hallam[19] in two dimensions and extended to three dimensions by Ashby and Sammis.[20] For simplicity, the flaws are assumed to have a uniform spacing, $1/f^{1/3}$, and uniform initial diameter, $2a$. Furthermore, the flaws are assumed to be oriented at 45° to the direction of the maximum principal compressive stress: the orientation that yields the maximum value of K_I. The damage state is characterized by

$$D = \frac{4}{3}\pi(l+\alpha a)^3 f$$

(B-3)

where l is the length of the wing crack and $\alpha a = 1/\sqrt{2}$ is the length of the preexisting flaw in the direction of the wing crack. D increases monotonically from its initial value,

$$D_0 = \frac{4}{3}\pi(\alpha a)^3 f$$

(B-4)

to $D = 1$ when neighboring cracks coalesce.

There are three contributions to the stress intensity at the wing crack tip[17] (i) a sliding stress across the crack faces that wedges open the wing cracks, (ii) a closing stress on the wing crack due to the global compressive hydrostatic stress, and (iii) an internal stress in the uncracked ligament between wing cracks

that balances the wedging stress. The net stress intensity factor at the tip of the wing crack can be expressed in terms of the first two stress invariants as

$$\frac{K_I}{\sqrt{\pi a}} = A\sigma_m + B\sigma_e$$

(B-5)

where

$$A \equiv c_1(c_2 A_3 - c_2 A_1 + c_3)$$

(B-6)

$$B \equiv \frac{c_1}{\sqrt{3}}(c_2 A_3 + c_2 A_1 + c_3)$$

(B-7)

$$c_1 = \frac{1}{\pi^2 \alpha^{3/2}\left[\left(\frac{D}{D_0}\right)^{1/3}-1+\frac{\beta}{\alpha}\right]^{3/2}}$$

$$c_2 = 1 + 2\left[\left(\frac{D}{D_0}\right)^{1/3}-1\right]^2\left(\frac{D_0^{2/3}}{1-D^{2/3}}\right)$$

(B-8)

$$c_3 = 2\alpha^2\pi^2\left[\left(\frac{D}{D_0}\right)^{1/3}-1\right]^2$$

$$A_1 = \pi\sqrt{\frac{\beta}{3}}\left[(1+\mu^2)^{1/2}-\mu\right]$$

(B-9)

$$A_3 = \pi\sqrt{\frac{\beta}{3}}\left[(1+\mu^2)^{1/2}+\mu\right]$$

These results apply when the crack faces remain in contact but the shear stress is sufficient to overcome friction and cause sliding of the crack faces. This is denoted Regime II.

Two other behavioral regimes are obtained. When the stress triaxiality is sufficiently low (negative), sliding does not occur across the crack faces and hence $K_I = 0$ (regime I). This condition is obtained when

$$\lambda \equiv \frac{\sigma_m}{\sigma_e} \leq -\frac{B}{A}$$

(B-10)

At the other extreme, where the triaxiality is high, the crack faces lose contact with one another (regime III). In this domain, K_I is given approximately by[21,22]

$$\frac{K_I}{\sqrt{\pi a}} = \left(C^2\sigma_m^2 + E^2\sigma_e^2\right)^{1/2}$$

(B-11)

The coefficients C and E are selected to ensure the elastic strains are continuous across the transition between regimes II and III. This requires that

$$E^2 = \frac{B^2 C^2}{C^2 - A^2}$$

(B-12)

and the regime transition occurs when

$$\lambda = \frac{AB}{C^2 - A^2}$$

(B-13)

where C is defined as

$$C \equiv A + \gamma\sqrt{\alpha\left(\frac{D}{D_0}\right)^{1/3}}$$

Crack growth is assumed to occur at a rate $i/i_0 = (K_I/K_{IC})^m$, but limited by the shear wave speed of the uncracked material.

Thus

$$i = \min\left[i_0 (K_I/K_{IC})^m, \sqrt{\frac{G}{\rho}}\right] \qquad \text{(B-14)}$$

where G and ρ are the shear modulus and the density of the uncracked ceramic. The sensitivity exponent, $m = 34$, and reference crack growth rate, $i_0 = 0.01$ m/s, were obtained by fitting data reported by Aeberli and Rawlings.[11]

For numerical implementation, full comminution was deemed to have occurred when the damage reaches a value $D = 1 - 0.1D_0$. (Allowing $D = 1$ results in division by 0 in the calculation of c_2.)

(3) Granular Plasticity

Granular plasticity of the fully comminuted ceramic is taken to follow a linear Drucker–Prager law,[23] with an effective stress defined as

$$\hat{\sigma} \equiv \frac{\sigma_e + (\tan\beta)\sigma_m}{1 - \tan\omega/3} \qquad \text{(B-15)}$$

where ω is the friction angle. The granular strain rate is given by

$$\frac{\dot{\varepsilon}_e^g}{\dot{\varepsilon}_s} = \begin{cases} \left(\frac{\hat{\sigma}}{\Sigma_c} - 1\right)^s & \hat{\sigma} > \Sigma_c \\ 0 & \text{otherwise} \end{cases} \qquad \text{(B-16)}$$

where Σ_c is the uniaxial compressive strength of the comminuted ceramic (assumed to be 1 MPa) and $\dot{\varepsilon}_s$ is the reference strain rate. For compressive mean stress states, the effective stress scales with the square of the strain rate in a Bagnold-type flow law giving $s = 1/2$. For $\sigma_m > 0$, s is large (~ 10) to prevent the granular medium from supporting large tensile stresses.

In general, granular flow is expected to occur with some dilatation, dictated by the value of ψ. The granular flow surface is G_g is defined as

$$G_g \equiv \begin{cases} \sigma_e + (\tan\psi)\sigma_m & \sigma_m < 0 \\ \sqrt{(\zeta\Sigma_t)^2 + \sigma_e^2} + \frac{\Sigma_c}{\Sigma_t}\sigma_m & \text{otherwise} \end{cases} \qquad \text{(B-17)}$$

where Σ_t is the hydrostatic tensile strength of the comminuted ceramic and ζ is a nondimensional regularization parameter to ensure purely dilatational flow under hydrostatic tension.

References

[1]V. S. Deshpande and A. G. Evans, "Inelastic Deformation and Energy Dissipation in Ceramics: A Mechanism-Based Constitutive Model," *J. Mech. Phys. Solids*, **56**, 3077–100 (2008).

[2]V. S. Deshpande, E. A. Gamble, B. G. Compton, R. M. McMeeking, A. G. Evans, and F. W. Zok, "A constitutive description of the inelastic response of ceramics," *J. Am. Ceram. Soc.*, 2011 (this issue).

[3]B. R. Lawn, "Indentation of Ceramics with Spheres: A Century after Hertz," *J. Am. Ceram. Soc.*, **81** [8] 1977–94 (1998).

[4]F. Guiberteau, N. P. Padture, and B. R. Lawn, "Effect of Grain Size on Hertzian Contact Damage in Alumina," *J. Am. Ceram. Soc.*, **77** [7] 1825–31 (1994).

[5]J. C. LaSalvia and J. W. McCauley, "Inelastic Deformation Mechanisms and Damage in Structural Ceramics Subjected to High-Velocity Impact," *Int. J. Appl.Ceram. Technol.*, **7** [5] 595–605 (2010).

[6]C. A. Powell-Dogan and A. H. Heuer, "Devitrification of the Grain Boundary Glassy Phase in a High-Alumina Ceramic Substrate," *J. Am. Ceram. Soc.*, **77** [10] 2593–8 (1994).

[7]Q. Ma and D. R. Clarke, "Stress Measurement in Single-Crystal and Polycrystalline Ceramics Using Their Optical Fluorescence," *J. Am. Ceram. Soc.*, **76** [6] 1433–40 (1993).

[8]J. He and D. R. Clarke, "Determination of the Piezospectroscopic Coefficients for Chromium-Doped Sapphire," *J. Am. Ceram. Soc.*, **78** [5] 1347–53 (1995).

[9]H. J. Frost and M. F. Ashby, *Deformation Mechanism Maps*. Pergamon Press, Oxford, U.K., 1982.

[10]K. E. Aeberli and R. D. Rawlings, "Effect of Simulated Body Environments on Crack Propagation in Alumina," *J. Mater. Sci. Lett.*, **2**, 215–20 (1983).

[11]T. W. Lambe and R. V. Whitman, *Soil Mechanics*. John Wiley & Sons Inc, New York, 1969.

[12]A. Nash, "Compressive Failure Modes of Alumina in Air and Physiological Media," *J. Mater. Sci.*, **18**, 3571–7 (1983).

[13]M. Adams and G. Sines, "Determination of Biaxial Compressive Strength of a Sintered Alumina Ceramic," *J. Am. Ceram. Soc.*, **59** [7] 300–4 (1976).

[14]J. Lankford, "Compressive Strength and Microplasticity in Polycrystalline Alumina," *J. Mater. Sci.*, **12**, 791–6 (1977).

[15]D. A. Shockey, A. H. Marchand, S. R. Skaggs, G. E. Cort, M. W. Burkett, and R. Parker, "Failure Phenomenology of Confined Ceramic Targets and Impacting Rods," *J. Impact Eng.*, **9** [3] 263–75 (1990).

[16]D. A. Shockey, J. W. Simons, and D. R. Curran, "The Damage Mechanism Route to Better Armor Materials," *Int. J. Appl. Ceram. Technol.*, **7** [5] 566–73 (2010).

[17]D. M. Lipkin and D. R. Clarke, "Sample-Probe Interactions Inspectroscopy: Sampling Microscopic Property Gradients," *J. Appl. Phys.*, **77** [5] 1855–63 (1995).

[18]K. Wan, W. Zhu, and G. Pezzotti, "Determination of In-Depth Probe Response Function Using Spectral Perturbation Methods," *J. Appl. Phys.*, **98**, 1–7 (2005).

[19]M. F. Ashby and S. D. Hallam, "The Failure of Brittle Solids Small Cracks Under Compressive Containing Stress States," *Acta Metall.*, **34** [3] 497–510 (1986).

[20]M. F. Ashby and C. G. Sammis, "The Damage Mechanics of Brittle Solids in Compression," *Pure Appl. Geophys.*, **133** [3] 489–521 (1990).

[21]J. R. Bristow, "Microcracks, and the Static and Dynamic Elastic Constants of Annealed and Heavily Cold-Worked Metals," *J. Phys. D Appl. Phys.*, **11** [2] 81–5 (1960).

[22]B. Budiansky and R. J. O'Connell, "Elastic Moduli of a Cracked Solid," *Int. J. Solids Struct.*, **12** [2] 81–97 (1976).

[23]D. C. Drucker and W. Prager, "Soil Mechanics and Plastic Analysis or Limit Design," *Q. Appl. Mathematics*, **10** [2] 157–65 (1952).   □

J. Am. Ceram. Soc., **94** [S1] S226–S235 (2011)

DOI: 10.1111/j.1551-2916.2011.04432.x

© 2011 The American Ceramic Society

Journal

Large Plastic Deformation in High-Capacity Lithium-Ion Batteries Caused by Charge and Discharge

Kejie Zhao, Matt Pharr, Shengqiang Cai, Joost J. Vlassak, and Zhigang Suo[†]

School of Engineering and Applied Sciences, Kavli Institute for Bionano Science and Technology, Harvard University, Cambridge, Massachusetts 02138

Evidence has accumulated recently that a high-capacity electrode of a lithium-ion battery may not recover its initial shape after a cycle of charge and discharge. Such a plastic behavior is studied here by formulating a theory that couples large amounts of lithiation and deformation. The homogeneous lithiation and deformation in a small element of an electrode under stresses is analyzed within nonequilibrium thermodynamics, permitting a discussion of equilibrium with respect to some processes, but not others. The element is assumed to undergo plastic deformation when the stresses reach a yield condition. The theory is combined with a diffusion equation to analyze a spherical particle of an electrode being charged and discharged at a constant rate. When the charging rate is low, the distribution of lithium in the particle is nearly homogeneous, the stress in the particle is low, and no plastic deformation occurs. When the charging rate is high, the distribution of lithium in the particle is inhomogeneous, and the stress in the particle is high, possibly leading to fracture and cavitation.

I. Introduction

LITHIUM-ION batteries are being developed to achieve safe operation, high capacity, fast charging, and long life. Each electrode in a lithium-ion battery is a host of lithium.[1] Lithium diffuses into and out of the electrode when the battery is charged and discharged. If a particle of the electrode material is charged or discharged slowly and is unconstrained by other materials, the distribution of lithium in the particle is nearly homogeneous, and the particle expands or contracts freely, developing no stress. In practice, however, charge and discharge cause a field of stress in the particle when the distribution of lithium is inhomogeneous,[2–5] when the host contains different phases,[6] or when the host is constrained by other materials.[7] The stress may cause the electrode to fracture,[6–10] which may lead the capacity of the battery to fade.[11–14]

Lithiation-induced fracture not only occurs in current commercial lithium-ion batteries, but is also a bottleneck in developing future lithium-ion batteries.[15,16] For example, of all known materials for anodes, silicon offers the highest theoretical specific capacity—each silicon atom can host up to 4.4 lithium atoms. By comparison, in commercial anodes of graphite, every six carbon atoms can host up to one lithium atom. However, silicon is not used in anodes in commercial lithium-ion batteries, mainly because the capacity of silicon fades after a small number of cycles—a mode of failure often attributed to lithiation-induced fracture.

Recent experiments, however, have shown that the capacity can be maintained over many cycles for silicon anodes of small

feature sizes, such as thin films,[17] nanowires,[18] and porous structures.[19] When silicon is fully lithiated, the volume of the material swells by ~300%.[15] For anodes of small feature sizes, evidence has accumulated recently that this lithiation-induced strain can be accommodated by plastic deformation. For instance, cyclic lithiation causes 50-nm-thick silicon films to develop undulation without fracture.[17] Likewise, cyclic lithiation causes surfaces of silicon nanowires to roughen.[18] These studies suggest that lithiated silicon undergoes plastic deformation. Furthermore, during cyclic lithiation of an amorphous silicon thin film attached to a substrate, the measured relation between the stress in the film and the state of charge exhibits pronounced hysteresis.[20] This observation indicates that the lithiated silicon film deforms plastically when the stress exceeds a yield strength. At this writing, the atomistic mechanism of plastic flow of lithiated silicon is unclear, but is possibly due to a "lubricating effect" of lithium. Cyclic lithiation and delithiation can also cause silicon to grow cavities.[21]

This paper develops a theory of finite plastic deformation of electrodes caused by charge and discharge. Plastic deformation has not been considered in most existing theories of lithiation-induced deformation.[2–10] The works that do include plasticity have been limited to small deformation.[22,23] Here, we analyze the homogeneous lithiation and deformation in a small element of an electrode within nonequilibrium thermodynamics, stipulating equilibrium with respect to some processes, but not others. The element is assumed to undergo plastic deformation when the state of stress reaches a yield condition. We then combine large plastic deformation and diffusion to analyze the lithiation of a spherical particle of an electrode. When a small particle is charged and discharged slowly, the stress is small. When the particle is charged and discharged quickly, the stress is high, potentially leading to fracture or cavitation.

II. Nonequilibrium Thermodynamics of Coupled Lithiation and Deformation

Figure 1 illustrates an element of an electrode subject to a cycle of charge and discharge. The element is small in size, so that fields in the element are homogeneous. In the reference state (Fig. 1(a)), the element is a unit cube of a host material, free of lithium and under no stress. When the element is connected to a reservoir of lithium at chemical potential μ and is subject to stresses s_1, s_2, and s_3, as illustrated in Fig. 1(b), the element absorbs a number C of lithium atoms, becomes a block of sides λ_1, λ_2, and λ_3, and gains in free energy W.

By definition, s_1, s_2, and s_3 are nominal stresses—forces acting on the element in the current state divided by the areas of the element in the reference state. The true stresses, σ_1, σ_2, and σ_3, are forces per unit areas of the element in the current state. The true stresses relate to the nominal stresses by $\sigma_1 = s_1/(\lambda_2\lambda_3)$, $\sigma_2 = s_2/(\lambda_3\lambda_1)$, and $\sigma_3 = s_3/(\lambda_1\lambda_2)$.

Associated with small changes in the stretches, $\delta\lambda_1$, $\delta\lambda_2$, and $\delta\lambda_3$, the forces do work $s_1\delta\lambda_1 + s_2\delta\lambda_2 + s_3\delta\lambda_3$. Associated with small change in the number of lithium atoms, δC, the chemical potential does work $\mu\delta C$. Thermodynamics dictates that the

---

R. McMeeking—contributing editor

Manuscript No. 28712. Received October 2, 2010; approved January 4, 2011.

This work is supported by the National Science Foundation through a grant on Lithium-ion Batteries (CMMI-1031161).

[†]Author to whom correspondence should be addressed. e-mail: suo@seas.harvard.edu

combined work should be no less than the change in the free energy:

$$s_1\delta\lambda_1 + s_2\delta\lambda_2 + s_3\delta\lambda_3 + \mu\delta C \geq \delta W \tag{1}$$

The work done minus the change in the free energy is the dissipation. The inequality (1) means that the dissipation is non-negative with respect to all processes. The object of this section is to examine the large amounts of lithiation and deformation by constructing a theory consistent with the thermodynamic inequality (1).

As illustrated in Fig. 1(b), when the unit cube of the host is lithiated under stresses, the deformation is anisotropic: the unit cube will change both its shape and volume. For example, a thin film of an electrode constrained on a stiff substrate, upon absorbing lithium, deforms in the direction normal to the film, but does not deform in the directions in the plane of the film.

The material deforms by mechanisms of two types: inelastic and elastic. Inelastic deformation involves mixing and rearranging atoms. Elastic deformation involves small changes of the relative positions of atoms, retaining the identity of neighboring atoms as well as the concentration of lithium. When the stresses are removed and the reservoir of lithium is disconnected, the material element will retain part of the anisotropic deformation (Fig. 1(c)). The phenomenon is reminiscent of plasticity of a metal. The remaining deformation is characterized by three stretches λ_1^i, λ_2^i, and λ_3^i, which we call inelastic stretches. The part of deformation that disappears upon the removal of the stresses is characterized by three stretches λ_1^e, λ_2^e, and λ_3^e, which we call elastic stretches. The total stretches are taken to be the products of the two types of the stretches

$$\lambda_1 = \lambda_1^e\lambda_1^i, \quad \lambda_2 = \lambda_2^e\lambda_2^i, \quad \lambda_3 = \lambda_3^e\lambda_3^i \tag{2}$$

Similar multiplicative decomposition is commonly used to describe elastic–plastic deformation of metals[24] and polymers,[25] as well as growth of tissues.[26]

We characterize the state of the material element by a total of seven independent variables: λ_1^e, λ_2^e, λ_3^e, λ_1^i, λ_2^i, λ_3^i and C.

Plasticity of a metal and inelasticity of an electrode differ in a significant aspect. While plastic deformation of a metal changes shape but conserves volume, inelastic deformation of an electrode changes both volume and shape. We decompose the inelastic stretches by writing

$$\lambda_1^i = \Lambda^{1/3}\lambda_1^p, \quad \lambda_2^i = \Lambda^{1/3}\lambda_2^p, \quad \lambda_3^i = \Lambda^{1/3}\lambda_3^p \tag{3}$$

Here, Λ is the volume of the material element after the removal of the stresses, namely,

$$\lambda_1^i\lambda_2^i\lambda_3^i = \Lambda \tag{4}$$

Inelastic shape change of the material element is described by λ_1^p, λ_2^p, and λ_3^p (Fig. 1(d)). By the definition of (3) and (4), the plastic stretches do not change volume, namely,

$$\lambda_1^p\lambda_2^p\lambda_3^p = 1 \tag{5}$$

We will call λ_1^p, λ_2^p, and λ_3^p the plastic stretches. In our terminology, inelastic deformation includes the changes in both volume and shape, while plastic deformation involves only the change in shape.

A combination of (2) and (3) gives $\lambda_1 = \lambda_1^e\lambda_1^p\Lambda^{1/3}$. Taking the logarithm of both sides of this equation, we write $\log\lambda_1 = \log\lambda_1^e + \log\lambda_1^p + \log\Lambda^{1/3}$. The quantity $\log\lambda_1$ is the natural strain, $\log\lambda_1^e$ the elastic part of the natural strain, and $\log\lambda_1^p$ the plastic part of the natural strain.

Most existing theories of lithiation-induced deformation do not consider plasticity. In effect, these theories assume that, after a cycle of lithiation and delithiation, the material element recovers its initial shape. Such an assumption disagrees with experimental observations of lithiation of large-capacity hosts, such as silicon, as discussed in the Introduction and in several recent papers.[22,23,27] This paper will allow plasticity, and will describe rules to calculate large plastic deformation.

The state of the element can be characterized by an alternative list of seven independent variables: λ_1^e, λ_2^e, λ_3^e, λ_1^p, λ_2^p, Λ, and C. To progress further, we make the following simplifying assumptions. The inelastic expansion of the volume is taken to be entirely due to the absorption of lithium, and is a function of the concentration of lithium:

$$\Lambda = \Lambda(C) \tag{6}$$

This function is taken to be characteristic of the material, and is independent of the elastic and plastic stretches. Equation (6) eliminates Λ from the list of independent variables, so that the state of the material element is characterized by six independent variables: λ_1^e, λ_2^e, λ_3^e, λ_1^p, λ_2^p, and C.

Following the theory of plasticity, we assume that the free energy of the material element is unaffected by the plastic stretches. Thus, the free energy is a function of four variables:

$$W = W(\lambda_1^e, \lambda_2^e, \lambda_3^e, C) \tag{7}$$

This assumption is understood as follows. The plastic stretches characterize inelastic shape change, involving rearranging atoms without changing the concentration of lithium. This rearrangement of atoms may dissipate energy, but does not alter the amount of free energy stored in the host materials. The situation is reminiscent of shear flow of a liquid of small molecules—the free energy is stored in molecular bonds, independent of the amount of flow. For a work-hardening metal, however, plastic strain will change the free energy stored in the material, for example, by creating more dislocations. Thus, assuming (7) amounts to a stipulation that plastic strains do not create such microstructural changes in lithiated electrodes.

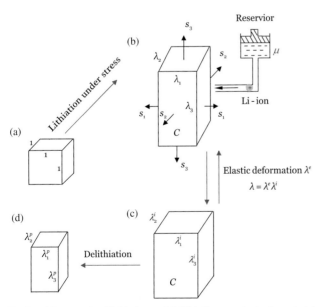

Fig. 1. After a cycle of lithiation and delithiation, an electrode material may not recover its initial shape. (a) In the reference state, an element of an electrode material is a lithium-free and stress-free unit cube. (b) Subject to forces and connected to a reservoir of lithium, the material element absorbs lithium, and undergoes anisotropic deformation. (c) When the stresses are removed and the reservoir of lithium is disconnected, the material element unloads elastically, and the remaining inelastic deformation is anisotropic. (d) After the material element desorbs lithium under no stress, the lithium-free host becomes a rectangular block.

Rewriting the inequality (1) in terms of the changes in the six independent variables, λ_1^e, λ_2^e, λ_3^e, λ_1^p, λ_2^p, and C, we obtain that

$$\left(\sigma_1\lambda_1\lambda_2\lambda_3 - \frac{\partial W}{\partial \log \lambda_1^e}\right)\delta \log \lambda_1^e + \left(\sigma_2\lambda_1\lambda_2\lambda_3 - \frac{\partial W}{\partial \log \lambda_2^e}\right)$$
$$\times \delta \log \lambda_2^e + \left(\sigma_3\lambda_1\lambda_2\lambda_3 - \frac{\partial W}{\partial \log \lambda_3^e}\right)\delta \log \lambda_3^e$$
$$+ \left(\mu - \frac{\partial W}{\partial C} + \Omega\sigma_m\right)\delta C$$
$$+ \lambda_1\lambda_2\lambda_3\left[(\sigma_1 - \sigma_3)\delta \log \lambda_1^p + (\sigma_2 - \sigma_3)\delta \log \lambda_2^p\right] \geq 0$$
$$(8)$$

where $\sigma_m = (\sigma_1 + \sigma_2 + \sigma_3)/3$ is the mean stress, and $\Omega = \lambda_1^e\lambda_2^e\lambda_3^e d\Lambda(C)/dC$ is the volume per lithium atom in the host.

Each of the six independent variables represents a process to evolve the material. The processes take place at different rates. We will adopt a commonly used simplifying approach. Say that we are interested in a particular time scale. Processes taking place faster than this time scale are assumed to be instantaneous. Processes taking place slower than this time scale are assumed to never occur. In the present problem, the particular time scale of interest is the time needed for a particle of an electrode material of a finite size to absorb a large amount of lithium. This time is taken to be set by the diffusion time scale.

Elastic relaxation is typically much faster than diffusion. We assume that the material element is in equilibrium with respect to the elastic stretches, so that in (8) the coefficients associated with $\delta\lambda_1^e$, $\delta\lambda_2^e$, and $\delta\lambda_3^e$ vanish:

$$\sigma_1 = \frac{\partial W}{\lambda_1\lambda_2\lambda_3\partial \log \lambda_1^e}, \quad \sigma_2 = \frac{\partial W}{\lambda_1\lambda_2\lambda_3\partial \log \lambda_2^e},$$
$$\sigma_3 = \frac{\partial W}{\lambda_1\lambda_2\lambda_3\partial \log \lambda_3^e}$$
$$(9)$$

We further assume that the material element is in equilibrium with respect to the concentration of lithium, so that in (8) the coefficient of δC vanishes:

$$\mu = \frac{\partial W(\lambda_1^e, \lambda_2^e, \lambda_3^e, C)}{\partial C} - \Omega\sigma_m \quad (10)$$

The free energy is adopted in the following form:

$$W = W_0(C) + \Lambda G\left[(\log \lambda_1^e)^2 + (\log \lambda_2^e)^2 + (\log \lambda_3^e)^2\right.$$
$$\left. + \frac{v}{1-2v}(\log \lambda_1^e\lambda_2^e\lambda_3^e)^2\right]$$
$$(11)$$

where G is the shear modulus, and v Poisson's ratio. Equation (11) can be interpreted as the Taylor expansion in terms of elastic strains. We have assumed that the elastic strains are small, and only retain terms up to those that are quadratic in strains. We have neglected any dependence of elastic moduli on the concentration of lithium. A combination of (9) into (11) shows that the stresses relate to the elastic strains as

$$\sigma_1 = 2G\left(\log \lambda_1^e + \frac{v}{1-2v}\log \lambda_1^e\lambda_2^e\lambda_3^e\right)$$
$$\sigma_2 = 2G\left(\log \lambda_2^e + \frac{v}{1-2v}\log \lambda_1^e\lambda_2^e\lambda_3^e\right)$$
$$\sigma_3 = 2G\left(\log \lambda_3^e + \frac{v}{1-2v}\log \lambda_1^e\lambda_2^e\lambda_3^e\right)$$
$$(12)$$

A combination (10) and (11) expresses the chemical potential of lithium as

$$\mu = \frac{dW_0(C)}{dC} - \Omega\sigma_m \quad (13)$$

In writing (12) and (13), we have neglected the terms quadratic in the elastic strains.

The material element, however, may not be in equilibrium with respect to plastic stretches. Consequently, the inequality (8) is reduced to

$$(\sigma_1 - \sigma_3)\delta \log \lambda_1^p + (\sigma_2 - \sigma_3)\delta \log \lambda_2^p \geq 0 \quad (14)$$

This thermodynamic inequality may be satisfied by many kinetic models—creep models that relate the rate of plastic strains to stresses. For simplicity, here we adopt a particular type of kinetic model: the model of time-independent plasticity.[28] A material is characterized by a yield strength. When the stress is below the yield strength, the rate of the plastic strain is taken to be so low that no additional plastic strain occurs. When the stress reaches the yield strength, the rate of plastic strain is taken to be so high that plastic strain increases instantaneously.

To calculate plastic deformation, we adopt the J_2 flow theory.[28] Recall that the plastic stretches preserve the volume, $\lambda_1^p\lambda_2^p\lambda_3^p = 1$. Consequently, (14) may be written in a form symmetric with respect to the three directions:

$$(\sigma_1 - \sigma_m)\delta \log \lambda_1^p + (\sigma_2 - \sigma_m)\delta \log \lambda_2^p$$
$$+ (\sigma_3 - \sigma_m)\delta \log \lambda_3^p \geq 0$$
$$(15)$$

The J_2 flow theory is prescribed as

$$\delta \log \lambda_1^p = \alpha(\sigma_1 - \sigma_m),$$
$$\delta \log \lambda_2^p = \alpha(\sigma_2 - \sigma_m),$$
$$\delta \log \lambda_3^p = \alpha(\sigma_3 - \sigma_m)$$
$$(16)$$

where α is a nonnegative scalar. This flow theory is symmetric with respect to the three directions, satisfies $\lambda_1^p\lambda_2^p\lambda_3^p = 1$, and is consistent with the thermodynamic inequality (15). In numerical calculations performed later, we will assume that the material is perfectly plastic. Let σ_Y be the yield strength measured when the material element is subject to uniaxial stressing. The yield strength is taken to be a function of concentration alone, $\sigma_Y(C)$. When the material element is under multiaxial stressing, the equivalent stress is defined by

$$\sigma_e = \sqrt{\frac{3}{2}\left[(\sigma_1 - \sigma_m)^2 + (\sigma_2 - \sigma_m)^2 + (\sigma_3 - \sigma_m)^2\right]} \quad (17)$$

The material element yields under the von Mises condition: $\sigma_e = \sigma_Y(C)$. The value of α is specified by the following rules

$$\begin{cases} \alpha = 0, & \sigma_e < \sigma_Y \\ \alpha = 0, & \sigma_e = \sigma_Y, \quad \delta\sigma_e < \delta\sigma_Y \\ \alpha > 0, & \sigma_e = \sigma_Y, \quad \delta\sigma_e = \delta\sigma_Y \end{cases} \quad (18)$$

III. A Spherical Particle of an Electrode

We now apply the theory to a spherical particle of an electrode material (Fig. 2). We have previously solved this problem using a theory of small plastic deformation.[22] Here we will allow large deformation. We model such an inelastic host of lithium by considering coupled lithium diffusion and large elastic–plastic deformation. In the following paragraphs, we specify the kinematics of large deformation, the kinetics of lithium diffusion, the flow rule of plasticity, and the thermodynamics of lithiation. The results of an amorphous silicon particle during charge and discharge are shown in Section IV. The detailed numerical procedure is included in Appendix A.

Before absorbing any lithium, the particle is of radius A, and is stress-free. This lithium-free particle is taken to be the reference configuration. At time t, the particle absorbs some lithium,

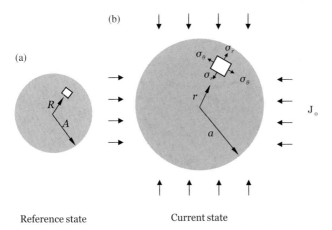

Fig. 2. (a) In the reference state, a spherical particle of an electrode is lithium-free and stress-free. (b) In the current sate, the particle is partially lithiated, and develops a field of stress.

the distribution of which may be inhomogeneous in the radial direction, but retains spherical symmetry. The inhomogeneous distribution of lithium induces in the particle a field of stress, and the particle swells to a radius a.

The kinematics of the large deformation is specified as follows. The spherical particle consists of a field of material elements. A material element a distance R from the center in the reference configuration moves, at time t, to a place a distance r from the center. The function $r(R, t)$ specifies the deformation of the particle. In representing a field, we may choose either r or R as an independent variable. One variable can be changed to the other by using the function $r(R, t)$. We will indicate our choice in each field explicitly when the distinction is important. The radial stretch is

$$\lambda_r = \frac{\partial r(R, t)}{\partial R} \tag{19}$$

The hoop stretch is

$$\lambda_\theta = \frac{r}{R} \tag{20}$$

The kinematics of lithium is specified as follows. Let C be the nominal concentration of lithium (i.e., the number of lithium atoms per unit volume of pure silicon in the reference state). The distribution of lithium in the particle is specified by the function $C(R, t)$. Because of the spherical symmetry, lithium diffuses in the radial direction. Let J be the nominal flux of lithium (i.e., the number of lithium atoms per unit reference area per unit time). The nominal flux is also a time-dependent field, $J(R, t)$. Conservation of the number of lithium atoms requires that

$$\frac{\partial C(R, t)}{\partial t} + \frac{\partial (R^2 J(R, t))}{R^2 \partial R} = 0 \tag{21}$$

Later we will also invoke the true concentration c (i.e., the number of lithium atoms per unit volume in the current configuration), and the true flux j (i.e., the number of lithium atoms per current area per time). These true quantities relate to their nominal counterparts by $J = j\lambda_\theta^2$ and $C = c\lambda_r\lambda_\theta^2$.

Write the radial stretch λ_r and the hoop stretch λ_θ in the form

$$\lambda_r = \lambda_r^e \lambda_r^p \Lambda^{1/3}, \quad \lambda_\theta = \lambda_\theta^e \lambda_\theta^p \Lambda^{1/3} \tag{22}$$

We will mainly consider high-capacity hosts that undergo large deformation by lithiation, and will neglect the volumetric change due to elasticity, setting $\lambda_r^e(\lambda_\theta^e)^2 = 1$. Consistent with this assumption, we set Poison's ratio $\nu = 1/2$, and set Young's mod-

ulus $E = 3G$. Recall that by definition the plastic stretches preserve volume, $\lambda_r^p(\lambda_\theta^p)^2 = 1$.

In the spherical particle, each material element is subject to a state of triaxial stresses, $(\sigma_r, \sigma_\theta, \sigma_\theta)$, where σ_r is the radial stress, and σ_θ the hoop stress. The state of elastic–plastic deformation is taken to be unaffected when a hydrostatic stress is superimposed on the element. In particular, as illustrated in Fig. 3(a), superimposing a hydrostatic stress $(-\sigma_\theta, -\sigma_\theta, -\sigma_\theta)$ to the state of triaxial stresses $(\sigma_r, \sigma_\theta, \sigma_\theta)$ results in a state of uniaxial stress $(\sigma_r-\sigma_\theta, 0, 0)$. The state of elastic–plastic deformation of the element subject to the triaxial stresses is the same as the state of plastic deformation of the element subject to a uniaxial stress $\sigma_r-\sigma_\theta$. We represent the uniaxial stress–stretch relation by the elastic and perfectly plastic model. Figure 3(b) sketches this stress–strain relation in terms of the stress $\sigma_r-\sigma_\theta$ and the elastic–plastic part of the true strain, $\log(\lambda_r^e\lambda_r^p) = \log(\lambda_r\Lambda^{-1/3})$. The yield strength in the state of uniaxial stress, σ_Y, is taken to be a constant independent of the plastic strain and concentration of lithium.

In the spherical particle the stresses are inhomogeneous, represented by functions $\sigma_r(R, t)$ and $\sigma_\theta(R, t)$. The balance of forces acting on a material element requires that

$$\frac{\partial \sigma_r(R, t)}{\lambda_r \partial R} + 2\frac{\sigma_r - \sigma_\theta}{\lambda_\theta R} = 0 \tag{23}$$

We specify a material model of transport as follows. We assume that each material element is in a state of local equilibrium with respect to the reaction between lithium atoms and host atoms, so that we can speak of the chemical potential of lithium in the material element. We further assume that the chemical potential of lithium in the material element takes the form:

$$\mu = \mu^0 + kT\log(\gamma c) - \Omega\sigma_m \tag{24}$$

where μ^0 is a reference value, γ the activity coefficient, and c the true concentration of lithium.

If the distribution of lithium in the particle is inhomogeneous, the chemical potential of lithium is a time-dependent field, $\mu(r, t)$, and the particle is not in diffusive equilibrium. The gradient of the chemical potential drives the flux of lithium. We adopt a linear kinetic model:

$$j = -\frac{cD}{kT}\frac{\partial \mu(r, t)}{\partial r} \tag{25}$$

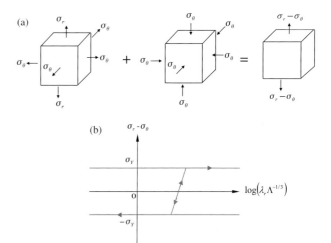

Fig. 3. (a) The state of elastic–plastic deformation of the element subject to the triaxial stresses $(\sigma_r, \sigma_\theta, \sigma_\theta)$ is the same as that of the element subject to a uniaxial stress $(\sigma_r-\sigma_\theta)$. (b) The uniaxial stress–strain relation in terms of the stress $(\sigma_r-\sigma_\theta)$ and the true elastic–plastic strain $\log(\lambda_r\Lambda^{-1/3})$.

This relation has been written in a conventional form, in terms of the true flux j and the true concentration c. Note that kT is the temperature in dimensions of energy, and that (25) may be regarded as a phenomenological definition of the diffusivity D. Recall that the flux relates to the drift velocity of lithium in the host by $j = c v_{drift}$. Thus, D/kT is the mobility of lithium in the host. The diffusivity may depend on concentration and stress.

The particle is subject to the following boundary conditions. Because of the symmetry, $r(0, t) = 0$ and $J(0, t) = 0$. On the surface of the particle, the radial stress vanishes at all time, $\sigma_r(A, t) = 0$. The particle is charged and discharged by prescribing on the surface of the particle a constant flux J_0, namely, $J(A, t) = \pm J_0$. The signs differ for charge and discharge.

IV. Numerical Results and Discussions

We solve the initial-boundary value problem numerically, as described in Appendix A. This section describes the numerical results, and discusses their implications. For simplicity, we set $\gamma = 1$ and assume that Ω, E, σ_Y, and D are constant. The initial-boundary value problem has three dimensionless parameters: $\Omega E/kT$, σ_Y/E, and $J_0 A \Omega/D$. For lithium in amorphous silicon, representative values are $\Omega = 1.36 \times 10^{-29}$ m^3, $E = 80$ GPa, $\sigma_Y = 1.75$ GPa,[20] and $D = 10^{-16}$ m^2/s,[29] giving $\Omega E/kT = 263$ and $\sigma_Y/E = 0.022$. The parameter $J_0 A \Omega/D$ is a dimensionless measure of the charging rate, and may be interpreted as follows. Let C_{max} be the maximum theoretical concentration of lithium. When the spherical particle of radius A is charged by a constant flux J_0, the nominal time τ needed to charge the particle to the theoretical maximum concentration is given by $4\pi A^2 J_0 \tau = (4\pi A^3/3) C_{max}$. For silicon, the volume of fully lithiated state swells by about 300%, so that $\Omega C_{max} \approx 3$. For a particle of radius $A = 1$ μm, $\tau = 1$ h corresponds to $J_0 A \Omega/D = 2.8$. In the following description, we use τ to represent the charge rate; for example, $\tau = 0.5$ h means that it needs 0.5 h to charge silicon to reach the theoretical maximum lithium concentration. Smaller values of τ represent a faster charge process. We normalize time as Dt/A^2. For a particle of radius $A = 1$ μm, the diffusion time scale is $A^2/D = 10^4$ s.

Figure 4 shows the evolution of the fields in the spherical particle charged at the rate of $\tau = 1$ h. The simulation is initiated when the particle is lithium-free, and is terminated when the concentration at the surface of the particle reaches the full capacity, $\Omega C_{max} = 3$, and the interior of the particle is still much below the full capacity (Fig. 4(a)). At all time, $r(R, t) > R$, indicating all material elements in the particle move away from the center of the particle (Fig. 4(b)). The ratio λ_r/λ_θ measures the anisotropy of the deformation (Fig. 4(c)). The deformation is highly anisotropic near the surface of the particle, but is isotropic at the center of the particle. Plastic deformation occurs near the surface of the particle, but is absent at the center of the particle (Fig. 4(d)). The chemical potential of lithium in the particle is inhomogeneous, driving lithium to diffuse from the surface of the particle toward the center (Fig. 4(e)).

Figures 4(f)–(h) show the distributions of the radial, hoop and equivalent stresses. The traction-free boundary condition requires that the radial stress at the surface of the particle to vanish at all times. Because the distribution of lithium in the particle is inhomogeneous, the particle expands more near the surface than at the center, resulting in tensile radial stresses inside the particle. The hoop stress is compressive near the surface, and tensile near the center. For the spherical particle, the equivalent stress is $\sigma_e = |\sigma_\theta - \sigma_r|$, which is bounded in the interval $0 \le \sigma_e \le \sigma_Y$. By symmetry, the center of the sphere is under equal-triaxial tensile stresses. Because of the triaxial constraint at the center, the radial stress and hoop stress can exceed the yield strength.

The high level of tensile stresses at the center of the sphere may generate cavities. An experimentally measured value of the yield strength is $\sigma_Y = 1.75$ GPa.[20] Our calculation indicates that the stress at the center of the particle can be several times the yield strength. Let ρ be the radius of a flaw and $\gamma_{surface}$ be the surface energy. For the flaw to expand, the stress needs to overcome the effect of the Laplace pressure, $2\gamma_{surface}/\rho$. Taking $\gamma_{surface} = 1$ J/m^2, we estimate that the critical radius $\rho = 1$ nm when the stress at the center of the particle is 2 GPa. Lithiation-induced cavitation has been observed in a recent experiment.[21]

To illustrate effects of yielding, Fig. 5 compares the two cases: $\sigma_Y = 1.75$ GPa and $\sigma_Y = \infty$ (no yielding). The fields are plotted at time $Dt/A^2 = 0.048$. As expected, yielding allows the particle to accommodate lithiation by greater anisotropic deformation, Fig. 5(b). Yielding also significantly reduces the magnitudes of the stresses, Figs. 5(c) and (d). This observation implies that fracture and cavitation may be avoided for electrodes with low yield strength.

As shown by (24), the chemical potential of lithium depends on the mean stress. This effect of stress on chemical potential has been neglected in some of the previous models. The representative values $\Omega = 1.36 \times 10^{-29}$ m^3 and $\sigma_Y = 1.75$ GPa give an estimate $\Omega \sigma_Y = 0.15$ eV, which is a value significant compared with the value of the term involving concentration in the expression of the chemical potential. Figure 6 compares results calculated by including or neglecting the term $\Omega \sigma_m$ in the expression of the chemical potential (24). The mean stress σ_m is compressive near the surface of the particle and tensile at the center. Consequently, the gradient of the mean stress also motivates lithium to migrate toward the center. Here we show the fields at the time when the surface of the particle attains the full capacity. The case with the mean stress included in the expression of the chemical potential absorbs more lithium at the center, Fig. 6(a). The stress gradient also decreases the chemical potential of lithium in the particle, Fig. 6(b). The contribution of the stress to the chemical potential helps to homogenize lithium distribution, and consequently reduces the stress.

Figure 7 compares the fields for three charging rates, $\tau = 0.5$ h, $\tau = 1$ h and $\tau = 2$ h. Each simulation is terminated when the surface of the particle attains the full capacity. The stress level is determined by the total amount of lithium inserted into the particle and the degree of inhomogeneity in the distribution of this amount of lithium. At a high charging rate, $\tau = 0.5$ h, the distribution of lithium is highly inhomogeneous, as in Fig. 7(a). However, the surface of the particle reaches its full capacity very rapidly. At this time, not much lithium has been inserted into the particle, so the stresses are fairly low and the deformation is anisotropic only near the surface of the particle. At such a fast charging rate, the particle will not store much lithium. At an intermediate charging rate, $\tau = 1$ h, lithium has some time to diffuse toward the center of the particle, but not enough time to fully homogenize the distribution of lithium. Consequently, relatively large stresses develop and the deformation is quite anisotropic. At a yet slower charging rate, $\tau = 2$ h, there is enough time for diffusion to nearly homogenize the distribution of lithium. At this slow charging rate, the particle is effective as an electrode in that it can store a large amount of lithium before the surface of the particle reaches the full capacity. This homogenization leads to relatively low stresses and nearly isotropic deformation.

Figure 8 shows the time evolution of the stress at the center of the particle for three charging rates. For a fast charging case ($\tau = 0.5$ h), the stress builds up until the outer surface reaches its full capacity. At an intermediate charging rate ($\tau = 1$ h), the stress builds up until it is large enough to significantly contribute to the chemical potential. At this point, both the stress gradient and concentration gradients tend to homogenize the concentration of lithium, and the stress decreases. However, the charging rate is still fairly fast relative to the time for diffusion. Thus, the concentration of lithium cannot be fully homogenized and the stress cannot be fully relaxed before the surface of the particle reaches the full capacity. By contrast, for the case of $\tau = 2$ h, the charging rate is slow enough for diffusion to homogenize the distribution of lithium before the surface of the particle reaches the full capacity, as shown in Fig. 7(a). As the distribution of

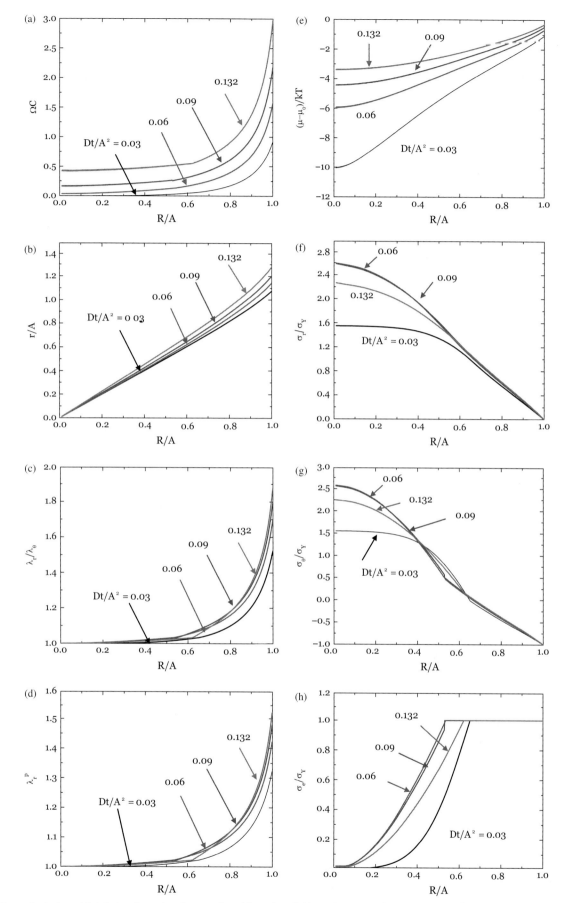

Fig. 4. As a spherical particle is being charged at the rate of $\tau = 1$ h, various fields evolve: (a) concentration of lithium, (b) deformation field, (c) ratio of the radial stretch to the hoop stretch, (d) plastic stretch in the radial direction, (e) chemical potential of lithium, (f) radial stress, (g) hoop stress, and (h) equivalent stress.

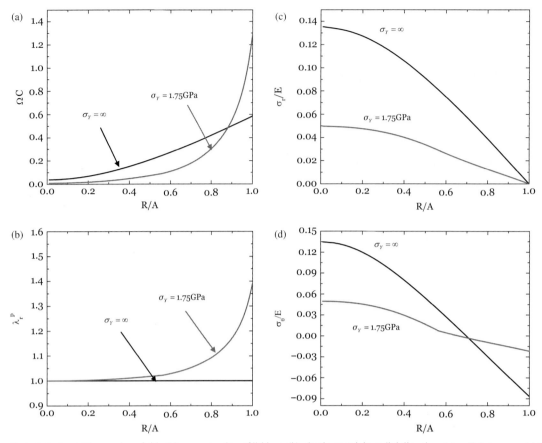

Fig. 5. The effect of plastic yield on various fields: (a) concentration of lithium, (b) plastic stretch in radial direction, (c) radial stress, and (d) hoop stress. The charging rate is $\tau = 1$ h, and the fields are given at time $Dt/A^2 = 0.048$.

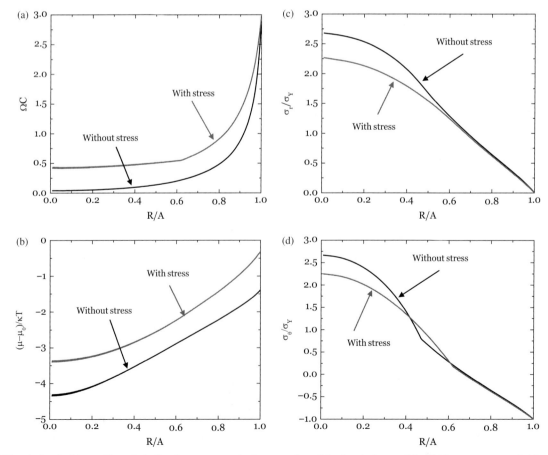

Fig. 6. Fields calculated with or without including the mean stress in the expression of the chemical potential of lithium are compared: (a) concentration of lithium, (b) chemical potential of lithium, (c) radial stress, and (d) hoop stress. The charging rate is $\tau = 1$ h, and both fields are given at the end of charge time, i.e., $Dt/A^2 = 0.132$ with stress calculation, $Dt/A^2 = 0.009$ without stress calculation.

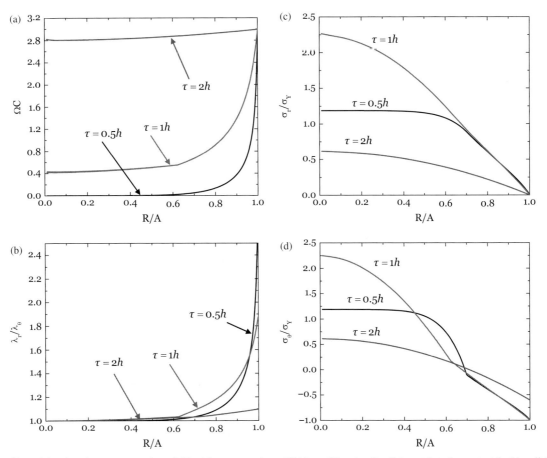

Fig. 7. The effect of the charging rate on various fields: (a) concentration of lithium, (b) ratio of radial stretch to hoop stretch, (c) radial stress, and (d) hoop stress.

lithium is homogenized, the stress is relaxed, as shown in Fig. 8. The effect of the charging rate on the triaxial tension at the center of the particle may be used to guide future experiments to study cavitation.

Figure 9 shows the time evolution of the lithium concentration and the stress fields as lithium desorbs from the spherical particle. The simulation begins when the particle is at full capacity of lithium and is stress-free, and is terminated when the concentration of lithium vanishes near the surface of the particle. As lithium desorbs, the concentration near the surface becomes lower than it is near the center of the particle, Fig. 9(a). This inhomogeneity causes the particle to contract more near the surface than at the center. Consequently, a compressive radial stress develops, Fig. 9(b). The hoop stress at the

surface becomes tensile with magnitude σ_Y, Fig. 9(c). This tensile stress may result in the propagation of surface flaws. Because the tensile stress is limited by the yield strength, fracture may be averted when the yield strength is low.

V. Conclusions

This paper formulates a theory that couples lithiation and large elastic–plastic deformation. The homogeneous lithiation and deformation in a small material element is analyzed using nonequilibrium thermodynamics. The material is assumed to undergo plastic deformation when the state of stress reaches the yield condition. A spherical particle subject to a constant rate of charge and discharge is analyzed by coupling diffusion and large plastic deformation. The effect of plastic yielding, stress on the chemical potential of lithium, and charging rates are studied. When the charging rate is low, the distribution of lithium in the particle is nearly homogeneous, and the stress is low. When the charging rate is high, the stress at the center of the particle can substantially exceed the yield strength. The developed stress gradient also greatly influences the diffusion of lithium, tending to homogenize the distribution of lithium in the particle. Plastic yielding can markedly reduce the magnitude of stress.

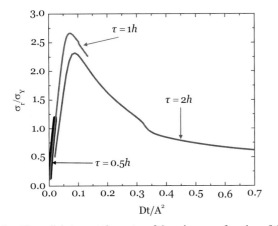

Fig. 8. The radial stress at the center of the sphere as a function of time at various charge rates.

Appendix A: Notes on Numerical Procedure

We rewrite the governing equations in the form used in our numerical simulation. It has been assumed that neither elastic nor plastic deformation causes any volumetric change. It has been also assumed that Ω is a constant. Consequently, the volumetric change of a material element is

$$\lambda_r \lambda_\theta^2 = 1 + \Omega C \qquad \text{(A-1)}$$

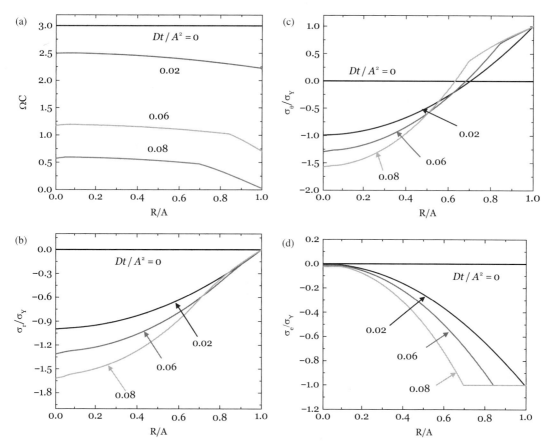

Fig. 9. As a spherical particle is being discharged at the rate of $\tau = 0.5$ h, various fields evolve (a) concentration of lithium, (b) radial stress, (c) hoop stress, and (d) equivalent stress.

A combination of (A-1), (19) and (20) gives that

$$r(R, t) = \left[3 \int_0^R (1 + \Omega C) R^2 \mathrm{d}R \right]^{1/3} \tag{A-2}$$

Here we have used a condition due to the spherical symmetry of the problem, $r(0, t) = 0$.

The kinetic model (25) can be written in terms of the nominal quantities:

$$J = -\frac{CD}{kT\lambda_r^2} \frac{\partial \mu(R, t)}{\partial R} \tag{A-3}$$

A combination of (A-3) and (24) gives

$$J = -\frac{CD}{(1 + \Omega C)^2} \left(\frac{r}{R} \right)^4 \frac{\partial}{\partial R}$$
$$\times \left[\log \frac{\gamma C}{1 + \Omega C} - \frac{\Omega(\sigma_r + 2\sigma_\theta)}{3kT} \right] \tag{A-4}$$

Recall that the flux satisfies the boundary conditions $J(0, t) = 0$ and $J(A, t) = \pm J_0$. The sign of the latter condition depends on whether the particle is being charged or discharged.

The stress $(\sigma_\theta - \sigma_r)$ and the plastic stretch evolve according to the ideal elastic–plastic model, Fig. 3(b). When $|\sigma_r - \sigma_\theta| < \sigma_Y$, the plastic stretch λ_r^p remains fixed, and the elastic stretch is given by $\log \lambda_r^e = (\sigma_r - \sigma_\theta)/E$, which is written as

$$\sigma_r - \sigma_\theta = E \log(\Lambda^{-1/3} \lambda_r / \lambda_r^p) \tag{A-5}$$

when $\sigma_r - \sigma_\theta = \pm \sigma_Y$, the elastic stretch is fixed at $\log \lambda_r^e = \pm \sigma_Y / E$, and the plastic stretch adopts a value

$$\lambda_r^p = \lambda_r \Lambda^{-1/3} \exp\left(\mp \frac{\sigma_Y}{E} \right) \tag{A-6}$$

Integrating (23), we obtain that

$$\sigma_r(R, t) = 2 \int_A^R \frac{(\sigma_r - \sigma_\theta)(1 + \Omega C) R^2}{r^3} \mathrm{d}R \tag{A-7}$$

Here we have used the boundary condition $\sigma_r(A, t) = 0$.

We use the finite difference method, and divide the interval $0 \leq R \leq A$ into small elements. The initial condition $C(R, 0)$ is prescribed; for example, we set $C(R, 0) = 0$ to simulate the process of lithiation, and set $C(R, 0) = C_{max}$ to simulate the process of delithiation. The initial values of the function $J(R, 0)$ are set with $J(0, 0) = 0$ and $J(A, 0) = \pm J_0$ at the boundaries, and $J(R, 0) = 0$ at the interior points. We then evolve all functions with a time step Δt. At a given time t, the functions $C(R, t)$ and $J(R, t)$, along with the boundary conditions $J(0, t) = 0$ and $J(A, t) = \pm J_0$, are inserted into (21) to calculate $C(R, t+\Delta t)$. The result is then inserted into (A-2) to calculate $r(R, t+\Delta t)$. The field $\sigma_r(R, t+\Delta t)$ is calculated by the integrating (A-7), where $(\sigma_\theta - \sigma_r)$ is determined by the uniaxial stress–strain relation. We then calculate $J(R, t+\Delta t)$ by using (A-4). This procedure is repeated for the next time step to evolve the fields.

Acknowledgments

Matt Pharr acknowledges the support of the Department of Defense (DoD) through the National Defense Science & Engineering Graduate Fellowship (NDSEG) Program.

This paper is submitted to the *Journal of the American Ceramic Society* for a special issue dedicated to the memory of A. G. Evans.

References

[1]R. A. Huggins, *Advanced Batteries: Materials Science Aspects.* Springer, New York, 2009.

[2]J. Christensen and J. Newman, "Stress Generation and Fracture in Lithium Insertion Materials," *J. Solid State Electrochem.*, **10** [5] 293–319 (2006).

[3]X. C. Zhang, W. Shyy, and A. M. Sastry, "Numerical Simulation of Intercalation-Induced Stress in Li-Ion Battery Electrode Materials," *J. Electrochem. Soc.*, **154** [10] A910–6 (2007).

[4]S. Golmon, K. Maute, S. H. Lee, and M. L. Dunn, "Stress Generation in Silicon Particles During Lithium Insertion," *Appl. Phys. Lett.*, **97** [3] 033111, 3pp (2010).

[5]Y. T. Cheng and M. W. Verbrugge, "Evolution of Stress within a Spherical Insertion Electrode Particle Under Potentiostatic and Galvanostatic Operation," *J. Power Sources*, **190** [2] 453–60 (2009).

[6]Y. H. Hu, X. H. Zhao, and Z. G. Suo, "Averting Cracks Caused by Insertion Reaction in Lithium-Ion Batteries," *J. Mater. Res.*, **25** [6] 1007–10 (2010).

[7]R. A. Huggins and W. D. Nix, "Decrepitation Model for Capacity Loss During Cycling of Alloys in Rechargeable Electrochemical Systems," *Ionics*, **6**, 57–64 (2000).

[8]T. K. Bhandakkar and H. J. Gao, "Cohesive Modeling of Crack Nucleation Under Diffusion Induced Stresses in a Thin Strip: Implications on the Critical Size for Flaw Tolerant Battery Electrodes," *Int. J. Solids Struc.*, **47** [10] 1424–34 (2010).

[9]W. H. Woodford, Y. M. Chiang, and W. C. Carter, "Electrochemical Shock of Intercalation Electrodes: A Fracture Mechanics Analysis," *J. Electrochem. Soc.*, **157** [10] A1052–9 (2010).

[10]K. J. Zhao, M. Pharr, J. J. Vlassak, and Z. G. Suo, "Fracture of Electrodes in Lithium-Ion Batteries Caused by Fast Charging," *J. Appl. Phys.*, **108** [7] 073517, 6pp (2010).

[11]Y. Itou and Y. Ukyo, "Performance of LiNiCoO$_2$ Materials for Advanced Lithium-Ion Batteries," *J. Power Sources*, **146** [1–2] 39–44 (2005).

[12]P. Arora, R. E. White, and M. Doyle, "Capacity Fade Mechanisms and Side Reactions in Lithium-Ion Batteries," *J. Electrochem. Soc.*, **145** [10] 3647–67 (1998).

[13]D. Y. Wang, X. D. Wu, Z. X. Wang, and L. Q. Chen, "Cracking Causing Cyclic Instability of LiFePO$_4$ Cathode Material," *J. Power Sources*, **140** [1] 125–8 (2005).

[14]K. E. Aifantis and J. P. Dempsey, "Stable Crack Growth in Nanostructured Li-Batteries," *J. Power Sources*, **143** [1–2] 203–11 (2005).

[15]L. Y. Beaulieu, K. W. Eberman, R. L. Turner, L. J. Krause, and J. R. Dahn, "Colossal Reversible Volume Changes in Lithium Alloys," *Electrochem. Solid State Lett.*, **4** [9] A137–40 (2001).

[16]W. J. Zhang, "A Review of the Electrochemical Performance of Alloy Anodes for Lithium-Ion Batteries," *J. Power Sources*, **196** [1] 13–24 (2011).

[17]T. Takamura, S. Ohara, M. Uehara, J. Suzuki, and K. Sekine, "A Vacuum Deposited Si Film having a Li Extraction Capacity Over 2000 mAh/g with a Long Cycle Life," *J. Power Sources*, **129** [1] 96–100 (2004).

[18]C. K. Chan, H. L. Peng, G. Liu, K. McIlwrath, X. F. Zhang, R. A. Huggins, and Y. Cui, "High-Performance Lithium Battery Anodes Using Silicon Nanowires," *Nature Nanotechnol.*, **3** [1] 31–5 (2008).

[19]H. Kim, B. Han, J. Choo, and J. Cho, "Three-Dimensional Porous Silicon Particles for Use in High-Performance Lithium Secondary Batteries," *Angew. Chem.-Int. Ed.*, **47** [52] 10151–4 (2008).

[20]V. A. Sethuraman, M. J. Chon, M. Shimshak, V. Srinivasan, and P. R. Guduru, "In situ Measurements of Stress Evolution in Silicon Thin Films During Electrochemical Lithiation and Delithiation," *J. Power Sources*, **195** [15] 5062–6 (2010).

[21]J. W. Choi, J. McDonough, S. Jeong, J. S. Yoo, C. K. Chan, and Y. Cui, "Stepwise Nanopore Evolution in One-Dimensional Nanostructures," *Nano Lett.*, **10** [4] 1409–13 (2010).

[22]V. A. Sethuraman, V. Srinivasan, A. F. Bower, and P. R. Guduru, "In situ Measurements of Stress-Potential Coupling in Lithiated Silicon," *J. Electrochem. Soc.*, **157** [11] A1253–61 (2010).

[23]K. J. Zhao, M. Pharr, J. J. Vlassak, and Z. G. Suo, "Inelastic Hosts as Electrodes for High-Capacity Lithium-Ion Batteries," *J. Appl. Phys.*, **109** [1] 016110, 3pp (2011).

[24]E. H. Lee, "Elastic–Plastic Deformation at Finite Strains," *J. Appl. Mech.*, **36** [1] 1–8 (1969).

[25]M. N. Silberstein and M. C. Boyce, "Constitutive Modeling of the Rate, Temperature, and Hydration Dependent Deformation Response of Nafion to Monotonic and Cyclic Loading," *J. Power Sources*, **195** [17] 5692–706 (2010).

[26]M. Ben Amar and A. Goriely, "Growth and Instability in Elastic Tissues," *J. Mech. Phys. Solids*, **53** [10] 2284–319 (2005).

[27]T. Song, J. L. Xia, J. H. Lee, D. H. Lee, M. S. Kwon, J. M. Choi, J. Wu, S. K. Doo, H. Chang, W. Il Park, D. S. Zang, H. Kim, Y. G. Huang, K. C. Hwang, J. A. Rogers, and U. Paik, "Arrays of Sealed Silicon Nanotubes as Anodes for Lithium Ion Batteries," *Nano Lett.*, **10** [5] 1710–6 (2010).

[28]R. Hill, *The Mathematical Theory of Plasticity*. Oxford University Press, Oxford, 1950.

[29]N. Ding, J. Xu, Y. X. Yao, G. Wegner, X. Fang, C. H. Chen, and I. Lieberwirth, "Determination of the Diffusion Coefficient of Lithium Ions in Nano-Si," *Solid State Ionics*, **180** [2–3] 222–5 (2009). ☐

J. Am. Ceram. Soc., **94** [S1] S236–S240 (2011)
DOI: 10.1111/j.1551-2916.2011.04492.x
© 2011 The American Ceramic Society

journal

Searching for a Biomimetic Method of Fabricating Network Structures

Brian Cox[†]

Teledyne Scientific Co. LLC, Thousand Oaks, California 91360

A recent theory has suggested that populations of migrating and proliferating cells can create network structures by responding to strain cues associated with the cell density variations that arise during network formation. Unlike prior theories of network formation, the strain-cue mechanism leads to nonfractal networks, consisting of closed loops rather than tree-like morphologies. In this paper, the possibility is suggested of developing a synthetic process for fabricating networks that mimic the mechanisms present in the theory. Such a biomimetic process should replicate three phenomena: (1) the extension of the existing network domain should be governed by the strain field that exists just outside the existing domain, rather than the strains within the domain; (2) the process should be stochastic; and (3) a relaxation mechanism must be present by which the strains that induce network extension at any location will fade with time. Simulations imply values for the key parameters of these mechanisms to achieve useful networks. Possible routes to realizing a biomimetic system are discussed.

I. Introduction

MULTIPHASE materials in which one phase constitutes a connected network can be desirable for a number of reasons. (1) A connected phase can provide pathways for thermal or electronic heat conduction. If the conducting phase is introduced as spatially random particles, conduction relies on achieving the percolation threshold and, if conductivity is required in three dimensions, percolation occurs at quite high volume fractions of the conducting phase. To reduce the volume fraction, some process is needed by which the particles can be assembled into line aggregates rather than being spatially random, which is not an easy feat. If conducting lines are achieved instead by using continuous fibers as the conducting phase, challenges arise with deploying the fibers in three directions—once again, isotropic conductivity is difficult to achieve—and with handling fibers of very small diameter. (2) Interconnected ligaments of a tough and strong material can provide mechanical bridging and therefore impart macroscopic toughness to an otherwise brittle material. Achieving isotropy is again a challenge; current methods based on chopped fibers or whiskers, for example, tend to produce approximately two-dimensional (2D) felt-like reinforcement (because of particle settling during processing), with relatively weak mechanical effect, for a given volume fraction of reinforcement, in the direction perpendicular to the layers of fibers or whiskers. An isotropic interconnected network of reinforcing ligaments could be significantly tougher than long-particle-reinforced materials for a given volume fraction. In a chopped fiber or whisker-reinforced material, a crack can run around the ends of the particles, which it cannot do if it propagates through a connected network of reinforcing ligaments. (3) If a connected network is formed in a solid body, simple

chemical processing (e.g., etching) can remove the matrix material, leaving the network as a free-standing structure. A free-standing three-dimensional (3D) network with open cell faces has numerous potential applications, including filtering, biological cell culture, catalysis, etc. The fact that it is a connected network gives it useable structural integrity, while control of the strut diameter and length and the network density would allow the surface-to-volume ratio and the permeability of the network to be tailored to a particular purpose.

While forming networks is a poorly developed art in materials engineering, nature forms a great variety of networks (especially nerve and vascular) during organogenesis with a wide range of forms and at spatial scales down to that of a single cell (∼ 10 μm). Many of these networks are tree-like structures, often with fractal scaling properties.[1] But others are nonfractal, closed networks, characterized by roughly periodic repetitions of truss-like strut-and-node assemblies, a pattern that is more desirable for many engineering materials applications. Given the lack of processing routes to such networks in materials engineering, the questions of just how nature does it, and how nature's methods might be replicated, deserve some attention.

In a recent paper,[2] a theory was presented that can account for the formation of a 2D network by neural cells invading the gut of a mouse. The direction of migration of a neural cell invading the host population of the gut was assumed to be controlled by the magnitude of the strains in the host medium (cells plus extracellular matrix) that arise as the host medium deforms to accommodate the invader (Fig. 1). This single assumption of a strain cue that references strains external to the invader is sufficient to generate network structures. The strain induced by a line of invaders is greatest at the extremity of the line and thus the strain field breaks symmetry, stabilizing branch formation. The strain cue also triggers sprouting from existing branches, with no further model assumption. The characteristics of networks predicted by the theory depend primarily on the ratio of the rate of advance of the invaders to the rate of relaxation of the host cells after their initial deformation. A second model parameter arises from the assumption that cell advance is a stochastic process.

Prior models of branching structures and other shape instabilities that mark organogenesis have exploited concepts that originated in the physical sciences. Viscous fingering (a problem first identified in the oil industry and explained in Saffman and Taylor[3]) arises when a less viscous fluid is injected into a more viscous fluid in the confines of a porous medium. Repeated finger formation can result in fractal tree-like structures. While the formative shape instability depends on the viscosity ratio and the local pressure at branch tips, the driving force is ultimately the far-field pressure, whose transmission and diminution along the existing structure is important to the morphological outcome. Diffusion-limited aggregation (DLA) (first conceived to account for the aggregation of colloids[4]) generates shape instabilities in an expanding domain whose growth depends on the availability of some species that diffuses toward it from surrounding material. Enhanced particle collection by nascent protrusions favors their extension while shadowing the adjacent growth front, which therefore tends not to advance, resulting in fractal tree structures very similar to those caused by viscous fingering.

R. McMeeking—contributing editor

Manuscript No. 28613. Received September 21, 2010; approved February 11, 2011.
[†]Author to whom correspondence should be addressed. e-mail: bcox@teledyne.com

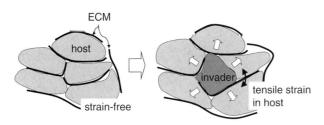

Fig. 1. When invaded, the host cells and their associated extracellular matrix must deform to accommodate the invader.

As pointed out by Halsey,[5] theories of viscous fingering and DLA are mathematically isomorphic, at least in simple renditions, in spite of the very different physics involved. Both can be framed as solutions to the Laplace equation, one using pressure as a dependent variable, the other using a probability field; in viscous fingering, the boundary between the two fluids moves in response to the gradient of the pressure across the boundary, while in DLA, the boundary of the accretion domain moves in response to the gradient of the probability variable across the boundary; and in viscous fingering, the boundary marks an isobar for the pressure variable, while in DLA, it marks an isocontour of the probability variable. Among other implications, this curious result issues a warning: finding a mechanism that replicates an observed morphology does not in itself prove that the mechanism is the operative one in nature. In the case of fractal tree-like networks, the observed morphology can be fitted equally well by two completely different physical phenomena. In the strain-cue theory of nonfractal network formation, a number of additional factors fall into agreement with experimental observations, which adds further credence to the strain-cue hypothesis[2]; but a strict test of the hypothesis remains for the future.

Nevertheless, with what belongs to the future left to the future, the strain-cue theory of Cox[2] does generate nonfractal networks and therefore one can pursue the following question: is it possible to mimic the phenomena identified in the theory by some invented laboratory process and thus replicate the method by which nature is hypothesized to form nonfractal networks with a view to applications in materials science? If so, this would constitute a significant advance in what is meant by the term "biomimetic." In the prior literature (to the author's knowledge), biomimicry has consisted of replicating morphologies from nature by whatever process can make them; in the present paper, the idea is proposed that the process itself might parallel the specific steps taken by living cells during natural morphogenesis. Thus, the entity being mimicked is not just the morphology, but the process by which a class of morphology is achieved in nature.

To open up a possible search for a biomimetic approach to fabricating nonfractal networks, this paper reviews the relevant details of the strain-cue theory of network formation and highlights the phenomena that must be replicated to achieve a biomimicking process.

II. The Strain-Cue Theory of Network Formation

The following is a summary of the theory presented by Cox.[2] The theory deals with the penetration of a 3D domain of host cells by an army of invading cells, which form a network. The invasion causes spatial variations in the density of cells as invaders intercalate themselves among the hosts. The density variations create spatially varying and time-dependent strain and stress fields. The fields are recomputed continually as the invasion progresses. The local values of strain components are used as cues to control the migration of invaders from domains that have already been invaded. The strain-cue rule is applied equally to all invaders on every iteration; all variations in migration patterns result solely from variations in the strain fields. No explicit control over branch stability, sprouting, or any other network characteristic is introduced.

Exploiting intuition from the mechanics of inclusions, the strain cues needed to model network formation by cells are calculated as an eigenstrain problem. The problem is solved numerically on a simple cubic mesh with element size a, aligned with a Cartesian coordinate system (x, y, z). Each cubic element may be occupied by a combination of hosts and invaders, with number n_h and n_i, respectively, which need not be integers. An element is defined to be somewhat larger than a single cell. The cell number densities in any element are $\rho_h = n_h/a^3$ and $\rho_i = n_i/a^3$, respectively. The cell densities are smeared out to be uniform within each element, but they differ from element to element.

In the absence of mechanical stress, a single host and a single invader cell are assumed to have volumes V_h and V_i, respectively. The volume that the cells in a single element would occupy in the absence of stress is $V = n_h V_h + n_i V_i$. If this volume is larger than that of the occupied element, a mechanical stress will arise if the element is confined among other elements. The stress field is computed using analytical formulae derived for an "inclusion" in an elastic medium that contains a "stress-free eigenstrain."[6–8] The volume eigenstrain, $\bar{\varepsilon}$, measures the difference between the stress-free volume of the cells that occupy an element and the volume of the element. The system of cubic elements containing various eigenstrains generates spatially varying strain fields, which are computed by methods detailed in Cox.[2]

A critical element of the hypothesis for natural network formation is a strain-based criterion for cell migration. This criterion must appear in a successful biomimicking process. Of special interest is the quantity $\varepsilon_s = \varepsilon_1 + \varepsilon_2$, evaluated at a point P just outside any element that has already been invaded, where the directions 1 and 2 are the directions normal to the direction of a candidate migration step that passes through that point. The spatial variations in ε_s break symmetry, presenting a strain concentration when P lies at the extremity of any branch, which stabilizes branches. When a branch bends by a random variation in its direction to form an elbow, a strain enhancement also arises at the point of the elbow. This trend generates sprouting. These two effects together generate networks.

A second key element for which a biomimicking analogue must be sought is the assumed stochastic nature of cell migration (or cell proliferation) in nature. Networks only arise from a stochastic process. In the theory of Cox,[2] the stochastic process was implemented in the criterion for cell advance (migration). If element i has been invaded before the m-th time step, a strain-cue factor, S_{ij}, for migration into a contacting element j (i.e., an element sharing an entire face with the i-th element) during the m-th time step is defined to include a stochastic parameter that is determined by a pseudorandom number generator. The probabilistic nature of the migration criterion represents the effects of random variations in the local character of the host/invader system that are not represented by variations in the local strain fields. The random fluctuations have no spatial order; they introduce no spatial bias that might favor branch formation or stabilization, but they do influence network morphology and density.

The third key element in the theory of biological network formation is the introduction of strain relaxation. Where invading cells have migrated into regions already occupied by host cells, the local cell density rises and the host cells will tend to relax away from the invaders to restore uniformity. This motion is introduced into the simulations as a diffusion-like process in which only the host cells relax.

Relaxation reduces the stress and strain fields around an advancing branch of invaders and introduces a second rate process that competes with the rate of cell advance. The ratio of these two rates is critical to the outcome. If the branch does not advance during a time increment, further strain reduction will occur; and therefore, the longer a branch lingers at a given length, the lower the probability of its further extension. One possible outcome is that the branch will arrest permanently as the strains become evanescent. A network consisting of many branches can only continue to advance with undiminished density if new branches are formed from existing branches by sprouting at a rate per unit length of advance of a branch that at least equals the probability of arrest.

III.   Characteristics of Networks Predicted by the Strain-Cue Theory

The simulations reported by Cox[2] are of the formation of planar networks; the invading cells are restricted to migrate within a single layer of elements whose centers lie on $z = 0$ in the Cartesian system (x, y, z). This restriction is appropriate for the formation of some nerve networks, such as in the gut and retina, and some instances of angiogenesis. For a biomimetic materials process, both 2D and 3D networks are of interest. While not addressed by Cox,[2] 3D networks would be created by the same mechanism if the constraint were removed that cells migrate in a single plane.

The following notes summarize events in network formation that should be part of a biomimetic process. They also highlight the local nature of the network forming mechanism.

(1)   Networks Formed by Invasion from an Initial Band of Cells

In the simulations of Cox,[2] invasion begins from some initial set of invaders with an assigned spatial distribution. Consider first a relatively dense population of invaders distributed randomly in space, with statistically uniform density, over a narrow band $-5a < x < 0$. Given this starting population, the invasion tends to progress across a diffuse front toward positive large x, yielding networks exemplified by Fig. 2(a).

Figure 2(a) contains a number of features similar to those formed by neural crest cells invading the ileum and cecum of a mouse (Fig. 2(b)). The proportion of the width of a branch of the network to the width of typical voids are approximately equal in Figs. 2(a) and (b). The front of the invasion in Figs. 2(a) and (b) is comparable with the width of the larger voids (areas enclosed by branches) in the network. Comparison of other images from both nature and simulations shows some backfilling as the invasion proceeds, i.e., densification of the branch structure behind the leading extremity of the invasion. Branches occasionally appear to terminate permanently in the image of Fig. 2(b). Terminated branches in Fig. 2(a) occur at an approximately similar frequency, in a small percentage of enclosed void areas. Invading cells occasionally appear in clusters, which arise in simulations as fluctuations in the outcome of inserting the

local strain field values into the rule for advance. Where sprouting forms three-branch junctions in either Figs. 2(a) or 2(b), no two of the three branches involved are collinear. This is consistent with the prediction of the simulations that sprouting is triggered by the strain concentration that arises at an elbow when a branch's extension randomly changes direction.

(2)   Network Development from a Single Invader

The characteristics of a predicted network do not depend on the initial spatial configuration assigned to the invaders. Figure 3 shows network development from a single centrally located invader cell. The problem definition in this simulation is similar to that in Fig. 2(a), the main difference being the initial distribution of invaders. Snapshots at increasing times show a network expanding in a roughly circular shape from the single initial invader, eventually filling the host domain. In the fully developed network, the branch density, sprout density, and distribution of enclosed void areas are statistically similar to those calculated from Fig. 2(a). Thus a biomimetic system can use various seeding or nucleation strategies to achieve a given network.

(3)   Parameters that Control the Network

While a number of parameters were defined in the model of Cox,[2] the scaling properties of the governing equations together with statistical analysis show that most of these parameters do not affect network morphology independently. In fact, network morphology depends only on two parameters: one is the ratio of the rate of cell migration to the relaxation rate of the strains induced in the host by a migrating invader; the second is a parameter describing how sensitive the probability of cell migration is to fluctuations in the magnitudes of the strains around an invader.

(4)   Physical Characteristics of the Mechanism of Network Formation

The key strain variations and the host relaxation mechanisms involve the interactions of 100 cells to order of magnitude; the influence on strain fields of cells beyond the nearest 100 cells in a branch falls rapidly with their distance. Thus, the network-forming mechanism has the robust characteristic of not depending on the maintenance of long-range order in the population.

Network formation requires that the strain stimulus mechanism be switched off at some point. If strain stimulus continues to operate indefinitely around a newly formed network branch, the branch disappears quite quickly due to the formation of an ever-increasing density of newly formed branches emerging from

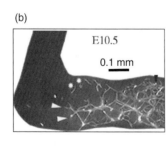

Fig. 2.   (a) A network grown from an initial band of invaders. Branches of invader cells (gray) are punctuated by occasional bifurcation or sprout points (black). Permanently arrested branches and clusters of invaders occur occasionally. (b) Nerve networks (white) formed in the mouse ileum and cecum, imaged at two different times during the invasion by neural crest cells (from Druckenbrod and Epstein[12]). The invasion is progressing from left to right in (a) and right to left in (b).

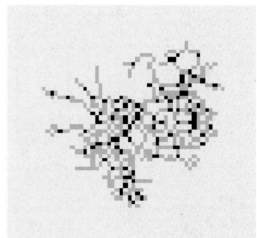

Fig. 3.   A network grown from a single invader located near the center of the image. Branches of invader cells (gray) are punctuated by occasional bifurcation or sprout points (black). (For parameters similar to those used in Cox.[2])

it. The switching off is effected by relaxation due to host cell motions, which removes the signaling strains.

IV. An Inferred Template for Biomimetic Network Fabrication

(1) Strain Cues

The first key characteristic of the theory is that extension of the spatial domain occupied by the existing network must be triggered by values of a mechanical strain field (or stress field) at points exterior to, but immediately adjacent to, the existing domain. For successful mimicry, the strain field must first be generated and a mechanism then created for domain extension to be guided by the strain field. Following nature, the strain field can be generated if the process of network extension temporarily raises the local material density in the network domain.

Nature suggests several mechanisms by which domain extension can be guided by strain fields external to the existing network domain. Cells themselves can probe the space around them using filipodia and lamellipodia, which are thin protrusions of the cell membrane that can sense either chemical or strain signals. However, analogues of filipodia or lamellipodia in a nonanimate, biomimetic laboratory process require some imagination to conceive. Alternative candidate natural mechanisms whose mimicry may be more practicable include the formation of some species within the host domain under the stimulus of high strain levels, which can diffuse back to the network domain, where it could stimulate mass transport; and the transformation of host domain material into network domain material via a phase transformation triggered by strain.

(2) Stochastic Character

The strain-cue theory uses a stochastic algorithm to predict network extension. The stochastic character plays a pivotal role in introducing random sprouting from existing branches and in the occasional change of direction of an existing branch.

In nature, a number of mechanisms can be conjectured that would lead to the process of cell migration from the existing network domain being stochastic, including randomness in the locations of the host cells between which the invader must intercalate itself to effect migration. In the simulations described by Cox,[2] randomness was introduced by the Monte Carlo method using a pseudorandom number generator.

Random numbers were chosen each time migration from an already-invaded element was checked for advance in any direction. Thus, different cells with equal strain environments might differ in whether they advance or not on a given iteration. The fact that not all invaders subject to conducive strain conditions advance is one factor that controls the density of the network that is formed.

To assess prospective advance from the same already-invaded element on successive iterations, different random numbers were used. This can be interpreted as an expected result of the continuing motion of the host cells surrounding an invader, which changes the physical environment of the invader in a stochastic manner.

Thus, an analogue of a stochastic natural system would use a host medium with two features: random granularity (in correspondence with the discreteness of host cells) and continual motion, for example under thermal activation.

(3) Replenishment

The networks formed by migration from a planar reservoir are similar to those in the gut in general appearance and also in the fact that the entire network is formed by a cell population that is drawn initially from a reservoir on one side or end of the network (from the neural crest in nature). For a network to propagate indefinitely away from the reservoir of invaders, the invader population must be replenished. In nature, this occurs by a combination of new invaders migrating from the reservoir along existing network branches and the proliferation of invaders.

Replenishment by migration in a biomimetic system might be accomplished by, for example, ion diffusion. Replenishment by an analogue of proliferation could occur, for example, by a volume-enhancing phase transformation that draws on a continuously distributed source material.

(4) Relaxation Mechanism

In the strain-cue theory, relaxation of strains governs the probability of branch arrest and therefore the density of the network. Relaxation is represented as a mass transport process, occurring at a rate that is proportional to the spatial gradient of the cell density.

Any mass transport mechanism could serve as an analogue in a biomimetic system.

(5) Parameters Whose Values must be Maintained

Only two parameters control the outcome of simulations in the strain-cue theory: the ratio of the rates of the advance of the invasion front and the relaxation of induced strains, v_f/v_r, and the parameter, c_1, that defines the sensitivity of the probability of migration of an invader to deviations in its strain environment.

The value of c_1 should be selected to correspond to the expected variations in strain due to shape variations around a network branch. The mechanics of inclusions show that variations of 30% occur in the relative magnitude of the strain components computed in the strain-cue theory of Cox.[2] In nature, one can argue that cells are most likely to have evolved to sense strain variations of this magnitude, because that would make their action in forming networks most robust; if cells were substantially more sensitive to strain variations, slight aberrations in the spatial configuration of the cell population would cause spurious cell responses, while if they were substantially less sensitive, they would not sense the available strain cues. Consistently, the laboratory process used for network extension should occur at a rate that changes substantially when the guiding strain fields vary by approximately 30% in magnitude.

The acceptable range of values of v_r/v_f is determined fairly closely by the outcome of simulations. Depending on the value of c_1, v_r/v_f must lie approximately in the range of 0.03–0.06. Control of this ratio in nature appears to be affected mainly by the variation of v_f, the migration speed of the invading cells. For example, the relaxation velocity deduced for ameloblasts during the formation of human dental enamel is within 50% of the relaxation velocity that is required in simulations of the innervation of the mouse gut to obtain agreement with nature.[2] In contrast, the migration speeds of ameloblasts during human amelogenesis and neural cells during innervation of the mouse gut differ by three orders of magnitude (amelogenesis being very slow). One might surmise that the relaxation velocity is tied to rate processes in cells, which tend to occur at an invariant rate for different cell types under the stimulus of equal density gradients; whereas, a cell's migration speed might be varied more widely by switching its motility on or off. In a biomimetic system, either rate could feasibly be manipulated.

(6) Parameters Whose Values are Immaterial

The simulations are invariant with regard to the morphology of the predicted networks under changes of certain parameter values. The absolute migration and relaxation velocities are immaterial; only their ratio matters. The elasticity of the materials does not enter; the morphology is determined by the strain fields arising around a system of inclusions bearing eigenstrain, which are independent of the elasticity of the medium, provided the network and host domains have similar elasticity. The width of a network branch does not affect the network morphology; the network will simply scale in a geometrically similar manner. Neither does the choice of element size in the theory affect network morphology; the outcome is invariant under changes in spatial scale. Thus, biomimetic systems can target networks possessing any characteristic length scale (or characteristic area of the voids enclosed by branches).

(7) Networks in Three Dimensions

If the restriction that migration can occur only on a plane is relaxed in simulations, the rules used by Cox,[2] without modification, will generate 3D networks. The strain concentrations around either branch tips or branch elbows are independent of permutations of the global directions (x, y, z). Therefore, both the sprouting rate and the probability of branch arrest per unit length of advance of a branch will be the same in 2D and 3D networks for given model parameters. The rate of branch coalescence will drop, implying a tendency toward open-tree rather than closed network structures.

V. Some Material Concepts on which a Biomimetic System could be Based

Two material processes might offer developmental pathways toward biomimetic network formation. The first is strain-induced (or stress-induced) phase transformation in a solid material. Consider, for example, the phase transformation seen in partially stabilized zirconia, a phenomenon with a long presence in the ceramics literature. The image shown in Fig. 4 shows fingers or branches of transformed material (transformation of tetragonal zirconia to monoclinic zirconia at a critical value of local stress) in a specimen subject to applied mechanical load.[9] This phenomenon shows the potential of strain fields induced by existing transformed material to generate geometrically stable branch structures; the untransformed material around the tip of a nascent branch possesses spatial strain variations that favor the continued extension of the branch in its original direction of propagation,[10] in close analogy with the strain fields calculated in the simulations of Cox.[2] (The problem is also analogous to crack extension along a straight fracture path due to the strain concentration caused by the existing crack.) Presumably because the transforming material has a random microstructure, occasional bifurcation of transformation branches occurs, leading to the formation of closed void areas of Fig. 4, a key element of a network.

However, unlike the simulations of Cox,[2] no mechanism exists in the material of Fig. 4 to turn off the transformation process in the wake of the advancing front of branch tips. Therefore, the branches broaden indefinitely, merging and washing out any network morphology. This is analogous to results in Cox[2] for very low values of the ratio of rates, v_r/v_f. To achieve network morphology, a relaxation mechanism must be added to the transformation phenomenon. A simple method for achieving this would be to suspend transforming particles of zirconia (or other material) in a viscous matrix. The rate of branch extension can be controlled by varying the rate of the far-field load displacement in an experiment such as that of Fig. 4; the rate of strain relaxation around branches can be controlled by varying the viscosity of the matrix.

A second material phenomenon of possible interest is ion diffusion. Consider a material containing cavities into which ions might diffuse, e.g., dendrimer cages, selected to possess a chemical potential for the ions that varies when mechanical strain is applied to the cavities. If occupancy of the cavities by ions causes volume expansion, then occupied cavities will induce a strain field around themselves that will promote diffusion of ions into adjacent cavities that are not yet occupied. Mechanically, such a system is again analogous to the inclusion problem analyzed by Cox[2]; the strain fields will replicate the symmetry-breaking properties of the strains around invading cells in biological network formation. The domains of material in which cavities are occupied might therefore extend in geometrically stable branch structures.

VI. Concluding Remarks

A laboratory process that mimics the way in which cell populations create networks must include a network extension mechanism cued by strains outside the existing network domain, a stochastic factor in network extension, and a relaxation process that causes the driving force for extension at any point to diminish in time. Given the presence of these phenomena, the key parameter that must be controlled is the ratio of the extension (migration) and relaxation velocities.

Acknowledgments

The author is indebted to Drs. Tim Bromage, Luisa Iruela-Arispe, Rodrigo Lacruz, Kerry Landman, Don Newgreen, C. E. Smith, and Mal Snead for conversations about biological aspects of networks and dental enamel. On aspects of physical and mechanical modeling, the author has accumulated much knowledge from old friends who worked at or visited the University of California, Santa Barbara, over the last 25 years, including Drs. John Hutchinson, Norman Fleck, Dave Marshall, Bob McMeeking, Zhigang Suo, and many others. He owes special debts to Dr. Fred Leckie, who, with characteristic charm, gently persuaded him of the value of seeking the simplest possible explanations of phenomena, even when the mathematics involved could easily run off into its own life; and Dr. Tony Evans, who brought us all together and made everything each one of us did seem an order of magnitude more important than we could otherwise have believed.

References

[1]V. Fleury, J.-F. Gouyet, and M. Leonetti (eds). *Branching in Nature*. Springer-Verlag, Berlin, 2001.

[2]B. N. Cox, "A Strain-Cue Hypothesis for Biological Network Formation," *J. R. Soc. Interface*, **8**, 377–94 (2011).

[3]P. G. Saffman and G. I. Taylor, "The Penetration of a Fluid into a Medium of Hele-Shaw Cell Containing a More Viscous Liquid," *Proc. R. Soc. London, A*, **245**, 312–29 (1958).

[4]T. A. Jr. Witten and L. M. Sander, "Diffusion-Limited Aggregation, a Kinetic Critical Phenomenon," *Phys. Rev. Lett.*, **47**, 1400–3 (1981).

[5]T. C. Halsey, "Diffusion-Limited Aggregation: A Model for Pattern Formation," *Physics Today*, (2000).

[6]T. Mura, *Micromechanics of Defects in Solids*. Martinus Nijhoff, The Hague, 1982.

[7]Q. Li and P. M. Anderson, "A Compact Solution for the Stress Field from a Cuboidal Region with a Uniform Transformation Strain," *J. Elasticity*, **64**, 237–45 (2001).

[8]J. D. Eshelby, "The Determination of the Elastic Field of an Ellipsoidal Inclusion, and Related Problems," *Proc. R. Soc. A*, **241**, 376–96 (1957).

[9]P. E. Reyes-Morel and I.-W. Chen, "Transformation Plasticity of CeO$_2$-Stabilized Tetragonal Zirconia Polycrystals: I, Stress Assistance and Autocatalysis," *J. Am. Ceram. Soc.*, **71** [5] 343–53 (1988).

[10]R. M. McMeeking and A. G. Evans, "Mechanics of Transformation-Toughening in Brittle Materials," *J. Am. Ceram. Soc.*, **65** [5] 242–6 (1982).

[11]B. N. Cox, "A Multi-Scale, Discrete-Cell Simulation of Organogenesis: Application to the Effects of Strain Stimulus on Collective Cell Behavior During Ameloblast Migration," *J. Theor. Biol.*, **262**, 58–72 (2010).

[12]N. R. Druckenbrod and M. L. Epstein, "The Pattern of Neural Crest Advance in the Cecum and Colon," *Dev. Biol.*, **287**, 125–33 (2005). □

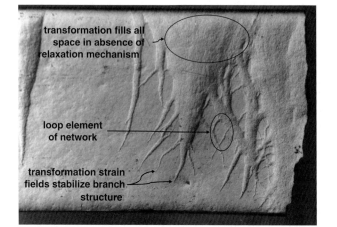

Fig. 4. Strain-induced transformation in partially stabilized zirconia exhibits branch structure and elements of network formation (original photograph provided by Dr. I-Wei Chen).

Call for Papers
Abstracts Due July 20, 2011

36TH INTERNATIONAL CONFERENCE AND EXPOSITION ON
ADVANCED CERAMICS AND COMPOSITES

January 22-27, 2012
Hosted at Hilton Daytona Beach Resort and Ocean Center
Daytona Beach Florida, USA

www.ceramics.org/daytona2012

Organized by The American Ceramic Society and The American Ceramic Society's Engineering Ceramics Division